D1087983

Disease Mapping and Risk Assessment for Public Health

Disease Mapping and Risk Assessment for Public Health

WITHDRAWN

Edited by

Andrew Lawson
University of Aberdeen, UK

and

Annibale Biggeri
University of Florence, Italy

Dankmar Böhning
Free University, Berlin, Germany

Emmanuel Lesaffre
Catholic University, Leuven, Belgium

Jean-François Viel
University of Besancon, France

Roberto Bertollini
WHO, Italy

JOHN WILEY & SONS, LTD

Chichester · New York · Weinheim · Brisbane · Singapore · Toronto

Copyright © 1999 by John Wiley & Sons Ltd,
Baffins Lane, Chichester,
West Sussex, PO19 1UD, England

National (01243) 779777
International (+44) 1243 779777

e-mail (for orders and customer service enquiries): cs-books@wiley.co.uk
Visit our Home Page on http://www.wiley.co.uk or http://www.wiley.com

Reprinted August 2000, September 2001, July 2002

All Rights Reserved. No part of this publication may be reproduced, stored in a retrieval system, or transmitted, in any form or by any means, electronic, mechanical, photocopying, recording, scanning or otherwise, except under the terms of the Copyright, Designs and Patents Act, 1988 or under the terms of a licence issued by the Copyright Licensing Agency, 90 Tottenham Court Road, London W1P 9HE, UK, without the permission in writing of the Publisher.

Designations used by companies to distinguish their products are often claimed as trademarks. In all instances where John Wiley & Sons is aware of a claim, the product names appear in initial capital or all capital letters. Readers, however, should contact the appropriate companies for more complete information regarding trademarks and registration.

Other Wiley Editorial Offices

John Wiley & Sons, Inc., 605 Third Avenue,
New York, NY 10158-0012, USA

Wiley-VCH Verlag GmbH, Pappelallee 3,
D-69469 Weinheim, Germany

Jacaranda Wiley Ltd, 33 Park Road, Milton,
Queensland 4064, Australia

John Wiley & Sons (Asia) Pte Ltd, 2 Clementi Loop #02-01,
Jin Xing Distripark, Singapore 129809

John Wiley & Sons (Canada) Ltd, 22 Worcester Road,
Rexdale, Ontario, M9W 1L1, Canada

Library in Congress Cataloging-in-Publication Data

Disease mapping and risk assessment for public health / edited by
Andrew Lawson . . . [et al.].
p. cm.
Includes bibliographical references and index.
ISBN 0-471-98634-8
1. Medical geography—Congresses. 2. Epidemiology—Congresses.
3. Risk assessment—Congresses. I. Lawson, Andrew (Andrew B.)
[DNLM: 1. Topography, Medical congresses. 2. Epidemiology
congresses. 3. Risk Assessment congresses. WB 700 D611 1999]
RA791.2.D56 1999
614.4'2—dc21
DNLM/DLC
for Library of Congress 98-40964
CIP

British Library Cataloguing in Publication Data

A catalogue record for this book is available from the British Library

ISBN 0-471-98634-8

Typeset in 10/12pt Photina by Thomson Press (India) Ltd, New Delhi
Printed and bound in Great Britain by Antony Rowe Ltd., Chippenham, Wiltshire.
This book is printed on acid-free paper responsibly manufactured from sustainable forestry in which at least two trees are planted for each one used for paper production.

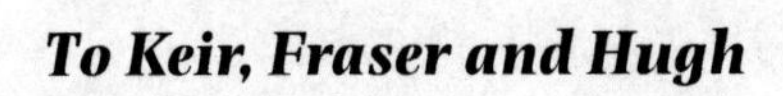

To Keir, Fraser and Hugh

Contents

Editors' Preface

Disease mapping and risk assessment is now a major focus of interest in the area of public health. The geographical distribution of the incidence of disease has an important role to play in the development of understanding the origins and causes of many diseases, and its role should not be underestimated. One of the earliest examples of the important role of geographical analysis of disease was the analysis of cholera outbreaks in the east and of London by John Snow in 1854. Snow constructed maps of the locations of cholera deaths and noted the particular elevated incidence around the Broad Street water pump, a source of water supply for the local area. Subsequently the local water company was tasked with improving the supply quality. More recent examples of the use of geographical analysis can be found. The incidence of asbestos-related lung cancer amongst shipyard workers in Georgia, USA, was established by large-scale comparative mapping of the geographical distribution of the disease. Only once the mapped incidence had been examined did the link between shipyard employment and asbestos exposure risk become established. More recently, outbreaks of asthma in areas of Barcelona during the 1980s have been traced by geographical analysis to the unloading of soybean cargo in Barcelona harbour. Some examples of geographical analysis of clusters are provided later in this book.

Besides the application of a geographical approach to the assessment of local excesses of disease incidence, there are now many branches in the study of geographical distribution of disease incidence. These branches reflect the varied needs of public health analysts and epidemiologists in their quest for the assessment of disease aetiology and the relationship between disease and factors contributing to its occurrence. First of all, there is a need to produce accurate maps of disease incidence so that map-users can, with confidence, assess the true underlying distribution of disease. This branch of the subject is usually termed *disease mapping*, and there is now an established range of methods which address this area of concern. Part I of this volume contains a range of contributions which focus on this subject area. Second, assessment of local aggregations of disease on maps is usually termed *disease clustering*, and this has become an important area of concern for public health. Part II of this volume provides a range of contributions which assess the variety of methods currently available to deal with these problems.

The assessment of the relations between disease incidence and variables or factors which could affect that incidence is the subject of spatial regression and *ecological analysis*. This subject area encompasses the situation where explanatory variables are measured at spatial locations or averaged over areas and are to be related to the distribution of the disease of interest. Ecological analysis usually also involves the problem of matching disease incidence given at one resolution level to variables possibly only available at aggregated levels of resolution. Part III of this volume focuses on issues related to this aspect. Health risk assessment around specific known putative (supposed) locations of hazard is now of considerable importance due to public demands on local health authorities to assess such risks. The highly publicised studies of the incidence of childhood leukaemia around nuclear facilities in the United Kingdom and more recently in France, are particular examples of such types of study. These studies represent a form of clustering study involving ecological analysis where the location centre is known and emphasis is placed on the exposure modelling around the putative source. Part IV of this volume focuses on methods which can be applied in these types of study. Finally, Part V of the volume focuses on a range of case studies on the application of methods to particular problems in analysis of the distribution of disease over geographical regions and their possible use for public health decision making.

This book has arisen from a major initiative of the European Union within the Biomed 2 programme of the 4th framework on Disease Mapping and Risk Assessment (contract number: BHM4-CT96-0633), in conjunction with the Rome Division of the European Center for Environment and Health, World Health Organisation (WHO), Regional Office for Europe. The culmination of this initiative was a WHO workshop which took place in October 1997 in Rome on Disease Mapping and Risk Assessment for Public Health Decision Making. This book follows from the ideas and contributions discussed during the workshop and it is hoped that it will form a marker of the state of the subject area at that time. The final report of the workshop appears as an appendix to this book.

Andrew B. Lawson (Aberdeen)

Annibale Biggeri (Florence)

Dankmar Böhning (Berlin)

Emmanuel Lesaffre (Leuven)

Jean-Francois Viel (Besancon)

Roberto Bertollini (WHO, Rome)

Contributors

Dr B. Armstrong
Environmental Epidemiology Unit
Department of Public Health and Policy
London School of Hygiene and Tropical Medicine, Keppel Street
London
UK

Professor Olav Axelson
Division of Occupational and Environmental Medicine
Department of Health and the Environment
Linköping University
Linköping
Sweden

Professor Luisa Bernardinelli
Dipartimento di Scienze Sanitarie
Applicate e Psicocomportamentali
Universitá di Pavia
Via Bassi 21, Pavia
Italy

Dr Nicola Best
Department of Epidemiology & Public Health
Imperial College School of Medicine
St Mary's Campus
London, W2 1PG
UK

Professor Annibale Biggeri
Dipartimento di Statistica 'G. Parenti'
Università di Firenze
Viale Morgani 59, Firenze
Italy

Dr John F. Bithell
Department of Statistics
University of Oxford
1 South Parks Road
Oxford
UK

Professor Dankmar Böhning
Department of Epidemiology
Free University Berlin
Fabeckstr 60–62
Haus 562
Berlin
Germany

Mr Allan Clark
Department of Mathematical Sciences
University of Aberdeen
Meston Building
Aberdeen
UK

Dr Erin Conlon
Department of Biostatistics
University of Minnesota
A460 Mayo Building Box 303
420 Delaware Street SE
Minneapolis
USA

Professor Noel A. C. Cressie
Department of Statistics
404 Cockins Hall
Ohio State University
Columbus
Ohio
USA

Ms Rosemary Day
Centre for Economic and Social Research on the Global Environment (CSERGE)
University of East Anglia
Norwich
UK

Dr Fabio Divino
Dipartimento di Statistica 'G. Parenti'
Universita di Firenze
Viale Morgagni 59
Firenze
Italy

Dr Helen Dolk
Environmental Epidemiology Unit
Department of Public Health and Policy
London School of Hygiene and Tropical Medicine
Keppel Street, London
UK

Ms Emanuela Dreassi
Dipartimento di Statistica 'G. Parenti'
Universita di Firenze
Viale Morgagni 59
Firenze
Italy

Professor Juan Ferrandiz
Department d'Estadistica i I. O.,
Universitat de València
Dr Moliner 50
Burjassot
Spain

Professor Arnoldo Frigessi
Norwegian Computing Center
Gaudstadalleen, 23
PO Box 114
Blindern
Oslo
Norway

Dr W. Gilks
MRC Biostatistics Unit
Institute of Public Health
Robinson Way
Cambridge
UK

Professor Harvey Goldstein
Multilevel Models Project
Institute of Education
University of London
UK

Mr Michael Hills
London School of Hygiene and Tropical Medicine (retired)
London
UK

Dr Wolfgang Hoffmann
Bremen Institute for Prevention Research and Social Medicine (BIPS)
Gruenenstrasse 120
Bremen
Germany

Professor John Jacquez
Departments of Biostatistics and Physiology
The University of Michigan
Ann Arbor
Michigan
USA

Dr Geoffrey Jacquez
Biomedware Inc,
516 North State Street
Ann Arbor
Michigan
USA

Dr Alan Kelly
Small Area Health Research Unit
Department of Community Health & General Practice
Trinity College
Dublin

Dr J. Komakec
Dipartimento di Scienze Sanitarie Applicate e Psicocomportamentali
Universitá di Pavia
Via Bassi 21
Pavia
Italy

Dr Martin Kulldorff
Division of Biostatistics
Department of Community Medicine and Health Care
University of Connecticut
School of Medicine
Farmington, Connecticut
USA

Dr Corrado Lagazio
Dipartimento di Statistica 'G. Parenti'
Universita di Firenze
Viale Morgagni 59
Firenze
Italy

Dr Ian H. Langford
CSERGE
University of East Anglia
Norwich
UK

Dr Andrew B. Lawson
Department of Mathematical Sciences
University of Aberdeen
Meston Building
Aberdeen,
UK

Professor Emmanuel Lesaffre
Epidemiology and Biostatistics
Catholic University
Leuven
Belgium

Dr Alistair H. Leyland
Public Health Research Unit
University of Glasgow
Glasgow
UK

Dr A. López
Department d'Estadistica i I.O.
Universitat deValència
Dr Moliner 50
Burjassot
Spain

Professor Thomas A. Louis
Department of Biostatistics
University of Minnesota
A460 Mayo Building Box 303
420 Delaware Street SE
Minneapolis
USA

Ms Anné-Lise McDonald
CSERGE
University of East Anglia
Norwich
UK

Dr L. Marrett
Department of Preventive Oncology
Cancer Care Ontario
Toronto, Ontario
Canada

Dr Marco Martuzzi
WHO European Center for Environment and Health
Via Francesco Crispi, 10
Rome
Italy

Dr Annie Mollié
Institut National de la Santé et de la Recherche Médicab U170
16 Avenue Paul Vaillant Couturier
Villejuif Cedex
France

Dr Cristina Montomoli
Dipartimento di Scienze Sanitarie Applicate e Psicocomportamentali
Universitá di Pavia
Via Bassi 21
Pavia
Italy

Dr Raymond Neutra
California Department of Health Sciences
5801 Christie Avenue
Suite 600
Emeryville
California
USA

Dr C. Pascutto
Dipartimento di Scienze Sanitarie Applicate e Psicocomportamentali
Universitá di Pavia
Via Bassi 21
Pavia
Italy

Professor Goran Pershagen
Institute of Environmental Medicine
Karolinska Institute
PO Box 210
Stockholm
Sweden

Dr Catherine Poquette
St Jude Children's Research Hospital
Memphis
Tennessee
USA

Mr Jon Rasbash
Multilevel Models Project
Institute of Education
University of London
London
UK

Dr Pilar Sanmartín
Department de Matemàtica Aplicada,
Universitat Jaume I
Spain

Dr Peter Schlattmann
Department of Psychiatry
Lab for Clinical Psychophysiology
Free University
Berlin
Eschenallee
Berlin
Germany

Dr Irene Schmidtmann
Institut für Medizinische Statistik und Dokumentation
Johannes Gutenberg-Universitat
Mainz
Germany

Professor Hal Stern
Department of Statistics
Iowa State University
Snedecor Hall
Ames, IA
USA

Professor Toshiro Tango
Division of Theoretical Epidemiology
National Institute of Public Health
4-6-1 Shirokanedai
Minato-Ku, Tokyo
Japan

Dr S. Martin Taylor
Department of Geography
McMaster University
Hamilton
Ontario
Canada

Professor Jean-François Viel
Department of Public Health
Faculty of Medicine
2 Place Saint Jacques
Besancon
France

Dr M. Vrijheid
Environmental Epidemiology Unit
Department of Public Health and Policy
London School of Hygiene and Tropical Medicine
London
UK

Dr Lance Waller
Department of Biostatistics
Emory University
Atlanta, Georgia
USA

Professor Stephen Walter
Department of Clinical Epidemiology and Biostatistics
HSC-2C16
1200 Main St W
Hamilton, Ontario
Canada

Dr Fiona L. R. Williams
Department of Epidemiology and Public Health
University of Dundee
Ninewells Hospital
Dundee
UK

Professor Steven Wing
Department of Epidemiology
School of Public Health
CB 7400 University of North Carolina
Chapel Hill
North Carolina
USA

Dr Iris Zöllner
Landesgesundheits amt
BadenWurttemberg
Abteilung IV
Hoppenlaustrasse 7
Stuttgart
Germany

PART I

Disease Mapping

1

Disease Mapping and Its Uses

Andrew B. Lawson

University of Aberdeen

D. Böhning

Free University, Berlin

A. Biggeri

University of Florence

E. Lesaffre

Katholic University, Leuven

J.-F. Viel

University of Besançon

1.1 INTRODUCTION

The representation and analysis of maps of disease incidence data is now established as a basic tool in the analysis of regional public health. The development of methods for mapping disease incidence has progressed considerably in recent years.

One of the earliest examples of disease mapping is the map of the addresses of cholera victims related to the locations of water supplies, by John Snow in 1854 (Snow, 1854). In that case, the street addresses of victims were recorded and their proximity to putative pollution sources (water supply pumps) was assessed.

The uses made of maps of disease incidence are many and various. Disease maps can be used to assess the need for geographical variation in health resource allocation, or could be useful in research studies of the relation of incidence to explanatory variables. In the first case, the purpose of mapping is to produce a map 'clean' of any random noise

Disease Mapping and Risk Assessment for Public Health. Edited by A.B. Lawson *et al.*
© 1999 John Wiley & Sons Ltd.

and any artefacts of population variation. This can be achieved by a variety of means. In the second case, specific hypotheses concerning incidence are to be assessed and additional information included in the analysis (for example, covariates). The first approach is close to image processing, and the second approach can be regarded as a spatial regression approach. This latter area is sometimes termed 'ecological analysis' and includes the analysis of the relationship between incidence and explanatory variables, which are available at an aggregated level. This includes the area of the analysis of putative sources of health hazard and a more general analysis of the relation of disease incidence to explanatory covariates.

The types of data that arise in mapping exercises can vary from the locations (usually residential) of cases of disease, to counts of disease within small areas (e.g. census tracts), and could involve functions of their distribution (e.g. rates or relative risks). Each data type requires the choice of a geographic resolution or scale at which the map is to be constructed. This choice may be forced on the analyst by existing geographic boundaries, or can be freely defined by the purpose of the mapping exercise. For example, a study of disease incidence within the area of a town will, per force, have the town limits as its boundary. However, when a free choice is available the analyst must make a trade-off between resolution of the data and the precision of measurement. Study area design is discussed more fully by Lawson and coworkers (Lawson, 1998b; Lawson and Waller, 1996). If rates or relative risks are to be mapped then there is a need to choose a reference expected rate or comparison group. These data must be geographically and temporally conformable, or a statistical model will be needed to allow correct matching of expected to observed data. Even if the data are conformable, models are effective in smoothing geographic variation. Adjustments for age or gender, for example, can be accomplished by pre-processing the data or by building such factors into the model.

In what follows we attempt to review a wide range of issues related to current developments in disease mapping. In other chapters of this volume there are examples of the application of mapping methods. In particular, detailed discussions of Bayesian and empirical Bayesian methods are provided by Mollié in Chapter 2, and some newer methodologies are examined in Chapters 3–5. Edge effects are examined in Chapter 6. A case study of the use of disease mapping in application to lung cancer in women is provided in Chapter 32.

1.2 SIMPLE STATISTICAL REPRESENTATION

The representation of disease incidence data can vary from simple point maps for cases, and pictorial representation of counts within tracts, to the mapping of estimates from complex models purporting to describe the structure of the disease events. In the following sections we describe the range of mapping methods from simple representation to model-based forms. Figure 1.1 displays a typical example of the type of data commonly found in disease mapping studies. In this example the address locations of cases are not available and the count of cases within census tracts are portrayed on the map.

Two fundamental forms of data are usually found in these studies. First, the residential addresses of cases of a disease of interest are available. In this case, we are concerned about the analysis of the coordinate locations of events. Note that the use of residential location as a factor in disease risk must be evaluated in any application.

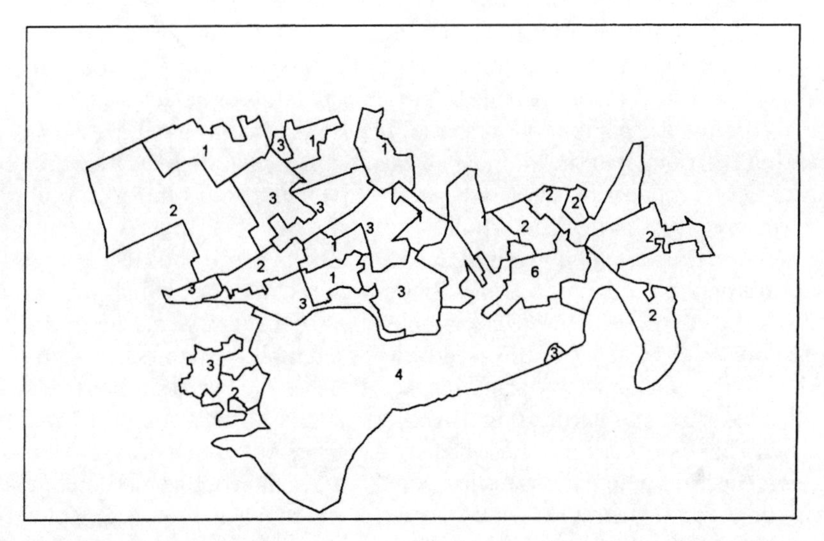

Figure 1.1 Falkirk, central Scotland: counts of respiratory cancer cases within census tracts for a fixed time period

Alternative exposure locations, such as workplace, may be more relevant in some disease studies. The second form of data, which is more commonly encountered, is the tract count, where small areas of a region (tracts) are the basis for aggregating the case events into counts. In this case the exact residential locations are unknown. These data are often more readily available as the use of case addresses may be restricted due to medical confidentiality. Figure 1.1 is an example of the latter data type.

1.2.1 Standardised mortality/morbidity ratios and standardisation

To assess the status of an area with respect to disease incidence it is convenient to attempt to first assess what disease incidence should be locally 'expected' in the area and then to compare the observed incidence with the 'expected' incidence. This approach has been traditionally used for the analysis of counts within tracts and can also be applied to case event maps.

Case events

Case events can be mapped as a map of point event locations. Define a realisation of m case locations in a study window W, as $\{\boldsymbol{x}_i\}, i = 1, \ldots, m$. For the purposes of assessment of differences in local disease risk it is appropriate to convert these locations into a continuous surface describing the spatial variation in *intensity* of the cases. Once this surface is computed then a measure of local variation is available at any spatial location within the observation window. Denote the intensity surface as $\lambda(\boldsymbol{x})$, where $\boldsymbol{x}$ is a spatial location. This surface can be formally defined as the first-order intensity of a point process on $\Re^2$ (see, for example, Lawson and Waller, 1996). This surface can be estimated by a variety of methods including density estimation (Härdle, 1991).

Since the case events occur within a population that is spatially varying in its density and composition, it is important to make some allowance within any study for this variation. This population can be termed the 'at risk' *population* or *background*. To provide an estimate of the 'at risk' population at spatial locations, it is necessary to first choose a measure that will represent the intensity of cases 'expected' at such locations. Define this measure as $g(\boldsymbol{x})$. Two possibilities can be explored. First, it is possible to obtain rates for the case disease from either the whole study window or a larger enclosing region. Usually these rates are available only aggregated into larger regions (e.g. census tracts). The rates are obtained for a range of subpopulation categories which are thought to affect the case disease incidence. For example, the age and sex structure of the population or the deprivation status of the area (see, for example, Carstairs, 1981) could affect the amount of population 'at risk' from the case disease. The use of such external rates is often called 'external standardisation' (Inskip *et al.*, 1983). Rates computed within census tracts will be smoother than those based on density estimation of case events, due to the aggregate level at which the rates are collected. An alternative method of assessing the 'at risk' population structure is to use a case event map of a disease which represents the background population but is not affected by the aetiological processes of interest in the case disease. For example, the spatial distribution of coronary heart disease (CHD) could provide a *control* representation for respiratory cancer when the latter is the case disease in a study of air pollution effects, as CHD is less closely related to air pollution insult. While exact matching of diseases in this way will always be difficult, there is an advantage in the use of control diseases in case event examples. If a realisation of the control disease is available in the form of a point event map, then it is possible to also compute an estimate of the first-order intensity of the control disease. This estimate can then be used to compare the intensity of case events with the intensity of the background.

The comparison of estimates of $\lambda(\boldsymbol{x})$ and $g(\boldsymbol{x})$ can be made in a variety of ways. First, it is possible to map the ratio form:

$$R(\boldsymbol{x}) = \frac{\hat{\lambda}(\boldsymbol{x})}{\hat{g}(\boldsymbol{x})} \tag{1.1}$$

(see, for example, Bithell, 1990; Kelsall and Diggle, 1995; Lawson and Williams, 1993).

Apart from ratio forms it is also possible to map transformations of ratios (e.g. log $R(\boldsymbol{x})$) or to map

$$D(\boldsymbol{x}) = \hat{\lambda}(\boldsymbol{x}) - \hat{g}(\boldsymbol{x}). \tag{1.2}$$

In all the above approaches to the mapping of case event data some smoothing or interpolation of the event or control data has to be made and the optimal choice of smoothing is of considerable importance given that two functions are being estimated. It is also possible to directly estimate $R(\boldsymbol{x})$ or $D(\boldsymbol{x})$ (Kelsall and Diggle, 1995).

Tract counts

As in the analysis of case events, it is usual to assess maps of count data by comparison of the observed counts with those counts 'expected' to arise given the tracts' 'at risk' population structure. Traditionally, the ratio of observed to expected counts within tracts is called a Standardised Mortality/Morbidity Ratio (SMR) and this ratio is an estimate of *relative risk* within each tract (i.e. the ratio describes the relative risk of being in the disease group rather than the background group). The justification for the use of SMRs can be supported by the analysis of likelihood models with multiplicative expected risk (see, for example, Breslow and Day, 1987).

Define $\boldsymbol{o}_i$ as the observed count of the case disease in the ith tract, and $\boldsymbol{e}_i$ as the expected count within the same tract. Then the SMR is defined as

$$R_i = \frac{\boldsymbol{o}_i}{\boldsymbol{e}_i}. \tag{1.3}$$

The alternative measure of relation between observed and expected counts, which is related to an additive risk model, is the difference

$$D_i = \boldsymbol{o}_i - \boldsymbol{e}_i. \tag{1.4}$$

In both cases the comments made above about mapping counts within tracts apply. In this case it must be decided whether to express the R_i or D_i as fill patterns in each region or across regions, or to locate the result at some specified tract location, such as the centroid. If it is decided that these measures should be regarded as continuous across regions, then some further interpolation of R_i or D_i must be made (see, for example, Breslow and Day, 1987, pp. 198–199).

1.2.2 Interpolation and smoothing

In many of the mapping approaches mentioned above, use must be made of interpolation methods to provide estimates of a surface measure at locations where there are no observations. Smoothing of SMRs has been advocated by Breslow and Day (1987). Those authors employ kernel smoothing to interpolate the surface (in a temporal application). One advantage of such smoothing is that the method preserves the positivity condition of SMRs: that is, the method does not produce negative interpolants. Geostatistical smoothing methods such as Kriging (Carrat and Valleron, 1992), which are designed to provide interpolation and prediction of continuous surfaces, do not preserve positivity, at least in their standard form. Kriging can be generalised to non-Gaussian models such as the lognormal, or applied to log transformed observations, thus assuring non-negative interpolation. Other interpolation methods also suffer from this problem (see, for example, Lancaster and Salkauskas, 1986; Ripley, 1981). Many mapping packages utilise interpolation methods to provide gridded data for further contour and perspective view plotting (e.g. AXUM, SPLUS). However, often the methods used are not clearly defined or based on mathematical interpolants (e.g. the Akima interpolator).

Note that the above comments also apply directly to case event density estimation. The use of kernel density estimation is recommended, as it provides a parsimoneous description of the intensity surface. However, as with many smoothing operators, edge effects can occur due to the censoring of observations outside the study region (i.e. the smoother uses only data from within the region and so at locations close to the study region boundary there can be an error induced by censoring. Corrections for edge effects are available (see Chapter 6 in this volume). For ratio estimation, Kelsall and

Diggle (1995) recommend the joint estimation of a common smoothing parameter for the numerator and denominator of $R(\boldsymbol{x})$ when a control disease realisation is available.

1.3 MODEL-BASED APPROACHES

It is possible to proceed by further assuming that a parametric model underlies the map of disease. The models described here can lead to the construction of likelihoods and to the specification of prior distributions for parameters, if a Bayesian approach is to be adopted.

1.3.1 Likelihood models

Case event data

Usually the basic model for case event data is derived from the following assumptions:

(i) Individuals within the study population behave independently with respect to disease propensity, after allowance is made for observed or unobserved confounding variables. That is, if the factors affecting disease risk were fully specified for each person, then they would have an independent chance of contracting the disease of interest.
(ii) The underlying 'at risk' population has a continuous spatial distribution, within the specified study region. This assumption allows the specification of a continuous 'at risk' background for modelling purposes. If this were not the case, for example when areas within the study region contained no population, then the study region would require modification.
(iii) The case events are unique, in that they occur as single spatially separate events. This assumption underlies the specification of point process models.

Given the above assumptions, it is justified to assume that the case events arise as a realisation of a Poisson point process, modulated by $g(\boldsymbol{x})$, a background function representing the 'at risk' population. This process is governed by a first order intensity:

$$\lambda(\boldsymbol{x}) = \rho \cdot g(\boldsymbol{x}) \cdot f(\boldsymbol{x}; \theta). \tag{1.5}$$

In this definition the local intensity of cases consists of an overall region-wide rate ρ, modified by two spatially-dependent components: the $g(\boldsymbol{x})$ function and $f(\cdot)$ representing a function of spatial location and possibly other parameters and associated covariates as well. These covariates could be observed explanatory variables (confounder variables) or could be unobserved effects. For example, a number of random effects representing unobserved heterogeneity could be included as well as observed covariates, as could functions of other locations. The likelihood associated with the Poisson process model is given by

$$L = \frac{1}{m!} \prod_{i=1}^{m} [\lambda(\boldsymbol{x}_i)] \cdot \exp\left\{ - \int_W \lambda(\boldsymbol{u}) \mathrm{d}\boldsymbol{u} \right\}. \tag{1.6}$$

For suitably specified $f(\cdot)$, a variety of models can be derived. In the case of disease mapping, where only the background is to be removed without further model assumptions, then a reasonable approach to intensity parameterisation is $\lambda(\boldsymbol{x}) = \rho \cdot g(\boldsymbol{x}) \cdot f(\boldsymbol{x})$.

Count event data

In the case of observed counts of disease within tracts, then given the above Poisson process assumptions it can be assumed that the counts are Poisson distributed with, for each tract, a different expectation: $\int_{W_i} \lambda(\boldsymbol{u})\mathrm{d}\boldsymbol{u}$, where W_i denotes the area of the ith tract. The log likelihood based on a Poisson distribution is then, bar a constant only depending on the data, given by

$$l = \sum_{i=1}^{m} \left\{ \boldsymbol{o}_i \log \int_{W_i} \lambda(\boldsymbol{u})\mathrm{d}\boldsymbol{u} - \int_{W_i} \lambda(\boldsymbol{u})\mathrm{d}\boldsymbol{u} \right\}. \tag{1.7}$$

Often a parameterisation in (1.7) is assumed where, as in the case event example, the intensity is defined as a simple multiplicative function of the background $g(\boldsymbol{x})$.

1.3.2 Random effects and Bayesian models

In the above sections some simple approaches to mapping intensities and counts within tracts have been described. These methods assume that once all known and observable confounding variables are included within the $g(\boldsymbol{x})$ estimation then the resulting map will be clean of all artefacts and hence depicts the true excess risk surface. However, it is often the case that unobserved effects could be thought to exist within the observed data and that these effects should also be included within the analysis. These effects are often termed *random* effects, and their analysis has provided a large literature both in statistical methodology and in epidemiological applications (see, for example, Breslow and Clayton, 1993; Cislaghi *et al.*, 1995; Ghosh *et al.*, 1998; Lawson *et al.*, 1996; Manton *et al.*, 1981; Marshall, 1991b). Within the literature on disease mapping there has been a considerable growth in recent years in modelling random effects of various kinds. In the mapping context, a random effect could take a variety of forms. In its simplest form a random effect is an extra quantity of variation (or variance component) which is estimable within the map and which can be ascribed a defined probabilistic structure. This component can affect individuals or can be associated with tracts or covariables. A simple general specification has been suggested by Besag *et al.* (1991) in application to count data, and further developed to include additional effects (Lawson *et al.*, 1996; Lawson, 1997). Define the intensity of the process at the ith location as

$$\lambda(\boldsymbol{x}_i) = \rho \cdot g(\boldsymbol{x}_i) \cdot \zeta_i \cdot \gamma_i,$$

where $\zeta_i = \exp(t_i + u_i + v_i)$ and $\gamma_i = f\{\sum_{j=1}^{k} h(\boldsymbol{x}_i - \boldsymbol{y}_j)\}$. This specification allows the inclusion of a variety of random effects at the ith observation level. Here, t is the trend component, and u and v are components that describe *extra* variation. This extra variation could arise from unobserved heterogeneity in the data, due to our inability to completely explain the individual or regional response to disease. These are non-specific random effects in that they do not *locate* random structures in the map but simply describe the overall random variation. Usually v is assumed to represent uncorrelated (overdispersion) heterogeneity and u represents the spatially correlated form.

The final term in this general model (γ) is a specific random effect, commonly known as a random *object* effect. This effect relates the observed locations to objects that have a random location. In this case the objects $\{\boldsymbol{y}\}$ are k cluster centres, and both $\{\boldsymbol{y}, k\}$ are unknown. The aim of an analysis which includes such effects would be to estimate the locations of clusters/cluster centres within the data. Other types of effect could be included within a particular analysis, depending on the level of aggregation and approximation used in the study.

A Bayesian approach

It is natural to consider modelling random effects within a Bayesian framework, although it is also possible to integrate over random effects and use a marginal likelihood approach (Aitken, 1996). First, random effects naturally have prior distributions. The prior distributions for these parameters have hyperparameters which can have hyperprior distributions also. In the full Bayesian approach, inference is usually based on samples of parameters from the joint posterior distribution. However, it is also possible to adopt an intermediate approach where the parameters of the prior distributions are estimated and further inference is made conditional on these estimated parameters. This type of empirical Bayes approach has often been applied in disease mapping (Manton *et al.*, 1981; Tsutakawa, 1988).

Few examples exist of simple Bayesian approaches to the analysis of case event data in disease mapping. The approach of Lawson *et al.* (1996) can be used with simple prior distributions for parameters. Their method relied on approximate maximum a posteriori (MAP) estimation of local intensities based on Dirichlet tile area approximations. They also compared these MAP estimates with full Bayesian modal estimates. For count data a number of examples exist where independent Poisson distributed counts, with constant tract rate (λ_i), are associated with prior distributions of varying complexities. The earliest examples of such a Bayesian approach are Manton *et al.* (1981) and Tsutakawa (1988). Clayton and Kaldor (1987) also developed a Bayesian analysis of a Poisson likelihood model where $\boldsymbol{o}_i$ has expectation $\theta_i \boldsymbol{e}_i$ and found that with the prior distribution given by $\theta_i \sim \text{Gamma}(\alpha, \beta)$ then the posterior estimate of θ_i is

$$\frac{\boldsymbol{o}_i + \alpha}{\boldsymbol{e}_i + \beta}$$

for estimates of the hyperparameters obtained from considering the negative binomial likelihood which is the unconditional distribution of $\{\boldsymbol{o}_i\}$. Hence, it would be possible to map directly the θ_i estimates as posterior means. On the other hand, the distribution of θ_i conditional on $\boldsymbol{o}_i$ is $\text{Gamma}(\boldsymbol{o}_i + \alpha, \boldsymbol{e}_i + \beta)$ and a full Bayesian approach would require the sampling of θ_i from this distribution, possibly with suitable sample summarisation. Other approaches and variants in the analysis of simple mapping models have been proposed by Devine and Louis (1994) and Marshall (1991b).

1.3.3 Mixture models and hidden structure

The above model-based mapping methods have a common theme, namely the assumption that the structure of the map can be described by a model with a global structure. An alternative approach is to assume that the map consists of a number of components, and that the task is to identify these components. Each component could potentially have a complex structure and so the model could be regarded as a composite of simple global models. In some work (e.g. Schlattmann and Böhning, 1993) each area of the map is classified as belonging to one component. However, it is not essential to assume this feature and broader models are possible.

Böhning and coworkers (Böhning *et al.*, 1992, 1998; Schlattmann and Böhning, 1993; Schlattmann *et al.*, 1996) first proposed a mixture model for a disease map, where the count is considered to be governed by a Poisson distribution with intensity $\lambda_i = \boldsymbol{e}_i \cdot \sum_{j=1}^{k} w_j \cdot \lambda_j^*$, where there are k components (λ_j^*) and weights (w_j). Note that this model can be formulated as a Bayesian model with a discrete prior distribution for the number of components. These models can be sampled via iterative simulation methods such as Markov chain Monte Carlo (MCMC) (Gilks *et al.*, 1996a) or by EM methods. More complex models with not only the *marginal* count distributions as mixtures but also the spatial distribution could be fruitfully explored.

1.4 SPATIO-TEMPORAL MODELLING

The development of disease incidence in the spatial domain has been paralleled by the development of methods where time is implicitly included in the analysis. This extension can take a variety of forms depending on how the sampling of the process in time is made. For example, at one extreme, a date of diagnosis of a case may be known as well as the residential location, whereas at another, only the count of cases within a small area tract within a time period may be available. In addition, many other forms of time-based information may arise: duration of periods of illness, longitudinal case history, intervention, or more general cohort effects. There could also be differences in approach depending on whether an evolutionary/sequential method is to be used (e.g. for screening of events as they occur), or an analysis is based on the full time span studied. Furthermore, there may also be some need to gain greater ancillary or covariate information in such studies due to the extension of study over periods where such variables are likely to change.

There has been some development of methods pertaining to small area counts in fixed time periods (Bernardinelli *et al.*, 1995b; Knorr-Held and Besag, 1998; Waller *et al.*, 1997a; Xia *et al.*, 1997). These approaches define either a Poisson likelihood for the counts with a log linear model for the spatial and spatio-temporal components, or a binomial model for the count within a finite population. Different authors have examined different parameterisations of this model. For example, Bernardinelli *et al.* (1995b) have specified an area-specific factorial effect and space–time factor interaction. There has been only limited interest in models for case-event data. However, it is possible to propose a general approach to case-event space–time modelling by extension of the basic unconditional intensity (1.5) to include temporal dependence. This general approach can be extended to small area counts also and so forms a general model.

The intensity specification at time t is

$$\lambda(\boldsymbol{x}, t) = \rho \cdot g(\boldsymbol{x}, t) \cdot f_1(\boldsymbol{x}; \theta_x) \cdot f_2(t; \theta_t) \cdot f_3(\boldsymbol{x}, t; \theta_{xt}),$$

where ρ is a constant background rate (in space $\times$ time units), $g(\boldsymbol{x}, t)$ is a modulation function describing the spatio-temporal 'at-risk' population background in the study region, f_k are appropriately defined functions of space, time and space–time, and θ_x, θ_t, and θ_{xt} are parameters relating to the spatial, temporal, and spatio-temporal components of the model.

Here each component of the f_k can represent a *full* model for the component, i.e. f_1 can include spatial trend, covariate and covariance terms, and f_2 can contain similar terms for the temporal effects, while f_3 can contain *interaction* terms between the components in space and time. Note that this final term can include *separate* spatial structures relating to interactions that are not included in f_1 or f_2. The exact specification of each of these components will depend on the application, but the separation of these three components is helpful in the formulation of components.

The above intensity specification can be used as a basis for the development of likelihood and Bayesian models for case events. If it can be assumed that the events form a modulated Poisson process in space–time, then a likelihood can be specified as in the spatial case.

Note that the above case-event intensity specification can be applied in the space–time case where small area counts are observed within fixed time periods $\{t_j\}, j = 1, \ldots, l$, by noting that

$$\mathrm{E}\{\boldsymbol{o}_{it_j}\} = \int_{t_j} \int_{W_i} \lambda(\boldsymbol{x}, t)\, \mathrm{d}\boldsymbol{x}\, \mathrm{d}t,$$

under the usual assumption of Poisson process regionalisation. In addition, the counts are independent conditional on the intensity given, and this expectation can be used within a likelihood modelling framework or within Bayesian model extensions. In previous published work in this area, cited above, the expected count is assumed to have constant risk within a given small-area/time unit, which is an approximation to the continuous intensity defined for the underlying case events. The appropriateness of such an approximation should be considered in any given application. If such an approximation is valid, then it is straightforward to derive the minimal and maximal relative risk estimates under the Poisson likelihood model assuming $\mathrm{E}\{\boldsymbol{o}_{it_j}\} = \lambda_{it_j} = e_{it_j} \cdot \theta_{it_j}$, where e_{it_j} is the expected rate in the required region/period. The maximal model estimate is $\hat{\theta}_{it_j} = \boldsymbol{o}_{it_j}/e_{it_j}$, the space–time equivalent of the SMR, while the minimal model estimate is

$$\hat{\theta} = \frac{\sum_i \sum_j \boldsymbol{o}_{it_j}}{\sum_i \sum_j e_{it_j}}.$$

Smooth space–time maps, e.g. empirical Bayes or full Bayes relative risk estimates, will lie between these two extremes. If the full integral intensity is used, then these estimates have the sums in their denominators replaced by integrals over space–time units.

1.5 CONCLUSIONS

In this brief review we have attempted to highlight important areas of development in methods in disease mapping and to suggest areas of potential new work. It is clear that approaches based on Bayesian methodology provide a flexible paradigm for the assessment of variability of parameter samples and sensitivity, and MCMC methods are likely to facilitate this approach. Potentially fruitful areas which may benefit from development in the future are: the assessment of different scales of pattern or aggregation within a study, the use of hidden structure models, and finally, space-time modelling of case event and count data.

As in many areas of statistical application, the use of models for disease mapping should be accompanied by the assessment of the goodness-of-fit of the model, and where appropriate the sensitivity of the model to specification/variation in assumptions. For goodness-of-fit many global measures exist, e.g. for likelihood models the AIC (Akaike Information Criterion) can be employed. This measure combines the assessment of likelihood of a model with the number of parameters employed. Similarly, for Bayesian models the BIC (Bayesian Information criterion) can be employed. Residual analysis can also be employed to assess pointwise goodness-of-fit.

2

Bayesian and Empirical Bayes Approaches to Disease Mapping

Annie Mollié

INSERM

2.1 INTRODUCTION

The analysis of geographical variations in rates of disease mortality or incidence is useful in the formulation of aetiological hypotheses. Disease mapping aims to elucidate the geographical distribution of underlying disease rates, and to identify areas with low or high rates. The two main conventional approaches are maps of standardised rates based on Poisson inference and maps of statistical significance. The former has the advantage of providing estimates of the parameters of interest, namely the disease rates, but raises two problems.

First, for rare diseases and for small areas, variation in the observed number of events exceeds that expected from Poisson inference. In a given area, variation in the observed number of events is due partly to Poisson sampling, but also to extra-Poisson variation.

To overcome this problem, Bayesian approaches have been developed in disease mapping. They consist of considering, in addition to the observed events in each area, prior information on the variability of disease rates in the overall map. Bayesian estimates of area-specific disease rates integrate the two types of information. Bayesian estimates are close to the standardised rates, when based upon a large number of events or person-years of exposure. However, with few events or person-years, prior information on the overall map will dominate, thereby shrinking standardised rates towards the overall mean rate.

The second problem in using the conventional approach based on Poisson inference is that it does not take account of any spatial pattern in disease, i.e. the tendency for geographically close areas to have similar disease rates. Bayesian approaches with prior

Disease Mapping and Risk Assessment for Public Health. Edited by A.B. Lawson *et al.*
© 1999 John Wiley & Sons Ltd.

information on the rates allowing for local geographical dependence are then pertinent. With this prior information, a Bayesian estimate of the rate in an area is shrunk towards a local mean, according to the rates in the neighbouring areas. A parallel can be drawn with the image restoration problem, our goal being to reconstruct the true image (disease rates) from noisy observed data (event counts).

Empirical Bayes method yields acceptable point estimates of the rates but underestimates their uncertainty. A direct fully Bayesian approach is rarely tractable with the non-conjugate distributions typically involved. However, a Monte Carlo technique called the Gibbs sampler has been used to simulate posterior distributions and produce satisfactory point and interval estimates for disease rates.

This application is part of the general theory of Bayesian analysis using generalised linear mixed models, developed and discussed in Breslow and Clayton (1993) and Clayton (1996).

2.2 MAXIMUM LIKELIHOOD ESTIMATION OF RELATIVE RISKS OF MORTALITY

The map is supposed to be divided into n contiguous areas labelled $i = 1, \ldots, n$. Let $\boldsymbol{y} = (y_1, \ldots, y_n)$ denote the number of deaths from the disease of interest during the study period. Expected number of deaths $\boldsymbol{e} = (e_1, \ldots, e_n)$ are supposed to be known, constant during the study period, and calculated by applying the overall age- and sex-specific death rates to the population at risk in an area.

Independently in each area, the number of deaths y_i is supposed to follow a Poisson distribution with mean $e_i r_i$, where $\boldsymbol{r} = (r_1, \ldots, r_n)$ are the unknown area-specific relative risks of mortality from the disease. The likelihood of the relative risk r_i is

$$[y_i | r_i] = \exp\{-e_i r_i\} \frac{(e_i r_i)^{y_i}}{y_i!}.$$

Then the maximum likelihood estimate (MLE) of r_i is the standardised mortality ratio (SMR) for the ith area: $\hat{r}_i = y_i / e_i$ with estimated standard error $s_i = \sqrt{y_i}/e_i$.

However, the most extreme SMRs are those based on only a few cases. On the contrary, the most extreme p-values of tests comparing SMRs to unity or confidence intervals excluding unity may simply identify areas with large populations. These two drawbacks are emphasised in studies on rare diseases or small areas, making the interpretation of maps of SMRs or of p-values difficult, and even misleading.

We consider two illustrations on cancer mortality in the 94 mainland French 'départements' during the period 1986–1993 (Rezvani *et al.*, 1997). The first example consists of 16 923 deaths from leukaemias among females (7.32 per 100 000). The SMRs (Figure 2.1(a)) vary around their mean 0.98 (s.d. = 0.11) from 0.58 for 'département' 05 to 1.20 for 'département' 23, both amongst the 'départements' with one of the smallest population size and expected number of deaths. The five highest SMRs and the five lowest are presented in Table 2.1. Owing to the wide differences in population sizes, the standard errors of the SMRs also have large variations: they range from 0.036 for 'département' 75 which has the second largest population size, to 0.194 for 'département' 48 which has the smallest population (about 32 times less than 'département' 75). Ninety-five

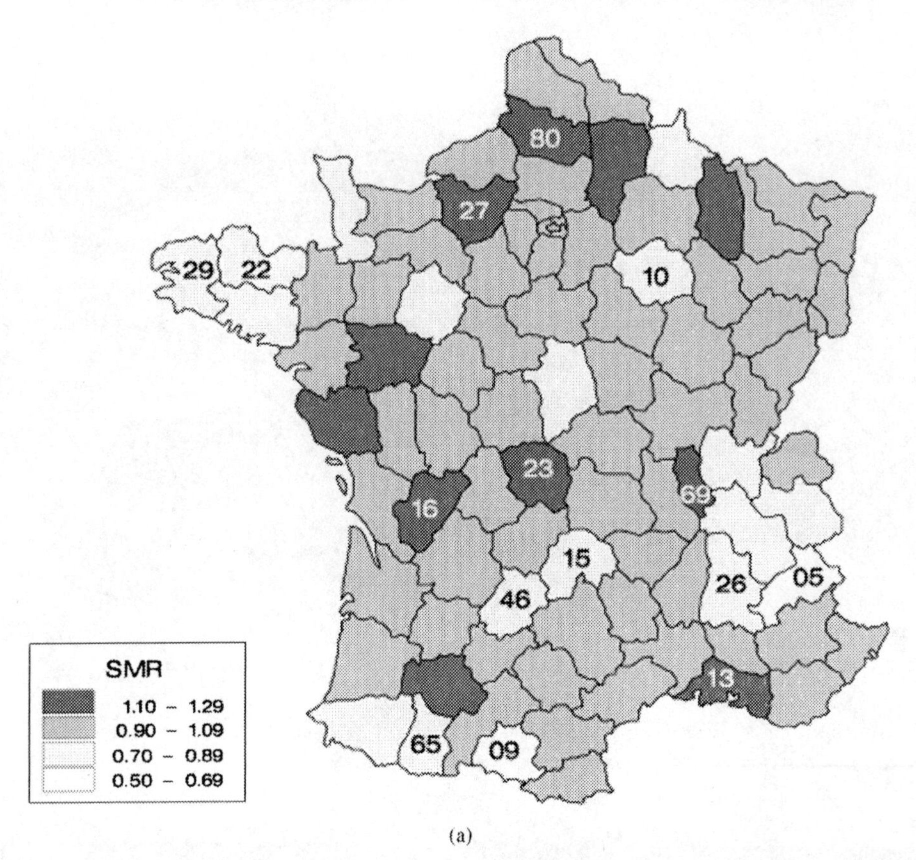

(a)

Figure 2.1 (a) SMRs for leukaemias for females in France, 1986–93. 'Départements' listed in Table 2.1 are labelled. (b) Posterior mean of relative risks of mortality from leukaemias for females in France, 1986–93 (based on 10 000 iterations of the Gibbs sampler, using a convolution Gaussian prior for the log relative risks). 'Départements' listed in Table 2.1 are labelled

percent confidence intervals based on Poisson distribution have been computed for each SMR. Those that exclude unity are presented in Table 2.1.

However, the SMRs highlighted on the map (Figure 2.1(a)) are often those based on the least reliable data and not specifically those significantly different from unity: among the 11 'départements' coloured in dark grey, only four have a significant SMR and among the 18 SMRs represented in white or light grey, only seven are significant. For instance, 'départements' 13, 16, 23 and 69 are all coloured in dark grey but 16 and 23 have a high SMR (1.18 and 1.20, respectively) based on a small population and not significantly different from unity, whereas 'départements' 13 and 69 have lower SMRs (1.13 and 1.15, respectively) based on two of the largest populations, significantly different from unity. Likewise 'départements' 10, 29 and 46 are all coloured in light grey but 10 and 46 have a low SMR (0.82 and 0.79, respectively) based on a small population, not significantly different from unity, whereas 'département' 29 has a higher SMR (0.87) based on a larger population, significantly different from unity.

The second example consists of 1122 deaths from testis cancer among males (0.51 per 100 000). The SMRs (Figure 2.2(a)) vary widely around their mean 1.01 (s.d. = 0.46)

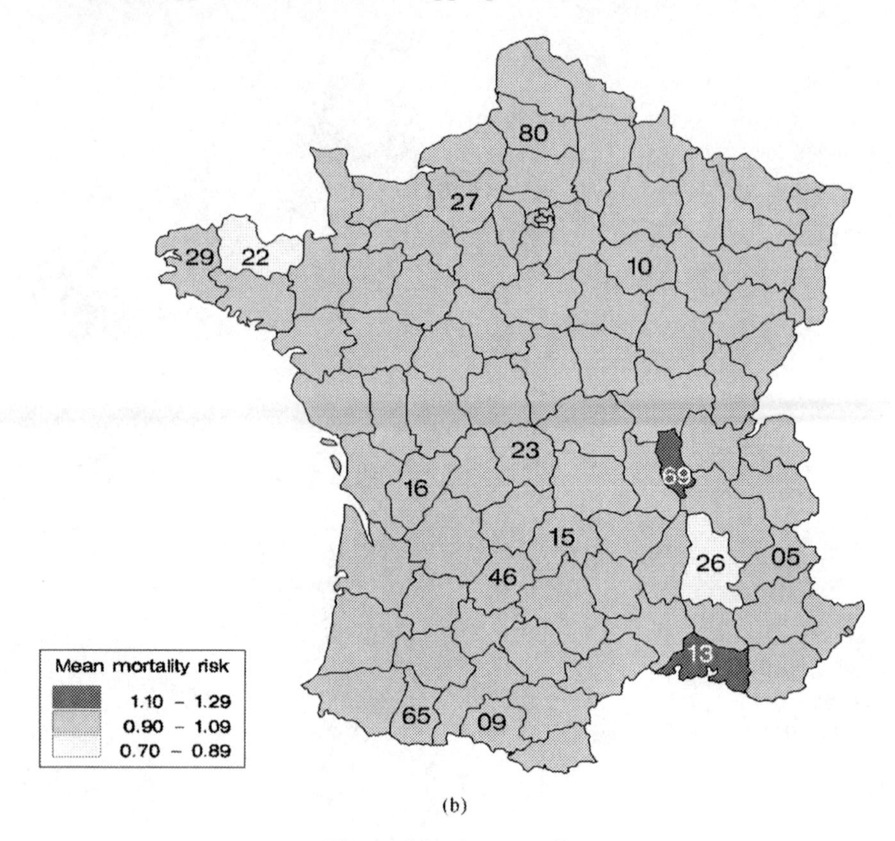

(b)

Figure 2.1 *(continued)*

from 0.15 for 'département' 47 to 2.42 for 'département' 48, which has the smallest population size and expected number of deaths. The five lowest SMRs and the five highest are presented in Table 2.2.

The standard errors of the SMRs range from 0.118 for 'département' 42 to 1.212 for 'département' 48 (Table 2.2) which has the smallest population, tenfold smaller than 'département' 42. Ninety-five percent confidence intervals based on Poisson distribution that exclude unity are presented in Table 2.2. Testis cancer being vary rare in France, the problem of the interpretation of the variations shown on the map of SMRs (Figure 2.2(a)) is emphasised. For instance, two (09, 48) of the six 'départements' coloured in black have SMRs not significantly different from unity, and two (32, 55) of the five 'départements' coloured in white have SMRs not significantly different from unity (Table 2.2) because they are based on very small populations.

Actually, for a rare disease and small areas, since individual risks are heterogeneous within each area, the variability of the average risk of the area exceeds that expected from a Poisson distribution. Extra-Poisson variation can be accommodated by allowing relative risks to vary within each area. Bayesian methods can be used for this, giving smoothed estimates of relative risks. Indeed, even if the SMR is the best estimate of the rate, for each area considered in isolation, Bayesian rules produce sets of estimates having smaller squared-errors loss (when $n \geq 3$) than the set of SMRs.

Table 2.1 Estimates of relative risks of mortality from leukaemias for females in France, 1986–93 (selected 'départements' shown, ordered by decreasing SMR)

i	'départements'	y_i	e_i	SMR	s_i	$CI_{95\%}$(SMR)	Mean	Median	$PI_{95\%}$
23	Creuse	76	63.3	1.20	0.138	(0.96–1.50)	1.03	1.03	(0.92–1.17)
27	Eure	158	133.4	**1.18**	**0.094**	**(1.01–1.39)**	1.06	1.06	(0.96–1.19)
16	Charente	142	120.7	1.18	0.099	(0.99–1.39)	1.06	1.05	(0.95–1.18)
80	Somme	183	156.4	**1.17**	**0.086**	**(1.01–1.36)**	1.08	1.08	(0.98–1.20)
69	Rhône	485	420.2	**1.15**	**0.052**	**(1.06–1.26)**	**1.10**	**1.10**	**(1.01–1.19)**
13	Bouches-du-Rhône	597	527.2	**1.13**	**0.046**	**(1.04–1.23)**	**1.10**	**1.10**	**(1.02–1.18)**
...	...	...	...	...	...	...	...	...	...
92	Hauts-de-Seine*	363	404.6	**0.90**	**0.047**	**(0.81–1.00)**	0.93	0.93	(0.86–1.01)
29	Finistère	245	280.8	**0.87**	**0.056**	**(0.77–0.99)**	**0.90**	**0.90**	**(0.81–0.99)**
10	Aube	75	91.1	0.82	0.095	(0.66–1.03)	0.94	0.95	(0.84–1.06)
22	Côtes-d'Armor	155	192.9	**0.80**	**0.065**	**(0.69–0.94)**	**0.88**	**0.88**	**(0.79–0.98)**
46	Lot	50	63.5	0.79	0.111	(0.58–1.04)	0.94	0.94	(0.82–1.06)
65	Hautes-Pyrénées	64	83.6	**0.77**	**0.096**	**(0.60–0.97)**	0.91	0.91	(0.79–1.03)
26	Drôme	94	127.5	**0.74**	**0.076**	**(0.60–0.91)**	0.89	0.89	(0.79–1.00)
15	Cantal	42	58.1	**0.72**	**0.112**	**(0.52–0.98)**	0.93	0.93	(0.82–1.05)
09	Ariège	39	56.2	**0.69**	**0.111**	**(0.50–0.94)**	0.91	0.91	(0.78–1.04)
05	Hautes-Alpes	21	36.1	**0.58**	**0.127**	**(0.36–0.89)**	0.91	0.92	(0.79–1.05)

i: number of the 'département'.
y_i: observed number of deaths in the ith 'département'.
e_i: expected number of deaths in the ith 'département'.
SMR: standardised mortality ratio for the ith 'département'.
s_i: estimated standard error of the SMR in the ith 'département'.
$CI_{95\%}$(SMR): 95% confidence interval of the SMR in the ith 'département' based on Poisson distribution.
Mean: posterior mean estimated from 10 000 cycles of the Gibbs sampler using a convolution Gaussian prior on the log relative risks.
Median: posterior median estimated from 10 000 cycles of the Gibbs sampler using a convolution Gaussian prior on the log relative risks.
$PI_{95\%}$: 95% posterior interval estimated from 10 000 cycles of the Gibbs sampler using a convolution Gaussian prior on the log relative risks.
*'Département' 92 is a small 'département' around Paris and is not labelled on the maps.

2.3 HIERARCHICAL BAYESIAN MODEL OF RELATIVE RISKS

2.3.1 Bayesian inference for relative risks

Bayesian approaches in this context combine two types of information: the information provided in each area by the observed deaths described by the Poisson likelihood $[\boldsymbol{y}|\boldsymbol{r}]$, and prior information on the relative risks specifying their variability in the overall map, summarised by their prior distribution $[\boldsymbol{r}]$.

Bayesian inference about the unknown relative risks $\boldsymbol{r}$ is based on the marginal posterior distribution $[\boldsymbol{r}|\boldsymbol{y}] \propto [\boldsymbol{y}|\boldsymbol{r}] \times [\boldsymbol{r}]$. The likelihood function of the relative risks $\boldsymbol{r}$ for the data (number of deaths) $\boldsymbol{y}$ is the product of n independent Poisson distributions, since y_i can be assumed to be conditionally independent given $\boldsymbol{r}$ and y_i depends only on r_i:

$$[\boldsymbol{y}|\boldsymbol{r}] = \prod_{i=1}^{n} [y_i|r_i].$$

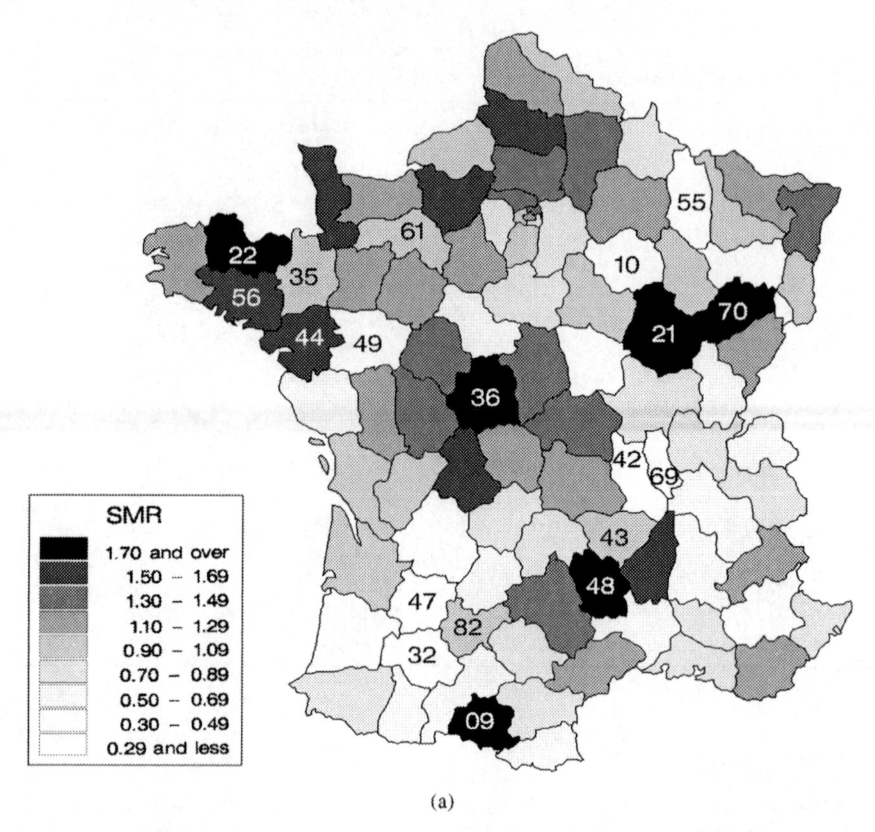

(a)

Figure 2.2 (a) SMRs for testis cancer for males in France, 1986–93. 'Départements' listed in Table 2.2 are labelled. (b) Posterior mean of relative risks of mortality from testis cancer for males in France, 1986–93 (based on 10 000 iterations of the Gibbs sampler, using a convolution Gaussian prior for the log relative risks). 'Départements' listed in Table 2.2 are labelled

The prior distribution $[\boldsymbol{r}]$ reflects prior belief about variation in relative risks over the map. It is supposed to be parameterised by hyperparameters γ and denoted $[\boldsymbol{r}|\gamma]$. The joint posterior distribution of the parameters $(\boldsymbol{r}, \gamma)$ is $[\boldsymbol{r}, \gamma|\boldsymbol{y}] \propto [\boldsymbol{y}|\boldsymbol{r}] \times [\boldsymbol{r}|\gamma] \times [\gamma]$. Thus the marginal posterior distribution for $\boldsymbol{r}$ given the data $\boldsymbol{y}$ is

$$[\boldsymbol{r}|\boldsymbol{y}] = \int [\boldsymbol{r}, \gamma|\boldsymbol{y}] \mathrm{d}\gamma.$$

A point estimate of the relative risks is given by a measure of location of this distribution: typically the posterior mean $\mathrm{E}[\boldsymbol{r}|\boldsymbol{y}]$ or the posterior median. However, direct evaluation of these parameters through analytic or numerical integration is not generally possible.

Another measure of location of this posterior distribution easier to compute and often used in image analysis applications (Besag, 1986, 1989) is the posterior mode or maximum a posteriori (MAP) estimate that maximises $[\boldsymbol{r}|\boldsymbol{y}, \gamma]$. MAP estimation can be

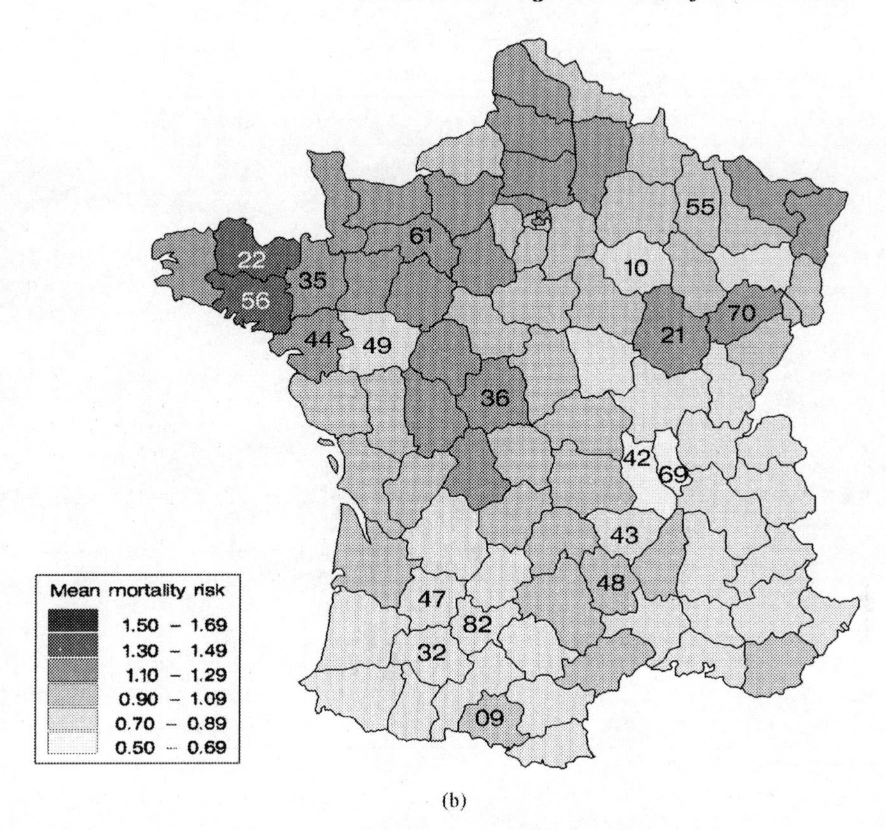

(b)

Figure 2.2 (*continued*)

performed using penalised likelihood maximisation (Clayton and Bernardinelli, 1992) and has been applied to disease mapping (Tsutakawa, 1985; Bernardinelli and Montomoli, 1992).

Standard Bayesian analysis considering a completely specified prior distribution $[\mathbf{r}|\gamma]$ with known hyperparameters γ is seldom used in practice. The empirical Bayes (EB) approach assumes that hyperparameters are unknown and are drawn from an unspecified distribution. The fully Bayesian formulation comprises a three-stage hierarchical model in which the hyperprior distribution $[\gamma]$ is specified.

2.3.2 Independent priors

If the prior belief suggests an unstructured heterogeneity of the relative risks, they will be considered to be independent given γ. In this case the simplest prior assuming exchangeable relative risks is the product of independent and identical distributions:

$$[\mathbf{r}|\gamma] = \prod_{i=1}^{n} [r_i|\gamma].$$

Table 2.2 Estimates of relative risks of mortality from testis cancer for males in France, 1986–93 (selected 'départements' shown, ordered by decreasing SMR)

i	'départements'	y_i	e_i	SMR	s_i	$CI_{95\%}$(SMR)	Mean	Median	$PI_{95\%}$
48	Lozère	4	1.65	2.42	1.212	(0.66–6.21)	0.97	0.96	(0.65–1.51)
36	Indre	12	5.33	**2.25**	**0.650**	**(1.16–3.94)**	1.24	1.22	(0.88–1.84)
70	Haute-Saône	10	4.64	**2.15**	**0.682**	**(1.03–3.96)**	1.14	1.13	(0.80–1.75)
09	Ariège	6	3.20	1.87	0.765	(0.69–4.09)	0.94	0.94	(0.60–1.52)
22	Côtes-d'Armor	19	10.86	**1.75**	**0.401**	**(1.05–2.73)**	1.39	1.39	(0.99–1.97)
21	Côte-d' Or	17	9.87	**1.72**	**0.418**	**(1.00–2.75)**	1.15	1.13	(0.83–1.66)
56	Morbihan	20	12.14	**1.65**	**0.368**	**(1.01–2.54)**	1.35	1.34	(0.98–1.87)
44	Loire-Atlantique	30	19.79	**1.52**	**0.277**	**(1.02–2.17)**	1.29	1.28	(0.97–1.71)
61	Orne	6	5.86	1.02	0.418	(0.38–2.23)	1.10	1.10	(0.76–1.55)
...	...	...	...	...	...	...	...	...	...
35	Ille-et-Vilaine	15	15.21	0.99	0.255	(0.55–1.63)	1.10	1.10	(0.78–1.48)
82	Tarn-et-Garonne	4	4.37	0.92	0.458	(0.25–2.35)	0.85	0.85	(0.57–1.26)
43	Haute-Loire	4	4.40	0.91	0.455	(0.25–2.33)	0.89	0.89	(0.60–1.30)
69	Rhône	14	29.01	**0.48**	**0.129**	**(0.26–0.81)**	**0.68**	**0.69**	**(0.47–0.92)**
49	Maine-et-Loire	6	13.52	**0.44**	**0.181**	**(0.16–0.97)**	0.89	0.91	(0.59–1.22)
55	Meuse	1	3.94	0.25	0.254	(0.01–1.41)	0.90	0.91	(0.58–1.31)
32	Gers	1	4.08	0.24	0.245	(0.01–1.36)	0.75	0.76	(0.48–1.12)
42	Loire	3	14.63	**0.21**	**0.118**	**(0.04–0.60)**	**0.69**	**0.71**	**(0.45–0.97)**
10	Aube	1	5.87	**0.17**	**0.170**	**(0.00–0.95)**	0.86	0.87	(0.55–1.22)
47	Lot-et-Garonne	1	6.57	**0.15**	**0.152**	**(0.00–0.85)**	0.74	0.75	(0.48–1.07)

i: number of the 'département'.
y_i: observed number of deaths in the ith 'département'.
e_i: expected number of deaths in the ith 'département'.
SMR: standardised mortality ratio for the ith 'département'.
s_i: estimated standard error of the SMR in the ith 'département'.
$CI_{95\%}$(SMR): 95% confidence interval of the SMR in the ith 'département' based on Poisson distribution.
Mean: posterior mean estimated from 10 000 cycles of the Gibbs sampler using a convolution Gaussian prior on the log relative risks.
Median: posterior median estimated from 10 000 cycles of the Gibbs sampler using a convolution Gaussian prior on the log relative risks.
$PI_{95\%}$: 95% posterior interval estimated from 10 000 cycles of the Gibbs sampler using a convolution Gaussian prior on the log relative risks.

The most convenient choice for $[r_i|\gamma]$ is the conjugate with the Poisson likelihood, the gamma prior Gamma(α, ν), with mean ν/α and variance ν/α^2:

$$[r_i|\gamma] = [r_i|\alpha, \nu] \propto \alpha^\nu r_i^{\nu-1} \exp\{-\alpha r_i\}.$$

Alternatively a normal prior distribution $N(\mu, \sigma^2)$ with mean μ and variance σ^2 on each log relative risk $x_i = \log(r_i)$ can be used:

$$[x_i|\gamma] = [x_i/\mu, \sigma] \propto \sigma^{-1} \exp\left\{-\frac{(x_i - \mu)^2}{2\sigma^2}\right\}.$$

To allow for area-specific covariates, each of these two priors can be easily generalised (Manton *et al.*, 1981, 1987, 1989; Tsutakawa *et al.*, 1985; Tsutakawa, 1988; Clayton and Kaldor, 1987).

Even if the choice of the conjugate prior can be justified as an appropriate distribution for modelling population risk processes (Manton *et al.*, 1981, 1987), and leads to estimates that have the best robustness properties in the class of all priors having the same mean and variance (Morris, 1983), gamma priors cannot be easily generalised to allow for spatial dependence, unlike normal priors.

2.3.3 Spatially structured priors

Prior knowledge may indicate that geographically close areas tend to have similar relative risks, i.e. there exists a local spatially structured variation in relative risks. To express this prior knowledge, nearest neighbour Markov random field (MRF) models are convenient. For this class of prior models, the conditional distribution of the relative risk in area i, given values for relative risks in all other areas $j \neq i$, depends only on the relative risk values in the neighbouring areas ∂i of area i. Thus relative risks have a locally dependent prior probability structure. Their joint distribution is determined (up to a normalising constant) by these conditional distribution (Besag, 1974). Gaussian MRF models on the log relative risks specify the conditional distribution of x_i to be normal with mean depending upon the mean of x_j in the neighbouring areas.

The usual forms of conditional Gaussian autoregression (Besag, 1974), first used on the log relative risks in Bayesian mapping by Clayton and Kaldor (1987), assume that the conditional variance is constant, and hence are not strictly appropriate for irregular maps where the number of neighbours varies. For irregular maps, intrinsic Gaussian autoregression is more suitable (Besag *et al.*, 1991) as the conditional variance of log relative risk x_i given all other x_j's is inversely proportional to the number of neighbouring areas w_{i+} of area i. The joint posterior distribution of $\boldsymbol{x}$ given the hyperparameter $\gamma(=\sigma)$ is then:

$$[\boldsymbol{x}\,|\,\gamma] = [\boldsymbol{x}\,|\,\sigma] \propto \sigma^{-n} \exp\left\{ -\frac{1}{2\sigma^2} \sum_{i=1}^{n} \sum_{i<j} w_{ij}(x_i - x_j)^2 \right\}.$$

The mean of $[\boldsymbol{x}\,|\,\gamma]$ is zero and its inverse variance–covariance matrix has diagonal elements w_{i+}/σ^2 and off-diagonal elements $-w_{i+}/\sigma^2$, where the w_{ij} are prescribed non-negative weights with:

$w_{ij} = 1$ if i and j are adjacent areas,
$w_{ij} = 0$ otherwise,
and $w_{i+} = \sum_{j=1}^{n} w_{ij}$.

Then, the normal conditional distribution of the log relative risk x_i, given all other x_j's and the hyperparameter σ, has mean and variance given by:

$$\mathrm{E}\,[x_i\,|\,x_j, j \neq i, \gamma] = \mathrm{E}[x_i\,|\,x_j, j \in \partial i, \sigma] = \bar{x}_i,$$

$$\mathrm{var}[x_i\,|\,x_j, j \neq i, \gamma] = \mathrm{var}[x_i\,|\,x_j, j \in \partial i, \sigma] = \frac{\sigma^2}{w_{i+}},$$

where ∂i denotes the set of areas adjacent to area i, w_{i+} its cardinal and $\bar{x}_i$ the mean of the x_j's for $j \in \partial i$.

In practice it is often unclear how to choose between an unstructured prior and a purely spatially structured prior. An intermediate distribution on the log relative risks that ranges from prior independence to prior local dependence, called a convolution Gaussian prior, has been proposed (Besag, 1989; Besag and Mollié, 1989). In this prior model the log relative risks are supposed to be the sum of two independent components:

$$\boldsymbol{x} = \boldsymbol{u} + \boldsymbol{v},$$

where $\boldsymbol{v}$ is a normal model with zero mean and variance λ^2, describing the unstructured heterogeneity of the relative risks, and $\boldsymbol{u}$ is an intrinsic Gaussian autoregression with conditional variances proportional to κ^2 representing local spatially structured variation.

The conditional variance of the log relative risk x_i, given all the other x_j's and the hyperparameters κ and λ, is the sum of the variances of the independent components $\boldsymbol{u}$ and $\boldsymbol{v}$:

$$\text{var}\,[x_i|x_j, j \neq i, \gamma] = \text{var}\,[x_i|x_j, j \in \partial i, \kappa, \lambda] = \frac{\kappa^2}{w_{i+}} + \lambda^2.$$

$\kappa^2 = 0$ corresponds to a total independence of the risks whereas $\lambda^2 = 0$ leads to a purely local dependence modelled by the intrinsic Gaussian autoregression.

The parameters κ^2 and λ^2 control the strength of each component: κ^2/λ^2 small reflects an unstructured heterogeneity, whereas κ^2/λ^2 large indicates that a spatially structured variation dominates.

This model can be generalised to allow for covariate effects (Mollié, 1990; Clayton and Bernardinelli, 1992; Clayton *et al.*, 1993).

2.4 EMPIRICAL BAYES ESTIMATION OF RELATIVE RISKS

As previously noted, a fully Bayesian analysis based on the marginal posterior distribution $[\boldsymbol{r}|\boldsymbol{y}]$ is often intractable. The EB idea consists in approximating $[\boldsymbol{r}|\boldsymbol{y}]$ by the marginal posterior distribution for $\boldsymbol{r}$ given the data $\boldsymbol{y}$ and the hyperparameters γ which is given by

$$[\boldsymbol{r}|\boldsymbol{y}, \gamma] \propto [\boldsymbol{y}|\boldsymbol{r}][\boldsymbol{r}|\gamma],$$

where the unknown hyperparameters γ are substituted by suitable estimates denoted $\hat{\gamma}$. This approximation is relevant if the distribution $[\gamma|\boldsymbol{y}]$ is very sharply concentrated at $\hat{\gamma}$. Generally these estimates are MLEs derived from the marginal likelihood of γ:

$$[\boldsymbol{y}|\gamma] = \int [\boldsymbol{y}|\boldsymbol{r}][\boldsymbol{r}|\gamma]\mathrm{d}\boldsymbol{r}$$

using the information relevant to the overall map structure, contained in $[\boldsymbol{y}|\gamma]$.

The EB estimate of the relative risks is then the posterior mean evaluated at the MLE $\hat{\gamma} : \mathrm{E}[\boldsymbol{r}|\boldsymbol{y}, \hat{\gamma}]$.

The conjugate gamma prior

For the conjugate gamma prior for the relative risks, the marginal posterior distribution $[\boldsymbol{r} \mid \boldsymbol{y}, \gamma]$ is the product of n marginal posterior distributions $[r_i \mid y_i, \alpha, \nu]$ which are also gamma distributed.

Thus the posterior mean of the relative risk for the ith area is a weighted average between the SMR for this area and the prior mean of the relative risks on the overall map, the weight being inversely related to the variance of the SMR. Since this variance is large for a rare disease and for small areas, a more important weight is given to the prior mean for every area, thereby producing a smoothed map.

Non-conjugate priors

However, in general cases where the prior distribution is not conjugate with the likelihood, the marginal posterior distribution $[\boldsymbol{r} \mid \boldsymbol{y}, \gamma]$ is non-standard and must be approximated to allow direct calculation of the posterior mean.

For multivariate normal priors on log relative risks, Clayton and Kaldor (1987) used a multinormal approximation for the marginal posterior distribution of $\boldsymbol{x}$. In addition, as the marginal likelihood of γ is rarely tractable, its maximisation requires the EM algorithm (Dempster *et al.*, 1977), even in the simplest case of independent normal priors. This method has been used by Clayton and Kaldor (1987) and Mollié and Richardson (1991) to smooth maps for rare diseases or small areas.

Disadvantages of EB estimation

The theoretical problem with EB methods is that even if the substitution of $\hat{\gamma}$ into $[\boldsymbol{r} \mid \boldsymbol{y}, \gamma]$ instead of using $[\boldsymbol{r} \mid \boldsymbol{y}]$, usually yields acceptable point estimates (Deely and Lindley, 1981), the variability in $\boldsymbol{r}$ is underestimated because no allowance is made for the uncertainty in γ. Thus 'naive' EB confidence intervals for $\boldsymbol{r}$ based on the estimated variance of the posterior $[\boldsymbol{r} \mid \boldsymbol{y}, \hat{\gamma}]$ are too narrow.

To allow for the uncertainty in γ two approaches, reviewed by Louis (1991), have been proposed in the EB context: adjustments based on the delta method (Morris, 1983) or bootstrapping the data $\boldsymbol{y}$ (Laird and Louis, 1987). In the context of disease mapping, Biggeri *et al.* (1993) followed the latter to the obtain confidence limit for $\boldsymbol{r}$ but using the conjugate gamma prior. However, these techniques are computationally difficult, especially in using non-conjugate priors.

Another disadvantage of EB estimation is that it may not provide an adequate description of the true dispersion in the rates. This problem has been addressed by Devine and Louis (1994) and Devine *et al.* (1994a,b) using a constrained EB estimator.

2.5 FULLY BAYESIAN ESTIMATION OF THE RELATIVE RISKS

The fully Bayesian approach gives a third way to incorporate variability in the hyperparameters γ in specifying the hyperprior distribution $[\gamma]$ and basing inference about $\boldsymbol{r}$ on the marginal posterior distribution $[\boldsymbol{r} \mid \boldsymbol{y}]$. However, except in the case of conjugate priors and hyperpriors, this distribution is often intractable.

Working directly with this posterior distribution $[\boldsymbol{r}|\boldsymbol{y}]$ will require analytic approximations or numerical evaluation of integrals (Tsutakawa, 1985; Tierney and Kadane, 1986). Otherwise Monte Carlo methods permit drawing samples from the joint posterior distribution $[\boldsymbol{r}, \gamma|\boldsymbol{y}]$ and hence from the marginal posteriors $[\boldsymbol{r}|\boldsymbol{y}]$ and $[\gamma|\boldsymbol{y}]$.

3.5.1 Hyperpriors

Classical choices for the hyperprior distribution $[\gamma]$ generally assume independence between the hyperparameters and may also assume a non-informative distribution for some hyperparameters. However, a non-informative uniform distribution $U(-\infty, +\infty)$ for the logarithm of the scale parameter σ^2 (or κ^2 and λ^2) results in an improper posterior with an infinite spike at $\sigma^2 = 0$ (or $\kappa^2 = 0$ and $\lambda^2 = 0$), forcing all relative risks to be equal.

For independent normal or multivariate normal priors for the log relative risks, a more general class of hyperpriors for the inverse variance $\theta(= \sigma^{-2}$ or k^{-2} and $\lambda^{-2})$ is conjugate gamma distributions with parameters assumed specified. With the convolution Gaussian prior on the log relative risks, in the absence of information about the relative importance of each component $\boldsymbol{u}$ and $\boldsymbol{v}$, it is reasonable to assume that they have the same strength and thus to choose vague gamma hyperpriors with means 2/var (log(SMR)) for λ^{-2} and $2/\bar{w}$ var (log(SMR)) for k^{-2}, where $\bar{w}$ is the mean of w_{i+}, and with large variances (Bernardinelli and Montomoli, 1992; Mollié, 1996).

A relative insensitivity of the relative risks estimates to the choice of hyperpriors has been found in a particular example of Bayesian mapping, with the convolution Gaussian prior and inverse chi-squared distributions on prior variances (Bernardinelli *et al.*, 1995a).

2.5.2 Monte Carlo methods: Gibbs sampling

The basic idea of the general Metropolis algorithm is to simulate a Markov chain whose equilibrium distribution is the desired distribution (Gilks *et al.*, 1996a). An adaptation of this algorithm, i.e. Gibbs sampling, is particularly convenient for MRF where the joint posterior distribution is complicated but full conditional distributions have simple forms and need to be specified only up to a normalising constant (Gilks, 1996).

For the hierarchical model discussed above, the joint posterior distribution of the log relative risks $\boldsymbol{r}$ and hyperparameters γ is (Mollié, 1996)

$$[\boldsymbol{x}, \gamma|\boldsymbol{y}] \propto [\boldsymbol{y}|\boldsymbol{x}][\boldsymbol{x}|\gamma][\gamma] = \prod_{i=1}^{n} [y_i|x_i][\boldsymbol{x}|\gamma][\gamma].$$

Gibbs sampling consists in visiting each parameter (log relative risks x_i, $i = 1, \ldots n$, and hyperparameters γ) in turn and simulating a new value for this parameter from its full conditional distribution given the current values for the remaining parameters. For instance, using the decomposition

$$[\boldsymbol{x}|\gamma] = [x_i|x_j, j \neq i, \gamma][x_j, j \neq i|\gamma]$$

a new value of the log relative risk x_i is drawn given the current values $(x'_j, j \neq i)$ and γ' from the full conditional distribution:

$$[x_i \mid x'_j, j \neq i, \gamma', \boldsymbol{y}] \propto [y_i \mid x_i][x_i \mid x'_j, j \in \partial i, \gamma']$$

and a new value of γ is drawn given the current values $\boldsymbol{x}'$ from the full conditional distribution:

$$[\gamma \mid \boldsymbol{x}'] \propto [\boldsymbol{x}' \mid \gamma][\gamma].$$

In theory, if the chain is irreducible, aperiodic and positive, recurrent, the joint distribution of the sample values of $(\boldsymbol{x}, \gamma)$ converges to the joint posterior distribution $[\boldsymbol{x}, \gamma \mid \boldsymbol{y}]$ and hence the distribution of the sample values of $\boldsymbol{x}$ (respectively of γ) converges to the marginal posterior distribution $[\boldsymbol{x} \mid \boldsymbol{y}]$ (resp. $[\gamma \mid \boldsymbol{y}]$) (Geman and Geman, 1984; Roberts and Polson, 1994; Roberts, 1996; Tierney, 1996). These distributions can be approximated by the empirical distributions of the sample values generated after convergence has been achieved.

In practice, after a sufficiently long burn-in of samples discarded for calculations, dependent samples approximately from the joint posterior are obtained. Several convergence diagnostics have been proposed (Cowles and Carlin, 1996).

Although the implementation of Gibbs sampling is quite easy, it may be very inefficient since it involves sampling from many non-standard but log concave distributions. The basic technique of rejection sampling has been improved by an adaptive rejection sampling method (Gilks and Wild, 1992).

When interest is in estimating the log relative risks $\boldsymbol{x}$, the marginal posterior distribution of $[\boldsymbol{x} \mid \boldsymbol{y}]$ can be approximated, ignoring the γ values. For each area, point estimates can be obtained from the simulated values, for example the posterior mean from the sample mean, the posterior median from the sample median, and interval estimation is also available by computing Bayesian credible intervals.

2.5.3 Examples

Using a fully Bayesian model with a convolution Gaussian prior for the log relative risks for each set of data, we performed a single run of the Gibbs sampler with a burn-in of 1000 iterations followed by 10 000 further cycles.

The mean of the number of neighbours being $\bar{w} = 5$ for 'départements' in France and var(log (SMR)) about 0.015 for leukaemias among females, gamma hyperpriors on the inverse variances have been set to have a mean of 27 for κ^{-2} and 135 for λ^{-2}, whereas var (log (SMR)) $\approx$ 0.31 for testis cancer in males led us to choose 1.3 for the mean of κ^{-2} and 6.5 for that of λ^{-2}. All variances have been set equal to 10^4.

For leukaemias among females, the heterogeneity component λ^2 was found to have a posterior mean of 0.0039, a posterior median of 0.0036 with a 95% Bayesian credible interval (0.0018–0.0070), and the spatially structured component κ^2 had a posterior mean of 0.0041, a posterior median of 0.0035 with a 95% Bayesian credible interval (0.0008–0.0115). The fully Bayesian estimates of relative risks in Figure 2.1 (b) show less variation than the SMRs (Figure 2.1 (a)). They vary from 0.88 for 'département' 22 to 1.10 for 'département' 69 with mean 0.99 (s.d. = 0.05). Extreme estimates based on small populations, for instance for 'départements' 05, 09, 10, 15, 23, 46 and 65 (Table 2.1) have disappeared and the map has been almost totally smoothed (Figure 2.1 (b)). On the other hand, extreme estimates based on very large populations for 'départements' 13 and 69 or on moderate populations for 'départements' 22, 26, 29, and 80 are maintained. Ninety-five percent Bayesian credible intervals excluding unity are only four, for 'départements' 13, 22, 29 and 69, which are nearly those highlighted in Figure 2.1 (b). Posterior medians are very close to posterior means (Table 2.1) and produce exactly the same map.

For testis cancer among males, the heterogeneity component λ^2 was found to have a posterior mean of 0.0272, a posterior median of 0.0228 with a 95% Bayesian credible interval (0.0079–0.0807), and the spatially structured component κ^2 had a posterior mean of 0.0546, a posterior median of 0.0450 with a 95% Bayesian credible interval (0.0036–0.1638). The fully Bayesian estimates of relative risks in Figure 2.2 (b) show much less variation than the SMRs (Figure 2.2 (a)): they vary from 0.68 for 'département' 69 to 1.39 for 'département' 22 with mean 0.98 (s.d. = 0.16). Extreme estimates based on very small populations for 'départements' 09, 48 and 55 have disappeared, like those having intermediate populations but neighbours with opposite estimate risks ('départements' 21 and 49), whereas estimate risks for 'départements' 32, 36, 47 and 70, although having small populations, are less changed because their neighbours also have similar risk estimates (Table 2.2). On the other hand, extreme estimates based on large populations for 'département' 69 are maintained, like those based on intermediate populations but neighbours with similar risk estimates for 'départements' 22, 42, 44 and 56. Moreover, risk estimates of 'départements' 35 and 61 that have neighbours with high estimated risks, have been raised, and that of 'départements' 43 and 82 have been slightly reduced, even if their populations are small (Table 2.2). Ninety-five percent Bayesian credible intervals for 'départements' 42 and 69 only exclude unity. Posterior medians are very close to posterior means (Table 2.1) and produce a very similar map. The spatial structure of the relative risks is shown in Figure 2.2 (b): risks increase from the south-east and south-west to the north and north-west.

2.5.4 Convergence study

Many convergence diagnostics have been proposed (Cowles and Carlin, 1996) based either on a single chain or multiple chains, or both, either quantitative or graphical, and either univariate or using the full joint distribution. According to these authors these diagnostics should be used with caution because many of them can fail to detect the sort of convergence failure they were designed to identify. They suggested using a variety of diagnostics and recommended that automated convergence monitoring should be avoided.

For each set of data, convergence has been checked running four parallel chains using different overdispersed starting points, each of them with a burn-in of 1000 iterations followed by 10 000 further cycles. Visual inspection of these chains has been done by plotting the sampled values for a subset of log relative risks and for both variance components. Convergence for the log relative risks was achieved very quickly (less than 10 iterations), whereas it could be slower for both variance parameters. We computed the 'estimated potential scale reduction' $\sqrt{\hat{R}}$ indicating the factor by which Bayesian credible intervals might be shrunk if iterations were continued indefinitely (Gelman and Rubin, 1992a; Gelman, 1996): for leukaemias in females $\hat{R} = 1.0054$ for κ^2,

$\hat{R} = 1.0022$ for λ^2 and for testis cancer in males $\hat{R} = 1.0057$ for κ^2, $\hat{R} = 1.0030$ for λ^2 and over the 94 log relative risks, the maximum value of $\hat{R}$ was 1.0015. In practice, values of $\hat{R}$ greater than 1.2 indicate poor convergence.

2.5.5 Sensitivity to hyperprior choices

Investigating the influence of hyperprior choices on leukaemias among females, we chose different means and variances for the hyperpriors on κ^{-2} and λ^{-2}: moderate changes in these choices did not affect the parameter estimates or the maps even when the ratio of the means is different to $\bar{w}$. On the other hand, very high hyperprior means and variances for κ^{-2} and λ^{-2} produced totally smoothed maps which do not really match with the fact that 95% Bayesian credible intervals still exclude unity for the same three 'départements' (13, 22 and 29).

Data on testis cancer in males being more sparse, the results are more sensitive to the hyperprior choice. Moderate changes also produced very similar estimates or maps, whereas larger means, resulting in stronger belief in the existence of geographical variation, tended to increase the smoothing. Whatever, 95% Bayesian credible intervals excluding unity remained for the same two 'départements' (42 and 69) and the visual impact given by the smoothed maps even more emphasised, is always the same.

2.6 CONCLUSION

For a rare disease and for small areas, the Bayesian approaches overcome the problem of overdispersion of the classical SMRs. Indeed, they smooth SMRs based on unreliable data but preserve those based on large populations, as shown in the examples. Bayesian estimates of the relative risks are then easier to interpret.

The fully Bayesian method, which consists of simulating the joint posterior distribution, has the great advantage over the EB method, in that it not only produces both point and interval estimates for the relative risks, but also permits computations of appropriate statistics to a specific problem.

The Bayesian approaches raise the problem of choosing an appropriate prior for the relative risks. It seems that the convolution Gaussian prior gives a satisfactory intermediate prior between independence and a pure local spatially structured dependence of the risks.

However, one should be careful in using Bayesian methods. In fact, assessing and monitoring convergence may be difficult particularly when convergence is slow: this may be either due to the sparseness of the data and the large number of geographical areas studied, or caused by model misspecification such as highly correlated parameter estimates.

Secondly, the results may be sensitive to the hyperprior choices, depending on the sparseness of the data. In the case of very sparse data, following Bernardinelli *et al.* (1995), stronger hyperpriors than non-informative ones, e.g. gamma hyperpriors, are to be recommended first to increase the reliability and the speed of the convergence of the Gibbs sampler, and second because in epidemiology, except in very specific circumstances, very high variations of risks between areas are unlikely to be encountered.

3

Addressing Multiple Goals in Evaluating Region-Specific Risk Using Bayesian Methods

Erin M. Conlon and Thomas A. Louis

University of Minnesota

3.1 INTRODUCTION

In a broad array of statistical models and applied contexts, Bayes and empirical Bayes approaches can produce more valid, efficient and informative statistical evaluations than traditional approaches (Carlin and Louis, 1996; Christiansen and Morris, 1997; Gelman *et al.*, 1995). The beauty of the Bayesian approach is its ability to structure complicated assessments, guide development of appropriate statistical models and inferences and produce summaries that properly account for all uncertainties. Computing innovations enable the implementation of complex, relevant models and applications burgeon (Besag and Green, 1993; Carlin and Louis, 1996; Gilks *et al.*, 1996a).

Valid analysis of spatially and temporally configured data requires an appropriate sampling model, an accounting for covariate effects and an accounting for spatio/temporal correlation. A principal goal is the stabilisation of estimated disease rates or relative risks in small areas while retaining sufficient geographic resolution for producing maps, conducting health assessments and developing health policy. Hierarchical Bayesian models have proven very effective in accomplishing these goals (Besag *et al.*, 1991; Clayton and Bernardinelli, 1992; Clayton and Kaldor, 1987; Cressie, 1992, 1993; Devine *et al.*, 1994a,b; Pickle *et al.*, 1996; Waller *et al.*, 1997a,b; Xia *et al.*, 1997). Stabilisation results from 'borrowing information' from other regions, usually with relatively higher weight given to nearby regions via a prior distribution that includes spatial correlation. This approach captures the influence of unmeasured or poorly measured exposures and other covariates that are spatially correlated.

Disease Mapping and Risk Assessment for Public Health. Edited by A.B. Lawson *et al.*
© 1999 John Wiley & Sons Ltd.

Policy-relevant environmental assessments involve synthesis of information (e.g. disease rates or relative risks) from a set of related geographic regions. Most commonly, individual rates are estimated and used for a variety of assessments. However, estimates of the histogram or empirical distribution function (edf) of the underlying rates (for example, to evaluate the number of rates above a threshold) or comparisons among the regions, for example by ranking to prioritise environmental assessments, can be of equal importance. While posterior means are the 'obvious' and effective estimates for region-specific parameters, the edf of the posterior means is underdispersed and never valid for estimating the edf of the true, underlying parameters. Also, effective estimates of the ranks of the parameters should target them directly. Ranking observed data usually produces poor estimates and ranking posterior means can be inappropriate.

Though no single set of values can be optimal for all goals, in many policy settings communication and credibility will be enhanced by reporting a single set of estimates with good performance for all three goals (Shen and Louis, 1998). In this chapter we present a case study on how inferential and descriptive goals determine appropriate summaries of the posterior distribution. We focus on how to use the posterior distribution for various goals and not on choice of or evaluation of the prior distribution and the likelihood. However, we briefly compare exchangeable and spatial correlation priors. After outlining goals, models and methods, we analyse data on lip cancer in Scotland using spatial correlation models to compare approaches to estimating region-specific parameters, the parameter edf/histogram and parameter ranks.

3.2 MODELS

Consider disease prevalence or incidence data that are available as summary counts or rates for a defined region such as a county, district or census tract for a single time period (e.g. a year). Denote an observed count by y_k, with $k = 1, \ldots, K$ indexing regions. The observation y_k is a count generated from a population base of size n_k and a sampling model (likelihood) parametrised by a baseline rate and a region-specific relative risk ψ_k. Within a region, data may be available for subgroups such as gender, race and exposure. Models for $\boldsymbol{\psi} = (\psi_1, \ldots, \psi_K)^{\mathrm{T}}$ should incorporate this information.

Bayesian modelling entails four stages:

1. Specification of the likelihood for the observed counts conditional on the base rate and relative risks. We use the Poisson distribution.
2. Specification of the distributions at each higher level of the sampling hierarchy, in our case the prior distribution for the vector of relative risks $\boldsymbol{\psi}$ and the hyperprior distribution for unknowns in this prior. We use a lognormal prior and appropriate hyperpriors.
3. Computation of the posterior distribution of $\boldsymbol{\psi}$ or other parameters of interest. We use the BUGS (Carlin and Louis, 1996; Gilks *et al.*, 1996a) software.
4. Use of this posterior for inference, possibly guided by a loss function.

We concentrate on steps 1, 2 and 4. Until recently, step 3 was difficult or impossible for all but the most basic Bayesian models. However, modern Markov chain Monte Carlo computational algorithms allow the use of realistic models.

3.2.1 The hierarchical model

With $\boldsymbol{X} = (\boldsymbol{X}_1, \ldots, \boldsymbol{X}_K)^{\mathrm{T}}$ the matrix of region-specific covariate vectors, we use:

$$\begin{aligned} \boldsymbol{\eta} &\sim h(\boldsymbol{\eta}), \\ \boldsymbol{\psi} &\sim g(\boldsymbol{\psi} \mid \mathbf{X}, \boldsymbol{\eta}), \\ Y_k \mid \psi_k &\sim \text{Poisson}(m_k \psi_k), \\ f(y_k \mid \psi_k) &= \frac{1}{y_k!}(m_k \psi_k)^{y_k} \mathrm{e}^{-m_k \psi_k}, \end{aligned} \tag{3.1}$$

where m_k is the expected count under a null, constant relative risk model and $\boldsymbol{\eta}$ is a vector of parameters that determines the prior. With the $\boldsymbol{\eta}$ known, the posterior for $\boldsymbol{\psi}$ is proportional to $f(\boldsymbol{y} \mid \boldsymbol{\psi}) g(\boldsymbol{\psi} \mid \boldsymbol{X}, \boldsymbol{\eta})$, but using it requires mathematical or Monte Carlo approximations. More generally, once the posterior distribution is available it can be used to make environmental assessments and to inform environmental policy.

For *internal standardisation*, the m_k are estimated by $n_k(\sum_k y_k / \sum_k n_k)$ and are assumed to be known in that their sampling variation is ignored. This approach is used by Clayton and Kaldor (1987), but models can be generalised to accommodate statistical uncertainty. If the expected counts are defined with respect to some external reference, then the model is *externally standardised* (see, for example, Bernardinelli and Montomoli, 1992). In either case, the directly estimated relative risks (via maximum likelihood, ML) are $\psi_k^{\mathrm{ml}} = y_k / m_k$.

To complete the model the prior (g) for $\boldsymbol{\psi}$ and hyperprior (h) for $\boldsymbol{\eta}$ must be specified. Models can incorporate a variety of *main effect* and *interaction* terms. The following provides a quite general form. Conditional on $\boldsymbol{\eta}$ in (3.1), let

$$\log(\psi_k) = \mathbf{X}_k \boldsymbol{\alpha} + \theta_k + \phi_k. \tag{3.2}$$

The vector $\boldsymbol{\eta}$ includes $\boldsymbol{\alpha}$ and parameters specifying the distribution of the θs and the ϕs. The $\mathbf{X}_k \boldsymbol{\alpha}$ term introduces covariate effects, the θ_k are iid random effects that produce an exchangeable model with extra-Poisson variation, and the ϕ_k are random effects that induce spatial correlation. The $\mathbf{X}_k \boldsymbol{\alpha}$ component of the model standardises and adjusts for age, gender, exposure and other potential confounders. Therefore, the θs and ϕs are adjusted random effects for region-specific log relative risks. Their variation can be viewed as compensation for model misspecification, for example failure to include important covariates. Including additional covariates can reduce the magnitude of these random effects.

A variety of popular models are contained within this structure (Besag *et al.*, 1991; Bernardinelli and Montomoli, 1992; Clayton and Kaldor, 1987; Clayton and Bernardinelli, 1992). Correlation structures can depend on inter-region distances (Devine *et al.*, 1994a), or on nearest neighbours (Besag *et al.*, 1991; Waller *et al.*, 1997a,b). Such neighbours can be defined as regions contiguous to region k, or perhaps as regions within a prescribed distance of region k.

We are interested in making inferences on region-specific, covariate-adjusted relative risks, $\rho_k = \mathrm{e}^{\theta_k + \phi_k}$. Similar issues and approaches apply to making inferences on the ψ_k or other parameters.

3.2.2 The exchangeable model

The pure *exchangeable model* sets

$$\begin{aligned} \phi_k &\equiv 0, \\ \theta_1, \ldots, \theta_K &\quad \text{iid} \quad \mathrm{N}(0, \tau^2). \end{aligned} \tag{3.3}$$

To provide some insight, assume that in the exchangeable model (3.3) the sampling distribution (f) is Gaussian (rather than Poisson) with conditional mean θ_k and variance σ_k^2. Then

$$\begin{aligned} \mathrm{E}\left[\theta_k | y_k\right] &= (1 - B_k)\, y_k, \\ \mathrm{var}\left[\theta_k | y_k\right] &= (1 - B_k)\, \sigma_k^2, \\ \mathrm{E}\left[\rho_k | y_k\right] &= \mathrm{E}\left[\mathrm{e}^{\theta_k} | y_k\right] = \mathrm{e}^{(1-B_k)(y_k + \frac{1}{2}\sigma_k^2)}, \\ B_k &= \sigma_k^2 / (\sigma_k^2 + \tau^2), \end{aligned} \tag{3.4}$$

and all region-specific posterior distributions have posterior means (PM) that are shrunken from the ML estimate towards a common value with the amount of shrinkage depending on the relationship between τ^2 and the region-specific variance σ_k^2. Setting $\tau = 0$ produces complete shrinkage to 0; $\tau = \infty$ leaves the ML estimates unchanged. With a Poisson sampling distribution similar shrinkage occurs, but there is no closed form for the PM.

3.2.3 The conditional autoregressive model

A pure *spatial correlation model* sets $\theta_k \equiv 0$ and builds a correlation model for the ϕs, generally with correlations that decrease with distance. The conditionally autoregressive (CAR) model is relatively easily implemented using BUGS and has proven effective. It builds the full joint distribution from complete conditional distributions for each ϕ_k given all others. The ϕs are Gaussian with conditional mean for ϕ_k given all other ϕs a weighted average of these others with weights decreasing with distance. The conditional variance depends on the weights. For weights $w_{kj}, k, j = 1, \ldots, K$:

$$\begin{aligned} \phi_k | \phi_{j \neq k} &\sim \mathrm{N}(\bar{\phi}_k, V_k), \quad i = 1, \ldots, K, \\ \bar{\phi}_k &= \frac{\sum_{j \neq k} w_{kj} \phi_j}{\sum_{j \neq k} w_{kj}}, \\ V_k &= \frac{1}{\lambda \sum_{j \neq k} w_{kj}}. \end{aligned} \tag{3.5}$$

The hyperparameter λ controls the strength of the spatial similarity induced by the CAR prior; larger values of λ indicate stronger spatial correlations between neighbouring regions. Setting $\lambda = \infty$ produces complete shrinkage to a common value; $\lambda = 0$ produces PM = ML. Importantly, these situations are both special cases of the exchangeable model ($\tau^2 = 0$ and $\tau^2 = \infty$, respectively) and it may be difficult to choose between the two prior structures.

The weights can be very general so long as they are compatible with a 'legal' joint distribution for the ϕs. They can depend on Euclidean distance, adjacency, or another metameter that carries information on statistical correlation. In the *adjacency model* the weights are $w_{kj} = 1$ if areas k and j share a common boundary; $w_{kj} = 0$ otherwise.

The posterior distribution of ϕ_k (or a transform of it) depends on all the observed data and on the weights. For example, the posterior mean for $\rho_k = e^{\phi_k}$ is determined by the region-specific ML relative risk (y_k/m_k), the overall relative risk, a value determined by 'nearby' regions and the covariate model $\boldsymbol{X\alpha}$.

3.3 GOALS AND INFERENCES

3.3.1 Mapping via posterior means

A crude map is produced by coding the region-specific ML rate or (adjusted) relative risk estimates. If some of the MLs have a high standard error or a high coefficient of variation (due to small n_k or m_k), then the ML map will be visually distorted because these estimates will tend to be at the extremes. Mapping MLs can produce sampling variation-induced 'hot-spots' and overdispersion of the histogram of estimated rates. If the MLs are stable (for example, when computed for entire countries; WHO, 1997), then distortions will be small and the ML map will give a valid display.

Plotting a feature of the posterior distribution (e.g. posterior mean, median, mode or other summary) *smooths* the crude map and reduces or eliminates many of these problems. A prior with spatial correlation locally stabilises the estimates by borrowing information from other regions with greater influence from nearby regions.

To see the potential advantage of a spatial correlation model, consider several rate estimates from contiguous regions, each estimate higher than the rate estimate pooled over all regions and each with a large standard error. As shown in (3.4), the pure exchangeable model will shrink each estimated relative risk quite substantially towards the overall mean (the B_k are near 1), ignoring the spatial information that the regions are contiguous. The spatial correlation model will shrink the estimated rates from the contiguous regions towards a cluster mean and shrink this cluster mean towards the overall mean by a relatively moderate amount. The resulting estimates will retain the local signal in the data. This same phenomenon occurs for each collection of regions and the Bayesian formalism, aided by considerable computation, is the only way to obtain the joint posterior distribution. Generally, PMs from a CAR model will be more spread out than those from an exchangeable model.

3.3.2 The edf/histogram and ranks

We consider situations where in addition to the region-specific relative risks, their histogram or edf and their ranks are of interest. One may wish to estimate the histogram of true, underlying region-specific relative risks and compare these histograms among countries. Or, one may want to estimate the fraction of regions with relative risks that exceed a threshold or to rank regions to prioritise environmental risk assessments.

Similar goals apply in education and medicine (Goldstein and Spiegelhalter, 1996; Laird and Louis, 1989) and other applications.

Estimating the edf of the underlying parameters by the edf of the PMs and producing ranks by ranking PMs are intuitively appealing. Unfortunately, this appeal is misguided. The histogram of the PMs is underdispersed relative to the desired edf and ranking the PMs can produce suboptimal ranks. Use of the coordinate-specific MLs solves neither of these problems. Their edf is overdispersed and ranks based on them generally perform very poorly. Valid inferences require structuring using loss functions designed for these estimation goals.

Estimating the histogram and edf

Using the formulation of Shen and Louis (1998), define the edf:

$$G_K(t) = \frac{1}{K}\,[\text{number of } \rho_k \le t\,] = I_{\{\rho_k \le t\}}, \tag{3.6}$$

where $I_{\{\cdot\}}$ is the indicator function. We use the integrated squared error loss (ISEL) to structure estimating G. With $A(t)$ a candidate estimate, ISEL $= \int [G_K(t) - A(t)]^2 \zeta(t)dt$, where $\zeta(t)$ is a weight function. For ISEL, the Bayesian formalism produces the posterior expected value of G_K as the optimal estimate:

$$\bar{G}_K(t) = E_G[G_K(t)\,|\,\boldsymbol{y}] = \frac{1}{K}\sum P_G[\,\rho_k \le t\,|\,\boldsymbol{y}\,], \tag{3.7}$$

which is the *posterior probability* that a ρ_k chosen at random from the collection of ρs that generated the data is less than or equal to t. If G is continuous, then so is $\bar{G}_K$.

To see the reason for not using PMs to produce the edf, consider the mean and variance computed from $\bar{G}_K$. The mean is the same as the average of the coordinate-specific posterior means ($\rho_{\cdot}^{\text{pm}} = (1/K)\sum \rho_k^{\text{pm}}$), but the variance induced by the edf is the sum:

$$\frac{1}{K}\sum(\rho_K^{\text{pm}} - \rho_{\cdot}^{\text{pm}})^2 + \frac{1}{K}\sum v_k, \tag{3.8}$$

sample variance of the posterior means   average posterior variance

where v_k is the posterior variance for region k. An edf based on the PMs produces only the first term—they are underdispersed. A similar development for the coordinate-specific MLs shows that they are overdispersed. Therefore, $\bar{G}_K$ or a discretised version of it should be used as the estimated edf of the ρs.

We need a discrete version and use the Shen and Louis (1998) optimal discretisation:

$$\hat{G}_K, \text{ has mass } (1/K) \text{ at} : \hat{U}_j = \bar{G}_K^{-1}\left(\frac{2j-1}{2K}\right), \quad j = 1, \ldots, K. \tag{3.9}$$

In an earlier approach, Louis (1984) and Ghosh (1992) produce constrained Bayes (CB) estimates by retaining the location and shape of the histogram produced by the posterior means, but adjusting its variance to equal (3.8):

$$\rho_k^{\text{cb}} = \mu_{\cdot} + C \times (\rho_k^{\text{pm}} - \rho_{\cdot}^{\text{pm}}),$$

$$C = \left[1 + \frac{\sum v_k}{\sum(\rho_k^{\text{pm}} - \rho_{\cdot}^{\text{pm}})^2}\right]^{1/2}.$$

Estimating ranks

In general, the optimal ranks for the ρ_k are neither the ranks of the observed data nor the ranks of the posterior means. Laird and Louis (1989) structured the approach as follows. Represent the ranks, $\mathbf{R}(= R_1, \ldots, R_K)$ by, $R_k = \text{rank}(\rho_k) = \sum_{j=1}^{K} I_{\{\rho_k \geq \rho_j\}}$. The smallest ρ has rank 1. Use squared error loss for estimating the ranks, producing the optimal estimates:

$$\bar{R}_k = E_G[R_k | \boldsymbol{y}] = \sum_{j=1}^{K} P_G[\rho_k \geq \rho_j | \boldsymbol{y}].$$

The $\bar{R}_k$ are shrunk towards the mid-rank $(K+1)/2$, and generally the $\bar{R}_k$ are not integers. Though this feature can be attractive (because integer ranks can over-represent distance and under-represent uncertainty), we will need integer ranks. To obtain them, rank the $\bar{R}_k$, producing:

$$\hat{R}_k = \text{rank}(\bar{R}_k). \tag{3.10}$$

See Laird and Louis (1989) for the posterior covariance of the ranks and Goldstein and Spiegelhalter (1996), Morris and Christiansen (1996), and Stern and Cressie (1996) for other approaches and examples.

3.3.3 Triple-goal estimates

No single set of values can be optimal for estimating the parameter histogram, parameter ranks and region-specific parameters, but communication and credibility will be enhanced by 'triple-goal' estimates with a histogram that is a good estimate of the parameter histogram, with ranks that are good estimates of the parameter ranks, and with values that are good estimates of region-specific parameters.

Shen and Louis (1998) develop 'GR' estimates (parameter estimates computed by combining estimates of G and of the R): $\rho_k^{\text{gr}} = \hat{U}_{\hat{R}_k}$. They compare the ML, PM, CB and GR estimates and show that the GR are optimal for estimating G_K and the R_K. Furthermore, although the CB are generally better than the GR for estimating region-specific parameters, the difference in performance is small and both the GR and CB pay only a modest price in estimating region-specific parameters compared with the use of the PM.

3.4 USING MONTE CARLO OUTPUT

Our models are estimated using MCMC via BUGS (see the Appendix). We outline, using MCMC output, how to compute estimates of the ρs, $\hat{G}_K$ and various estimated ranks. Assume that MCMC has been run and that we have an $I \times K$ array in which rows ($i = 1, \ldots, I$) index the draws from the chain (draws from the posterior distribution) and columns ($k = 1, \ldots, K$) index the region. The chains have been stripped of burn-in values. We have available $\rho_{ik} = e^{\phi_{ik}}$ and can compute a wide variety of summaries, including:

- The region-specific posterior means: $\rho_k^{\text{pm}} = \rho_{.k}$.
- The region-specific posterior variances, v_k.
- S_k = ranks of the ρ_k^{pm}.
- The posterior expected ranks: $\bar{R}_k = R_{.k}$, where R_{ik} is the rank of region k among the $\rho_{ik}, k = 1, \ldots, K$. The standard error of $\bar{R}_k$ is the sample standard error of the R_{ik}. The $\bar{R}_k$ are not integers.
- $\hat{R}_k$ = ranks of the $\bar{R}_k$.
- $\tilde{R}_k$ = the posterior modal ranks computed as the mode of the $R_{ik}, i = 1, \ldots, I$. (A very large I will be needed to produce valid modes.)
- The Louis–Ghosh constrained estimates ρ_k^{cb}.
- $\hat{G}_K$: The edf of $U_1, \ldots, U_K$. To find the Us, pool all IK MCMC output values (all ρ_{ik}) and for $\ell = 1, \ldots, K$ let: U_ℓ = the $[(\ell - 0.5)I]th$ smallest value in the pooled output. This is the $\{100[(\ell - 0.5)/K]\}th$ percentile in the pooled output.
- The Shen–Louis parameter estimates: $\rho_k^{\text{gr}} = U_{\hat{R}_k}$.

Having samples from the posterior distribution enables straightforward computation of most summaries. For example, for ranking one can compute median ranks and posterior intervals. For inference on the $\psi_k = \rho_k e^{\mathbf{X}_k \boldsymbol{\alpha}}$, one can directly use the MCMC output, including samples from the posterior distribution of $\boldsymbol{\alpha}$.

3.5 SCOTTISH LIP CANCER DATA ANALYSIS

3.5.1 The dataset

This dataset (IARC, 1985) includes information for the 56 countries of Scotland pooled over the years 1975–80. The dataset includes observed and expected male lip cancer cases, the male population-years-at-risk, a covariate measuring the fraction of the population engaged in agriculture, fishing or forestry (AFF), and the location of each county expressed as a list of adjacent counties. Expected cases are based on the male population count and the age distribution in each county, using internal standardisation.

We report in detail on the model:

$$\log(\phi_k) = \alpha \text{AFF}_k + \phi_k, \tag{3.11}$$

with ϕ following the CAR model in (3.5) using adjacency indicators for the ws and taking the ρ_k as the parameters of interest. Since the CAR model implicitly includes a non-zero intercept, an intercept is not needed in the regression model. We make limited comparisons with the pure exchangeable model (ϕ_k in (3.11) replaced by θ_k; the regression augmented by an intercept α_0).

3.5.2 Estimates of prior parameters

For both the exchangeable and CAR models we used 2000 samples from the posterior distribution of α and λ, after stripping 1000 burn-in values from the BUGS output (see the Appendix). For the exchangeable model the priors for α_0 and α are normal with mean 0 and variances 10^6 and 10^8, respectively. These are essentially flat priors. The reciprocal prior variance (τ^{-2}) is gamma with mean = 1, variance = 1000. Table 3.1 gives results for the exchangeable model.

For the CAR model the prior for α is normal with mean 0 and variance 10^8. The precision parameter λ is gamma with mean = 0.25 and variance = 1000. Table 3.2 gives results for the CAR model.

The posterior mean and standard deviation of α depend on the random effects structure. Estimates for it report the impact of changes in the fraction of AFF. For the exchangeable model, $\alpha^{\text{pm}} = 6.95$ implies a doubling of the relative risk for every 0.1 increase in AFF. For the CAR model, $\alpha^{\text{pm}} = 4.04$ implies a 50% increase in relative risk.

3.5.3 Posterior means

Figures 3.1 and 3.2 display the ML and PM estimates of relative risk and the s.d. of the ML. (The middle line displays the ML estimates with 'whiskers' with length proportional to the s.d. The bottom line displays the PMs.) Notice that in Figure 3.1 shrinkage towards the overall value is more pronounced for regions with relatively unstable MLs than for regions with relatively stable MLs. A comparison of Figures 3.1 and 3.2 shows that the

Table 3.1 Posterior moments for the exchangeable model parameters

Parameter	Posterior mean	Posterior s.d.
α_0	−0.51	0.16
α	6.95	1.33
τ	0.62	0.09

Table 3.2 Posterior moments for the CAR model parameters

Parameter	Posterior mean	Posterior s.d.
α	4.04	1.13
λ	2.54	1.29

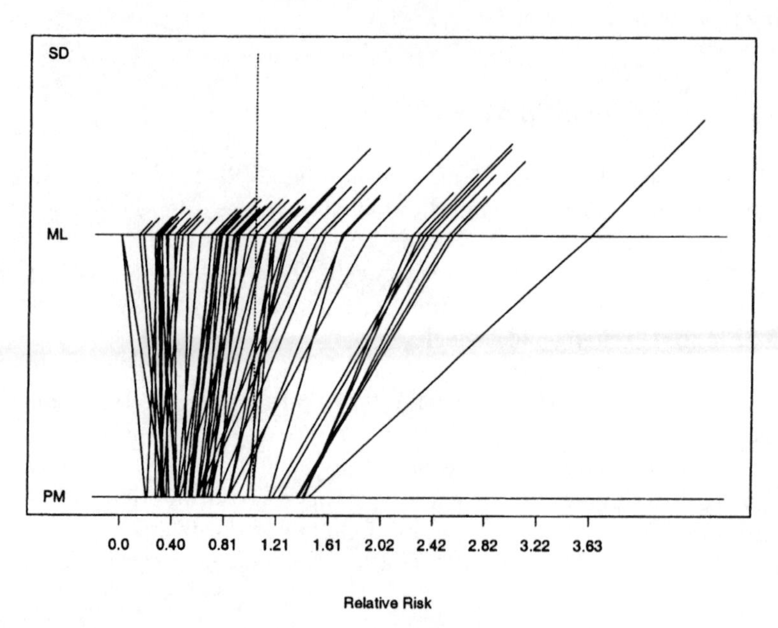

Figure 3.1 Maximum likelihood estimates (ML) of relative risk, their estimated standard errors (SD) and posterior mean (PM) relative risks computed from the exchangeable model. Note that when the MLE = 0, the estimated SD = 0

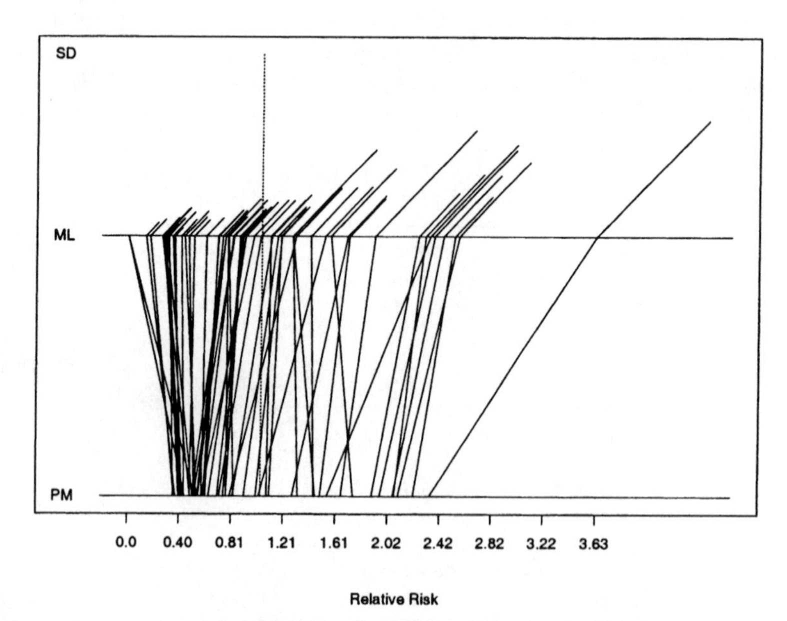

Figure 3.2 Maximum likelihood estimates (ML) of relative risk, their estimated standard errors (SD) and posterior mean (PM) relative risks computed from the CAR model. Note that when the MLE = 0, the estimated SD = 0

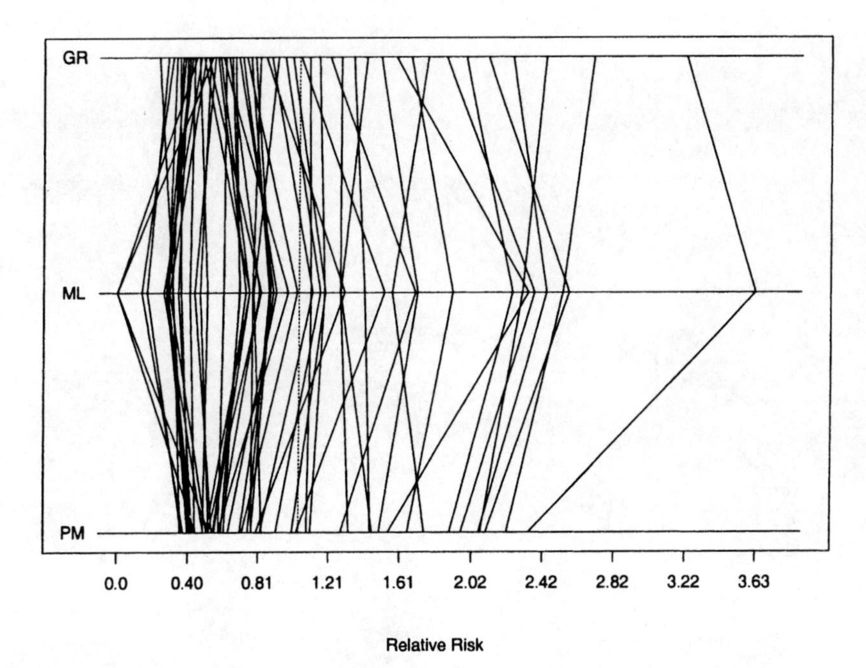

Figure 3.3 GR, ML, and PM estimates of relative risk for the CAR model

PMs from the CAR model are more spread out than are those for the exchangeable model. The CAR model preserves more of the local signal. Also, note that for both the exchangeable and CAR models the lines between the ML and PM axes cross, indicating that ranks based on PMs are different from those based on MLs.

A study of the regions with the fifth and sixth largest ML estimates (regions 8 and 9) provides a good comparison of the exchangeable and CAR models; the influence of including spatial correlation. These regions have approximately the same ML estimates (2.37 and 2.34, respectively) with approximately the same standard deviation. In Figure 3.1 the PMs also are similar and are moved substantially towards the value 1. In Figure 3.2 the PMs are quite different. Region 9 has six neighbours and a PM shrunken substantially towards 1 because the average of the ML estimates for its neighbours (1.69) is given considerable weight by the CAR prior. In contrast, region 8's only neighbour (region 6, actually separated by water) has ML = 1.41, but this value is given relatively little weight by the CAR prior but sufficient variance stabilisation to avoid the shrinkage produced by the exchangeable prior. The foregoing is only a partial explanation. The CAR prior produces very complicated relationships among observed and posterior distributions. Sorting these out requires an MCMC approach.

GR and PM

Henceforth, we consider only the CAR model. Figure 3.3 compares GR, ML and PM esitmates of relative risk. Note that the spread of the GR estimates lies between that for the ML and the PM, that the ranks of the GR estimates are different from those for the

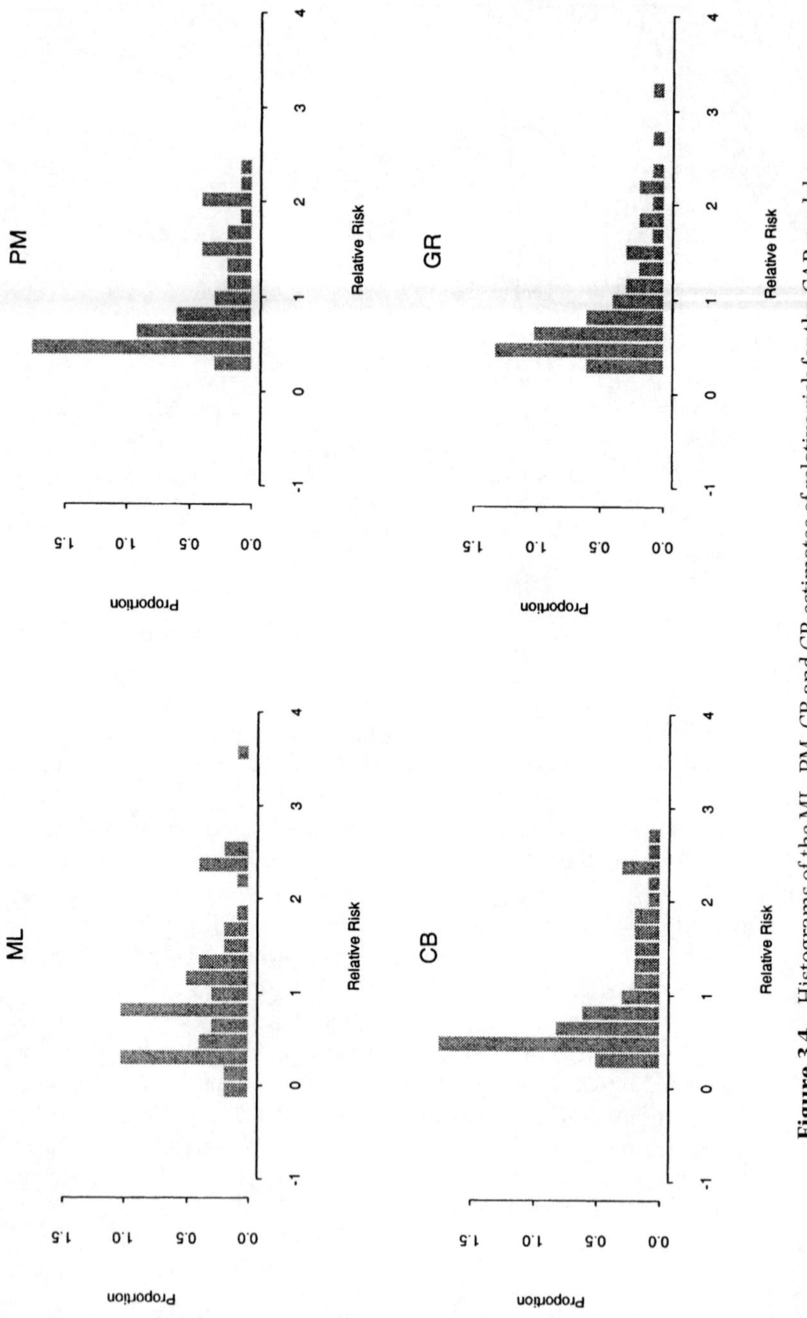

Figure 3.4 Histograms of the ML, PM, CB and GR estimates of relative risk for the CAR model

ML and the PM, that some of the GR estimates are farther from the shrinkage target than are the ML, and that lines cross, indicating different ranks from the different estimates. For each region, and in both the exchangeable and CAR models, the PM is closer to the shrinkage target than is the ML. For each model the GR estimate can be farther from the target than is the ML (the models 'stretch' away from the target). Stretching is possible whenever region-specific sampling variances (controlled by the m_k in our application) have a large relative range. In this situation, a low-variance ML with rank below but not at K can have $\hat{R} = K$ and GR estimate $\hat{U}_K$.

3.5.4 Histograms

Figure 3.4 displays histograms of the ML, PM, CB and GR estimates of relative risk. These show the spread relations noted in Figure 3.3 and allow a comparison of shape. The CB and GR estimates are similar for this data set.

Table 3.3 presents percentiles and moments for the ML, PM, CB, and GR estimates. Note that all have approximately the same mean and that the variances are ordered: PM $<$ CB $\approx$ GR $<$ ML. Quantiles document the shape differences of the histograms in Figure 3.4. These comparisons for the exchangeable model (not shown) are

Table 3.3 Percentiles and moments for the ML, PM, CB and GR estimates from the CAR model

Moments		
	ML	**PM**
Mean	1.04	0.92
s.d.	0.78	0.57
Mean	0.96	0.95
s.d.	0.67	0.67
	CB	**GR**

Quantiles		
	ML	**PM**
10%	0.28	0.41
25%	0.4	0.5
Median 50%	0.84	0.67
75%	1.33	1.29
90%	2.32	1.92
10%	0.38	0.37
25%	0.48	0.48
Median 50%	0.66	0.70
75%	1.37	1.23
90%	2.14	1.92
	CB	**GR**

more pronounced because PMs computed from it are less spread out than in the CAR model.

3.5.5 Ranks

Figure 3.5 reports on ranks using the various approaches, with $\bar{R}$ (the y-axis) taken as the gold standard. Note that the ranks based on the ML are very different from the best estimate, while the PM, CB and GR approaches produce ranks that have a monotone relation with $\bar{R}$. Of course, a monotonicity relation is assured for the GR estimates. Note that two ML estimates are tied at 0 though their $\bar{R}$ differ, producing two dots plotted for rank 1.

Tables 3.4 and 3.5 provide additional detail on ranking. They show how identification of extreme regions depends on the approach taken. Note that based on the MLs, region 8 is one of the five highest, but is not so categorised by PMs, CBs or GRs. Its standard error is sufficiently large relative to other regions with large MLs that its posterior distribution is pulled back toward 1 more than the other contenders. Similarly, based on the MLs, regions 43, 55 and 56 are three of the five lowest relative risk regions, but none achieves this status based on the PMs, CBs or GRs.

3.5.6 Transforming the parameter of interest

If we were to repeat the foregoing analyses with target parameter $\phi = \log(\rho)$ rather than ρ, then the GR estimates would map directly by the log transform (they are monotone-transform equivariant) and the $\bar{R}$ and $\hat{R}$ would not change (they are monotone-transform invariant). This transform equivariance and invariance are very attractive

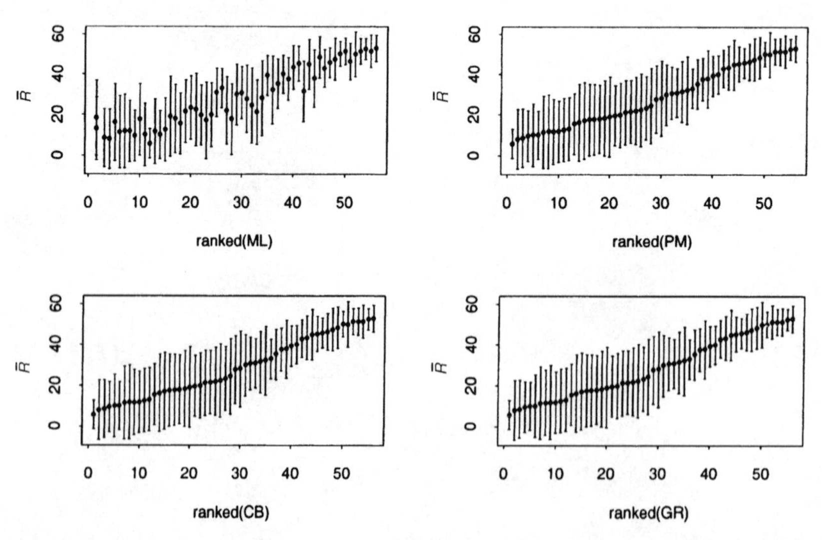

Figure 3.5 Conditional expected ranks ($\bar{R}$) versus ranked ML, PM, CB relative risk estimates for the CAR model

Table 3.4 Highest five relative risk regions. PM, CB and GR from the CAR model

Rank	ML region	PM region	CB region	GR region
56	1	1	1	1
55	3	2	2	2
54	2	3	3	11
53	5	11	11	5
52	8	5	5	3

Table 3.5 Lowest five relative risk regions. PM, CB and GR from the CAR model

Rank	ML region	PM region	CB region	GR region
1	55	49	49	49
2	56	53	53	53
3	54	54	54	54
4	53	42	42	42
5	43	48	48	45

properties not shared by the PM and CB estimates. Furthermore, direct logarithmic transformation of the MLs requires avoiding taking the log of 0 (usually by adding $a(< 1)$ to y_k and 1 to m_k). Hierarchical models avoid such *ad hoc* fixups.

3.6 CONCLUSION

In a fully Bayesian approach, all inferences are determined by the posterior distribution and a loss function. Statistical inferences generated by this formalism often have excellent frequentist properties. Hierarchical structuring and the Bayesian posterior computations are especially important in multi-dimensional settings and where goals are non-standard. As we illustrate, intuitive approaches such as using posterior means to produce a histogram or ranks can perform very poorly. For the triple goals of estimating individual parameters, the parameter histogram and ranks, the new GR estimates (which combine an estimate of G with an estimate of R) are preferred over maximum likelihood (ML), posterior means (PM) or constrained Bayes (CB). Though the PMs are optimal for estimating individual parameters, the GR or CB estimates produce a tolerable increase in squared error loss and generally outperform ML estimates. The advantages of reporting one set of estimates argue in favour of the GR approach. Of course, if estimating individual parameters is of dominant importance, then the PM or other parameter-specific estimates should be used.

Bayes and empirical Bayes hierarchical models with fixed effects for covariate influences and some combination of exchangeable and correlated region-specific random effects have the potential to reflect the principal features of environmental assessments. However, the approach must be used cautiously and models must be sufficiently flexible to incorporate important relations among exposure, geography and response and to accommodate the main stochastic features. Models must be robust. In addition to possible non-robustness to misspecification of the sampling distribution (f) (a potential problem shared by all methods), inferences from hierarchical models may not be robust to prior misspecification. A broadened class of priors (e.g. replace the Gaussian distribution by the *t*-family) can add robustness as can the use of semi- and non-parametric priors (Escobar, 1994; Magder and Zeger, 1996; Shen and Louis, 1997). Whatever choices are made for the prior and likelihood, goals and loss functions should determine how the posterior distribution is used for inference.

Considerable progress has been made in developing and implementing hierarchical models. Their important role in risk assessment and environmental policy energises and justifies accelerated methodologic research and development of applied insights.

ACKNOWLEDGEMENT

Support was provided by grant 1R01-ES-07750 from the US National Institute of Environmental Health Sciences.

APPENDIX: IMPLEMENTING BUGS

For the spatial CAR model, the posterior distribution of the slope on AFF (α), the precision parameter for the CAR variance (λ) and the ϕs are calculated as follows. The prior for α is normal with mean 0 and variance 10^8, essentially a flat prior. The precision parameter λ has a gamma prior with both mean and variance as free parameters. Several means and variances were tried; posterior distributions were insensitive to a wide range of choices. Reported results are based on a prior mean of 0.25 and variance of 1000. With Normal(*a*,*b*) denoting a normal distribution with mean *a* and variance *b* and similarly for Gamma(*a*,*b*), the CAR model (see equations 3.1 and 3.5) is programmed in BUGS as follows:

$$
\begin{aligned}
Y_k &\sim \text{Poisson}(m_k \psi_k), \\
\log(\psi_k) &= \alpha \text{AFF}_k + \phi_k, \\
\alpha &\sim \text{Normal}(0, 10^8), \\
\phi_k &\sim \text{Normal}(\bar{\phi}_k, V_k), \\
c_k &= \text{number of neighbours of } k, \\
\bar{\phi}_k &= \frac{1}{c_k} \sum_{j \in \text{neighbours}(k)} \phi_j, \\
V_k &= \frac{1}{\lambda c_k}, \\
\lambda &\sim \text{Gamma}(0.25, 1000), \\
\psi_k^{\text{ml}} &= Y_k / m_k.
\end{aligned}
$$

Running this model in BUGS gives the $\phi_{ik}, i = 1, \ldots, 2000, k = 1, \ldots, 56$, after dropping a burn-in of 1000. Inferences for a transform of ϕ (such as our parameter of interest $\rho_k = \mathrm{e}_k^{\phi}$) are based on the transformed ϕ_{ik}.

Implementation of the exchangeable model is similar.

4

Disease Mapping with Hidden Structures Using Mixture Models

D. Böhning and P. Schlattmann

Free University Berlin

4.1 INTRODUCTION

The analysis of the spatial variation of disease and its subsequent representation on a map has become an important topic in epidemiological research. Identification of spatial heterogeneity of disease risk gives valuable hints for possible exposure and targets for analytical studies.

Another important use of disease mapping may be seen in disease surveillance and health outcome research. Especially in cancer registries, maps are used to facilitate reporting of the public health situation and frequently maps are a starting point for cluster investigations.

The first step in the construction of disease maps is usually the choice of an epidemiological measure which shall be presented on the map. An often used measure is the Standardised Mortality Ratio, SMR $= O/E$, where the expected cases E are calculated based on a reference population, and O denotes the observed cases.

A common approach in map construction is the choropleth method (Howe, 1990). This method implies categorising each area and then shading or colouring the individual regions accordingly. Frequently the categorisation of the individual region is of particular importance.

Traditional approaches to categorisation are based on the percentiles of the SMR distribution. Most cancer atlases use this approach, usually based on quartiles, quintiles or sixtiles.

In our first example we present data from the former GDR within the time period from 1980 to 1989. The map of Figure 4.1 presents the regional distribution of childhood leukaemia in the former GDR.

Disease Mapping and Risk Assessment for Public Health. Edited by A.B. Lawson *et al.*
© 1999 John Wiley & Sons Ltd.

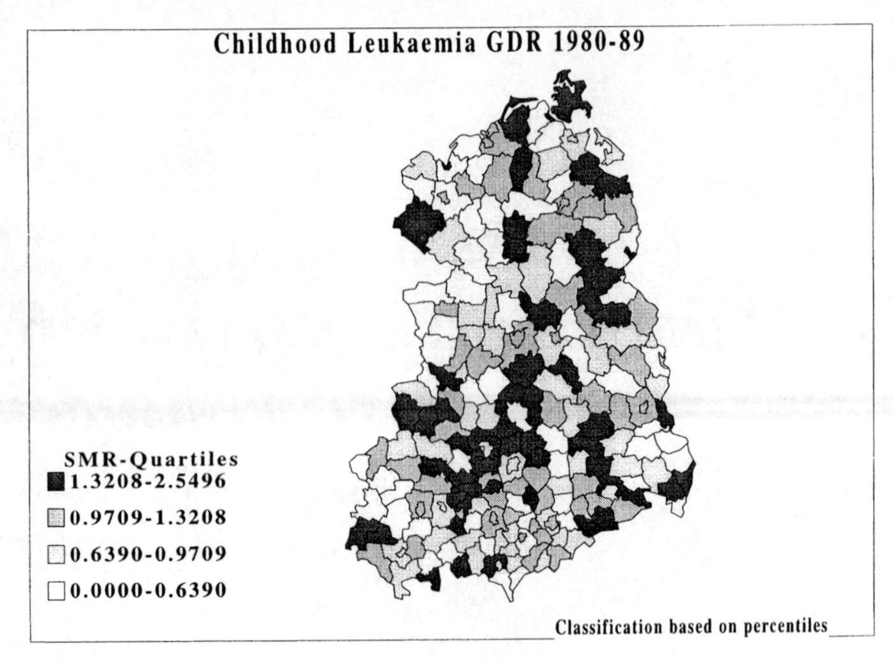

Figure 4.1 Map based on percentiles

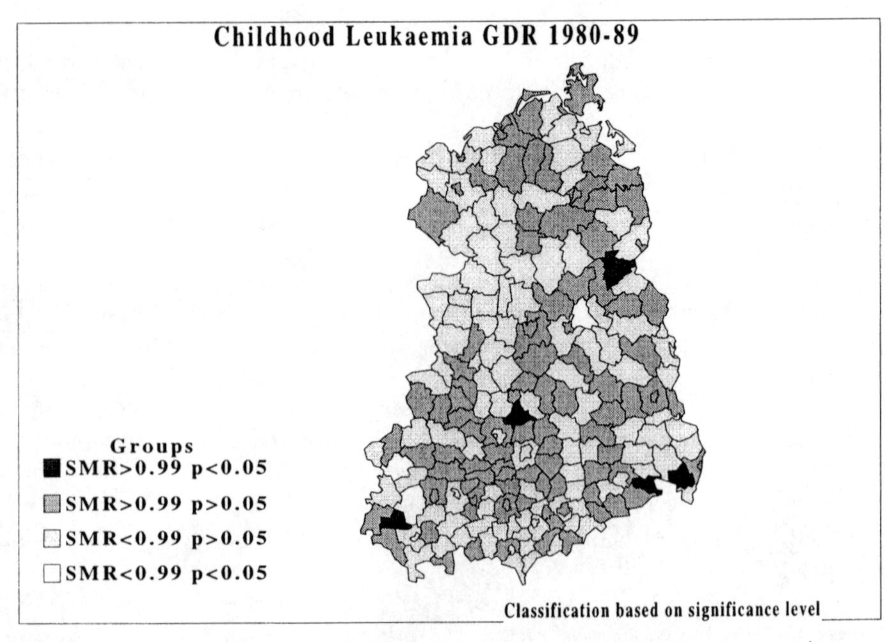

Figure 4.2 Map based on significance level with maximum likelihood estimate $\hat{\theta} = 0.99$

There has been a debate in the media about whether there is an excess of childhood leukaemia in the vicinity of the nuclear power plant Rossendorf close to Dresden in the south-east of the GDR. According to this percentile map there is an increased risk in the districts of Selbnitz and Dresden, both of which are close to the power plant. These districts are located in the south-east of the GDR, close the Czech border. This map seems to support the hypothesis of an increased risk of leukaemia in that area. But maps based on the percentiles of the SMR distribution are likely to reflect only random fluctuations in the corresponding small counts. The blank area in the following maps refers to the former western part of Berlin.

Thus another frequently used approach (Walter and Birnie, 1991) is based on the assumption that the observed cases O_i of the individual region follow a Poisson distribution with

$$O_i \sim \text{Po}(\theta E_i), \text{ with density } f(O_i, \theta, E_i) = \frac{e^{-(\theta E_i)}(\theta E_i)^{O_i}}{O_i!},$$

where again E_i denotes the expected cases in the region labelled i. Computation of the p-value is done under the null hypothesis $\theta = 1$ or based on the maximum likelihood estimator $\hat{\theta} = \sum_{i=1}^{n} O_i / \sum_{i=1}^{n} E_i$, where the latter is called the adjusted null hypothesis (n is the number of areas).

Again, based on this probability map (Figure 4.2) there is a significant excess in the district of Selbnitz. But probability maps based on a Poisson assumption face the problem of misclassification as well. Here regions with a large population tend to show significant results. Additionally, even if the null hypothesis of constant disease risk is true, misclassification occurs. It can be shown (Schlattmann and Böhning, 1993) that probability maps do not provide a consistent estimate of heterogeneity of disease risk. A false positive probability map may cause unnecessary public concern, especially if a disease map such as childhood leukaemia is presented, which is attached to highly emotional effects. Thus, the question remains whether the observed excess risk in the Dresden area is merely a methodological artefact.

4.2 THE EMPIRICAL BAYES APPROACH

4.2.1 The parametric model

A more flexible approach is given in random effects models, i.e. models where the distribution of relative risks θ_i between areas is assumed to have a probability density function $g(\theta)$. The O_i are assumed to be Poisson distributed conditional on θ_i with expectation $\theta_i E_i$.

Several parametric distributions like the gamma distribution or the lognormal distribution have been suggested for $g(\theta)$; for details see Clayton and Kaldor (1987) or Mollié and Richardson (1991). Among the parametric prior distributions the gamma distribution has been used several times for epidemiologic purposes (Martuzzi and Hills, 1995; see also Chapter 25 in this volume). In the case that the θ_i are assumed to be gamma distributed, with $\theta_i \sim \Gamma(\alpha, \nu)$, the parameters α and ν have to be estimated from the data. The marginal distribution $P(O_i = o_i) = \int_0^\infty \text{Po}(o_i, \theta, E_i) g(\theta) d\theta$, where $g(\cdot)$ follows a gamma distribution with parameters α and ν. Here we are led to a parametric mixture distribution. By applying Bayes' theorem we can estimate the posterior expectation for the relative risk of the individual area. The main distinction between empirical and full Bayesian methods can be seen in the fact that in the case of the empirical Bayes methodology the parameters of the prior distribution are estimated as point estimates $\hat{\alpha}$ and $\hat{\nu}$ from the data. Thus the posterior expectation of the relative risk are obtained conditional on these point estimates. In a full Bayesian approach a probability model for the whole set of parameters is specified (the prior distribution of α and ν included) and the posterior expectation of the relative risk is integrated over the posterior distribution of α and ν. The posterior expectation of the relative risk of the individual area is then given as an empirical Bayes estimate by

$$\widehat{\mathrm{SMR}}_i = \mathrm{E}(\hat{\theta}_i|\hat{\alpha}, \hat{\nu}, E_i, O_i) = \frac{O_i + \hat{\nu}}{E_i + \hat{\alpha}}.$$

First, in areas with a large population size the SMR_i based on this empirical Bayes approach change very little compared with the maximum likelihood estimates, whereas for areas with small population size the SMR_i shrinks to the global mean. Secondly, if the prior distribution is estimated to have small variance, then this is reflected in a large amount of shrinkage. Thus parametric empirical Bayes methods provide variance-minimised estimates of the relative risk of the individual area. But these methods still face the problem that they need a *post hoc* classification of the posterior estimate of the epidemiological measure in order to produce maps.

However, as can be seen in Figure 4.3 when using the same scale as in Figure 4.1, we obtain a homogeneous map of disease risk of childhood leukaemia in the former GDR.

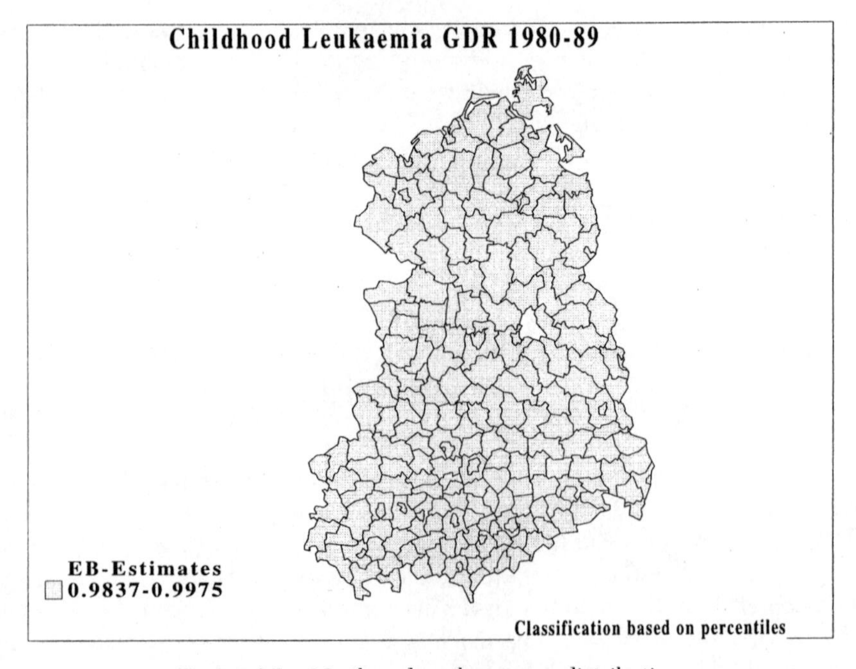

Figure 4.3 Map based on the gamma distribution

4.2.2 The non-parametric mixture model approach

Now let us assume that our population under scrutiny consists of subpopulations with different levels of disease risk θ_j. Each of these subpopulations with disease risk θ_j represents a certain proportion p_j of all regional units. Statistically, this means that the mixing distribution reduces to a finite mass point distribution. Here we face the problem of identifying the level of risk for each subpopulation and the corresponding proportion of the overall population. One can think of this situation as a *hidden* (or *latent*) structure, since the subpopulation to which each area belongs remains unobserved. These subpopulations may have different interpretations. For example, they could indicate that an important covariate has not been taken into account. Consequently, it is straightforward to introduce an unobserved or latent random vector Z of length k consisting of only 0s besides one 1 at some position (say the jth), which then indicates that the area belongs to the jth subpopulation. Taking the marginal density over the unobserved random variable Z we are led to a discrete semiparametric mixture model. If we assume a non-parametric parameter distribution

$$P = \begin{bmatrix} \theta_1 & \dots & \theta_k \\ p_1 & \dots & p_k \end{bmatrix}$$

for the mixing density $g(\theta)$ (which can be shown to be always discrete in its nature), then we obtain the mixture density as a weighted sum of Poisson densities for each area i:

$$f(O_i, P, E_j) = \sum_{j=1}^{k} p_j f(O_i, \theta_j, E_i), \quad \text{with} \sum_{j=1}^{k} p_j = 1 \text{ and } p_j \geq 0,\ j = 1, \dots, k.$$

Note that the model consists of the following parameters: the number of components k, the k unknown relative risks $\theta_1, \dots, \theta_k$ and $k-1$ unknown mixing weights $p_1, \dots, p_{k-1}$. To find the maximum likelihood estimates there are no closed form solutions available; suitable algorithms are given by Böhning *et al.* (1992). An overview of reliable algorithms may be found in Böhning (1995). Public domain software to estimate the parameters of the mixture is available with the package C.A. MAN (Böhning *et al.*, 1992). For the special case of disease mapping the package Dismap Win (Schlattmann, 1996) may be used. A general strategy implies calculating the non-parametric maximum likelihood estimator (NPMLE) and then applying a backward selection strategy to determine the number of components by means of the likelihood ratio statistic (Schlattmann and Böhning, 1993). Applying Bayes' theorem and using the estimated mixing distribution as a prior distribution we are able to compute the probability that each region belongs to a certain component:

$$\Pr(Z_{ij} = 1 | O_i, \hat{P}, E_i) = \frac{\hat{p}_j f(O_i, \hat{\theta}_j, E_i)}{\sum_{l=1}^{k} \hat{p}_l f(O_i, \hat{\theta}_l, E_i)}.$$

The ith area is then assigned to that subpopulation j for which it has the highest posterior probability of belonging. In terms of the latent vector Z, Bayes' theorem gives us its

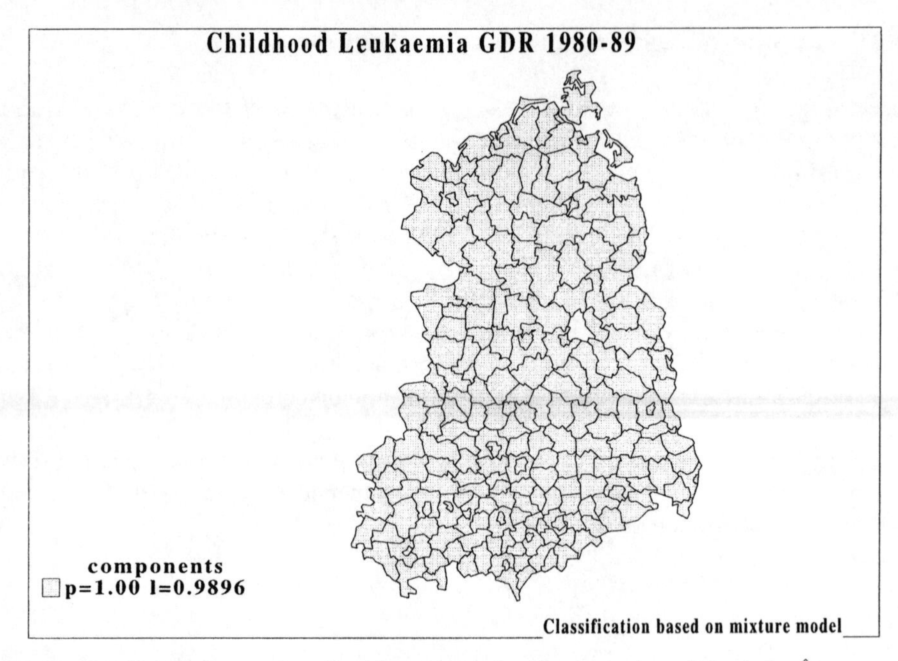

Figure 4.4 Map based on the mixture distribution, where l stands for $\hat{\theta}$

posterior distribution. For the leukaemia data we find a one-component or homogeneous model, i.e. with constant disease risk $\hat{\theta} = 0.99$ and common weight $\hat{p} = 1$. Clearly, in contrast to Figure 4.1 and 4.2 we obtain, in accordance with Figure 4.3, a homogeneous map for the leukaemia data (Figure 4.4). This could also be thought of as using the empirical Bayes estimate based on a posterior distribution which is a constant value for all regions equal to $\hat{\theta} = \sum O_i / \sum E_i$, in this case.

In general we can compute the posterior expectation for this model as follows:

$$\widehat{\mathrm{SMR}}_i = \mathrm{E}(\theta_i | O_i, \hat{P}, E_i) = \frac{\sum_{j=1}^{k} \hat{p}_j f(O_i, \hat{\theta}_j, E_i) \hat{\theta}_j}{\sum_{l=1}^{k} \hat{p}_l f(O_i, \hat{\theta}_l, E_i)}.$$

In this special case of a homogeneous solution the posterior expectation reduces to the maximum likelihood estimate of the relative risk $\hat{\theta}$. Table 4.1 contains the crude SMR, the Poisson probability and empirical Bayes estimates for regions in the Dresden area. Here EB stands for empirical Bayes estimates based on the gamma distribution as prior distribution and MIX-EB stands for the posterior expectation of the relative risk based on the mixture distribution. Clearly, we conclude that there is no excess risk in the Dresden area based on the spatial resolution of 'Landkreise'. Further investigations would need to refine the spatial resolution. However, in the case of routine maps produced by a cancer registry we would avoid a false positive result.

Table 4.1 Relative risk estimates for areas close to Rossendorf

Area	Cases	Expected cases	SMR	EB	MIX-EB	$\Pr(O = o_i)$
Dresden (City)	32	34.41	0.93	0.99	0.99	0.38
Dresden (area)	10	6.8	1.47	0.99	0.99	0.15
Selbnitz	9	3.53	2.55	0.99	0.99	0.01
Pirna	7	7.07	0.99	0.99	0.99	0.49
Bischofswerda	2	4.44	0.45	0.99	0.99	0.18

4.3 THE VALIDITY OF THE MIXTURE MODEL APPROACH FOR MAP CONSTRUCTION

The non-parametric mixture approach to map construction of the leukaemia data yields different results compared with traditional methods. These results are not necessarily more reliable. In simulation studies done by Schlattmann (1993) the mixture model approach of map construction was compared with traditional approaches of map construction such as using the percentiles of the SMR distribution or the approach based on the Poisson significance level. Various situations of heterogeneity have been simulated, assuming different levels of disease risk with two and three subpopulations. For each individual region therefore the true status of disease risk was known. The total number of different 'true' maps generated was 150. For each of these 'true' maps 2500 replications have been done in order to assess the percentage of correct classifications for each individual region and different approaches of map construction. Figure 4.5 shows the overall median percentage of correct classifications based on this simulation study, together with a 95% confidence interval.

As can be seen in the figure the mixture model approach provides by far the highest percentage of correct classifications compared with traditional methods. As a referee

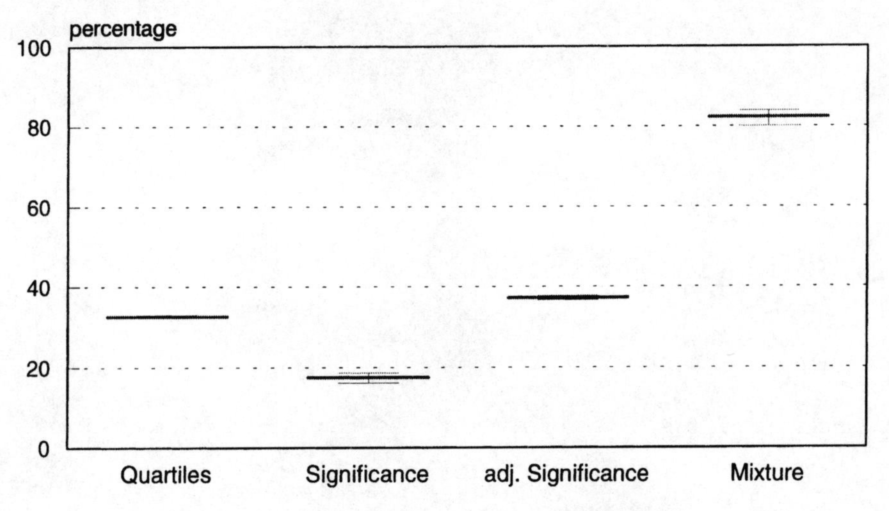

Figure 4.5 Percentage of correct classifications

pointed out, a more interesting exercise would be to compare the validity of empirical and full Bayesian methods. The authors are involved in a simulation study which seeks to investigate this point.

4.4 EXTENSIONS OF THE MIXTURE MODEL APPROACH

In a second example we investigate the regional distribution of lip cancer in males in the GDR from 1980 to 1989. When investigating the heterogeneity of these data we find a mixture model with three components as follows:

$$f(O_i, \hat{P}, E_i) = 0.097\,\mathrm{Po}(O_i, 0.449, E_i) + 0.680\,\mathrm{Po}(O_i, 0.970, E_i) + 0.223\,\mathrm{Po}(O_i, 2.176, E_i).$$

The log likelihood is -657.75 compared with -770.70 for a homogeneous model. Clearly heterogeneity is present for these data and three different levels of disease risk are identified: about 10% of all regions have a very low disease risk half the size of the standard, 68% have the same disease risk as the standard, and 22% of the regions have a disease risk which is twice as high as that of the standard. The map in Figure 4.6 is based on this mixture model.

Frequently one is interested only in residual heterogeneity, i.e. the question to be answered is whether heterogeneity remains after having adjusted for known covariates. In the homogeneous case covariates are included through Poisson regression (Breslow and Day, 1975). This leads to a loglinear model, where the Poisson parameter is given by $\theta_i = \exp(LP_i)$, with the linear predictor $LP_i = \alpha + \beta_1 x_{i1} + \ldots + \beta_M x_{iM} +$

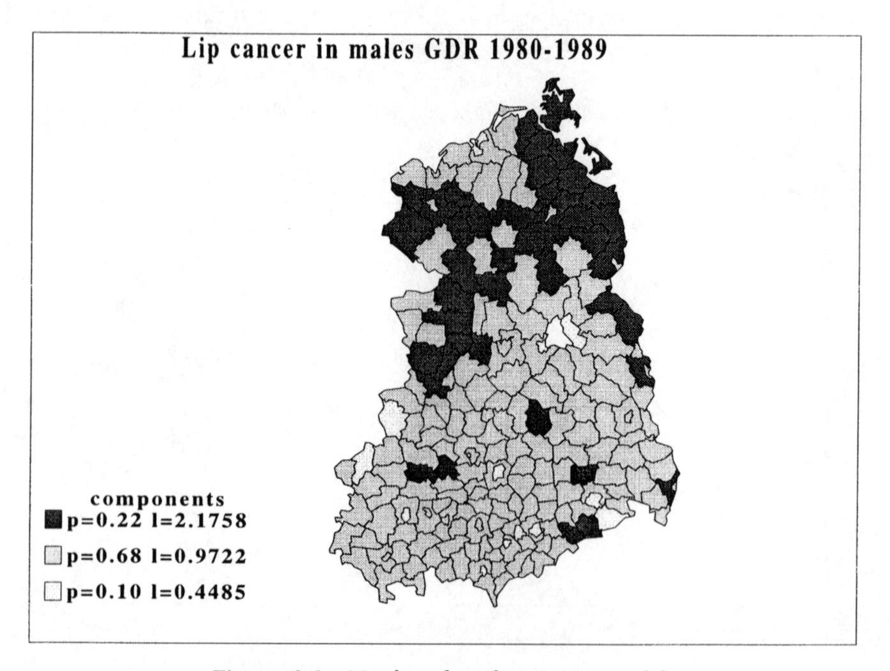

Figure 4.6 Map based on the mixture model

$\log E_i$. Exponentiating $\theta_i = \alpha + \beta_1 x_{i1} + \ldots + \beta_M x_{iM} + \log E_i$ we have a generalisation for the Poisson model $\theta_i E_i$.

Now the question arises: How do we include known covariates in the mixture model? The inclusion of covariates leads into the area of ecologic studies. For a detailed discussion see Chapter 13 in this volume. A natural extension of the homogeneous Poisson regression model is given by the mixed Poisson regression model (Dietz, 1992; Schlattmann *et al.*, 1996). An extension of the univariate Poisson mixture density $O_i \sim p_1 \text{Po}(O_i, \theta_1, E_i) + \ldots + p_k \text{Po}(O_i, \theta_k, E_i)$ is given by a random effects model where the random parameter P is discrete finite with

$$P = \begin{bmatrix} \boldsymbol{\beta_1} & \cdots & \boldsymbol{\beta_k} \\ p_1 & \cdots & p_k \end{bmatrix}, \quad \text{with } \boldsymbol{\beta_j} = (\alpha_j, \beta_{j1}, \ldots, \beta_{jM})^{\mathrm{T}} \quad \text{and} \quad j = 1, \ldots, k,$$

where M denotes the number of covariates in the Poisson regression model. The conditional distribution of O_i is given by $O_i \sim \sum_{j=1}^{k} p_j \text{Po}(O_i, \exp(LP_{ij}))$, with linear predictor $LP_{ij} = \alpha_j + \beta_{1j} x_{1i} + \ldots + \beta_{Mj} x_{Mi} + \log E_i$. The number ($M$+1) of parameters in the Poisson regression is the same for each subpopulation. The univariate mixture model approach may be considered as a special case with mixing only over the intercepts α_j and $\beta_{1j} = \ldots = \beta_{Mj} = 0, j = 1, \ldots, k$, where k denotes the number of components and M denotes the number of covariates. Again estimation may be done by maximum likelihood. If the indicator variables Z_{ij} were known, then the maximum likelihood estimators for the parameters would simply be the MLEs from each component groups. Again, there are no closed form solutions available for maximum likelihood estimates. An adaptation of the EM algorithm by Dempster *et al.* (1977) has been developed by Dietz (1992). See also Mallet (1986) for a discussion of maximum likelihood estimation. A detailed description can be found in Schlattmann *et al.* (1996) as well. The computations involved may be done with the program Dismap Win.

In our example we include an important covariate for lip cancer. Exposure to UV light is considered an important risk for lip cancer. Since there is no direct measure for exposure to UV light available, the surrogate measure AFF is applied. The covariate AFF describes the percentage of people working in agriculture, fisheries and forestry. Several models are fitted to the data. The first covariate-adjusted mixture model is obtained by using the adaptation of the EM algorithm, as described earlier in the text. This covariate-adjusted mixture model is a random effects model where a random effects model over the intercepts and a fixed effect for the covariate is modelled. The extension of this model assumes a full random effects model, i.e. the effect of the covariate may differ in each component of the mixture model. This second model provides a considerable improvement of the log likelihood, the value of the likelihood ratio statistic equals 8.08. The 95% bootstrapped critical value of the LRS is 7.5. Hence we conclude that the second covariate-adjusted mixture model is appropriate for the lip cancer data. Table 4.2 shows the estimates of the various models.

After adjusting for the proportion of people working in agriculture, forestry and fisheries, we still find residual heterogeneity, as indicated by a covariate-adjusted mixture model with three components and a random effects for the covariate AFF (Figure 4.7). If we compute the mean relative risk as the mean predicted value for each component, then the low-risk group has a weight of 28% and a relative risk estimate of 0.7, the intermediate risk group has a relative risk estimate of 1.22 with a corresponding weight of

Table 4.2 Parameter estimates of covariate-adjusted mixture models

Components $\hat{k}$	Weight $\hat{p}_j$ (S.E.)	Intercept $\hat{\alpha}_j$ (S.E.)	AFF $\hat{\beta}_j$ (S.E.)	Log L
$k = 1$	1.0	− 0.397 (0.03)	3.21 (0.19)	− 652.39
	0.25 (0.09)	− 0.80 (0.14)		
$k = 3$	0.67 (0.09)	− 0.24 (0.06)	2.88 (0.14)	− 629.00
	0.08 (0.04)	0.40 (0.10)		
	0.28 (0.09)	− 0.79 (0.17)	2.93 (0.94)	
$k = 3$	0.66 (0.09)	− 0.25 (0.06)	3.16 (0.33)	− 624.96
	0.06 (0.03)	0.80 (0.16)	0.61 (1.43)	

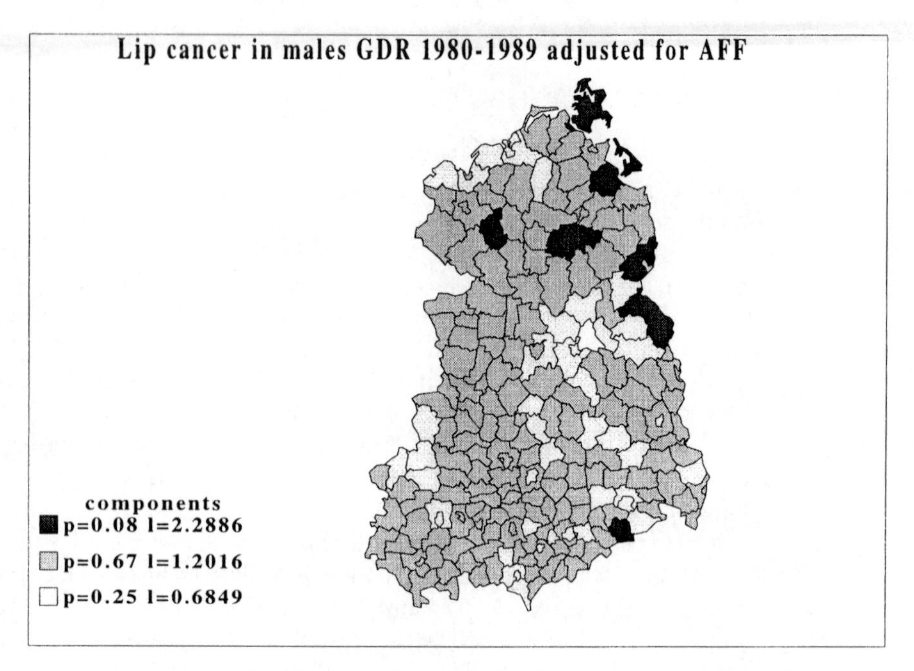

Figure 4.7 Map based on the covariate-adjusted mixture model AFF

66%. The high-risk group has mean relative risk of 2.43. Note that after the covariate is adjusted for a change in the appearance of the map can be observed, there is a drop from 22% to 6% in the proportion of the high-risk group. It is also noteworthy that the proportion of the medium-risk group remains almost the same, whereas the proportion of the low-risk group increases. Comparing Figures 4.4 and 4.6 we observe a shift from 'high-risk' areas to 'medium-risk' areas and a shift from 'medium-risk' areas to low-risk areas.

4.5 DISCUSSION AND CONCLUSIONS

Mixture models provide a valid method to detect and model heterogeneity of disease risk. The special and important case of homogeneous disease risk is detected in contrast to traditional methods of mapping. Heterogeneity of disease risk is detected as well and a valid model-based classification of the individual region is provided.

From a public health point of view these are quite desirable properties. A reliable map facilitates the reporting of data of a cancer registry. Especially in the case of a homogeneous disease risk there is less danger of a false positive cluster alarm induced by a reported map in contrast to traditional methods of disease mapping. In this circumstance the situation 'of a disease in search of exposure' is avoided. However, the primary goal of such a map is to separate signal from noise in the data. The absence of evidence is not necessarily evidence of absence. But disease maps avoid selection effects in the assessment of disease risk, since there is no a priori determined point source. If one is interested in the investigation of the hypothesis, for example of increased risk in the vicinity of nuclear power plants, then there could follow further analyses with a refined level of spatial aggregation as a second step and methods that address the effect of point sources.

One of the specific methodological attractive features of the non-parametric mixture approach consists in the fact that an estimate of the number of subpopulations is provided. This can be viewed as one example of inference for an object which is not fixed in its dimension (Richardson and Green, 1997). This paper investigated a Bayesian approach to address the uncertainty of the estimate of $\hat{k}$. A likelihood approach would use likelihood ratio tests. In this case it is important to keep in mind that the asymptotic null distribution of the likelihood ratio is non-standard (Böhning *et al.*, 1994). Alternatively, simulation or Bootstrap ideas (Schlattmann and Böhning, 1997, 1999) can be applied to address the variability of $\hat{k}$. These ideas are currently under investigation.

As with other empirical Bayes methods the posterior expectation of the epidemiological measure of the individual region can be obtained. Inclusion of the spatial structure such as an adjacency matrix into the model is easily achieved within a full Bayesian framework (Besag *et al.*, 1991; see also Chapter 2 in this volume. In contrast, inclusion of an adjacency matrix into mixture models is difficult and not straightforward. There have been some efforts to use auto models (Besag, 1974) for the purpose of disease mapping. Here the basic idea is in line with auto regressive modelling, namely that the distribution of the O_i is based on its adjacent regions. This model has been used by Ferrándiz *et al.* (1995) and Divino *et al.* (1998). An extension of these models within the mixture model framework has been proposed by Schlattmann and Böhning (1997, 1999). However, the use of auto models for the purpose of disease mapping is highly controversial (Lawson, 1996b) due to the fact that a normalising constant has to be taken into account which allows only negative values for the spatial parameters γ_{is}. As a result there is still a need for further research in order to address spatial autocorrelation into the mixture model framework.

However, for practical purposes of disease mapping with DismapWin available mixture models are a valuable tool for analysing the heterogeneity of disease risk. Mixture models are a methodological satisfactory method of classification of the individual region and thus may be a starting point in risk assessment based on disease maps.

4.6 APPENDIX: DETAILS ABOUT THE PROGRAM DismapWin

DismapWin has been developed to provide software for the mixture model approach to disease mapping described in this chapter. DismapWin runs MS-Windows requiring at least MS-Windows 3.x. DismapWin reads either ASCII or Dbase data files. Boundary files are read in the Epimap format. Thus each boundary file delivered or created with Epimap may be used with DismapWin. Maps may be constructed using traditional methods of map construction like percentiles, significance level or the mixture model and empirical Bayes framework as described in this chapter. This may be done either for rates, SMRs or continuous variables. The mixed Poisson regression approach is available as well. The individual maps are displayed on the screen, and paper copies may be obtained with any printer or device supported by MS-Windows. Maps may be copied to the clipboard and from there to word processors. Also implemented are some tests for autocorrelation like Moran's I, Ohno-Aoki, etc. and tests for heterogeneity like the test by Gail, the likelihood ratio statistic, etc. Empirical Bayes estimators for rates or SMRs are computed as well. Utility programs provide the computation of an adjacency matrix for a given boundary file. Another utility program provides parametric bootstrapping to obtain critical values for the likelihood ratio statistic, which is needed to determine the number of components in a mixture or to obtain standard error estimates. The program may be obtained from the Internet at the URL `http://www.medizin.fu-berlin.de/sozmed/DismapWin.html`.

5

Inference for Extremes in Disease Mapping

H. Stern and N. Cressie
Ohio State University

5.1 INTRODUCTION

Maps of disease-incidence or disease-mortality rates are often used to identify regions with unusually high rates. Such areas may then undergo detailed study to determine whether any specific risk factors can be ascertained. Bayes or empirical Bayes point estimates based on a form of squared error loss are commonly used to create the maps; see, for example, Tsutakawa *et al.* (1985), Clayton and Kaldor (1987), Manton *et al.* (1989), Mollié and Richardson (1991), Cressie (1992), Breslow and Clayton (1993), and Bernardinelli *et al.* (1995a). These estimates typically underestimate rates for those regions in the tails of the distribution of underlying rates, which are often the parts of the distribution of most interest of epidemiologists.

In this chapter we reconsider the Bayesian approach to inference about disease-incidence or disease-mortality rates, emphasising the power and flexibility of the posterior distribution for addressing a wide range of scientific questions. In particular, for inference about extremely high rates we consider loss functions that emphasise the extreme order statistic and its antirank (the ith antirank is the region corresponding to the ith order statistic). Although closed-form estimates are not available, simulation from the full posterior distribution can be used to obtain Bayes estimates for these loss functions.

In the next section the spatial model used to analyse disease-incidence or disease-mortality rates in n geographic regions is described. Section 5.3 describes a Bayesian simulation-based approach to drawing inferences from data using this model. Lip cancer mortality rates in the $n = 56$ districts of Scotland are used as an illustration (these geographic districts defined local governing district councils from 1974 until the 1995 reorganisation of local government). Section 5.4 introduces the need for summaries of the posterior distribution and considers Bayes estimates derived under a traditional, quadratic loss function. We also consider constrained Bayes estimates in this section.

Disease Mapping and Risk Assessment for Public Health. Edited by A.B. Lawson *et al.*
© 1999 John Wiley & Sons Ltd.

Section 5.5 introduces alternatives to the quadratic loss function. The various point estimates are demonstrated on Scotland lip cancer data in Section 5.6.

5.2 SPATIAL MODELS FOR DISEASE INCIDENCE OR MORTALITY

5.2.1 A Gaussian–Gaussian probability model

Let $\boldsymbol{O}$ represent the vector of observed disease-incidence counts or the number of deaths from a particular disease from n geographic regions or districts, and $\boldsymbol{E}$ represent the vector of expected counts adjusted for variation in demographic factors across the n regions. The ratio O_i/E_i is a disease-incidence rate relative to population norms, or a standardised mortality rate (SMR) when dealing with mortality data. Although $\boldsymbol{O}$ may be described best by a Poisson distribution, we assume here that, after a suitable transformation, disease-incidence or disease-mortality data can be thought of as approximately Gaussian. This approach was used successfully by Cressie and Chan (1989) to model sudden-infant-death rates in the 100 counties of North Carolina. In the example that follows, we take

$$Y_i \equiv \{O_i/E_i\}^{1/2} + \{(O_i+1)/E_i\}^{1/2}, \quad i = 1, \ldots, n, \tag{5.1}$$

which is the Freeman–Tukey transformation of the observed count divided by the square root of the expected count. The Freeman–Tukey transformation is similar to the somewhat simpler square-root transformation; it has the advantage that it is a better variance-stabilising transformation over a wider range of expected counts (Freeman and Tukey, 1950). Let $\theta_i = \mathrm{E}(Y_i)$ and $\lambda_i = \mathrm{E}(O_i)/E_i$. Then $\boldsymbol{\lambda}$ represents a vector of relative risk parameters. We obtain an approximate relationship between the parameters θ_i and λ_i by taking a Taylor series expansion of $\mathrm{E}(Y_i)$ around $\mathrm{E}(O_i)$:

$$\theta_i \approx 2\lambda_i^{1/2}\left(1 + \frac{1}{8\lambda_i E_i}\right). \tag{5.2}$$

This approximation is most accurate when $\mathrm{E}(O_i) = \lambda_i E_i$ is large, a condition that fails for some of the districts. Conditional on $\boldsymbol{\theta}$, the observed (transformed) data $\boldsymbol{Y}$ are modelled as Gaussian; that is,

$$\boldsymbol{Y}|\boldsymbol{\theta}, \sigma^2 \sim \mathrm{Gau}\left(\boldsymbol{\theta}, \sum = \sigma^2 D\right), \tag{5.3}$$

with D a known matrix. The results in this chapter apply for any known D, although we have taken D to be diagonal in our example.

The underlying rates $\boldsymbol{\theta}$ are modelled as Gaussian with mean depending on the $n \times p$ covariate matrix X and a variance matrix that incorporates spatial relationships among the districts. Specifically,

$$\boldsymbol{\theta}|\boldsymbol{\beta}, \tau^2, \phi \sim \mathrm{Gau}(X\boldsymbol{\beta}, \Gamma = \tau^2(I - \phi C)^{-1}M \equiv \tau^2\Phi), \tag{5.4}$$

where $\boldsymbol{\beta}$ is the unknown regression coefficient vector, $C = (c_{ij})$ is a matrix measuring spatial association, ϕ is a parameter measuring spatial dependence, M

is a known diagonal matrix chosen so that Γ is symmetric and positive-definite, and τ^2 is a scale parameter. The Gaussian model on $\boldsymbol{\theta}$ is an example of the conditional autoregressive model (see, for example, Besag, 1974; Cressie, 1993, Section 6.6). Let $N_i \equiv \{j : c_{ij} \neq 0\}$ represent the 'neighbours' of i and $\boldsymbol{\theta}_{-i} = (\theta_1, \ldots, \theta_{i-1}, \theta_{i+1}, \ldots, \theta_n)^{\mathrm{T}}$. Then we may equivalently write

$$\theta_i | \boldsymbol{\theta}_{-i} \sim \text{Gau}\left((X\boldsymbol{\beta})_i + \phi \sum_{j \in N_i} c_{ij}(\theta_j - (X\boldsymbol{\beta})_j), \tau^2 m_{ii} \right), \quad i = 1, \ldots, n, \tag{5.5}$$

where $(\boldsymbol{X\beta})_i$ is the ith element of the vector $\boldsymbol{X\beta}$. That is to say, the conditional autoregressive model assumes there is an association between the rate in region i and the rates in neighbouring regions. The parameter ϕ and the matrix C determine the degree of association.

Summing up, after a suitable transformation the data are fit to a hierarchical Gaussian–Gaussian model. Our results in this chapter, motivated by applications in disease mapping, can be used in any problem where a Gaussian–Gaussian model is used (e.g. Efron and Morris, 1973). In general, $\boldsymbol{\beta}$, σ^2, τ^2, and ϕ are unknown parameters. In the next few sections we consider several approaches to inference for these parameters and $\boldsymbol{\theta}$. As needed, we assume a (non-informative) prior distribution for these parameters:

$$p(\boldsymbol{\beta}, \sigma^2, \tau^2, \phi) \propto 1/\sigma^2 \tag{5.6}$$

where the distribution of ϕ is restricted to the range of values $(\phi_{\min}, \phi_{\max})$ for which Γ is positive-definite (Cressie, 1993, Section 7.6).

The Gaussian–Gaussian model is desirable because it permits us to compare a wide range of inference procedures, although negative values of $\boldsymbol{\theta}$ and $\boldsymbol{Y}$ are in the support of the joint distribution when in reality these quantities can never be negative for the Freeman–Tukey transformation. Given our interest in large values of $\boldsymbol{\theta}$, this was not a major problem for our analysis. A logarithmic transformation would avoid this problem. An alternative model, e.g. a Poisson–log Gaussian model, would also avoid the problem of negative rates but would require more computational effort to carry out all of the analyses considered here. The Poisson–log Gaussian has been used by a number of authors in this setting; see, for example, Besag *et al.* (1991).

5.2.2 Choosing the variance matrix

The choice of the variance matrix, Γ, of the underlying transformed rates, $\boldsymbol{\theta}$, is important because this is how the spatial association enters into the model. Recall from (5.4) that the variance matrix of the conditional autoregressive (CAR) Gaussian model is

$$\Gamma = \tau^2 (I - \phi C)^{-1} M \equiv \tau^2 \Phi,$$

where ϕ, C, and M are chosen so that Φ is symmetric and positive-definite and $M = \text{diag}(m_{11}, \ldots, m_{nn})$. The conditional variance must be positive so that

$$\text{var}\,(\theta_i | \boldsymbol{\theta}_{-1}) = \tau^2 m_{ii} > 0, \quad i = 1, \ldots, n,$$

and consideration of Φ^{-1} makes it apparent that Φ is symmetric if and only if

$$c_{ij}m_{jj} = c_{ji}m_{ii}, \quad i, j = 1, \ldots, n.$$

Furthermore, positive-definiteness is obtained if and only if

$$\phi \in (\phi_{\min}, \phi_{\max}),$$

where $\phi_{\min} = \eta_1^{-1}, \phi_{\max} = \eta_n^{-1}$, and $\eta_1 < 0 < \eta_n$ are, respectively, the smallest and largest eigenvalues of $M^{-1/2}CM^{1/2}$ (Cressie, 1993, p. 559).

There are a large number of choices for M, C, and ϕ that meet the conditions required for Γ to be a symmetric, positive-definite variance matrix. A common model for spatial association is the intrinsic conditional autoregression proposed by Besag *et al.* (1991) and obtained by choosing

(i) $m_{ii} = |N_i|^{-1}$, the inverse of the number of neighbours of the ith region, $i = 1, \ldots, n$;

(ii) $c_{ij} = \begin{cases} |N_i|^{-1}, & j \in N_i, \\ 0, & \text{elsewhere}, i = 1, \ldots, n; \end{cases}$ and

(iii) $\phi = 1 = \phi_{\max}$.

Besag *et al.* (1991) analysed data using a Poisson–log Gaussian probability model that incorporated this CAR model for spatial association and a vector of independent Gaussian distributed random effects to model general (non-spatial) heterogeneity of rates. That model has subsequently been used by Clayton and Bernardinelli (1992), Breslow and Clayton (1993), Bernardinelli *et al.* (1995), Besag *et al.* (1995), amongst others. The degree of spatial association in the Besag *et al.* (1991) model is determined by the relative magnitudes of the spatial-association variance component and the heterogeneity variance component.

Cressie and Chan (1989) proposed the following choice for a CAR variance matrix (which we use in subsequent sections):

(iv) $m_{ii} = E_i^{-1}$, the inverse of the expected count in the ith region, $i = 1, \ldots, n$;

(v) $c_{ij} = \begin{cases} (E_j/E_i)^{1/2}, & j \in N_i, \\ 0, & \text{elsewhere}, i = 1, \ldots, n; \end{cases}$ and

(vi) $\phi \in (\phi_{\min}, \phi_{\max})$.

Alternative specifications in Cressie and Chan (1989) incorporate a measure of the distance between two districts but we have not done so here to remain consistent with other analyses of the Scotland lip cancer data. The additional non-spatial heterogeneity used by Besag *et al.* (1991) enters our Gaussian–Gaussian probability model through the variance matrix $\Sigma = \sigma^2 D$.

The notable differences between the two proposals are the choice of the conditional (or partial) variances ($m_{ii} = |N_i|^{-1}$ versus $m_{ii} = E_i^{-1}$), the choice of spatial-dependence coefficients ($c_{ij} = m_{ii}$ versus $c_{ij} = \{m_{ii}/m_{jj}\}^{1/2}$), and the choice of the spatial-dependence parameter ($\phi = \phi_{\max}$ versus $\phi \in (\phi_{\min}, \phi_{\max})$). We now discuss the last two of these differences further. The spatial-dependence coefficients differ in that $\sum_j c_{ij} = 1$ under (ii) with no similar normalisation under (v). Both specifications appear reasonable in that they lead to similar variance matrices in the Scotland lip cancer example. The choice of ϕ represents another difference, with $\phi = \phi_{\max}$ in (iii) providing maximal spatial association. The variance matrix Γ is singular at $\phi_{\max}$, leading to an improper prior distribution of $\boldsymbol{\theta}$ which, as usual, should be considered carefully before implementation. By contrast, (vi) is straightforward to deal with. It allows the data to suggest likely values for ϕ from among those for which Γ is positive-definite because, a priori, it is allowed to vary over its parameter space, $(\phi_{\min}, \phi_{\max}\}$. With ϕ fixed at its maximum value, Besag *et al.* (1991) assess the degree of spatial association by looking at the relative contribution of the two variance components, the spatial-association-related variance τ^2 and a general (non-spatial) heterogeneity parameter (analogous to our σ^2). Another difference concerns the conditional correlation implied by the given choices of ϕ and $\{c_{ij}\}$. A result of Cressie (1993, Section 7.6) shows that $\phi^2 c_{ij} c_{ji} = \mathrm{corr}^2\{\theta_i, \theta_j | \boldsymbol{\theta}_{-i,-j}\}$ which we might call the partial or conditional correlation (it should be noted that the notation here differs slightly from that of Cressie's Section 7.6). Now, the choice of (ii) and (iii) yields $\phi^2 c_{ij} c_{ji} = |N_i|^{-1}|N_j|^{-1}$. That is, the model only allows weak partial correlations that vary according to the neighbourhood structure and behave like products of partial variances. By contrast, choice of (v) and (vi) yields $\phi^2 c_{ij} c_{ji} = \phi^2$ for $\phi \in (\phi_{\min}, \phi_{\max})$, regardless of the choice of partial variances m_{ii}, and therefore ϕ^2 is interpretable as a partial correlation squared that is invariant to the neighbourhood structure. In what is to follow, we assume (iv), (v), and (vi) for the model for $\mathbf{Y}$ and $\boldsymbol{\theta}$.

5.3 BAYESIAN INFERENCE VIA SIMULATION

5.3.1 The simulation approach

We describe the Bayesian approach to inference in the Gaussian–Gaussian model using lip cancer data from Scotland to demonstrate. Table 5.1 repeats the lip cancer data from Breslow and Clayton (1993) with district names provided in Cressie (1993, Section 7.5). A map showing the 56 districts of Scotland is also provided in Cressie (1993, p. 538). These 56 geographic districts helped define the local government structure of Scotland prior to the 1995 reorganization of local government. The table gives observed cases of lip cancer (O), expected cases (E), the standardised mortality rate (O/E), the Freeman–Tukey transformed data (Y), a single covariate measuring the percent of the population engaged in outdoor industry (agriculture, fishing, forestry, abbreviated to *AFF*) and the neighbours of each district. The identification numbers in the table represent the ranks of the districts according to the ratio O/E.

We apply the Gaussian–Gaussian model of Section 5.2 to the data with M, C, and ϕ given by (iv)–(vi) of that section. The covariate matrix X in the model for the underlying rates $\boldsymbol{\theta}$ contains a column of ones corresponding to the intercept and a column containing the variable *AFF*. The variance of $\mathbf{Y}$ in the model is $\sigma^2 D$, where D is assumed to be diagonal with ith element E_i^{-1}. This choice for D is motivated by a delta-method argument for the variance of the Freeman–Tukey transformed data. Specifically, if O_i were Poisson with mean $\lambda_i E_i$, then the delta-method suggests that the variance of Y_i is approximately $1/E_i$. Note that this argument would also suggest fixing $\sigma^2 = 1$ and, as we dicuss below, it is necessary to make some assumption about σ^2 for the lip cancer data because the data do not enable us to estimate τ^2 and σ^2 separately.

Table 5.1 Scotland lip cancer data

ID	District name	*Q*	*E*	*Q/E*	*Y*	*AFF*	Neighbours
1	Skye–Lochalsh	9	1.38	6.52	5.25	16	5, 9, 11, 19
2	Banff–Buchan	39	8.66	4.50	4.27	16	7, 10
3	Caithness	11	3.04	3.62	3.89	10	6, 12
4	Berwickshire	9	2.53	3.56	3.87	24	18, 20, 28
5	Ross–Cromarty	15	4.26	3.52	3.81	10	1, 11, 12, 13, 19
6	Orkney	8	2.40	3.33	3.76	24	3, 8
7	Moray	26	8.11	3.21	3.62	10	2, 10, 13, 16, 17
8	Shetland	7	2.30	3.04	3.61	7	6
9	Lochaber	6	1.98	3.03	3.62	7	1, 11, 17, 19, 23, 29
10	Gordon	20	6.63	3.02	3.52	16	2, 7, 16, 22
11	Western Isles	13	4.40	2.95	3.50	7	1, 5, 9, 12
12	Sutherland	5	1.79	2.79	3.50	16	3, 5, 11
13	Nairn	3	1.08	2.78	3.59	10	5, 7, 17, 19
14	Wigtown	8	3.31	2.42	3.20	24	31, 32, 35
15	NE Fife	17	7.84	2.17	2.99	7	25, 29, 50
16	Kincardine	9	4.55	1.98	2.89	16	7, 10, 17, 21, 22, 29
17	Badenoch	2	1.07	1.87	3.04	10	7, 9, 13, 16, 19, 29
18	Ettrick	7	4.18	1.67	2.68	7	4, 20, 28, 33, 55, 56
19	Inverness	9	5.53	1.63	2.62	7	1, 5, 9, 13, 17
20	Roxburgh	7	4.44	1.58	2.60	10	4, 18, 55
21	Angus	16	10.46	1.53	2.51	7	16, 29, 50
22	Aberdeen	31	22.67	1.37	2.36	16	10, 16
23	Argyll–Bute	11	8.77	1.25	2.29	10	9, 29, 34, 36, 37, 39
24	Clydesdale	7	5.62	1.25	2.31	7	27, 30, 31, 44, 47, 48, 55, 56
25	Kirkcaldy	19	15.47	1.23	2.25	1	15, 26, 29
26	Dunfermline	15	12.49	1.20	2.23	1	25, 29, 42, 43
27	Nithsdale	7	6.04	1.16	2.23	7	24, 31, 32, 55
28	East Lothian	10	8.96	1.12	2.16	7	4, 18, 33, 45
29	Perth–Kinross	16	14.37	1.11	2.14	10	9, 15, 16, 17, 21, 23, 25, 26, 34, 43, 50
30	West Lothian	11	10.20	1.08	2.12	10	24, 38, 42, 44, 45, 56
31	Cumnock–Doon	5	4.75	1.05	2.15	7	14, 24, 27, 32, 35, 46, 47
32	Stewartry	3	2.88	1.04	2.20	24	14, 27, 31, 35
33	Midlothian	7	7.03	1.00	2.06	10	18, 28, 45, 56
34	Stirling	8	8.53	0.94	2.00	7	23, 29, 39, 40, 42, 43, 51, 52, 54
35	Kyle–Carrick	11	12.32	0.89	1.93	7	14, 31, 32, 37, 46
36	Inverclyde	9	10.10	0.89	1.94	0	23, 37, 39, 41
37	Cunninghame	11	12.68	0.87	1.90	10	23, 35, 36, 41, 46
38	Monklands	8	9.35	0.86	1.91	1	30, 42, 44, 49, 51, 54
39	Dumbarton	6	7.20	0.83	1.90	16	23, 34, 36, 40, 41
40	Clydebank	4	5.27	0.76	1.85	0	34, 39, 41, 49, 52
41	Renfrew	10	18.76	0.53	1.50	1	36, 37, 39, 40, 46, 49, 53
42	Falkirk	8	15.78	0.51	1.47	16	26, 30, 34, 38, 43, 51
43	Clackmannan	2	4.32	0.46	1.51	16	26, 29, 34, 42
44	Motherwell	6	14.63	0.41	1.33	0	24, 30, 38, 48, 49
45	Edinburgh	19	50.72	0.37	1.24	1	28, 30, 33, 56
46	Kilmarnock	3	8.20	0.37	1.30	7	31, 35, 37, 41, 47, 53
47	East Kilbride	2	5.59	0.36	1.33	1	24, 31, 46, 48, 49, 53
48	Hamilton	3	9.34	0.32	1.22	1	24, 44, 47, 49

Table 5.1 *(continued)*

ID	District name	Q	E	Q/E	Y	AFF	Neighbours
49	Glasgow	28	88.66	0.32	1.13	0	38, 40, 41, 44, 47, 48, 52, 53, 54
50	Dundee	6	19.62	0.31	1.15	1	15, 21, 29
51	Cumbernauld	1	3.44	0.29	1.30	1	34, 38, 42, 54
52	Bearsden	1	3.62	0.28	1.27	0	34, 40, 49, 54
53	Eastwood	1	5.74	0.17	1.01	1	41, 46, 47, 49
54	Strathkelvin	1	7.03	0.14	0.91	1	34, 38, 49, 51, 52
55	Annandale	0	4.16	0.00	0.49	16	18, 20, 24, 27, 56
56	Tweeddale	0	1.76	0.00	0.75	10	18, 24, 30, 33, 45, 55

Applying the Bayesian paradigm for drawing inferences under the model of Section 5.2, we focus our attention on the joint posterior distribution of the model parameters:

$$p(\boldsymbol{\theta}, \sigma^2, \tau^2, \phi, \boldsymbol{\beta} \mid \mathbf{Y}) \propto p(\mathbf{Y} \mid \boldsymbol{\theta}, \sigma^2) p(\boldsymbol{\theta} \mid \boldsymbol{\beta}, \tau^2, \phi) p(\boldsymbol{\beta}, \sigma^2, \tau^2, \phi). \tag{5.7}$$

Inferences about $\boldsymbol{\theta}$ or the other model parameters can be obtained using suitable summaries of the posterior distribution. In analysing the Scotland lip cancer data we rely on a simulation approach and base our inferences on realisations from the posterior distribution $p(\boldsymbol{\theta}, \sigma^2, \tau^2, \phi, \boldsymbol{\beta} \mid \mathbf{Y})$. One approach to obtaining realisations from the full posterior distribution would apply Markov chain Monte Carlo (MCMC) methodology. However, because of the Gaussian–Gaussian hierarchical model, we are able to use analytic results and a discrete approximation to the marginal posterior distribution of τ^2 and ϕ to produce a direct simulation algorithm. The direct simulation approach does not require that we assess the convergence of a Markov chain, but it does rely on a discrete approximation. We describe the approach more fully in the next paragraphs.

To begin, we rewrite the joint posterior distribution (5.7) as

$$p(\boldsymbol{\theta}, \sigma^2, \tau^2, \phi, \boldsymbol{\beta} \mid \mathbf{Y}) \propto p(\boldsymbol{\beta} \mid \boldsymbol{\theta}, \sigma^2, \tau^2, \phi, \mathbf{Y})\, p(\boldsymbol{\theta} \mid \sigma^2, \tau^2, \phi, \mathbf{Y})\, p(\sigma^2, \tau^2, \phi \mid \mathbf{Y}),$$

with

$$\boldsymbol{\beta} \mid \boldsymbol{\theta}, \sigma^2, \tau^2, \phi, \mathbf{Y} \sim \mathrm{Gau}(\{X'\Gamma^{-1}X\}^{-1}\{X'\Gamma^{-1}\boldsymbol{\theta}\}, \{X'\Gamma^{-1}X\}^{-1}) \tag{5.8}$$

and

$$\begin{aligned} \boldsymbol{\theta} \mid \sigma^2, \tau^2, \phi, \mathbf{Y} \sim \mathrm{Gau}([&(\sigma^2 D)^{-1} + (I - P_\Gamma)'\Gamma^{-1}(I - P_\Gamma)]^{-1}(\sigma^2 D)^{-1}\mathbf{Y}, \\ &[(\sigma^2 D)^{-1} + (I - P_\Gamma)'\Gamma^{-1}(I - P_\Gamma)]^{-1}) \end{aligned} \tag{5.9}$$

and

$$\begin{aligned} &p(\sigma^2, \tau^2, \phi \mid \mathbf{Y}) \\ &\quad \propto p(\sigma^2, \tau^2, \phi)\sigma^{-n} |(\sigma^2 D)^{-1} + (I - P_\Gamma)'(\tau^2 \Phi)^{-1}(I - P_\Gamma)|^{-1/2} \\ &\quad \times |\Gamma|^{-1/2} |X'\Gamma X|^{-1/2} \exp(-0.5\mathbf{Y}'\{(\sigma^2 D)^{-1} - (\sigma^2 D)^{-1} \\ &\quad \times [(\sigma^2 D)^{-1} + (I - P_\Gamma)'\Gamma^{-1}(I - P_\Gamma)]^{-1}(\sigma^2 D)^{-1}\}\mathbf{Y}), \end{aligned} \tag{5.10}$$

where p is the rank of the matrix X, $\Gamma = \tau^2\Phi = \tau^2(I - \phi C)^{-1}M$, and $P_\Gamma = X(X'\Gamma^{-1}X)^{-1}X'\Gamma^{-1}$. Thus, draws from the posterior distribution (5.7) can be obtained by first drawing from the joint posterior distribution of σ^2, τ^2, ϕ, (5.10) and then sampling $\boldsymbol{\theta}$ and $\boldsymbol{\beta}$ from the relevant Guassian distributions ((5.9) and (5.8), respectively).

We examined the shape of the posterior distribution (5.10) for (σ^2, τ^2, and ϕ) on a grid of values for a number of prior distributions. Several interesting results are obtained. First, for $\phi = 0$ (corresponding to no spatial relationship), the model simplifies and, marginally, Y_i has variance proportional to $\sigma^2 + \tau^2$ (recall $D = M$ in our case). There is no information in the data to permit separate estimation of σ^2 and τ^2 in this case. For positive values of ϕ, the data appears to provide little information for separately estimating σ^2 and τ^2. For most of the prior distributions on σ^2 and τ^2 that we considered, the mode of the posterior distribution of (σ^2, τ^2, and ϕ) has σ^2 equal zero. For certain other choices of the prior distribution for σ^2 and τ^2 we obtain a mode with $\tau^2 = 0$. Given this difficulty in identifying the two variance parameters we choose to fix σ^2. There is no evidence of extra-Poisson variability in the counts once covariates and spatial associa-

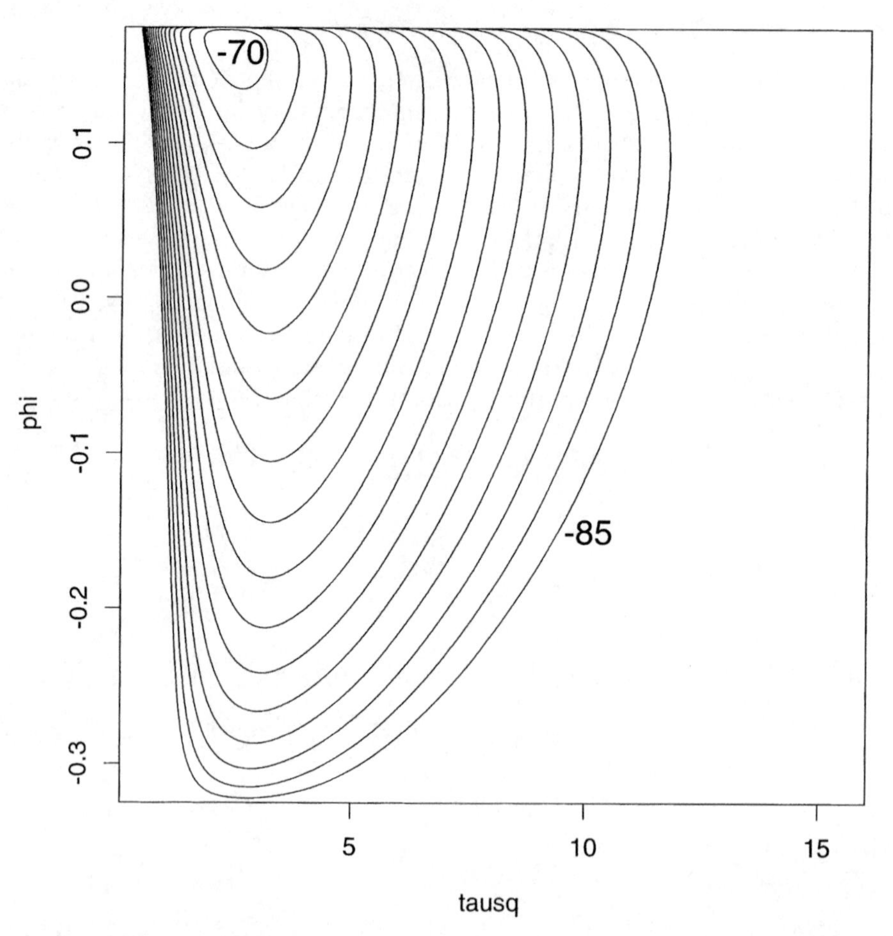

Figure 5.1 Contour plot of the logarithm of the marginal posterior distribution $p(\tau^2, \phi | \mathbf{Y}, \sigma^2 = 1)$

tion are included in the model, which leads us to choose $\sigma^2 = 1$, the value suggested by the Freeman–Tukey transformation. With $\sigma^2 = 1$, we construct a discrete approximation to the marginal posterior distribution, $p(\tau^2, \phi | \mathbf{Y}, \sigma^2 = 1)$ supported on a grid of values in $(0, 16] \times (\phi_{\min}, \phi_{\max})$. A contour plot of the logarithm of the marginal posterior distribution is shown in Figure 5.1. From this plot we see clearly that the data support values of the spatial-dependence parameter ϕ near, but below, its upper limit $\phi_{\max}$.

5.3.2 Simulation results

To illustrate the types of results produced under the Bayesian paradigm, we provide in Table 5.2 summaries of 2000 simulations from the posterior distribution. We transform the posterior simulations of $\boldsymbol{\theta}$ to $\boldsymbol{\lambda}$ using $\lambda_i = \theta_i^2/4$, which is an approximation to the inverse of the transformation (5.2). A more accurate inverse transformation can be derived but it relies on large expected counts. Values of θ_i less than zero are mapped into $\lambda_i = 0$; these are possible under the Gaussian–Gaussian model but occurred rarely. The posterior median and a 95% central posterior interval are provided for the λ_is, the two extrema $\lambda_{(56)}$ and $\lambda_{(1)}$, and the remaining model parameters. Figure 5.2 shows empirical posterior distributions based on 2000 simulations of the rates λ_i for four of the $n = 56$ districts in the Scotland lip cancer data along with the corresponding observed values of the SMR, O_i/E_i. We chose the districts with the highest (Skye–Lochalsh; 1) and lowest (Annandale; 55) value of the transformed counts $\{Y_i\}$ and the highest (Glasgow; 49) and lowest (Badenoch; 17) values of $\{E_i\}$. Two features are noteworthy. First, the variance of the posterior distributions are largest where there are few data (e.g. districts like Badenoch and Skye–Lochalsh where E_i is small) and smallest where there are substantial data (e.g. Glasgow). Secondly, there is a tendency for the rates corresponding to the extreme observed values to be shrunken towards the value that would be predicted by the regression surface $X\boldsymbol{\beta}$. Most of the posterior draws from Skye–Lochalsh are lower than the observed value and most of the draws for Annandale are higher than the observed value.

The effect of the covariate, *AFF*, can be assessed by considering the posterior distribution for the slope parameter β_2. On the basis of the posterior simulations displayed in Table 5.2, we estimate the posterior median of β_2 to be 0.07 and a central 95% posterior interval to be (0.04, 0.10). The posterior distribution provides strong support for a positive association between the covariate *AFF* and the underlying rates.

In this chapter, interest is in the largest rate and the identity of the region possessing the largest rate. Clearly, the simulations from the posterior distribution allow for a direct assessment of these quantities. Define

$$\lambda_{(n)} \equiv \max\{\lambda_i : i = 1, \ldots, n\},$$
$$i(n) \equiv \{i : \lambda_i = \lambda_{(n)}\}.$$

Figure 5.3 shows the results of a simulation from $p(\lambda_{(56)} | \mathbf{Y}, \sigma^2 = 1)$ for the Scotland lip cancer data, obtained directly from the simulations of $p(\boldsymbol{\theta} | \mathbf{Y}, \sigma^2 = 1)$ via transformation. The first two posterior moments, based on 2000 realizations, are computed to be

$$\mathrm{E}(\lambda_{(56)} | \mathbf{Y}, \sigma^2 = 1) = 6.28; \qquad \{\mathrm{var}(\lambda_{(56)} | \mathbf{Y}, \sigma^2 = 1)\}^{1/2} = 1.47.$$

Table 5.2 Posterior inference for $\boldsymbol{\lambda}$ and other parameters

	Posterior distribution				Posterior distribution		
Parameter	**2.5%**	**Median**	**97.5%**	**Parameter**	**2.5%**	**Median**	**97.5%**
λ_1	2.72	5.78	9.51	λ_{29}	0.78	1.24	1.81
λ_2	2.70	3.76	5.02	λ_{30}	0.58	1.07	1.69
λ_3	1.55	3.03	5.10	λ_{31}	0.42	1.05	1.99
λ_4	1.67	3.38	5.63	λ_{32}	0.50	1.50	3.01
λ_5	1.81	3.11	4.69	λ_{33}	0.52	1.06	1.79
λ_6	1.65	3.34	5.64	λ_{34}	0.44	0.90	1.51
λ_7	1.88	2.80	3.90	λ_{35}	0.50	0.92	1.43
λ_8	1.11	2.53	4.75	λ_{36}	0.39	0.81	1.39
λ_9	1.21	2.83	5.33	λ_{37}	0.56	0.97	1.52
λ_{10}	1.78	2.83	3.99	λ_{38}	0.34	0.72	1.26
λ_{11}	1.47	2.61	4.08	λ_{39}	0.52	1.10	1.85
λ_{12}	1.18	2.94	5.56	λ_{40}	0.22	0.70	1.44
λ_{13}	0.85	3.12	6.58	λ_{41}	0.29	0.53	0.86
λ_{14}	1.24	2.50	4.17	λ_{42}	0.43	0.79	1.21
λ_{15}	1.13	1.85	2.79	λ_{43}	0.21	0.80	1.71
λ_{16}	1.15	2.14	3.41	λ_{44}	0.20	0.45	0.80
λ_{17}	0.61	2.42	5.69	λ_{45}	0.30	0.44	0.62
λ_{18}	0.65	1.49	2.69	λ_{46}	0.19	0.52	1.05
λ_{19}	0.88	1.71	2.76	λ_{47}	0.09	0.42	1.05
λ_{20}	0.69	1.54	2.66	λ_{48}	0.14	0.41	0.82
λ_{21}	0.86	1.41	2.12	λ_{49}	0.26	0.37	0.50
λ_{22}	1.09	1.50	1.97	λ_{50}	0.21	0.42	0.70
λ_{23}	0.71	1.28	2.00	λ_{51}	0.03	0.42	1.22
λ_{24}	0.48	1.12	1.98	λ_{52}	0.04	0.41	1.15
λ_{25}	0.70	1.10	1.63	λ_{53}	0.05	0.29	0.79
λ_{26}	0.60	1.03	1.62	λ_{54}	0.04	0.27	0.73
λ_{27}	0.53	1.12	1.94	λ_{55}	0.01	0.28	0.90
λ_{28}	0.59	1.10	1.79	λ_{56}	0.00	0.26	1.38
$\lambda_{(56)}$	4.15	6.03	9.55	$\lambda_{(1)}$	0.00	0.08	0.26
τ^2	1.58	2.72	4.70	β_1	1.21	1.52	1.86
ϕ	0.016	0.142	0.173	β_2	0.037	0.067	0.098

From the same simulations, we obtain, in decreasing order of probability,

$$\Pr(i(n) = 1 \,|\, \mathbf{Y}, \sigma^2 = 1) = 0.674,$$
$$\Pr(i(n) = 13 \,|\, \mathbf{Y}, \sigma^2 = 1) = 0.078,$$
$$\Pr(i(n) = 4 \,|\, \mathbf{Y}, \sigma^2 = 1) = 0.050,$$
$$\Pr(i(n) = 6 \,|\, \mathbf{Y}, \sigma^2 = 1) = 0.044.$$

It may seem curious that district 13 (Nairn) has only the thirteenth largest value of $\{O_i/E_i\}$ yet has the second largest probability of being the district with maximum

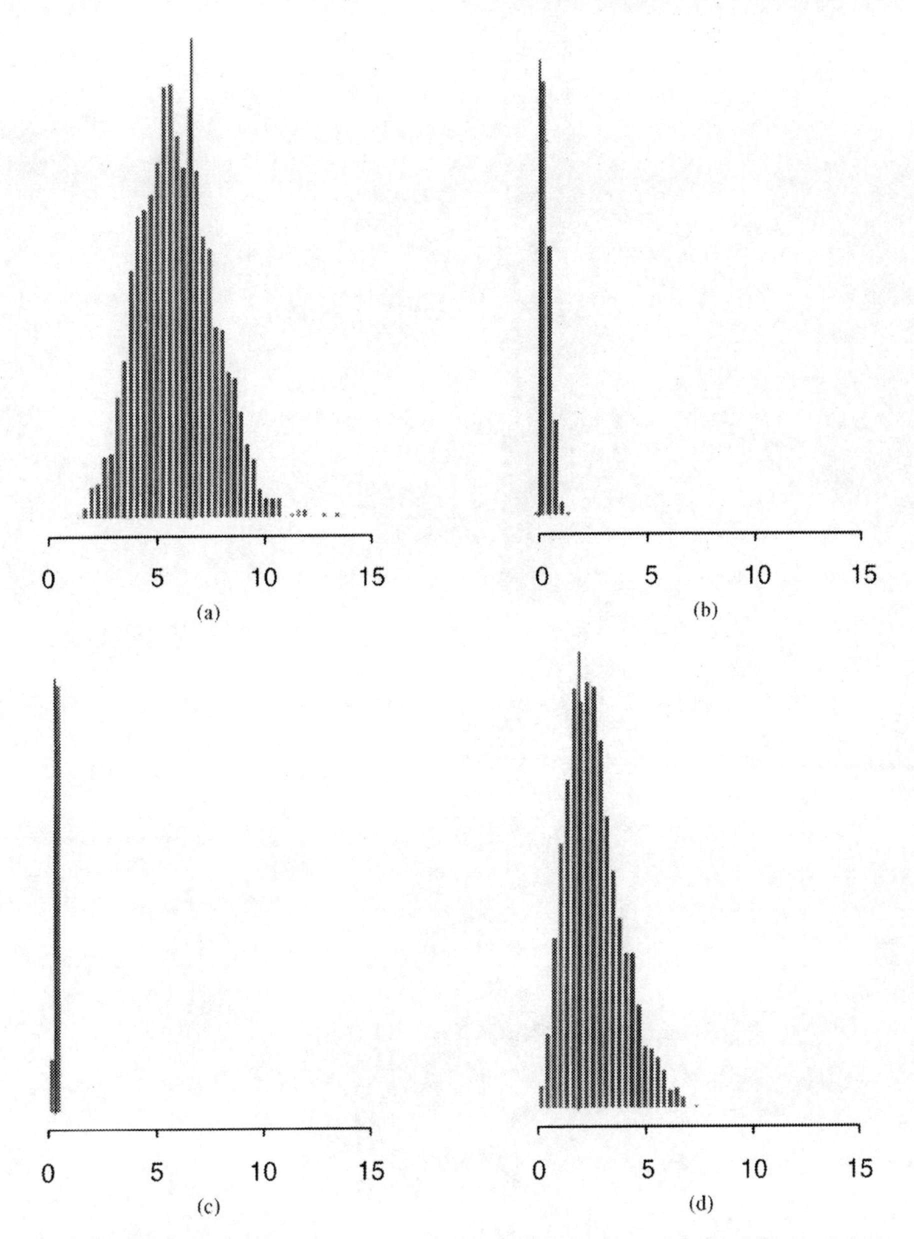

Figure 5.2 Posterior distributions of relative risk parameters, λ_i, for four districts estimated using 2000 realisations simulated from $p(\boldsymbol{\theta} \mid \mathbf{Y}, \sigma^2 = 1)$ and then transforming to the $\boldsymbol{\lambda}$ scale. Observed values of the standardised mortality rate, O_i/E_i, are indicated by a solid vertical line on each plot. (a) District 1, Skye–Lochalsh, has the largest Y_i. (b) District 55, Annandale, has the smallest, Y_i. (c) District 49, Glasgow, has the largest E_i. (d) District 17, Badenoch, has the smallest E_i.

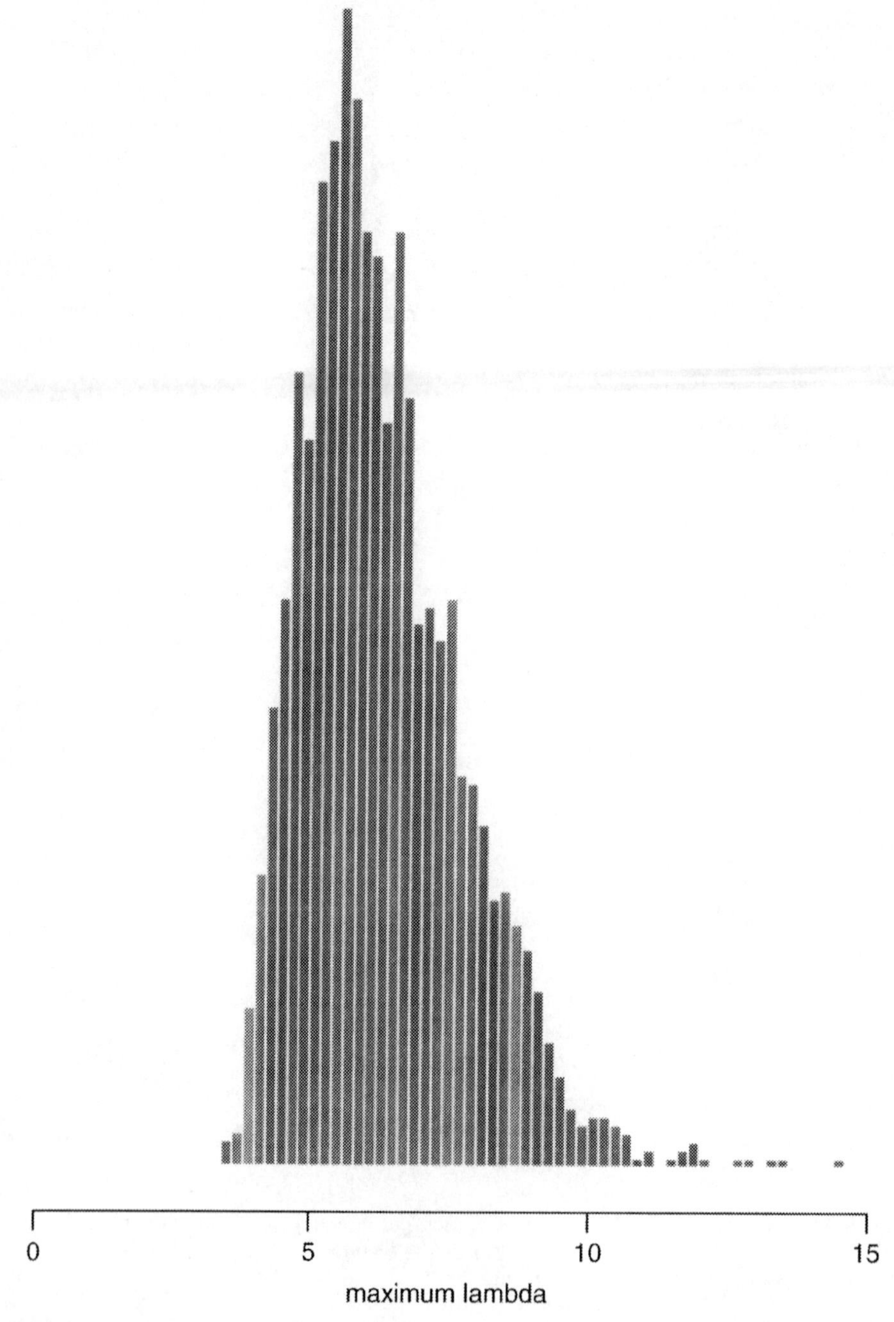

Figure 5.3 Posterior distribution $p(\lambda_{(56)} \mid \boldsymbol{Y}, \sigma^2 = 1)$, where $\lambda_{(56)}$ is the maximum value of $\boldsymbol{\lambda}$, estimated using 2000 realisations simulated from $p(\boldsymbol{\theta} \mid \boldsymbol{Y}, \sigma^2 = 1)$ and transformed to the $\boldsymbol{\lambda}$ scale

$\{\lambda_i\}$. There are several factors that help to explain this result. First, the Freeman–Tukey transformation used in modelling is not a monotonic function of O_i/E_i and districts like Nairn with low expected counts move up in the ranking as a result of the transformation. Secondly, the Bayesian approach filters the noisy $\{Y_i\}$ in order to obtain the presumably smoother underlying values $\{\lambda_i\}$. The relative degree of filtering depends on the value of the covariate in the district, the observed rates and covariate values for neighbouring districts, and the expected count in the district. Here, many of the higher-ranking districts based on O_i/E_i are estimated to have lower probability of being the extreme value because of their value of *AFF* and the information provided by neighbouring districts. Finally, the probability of an individual district having the extreme rate depends critically on the uncertainty concerning that district's underlying rate. Table 5.2 indicates that there is substantial uncertainty about λ_{13} with the posterior 95% interval including some extremely high values and some extremely low values. This is not surprising given the extremely low expected count in that district (second lowest out of the 56 districts).

5.4 BAYES AND CONSTRAINED BAYES ESTIMATES

The joint posterior distribution of all model parameters can, with the aid of simulation, address any scientific question of interest. Conceptually, each draw from the posterior distribution provides a 'plausible' map of disease-incidence rates. Our 2000 posterior simulations provide a reference set of 2000 plausible maps. Scientific questions, such as which district has the highest rate, can be addressed by reviewing the reference set of plausible maps as was done in the previous section. The flexibility provided by having this reference set (i.e. the posterior distribution) is a strength of the Bayesian approach. Despite, this, a single map or point estimate is often desirable for public policy reasons. However, it seems quite clear that there is not a single best set of small-area estimates for all purposes. A single map or point estimate can only be obtained by explicitly identifying a loss function that describes the scientific goal and the consequences of poor estimates. In this section and the next we consider a number of possible loss functions. We obtain optimal point estimates on the $\boldsymbol{\theta}$ scale, and then transform these estimates to the $\boldsymbol{\lambda}$ scale. An alternative would be to define loss functions directly on the $\boldsymbol{\lambda}$ scale and optimise on that scale. The various computational approaches used to derive the point estimates can be easily modified to accommodate such a change. We have not done so here because the transformation (5.2) defining $\boldsymbol{\lambda}$ is only an approximation, and moreover the approximation requires large expected counts for each district (a condition that is not satisfied in our illustrative example).

5.4.1 Bayes point estimates

In this section we consider the quadratic loss functions that have dominated statistical practice in many areas. A popular choice is the sum of squared errors, $L(\boldsymbol{\theta}, \mathbf{t}) = (\boldsymbol{\theta} - \mathbf{t})'(\boldsymbol{\theta} - \mathbf{t})$, but this ignores the spatial association that is built into our model. In this chapter we consider the quadratic loss function $L(\boldsymbol{\theta}, \mathbf{t}) = (\boldsymbol{\theta} - \mathbf{t})'\Phi^{-1}(\boldsymbol{\theta} - \mathbf{t})$ so that the spatial association is taken into account when assessing the fit of an estimate. Note that we use Φ rather than the complete variance matrix $\Gamma = \tau^2\Phi$ to define the loss function because we wish to express our loss in squared units of θ rather than in standardised units. The quadratic loss depending on Φ poses a problem because it depends on the spatial-dependence parameter, ϕ, and consequently it is difficult to evaluate the estimator (for reasons that are described below). Instead, in deriving point estimates, we consider ϕ to be fixed in our model (with the analysis perhaps repeated at different values). With ϕ fixed, the Bayes estimator for $\boldsymbol{\theta}$ is $\mathrm{E}(\boldsymbol{\theta}|\mathbf{Y}, \sigma^2 = 1, \phi)$. The Bayes estimator is obtained by averaging the full posterior distribution over the remaining parameters $\boldsymbol{\beta}$ and τ^2. In fact, as shown by Ghosh (1992), the estimate can be obtained by analytical integration followed by numerical integration over a single parameter. Here, we can integrate out $\boldsymbol{\beta}$ analytically to obtain

$$\mathrm{E}(\boldsymbol{\theta}|\mathbf{Y}, \tau^2, \sigma^2 = 1, \phi) = \{D^{-1} + (I - P_\Gamma)'\Gamma^{-1}(I - P_\Gamma)\}^{-1}D^{-1}\mathbf{Y}, \tag{5.11}$$

where, once again, $P_\Gamma \equiv X(X'\Gamma^{-1}X)^{-1}X'\Gamma^{-1}$. The Bayes estimate, $\mathrm{E}(\boldsymbol{\theta}|\mathbf{Y}, \sigma^2 = 1, \phi)$, is obtained by integrating the expectation (5.11) with respect to the marginal posterior distribution of τ^2,

$$\mathrm{E}(\boldsymbol{\theta}|\mathbf{Y}, \sigma^2 = 1, \phi) = \int_0^\infty \mathrm{E}(\boldsymbol{\theta}|\mathbf{Y}, \tau^2, \sigma^2 = 1, \phi)\, p(\tau^2|\mathbf{Y}, \sigma^2 = 1, \phi)\, \mathrm{d}\tau^2, \tag{5.12}$$

where

$$\begin{aligned} p(\tau^2|\mathbf{Y}, \sigma^2 = 1, \phi) \propto \tau^{-(n-2)}&|D^{-1} + (I - P_\Gamma)'\Gamma^{-1}(I - P_\Gamma)|^{-1/2} \\ &\exp(-0.5\mathbf{Y}'[D^{-1} - D^{-1}\{D^{-1} \\ &+ (I - P_\Gamma)'\Gamma^{-1}(I - P_\Gamma)\}^{-1}D^{-1}]\mathbf{Y}). \end{aligned} \tag{5.13}$$

This integration was performed numerically using Simpson's rule (with Fortran programs based on those in Press *et al.*, 1986, Section 4.2). The computation is simplified by the identity $(I - P_\Gamma)'\Gamma^{-1}(I - P_\Gamma) = (I - P_\Gamma)'\Gamma^{-1}$. Two modifications of this procedure are required if ϕ is treated as a random variable rather than a fixed quantity. First, the Bayes estimate for a weighted-squared error loss function that depends on parameters of the model is a ratio of posterior expectations rather than a single expectation. Secondly, the required numerical integrations would be over the two parameters τ^2 and ϕ.

The preceding discussion used hierarchical models to account for the unknown parameters $\boldsymbol{\beta}$ and τ^2 and hence might be called hierarchical Bayes (HB). An alternative is the empirical Bayes approach that obtains point estimates of $\boldsymbol{\beta}$ and τ^2 and then treats these parameters as fixed values for all subsequent analyses. See, for example, Kass and Steffey (1989) for a comparison of hierarchical Bayes and empirical Bayes methods.

5.4.2 Constrained Bayes estimates

Bayes point estimates based on some form of squared-error loss are commonly used to create maps of small-area rates, as described in Section 5.4.1. One positive feature of such maps is that they are smoother than maps of raw disease incidence rates because the impact of high-variance, low-population areas is lessened. Unfortunately, the use of posterior means creates an unanticipated difficulty because the resulting estimates for low population areas tend to be 'overshrunk' toward the overall mean. Louis (1984) shows that the variability of posterior means is an underestimates of the population variability of the parameters. Gelman and Price (1999) show that when posterior means are used, low-population areas are less likely to show up as extreme values than high-population areas.

One approach to improved inference for extremes is to adjust the estimates derived under quadratic loss so that these estimates no longer underestimate the variability of the ensemble of risk parameters. Louis (1984) constructs *constrained* Bayes estimates that are constrained to match the posterior expected values of the sample mean and sample variance of the parameters in a simple form of the Gaussian–Gaussian hierarchical model. Ghosh (1992) extends Louis's result by dropping the Gausian assumption for the data model. We provide a further extension to accommodate the covariates X and a general covariance structure. The constraints in this setting are that the regression function of the estimates on X match the expected posterior regression function of $\boldsymbol{\theta}$ on X, and that the residual variance of the estimates about the regression surface match the expected residual variance of the transformed rates $\boldsymbol{\theta}$ about the regression surface. Note that, as in the previous section, the variance matrix of $\boldsymbol{\theta}$ is $\Gamma = \tau^2\Phi$, whereas the loss function depends only on Φ.

Theorem 1. Suppose that $\mathbf{Y}|\boldsymbol{\theta} \sim p(\mathbf{Y}|\boldsymbol{\theta})$ and $\boldsymbol{\theta} \sim \text{Gau}(X\boldsymbol{\beta}, \Gamma)$, with $\Gamma = \tau^2\Phi$ positive definite, and β, τ^2, and Φ known. Let $Y_o = \{\mathbf{Y}: H_2(\mathbf{Y}) > 0\}$ with $H_2(\mathbf{Y})$ defined below, and let $P_\Phi = X(X'\Phi^{-1}X)^{-1}X'\Phi^{-1}$ denote the projection matrix that yields the predicted values for the generalised least squares regression on X with error vector that has variance matrix Φ. Then for $\mathbf{Y} \in Y_o$, the estimator $\mathbf{t}(\mathbf{Y})$ that minimises the posterior expected weighted squared error $\text{E}[(\boldsymbol{\theta} - \mathbf{t}(\mathbf{Y}))'\Phi^{-1}(\boldsymbol{\theta} - \mathbf{t}(\mathbf{Y}))|\mathbf{Y}]$ subject to

$$P_\Phi \text{E}[\boldsymbol{\theta}|\mathbf{Y}] = P_\Phi \mathbf{t} \tag{5.14}$$

and

$$\text{E}[\boldsymbol{\theta}'(I - P_\Phi)'\Phi^{-1}(I - P_\Phi)\boldsymbol{\theta}|\mathbf{Y}] = \mathbf{t}'(I - P_\Phi)'\Phi^{-1}(I - P_\Phi)\mathbf{t} \tag{5.15}$$

is given by

$$\mathbf{t}(\mathbf{Y}) = a\text{E}(\boldsymbol{\theta}|\mathbf{Y}) + (1 - a)P_\Phi \text{E}(\boldsymbol{\theta}|\mathbf{Y}) \tag{5.16}$$

with

$$a = a(\mathbf{Y}) = [1 + H_1(\mathbf{Y})/H_2(\mathbf{Y})]^{1/2} \tag{5.17}$$

and

$$H_1(\mathbf{Y}) = \text{tr}\{\text{var}(\Phi^{-1/2}(I - P_\Phi)\boldsymbol{\theta}|\mathbf{Y}), \tag{5.18}$$

$$H_2(\mathbf{Y}) = \text{E}(\boldsymbol{\theta}|\mathbf{Y})'(I - P_\Phi)'\Phi^{-1}(I - P_\Phi)\text{E}(\boldsymbol{\theta}|\mathbf{Y}). \tag{5.19}$$

The proof follows the same line of reasoning as Ghosh (1992); it is provided in an appendix. In the appendix we also show that it is possible to prove that the same estimator (5.16) is optimal if we require only the second constraint (5.15). Note that the optimal estimate (5.16) is a linear combination of $E(\boldsymbol{\theta}|\mathbf{Y})$ and $P_\Phi E(\boldsymbol{\theta}|\mathbf{Y})$ with weights that depend on $E(\boldsymbol{\theta}|\mathbf{Y})$ and $\text{var}(\boldsymbol{\theta}|\mathbf{Y})$ through (5.18) and (5.19). The optimal constrained estimates can be rewritten as $E(P_\Phi\boldsymbol{\theta}|\mathbf{Y})+ a(E(\boldsymbol{\theta}|\mathbf{Y}) - E(P_\Phi\boldsymbol{\theta}|\mathbf{Y}))$ with the weight a greater than one. Thus, the optimal constrained estimates are 'un' -shrunk to some degree; that is, they are on the same side of the regression surface (when the posterior means are regressed on X) but moved away from it in the direction of the vector of posterior means.

As with the traditional Bayes point estimates of Section 5.4.1 we consider ϕ to be fixed in the spatial model and the loss function, with the constrained analysis perhaps repeated at several different values. The constrained hierarchical Bayes estimates (CHB) require $E(\boldsymbol{\theta}|\mathbf{Y})$ and $\text{var}(\boldsymbol{\theta}|\mathbf{Y})$; in our case with Gaussian data $\mathbf{Y}$ and known σ^2 and ϕ, we should actually write $E(\boldsymbol{\theta}|\mathbf{Y},\sigma^2 = 1,\phi)$ and $\text{var}(\boldsymbol{\theta}|\mathbf{Y},\sigma^2 = 1,\phi)$. These quantities can be obtained by numerical integration over the single parameter τ^2. The computation of $E(\boldsymbol{\theta}|\mathbf{Y},\sigma^2 = 1,\phi)$ is described in Section 5.4.1. The $\text{var}(\boldsymbol{\theta}|\mathbf{Y},\sigma^2 = 1,\phi)$ is formally defined as

$$\text{var}(\boldsymbol{\theta}|\mathbf{Y},\sigma^2=1,\phi)=E(\boldsymbol{\theta}\boldsymbol{\theta}'|\mathbf{Y},\sigma^2=1,\phi) - E(\boldsymbol{\theta}|\mathbf{Y},\sigma^2=1,\phi)E(\boldsymbol{\theta}|\mathbf{Y},\sigma^2=1,\phi)'.$$

The first expression on the right-hand side can be evaluated by numerical integration after noting that

$$\begin{aligned}
E(\boldsymbol{\theta}\boldsymbol{\theta}'|\mathbf{Y},\sigma^2 = 1,\phi)&\\
&= E[E(\boldsymbol{\theta}\boldsymbol{\theta}'|\mathbf{Y},\tau^2,\sigma^2 = 1,\phi)|\mathbf{Y},\sigma^2 = 1,\phi]\\
&= E[\text{var}(\boldsymbol{\theta}|\mathbf{Y},\tau^2,\sigma^2 = 1,\phi)|\mathbf{Y},\sigma^2 = 1,\phi]\\
&\quad + E[E(\boldsymbol{\theta}|\mathbf{Y},\tau^2,\sigma^2 = 1,\phi)E(\boldsymbol{\theta}|\mathbf{Y},\tau^2,\sigma^2 = 1,\phi)'|\mathbf{Y},\sigma^2 = 1,\phi],
\end{aligned}$$

where the outer expectations in the last equality are expectations over $p(\tau^2|\mathbf{Y},\sigma^2 = 1,\phi)$ which is defined in (5.13). The quantities in the integrands, the mean and variance of $\boldsymbol{\theta}$ conditional on $\mathbf{Y},\tau^2,\sigma^2 = 1$, and ϕ, are obtained from (5.9) in Section 5.3.1.

5.5 LOSS FUNCTIONS FOR EXTREME VALUES

The most common use of epidemiological maps is to locate areas with unusually high incidence or mortality rates. Given this interest in extremes, reliance on quadratic loss functions that involve the parameters for all n geographical regions seems inappropriate. We consider in this section alternative loss functions that emphasise $\theta_{(n)}$, the maximum order statistic from $\boldsymbol{\theta}$, and $i(n)$, its antirank (the index of the region attaining the maximum). As in the previous section, to obtain estimates on the $\boldsymbol{\lambda}$ scale we just transform the optimal $\boldsymbol{\theta}$ estimates. The two loss functions considered here are easily modified to accommodate a change to the $\boldsymbol{\lambda}$ scale if we wish.

One obvious alternative to quadratic loss functions is the 0–1 loss function applied to these two quantities. Define $L_0(\{\theta_{(n)}, i(n)\}; \{\hat{\theta}_{(n)}, \hat{i}(n)\})$ as follows: $L_0 = 1$ unless $\hat{\theta}_{(n)} = \theta_{(n)}$ and $\hat{i}(n) = i(n)$, in which case $L_o = 0$; that is, the loss is one unless both quantities are estimated correctly. The resulting Bayes estimate of $\theta_{(n)}$ and $i(n)$ is the mode of their joint posterior distribution. We estimate this mode by combining the empirical posterior distribution for $i(n)$, which indicates the likelihood of each district being the maximum, and an empirical estimate of the posterior density of $\theta_{(n)}$ conditional on each district being the maximum. We multiply these two density estimates (one portion discrete and the other continuous) and find the mode by inspection. The density estimation for the continuous portion was carried out using the S-PLUS computing environment with Gaussian kernels of width 0.5 to provide sufficiently smooth densities.

The 0–1 loss seems somewhat difficult to apply because of the nature of the required calculations (including density estimation with the inherent uncertainty about kernel width). As a next approach, we proceed sequentially in building a loss function that is more appropriate to the two estimands, starting with the antirank. Define

$$L_a = k(n - j)/n, \quad \text{if } \hat{i}(n) = i(j), \tag{5.20}$$

where k is a constant to be determined. Thus, the loss associated with incorrectly selecting as our estimate the region that turns out to correspond to the jth order statistic, $\theta_{(j)}$, depends on the ordinal distance between j and n. For the maximum value itself, define a loss function based on the ratio of the estimated and true maxima:

$$L_m = \left(\frac{\max(\hat{\theta}_{(n)}, \theta_{(n)})}{\min(\hat{\theta}_{(n)}, \theta_{(n)})} - 1 \right)^2 \tag{5.21}$$

We can use expert opinion concerning the relative importance of the two types of errors to determine a value of k that calibrates the two loss functions. For example, if the θ_is are disease-incidence rates, then a ratio of 2 or more, corresponding to $\hat{\theta}_{(n)} > 2\theta_{(n)}$ or $\hat{\theta}_{(n)} < 0.5\theta_{(n)}$, might be considered cause for concern. An antirank estimate below the upper tenth percentile might cause a similar level of concern. Therefore, we can choose k to equate L_m and L_a at these values and thereby create a similar scale for the two loss functions. In our case the θ_is correspond approximately to twice the square root of a rate (see (5.2)) and thus a ratio of $2\sqrt{2}$ or more would be a cause for concern. This suggests $L_a(j = 0.9n) = L_m(\max/\min = 2\sqrt{2})$, which gives $k = 10(2\sqrt{2} - 1)^2 \approx 33$. Finally, we argue that simply adding together the two loss functions is inappropriate because it treats overestimates of the extreme and underestimates of the extreme symmetrically. Given our interest in obtaining accurate estimates of the extreme values it seems that an underestimate should be more heavily penalised, and the larger the antirank loss L_a, the higher the penalty. Thus, the extreme-value loss function we propose is

$$L(\{\theta_{(n)}, i(n)\}; \quad \{\hat{\theta}_{(n)}, \hat{i}(n)\}) = L_a + L_m + hL_aL_mI(\hat{\theta}_{(n)} < \theta_{(n)}), \tag{5.22}$$

where $h > 0$ is a constant chosen to control the underestimation penalty. In the analysis, we consider a range of values for h.

Naturally, it is not possible to derive in closed form the estimates $(\hat{\theta}_{(n)}, \hat{i}(n))$ that minimise the expected loss. We propose instead to minimise the expected loss numerically. Given a set of realisations from the posterior distribution of $\boldsymbol{\theta}$, we search over the two-dimensional space $(-\infty, \infty) \times \{1, \ldots, n\}$ for the values that minimise the (sample) expected loss. In practice, for each possible value of $\hat{i}(n)$ we search for the value of $\hat{\theta}_{(n)}$ that minimises the loss function. Then we compare the resulting pairs $\{\hat{\theta}_{(n)}, \hat{i}(n)\}$ to determine the single best estimate. Notice that neither the 0–1 loss function nor the extreme-value loss function involves the spatial dependence parameter ϕ in its definition and so ϕ can and should be integrated our of the posterior distribution. Also, it should be noted that both the 0–1 and extreme-value loss functions could easily be modified to focus on the extreme low value rather than the extreme high value.

5.6 RESULTS FOR THE SCOTLAND LIP CANCER DATA

In this section we return to the Scotland lip cancer data to demonstrate the ideas of Sections 5.4 and 5.5. Recall that the data are shown in Table 5.1. We have fixed $\sigma^2 = 1$ in all calculations, as described in preceding sections. All results for this section represent optimal estimates of the parameter vector $\boldsymbol{\theta}$, the expected value of the vector of Freeman–Tukey transformed counts **Y**, which are then transformed to obtain estimates of the relative risk parameter vector $\boldsymbol{\lambda}$.

Table 5.3 Hierarchical Bayes (HB) and constrained hierarchical Bayes (CHB) estimates

	HB		CHB			HB		CHB	
	ϕ		ϕ			ϕ		ϕ	
	0.00	0.14	0.00	0.14		0.00	0.14	0.00	0.14
λ_1	5.42	5.77	6.11	6.77	λ_{29}	1.17	1.24	1.16	1.25
λ_2	3.82	3.76	4.17	4.22	λ_{30}	1.15	1.06	1.14	1.04
λ_3	3.06	3.01	3.41	3.42	λ_{31}	1.11	1.05	1.13	1.07
λ_4	3.52	3.44	3.63	3.64	λ_{32}	1.52	1.50	1.36	1.36
λ_5	2.97	3.13	3.29	3.58	λ_{33}	1.10	1.07	1.09	1.05
λ_6	3.36	3.39	3.45	3.58	λ_{34}	0.99	0.89	0.99	0.88
λ_7	2.71	2.80	2.97	3.15	λ_{35}	0.94	0.91	0.94	0.90
λ_8	2.61	2.54	2.92	2.90	λ_{36}	0.83	0.81	0.88	0.85
λ_9	2.62	2.88	2.93	3.32	λ_{37}	0.98	0.97	0.94	0.93
λ_{10}	2.77	2.83	2.92	3.08	λ_{38}	0.82	0.73	0.86	0.74
λ_{11}	2.48	2.63	2.76	3.01	λ_{39}	1.09	1.10	1.00	1.01
λ_{12}	2.75	3.01	2.90	3.29	λ_{40}	0.76	0.69	0.81	0.71
λ_{13}	2.68	3.10	2.94	3.53	λ_{41}	0.56	0.54	0.56	0.52
λ_{14}	2.62	2.48	2.59	2.50	λ_{42}	0.77	0.78	0.65	0.65
λ_{15}	1.90	1.86	2.06	2.04	λ_{43}	0.80	0.78	0.69	0.66
λ_{16}	2.03	2.15	2.06	2.24	λ_{44}	0.46	0.45	0.45	0.43
λ_{17}	2.03	2.47	2.17	2.74	λ_{45}	0.42	0.44	0.40	0.41
λ_{18}	1.58	1.48	1.68	1.58	λ_{46}	0.53	0.52	0.48	0.45
λ_{19}	1.53	1.70	1.62	1.84	λ_{47}	0.47	0.42	0.46	0.39
λ_{20}	1.58	1.52	1.63	1.58	λ_{48}	0.41	0.41	0.39	0.37
λ_{21}	1.43	1.40	1.50	1.48	λ_{49}	0.36	0.37	0.34	0.34
λ_{22}	1.49	1.49	1.44	1.46	λ_{50}	0.38	0.42	0.36	0.39
λ_{23}	1.29	1.29	1.30	1.30	λ_{51}	0.45	0.41	0.44	0.38
λ_{24}	1.25	1.11	1.29	1.14	λ_{52}	0.43	0.39	0.41	0.36
λ_{25}	1.07	1.11	1.16	1.20	λ_{53}	0.31	0.29	0.28	0.24
λ_{26}	1.06	1.02	1.15	1.10	λ_{54}	0.27	0.27	0.24	0.22
λ_{27}	1.18	1.11	1.21	1.13	λ_{55}	0.25	0.29	0.14	0.16
λ_{28}	1.13	1.11	1.15	1.14	λ_{56}	0.30	0.26	0.22	0.16

Table 5.4 Estimates of maximum of λ and antirank, with estimated posterior expected extreme-value loss. Point estimates are derived on $\boldsymbol{\theta}$ scale and then transformed; expected loss is reported on the $\boldsymbol{\theta}$ scale

	Estimates		Estimated posterior expected extreme-value loss			
Estimation method	$\hat{\imath}(n)$	$\hat{\lambda}_{(n)}$	$h=0$	$h=1$	$h=10$	$h=100$
Hierarchical Bayes, $\phi=0.00$	1	5.42	0.640	0.642	0.658	0.816
Hierarchical Bayes, $\phi=0.14$	1	5.77	0.636	0.637	0.646	0.736
Constrained hierarchical Bayes, $\phi=0.00$	1	6.11	0.635	0.636	0.682	0.844
Constrianed hierarchical Bayes, $\phi=0.14$	1	6.77	0.638	0.638	0.640	0.658
Posterior median	1	5.78	0.636	0.637	0.646	0.735
Posterior mode (0–1 loss)	1	6.44	0.636	0.636	0.639	0.670
Bayes (extreme-value loss $h=0$)	1	6.15	0.635	–	–	–
Bayes (extreme-value loss $h=1$)	1	6.19	–	0.636	–	–
Bayes (extreme-value loss $h=10$)	1	6.47	–	–	0.639	–
Bayes (extreme-value loss $h=100$)	1	7.26	–	–	–	0.653

In Section 5.3 we introduced Table 5.2 which provides posterior medians and posterior intervals based on 2000 simulations for all of the districts' rates and the model parameters (except σ^2). Table 5.3 provides a variety of 'point estimates' or maps (although they are provided in tabular rather than graphical form). All are Bayes estimates for the quadratic loss function of Section 5.4, computed from a Bayesian analysis with fixed spatial-dependence parameter. There are two columns of hierarchical Bayes estimates ((5.12) in Section 5.4.1) and two columns of constrained hierarchical Bayes estimates ((5.16) in Section 5.4.2); the estimates are evaluated at two different values of the spatial dependence parameter ϕ, namely $\phi=0.00$ (which corresponds to no spatial dependence), and $\phi=0.14$ (the posterior median of ϕ in the full Bayesian analysis that includes ϕ as an unknown parameter). As we would expect, the medians of the marginal posterior distributions of the rates from the full Bayesian analysis (see Table 5.2) appear to be centred close to the hierarchical Bayes estimate with $\phi=0.14$ because the data support large values of ϕ. The constrained Bayes estimates exhibit more variation than the hierarchical Bayes estimates, as they were designed to do.

Suppose we now consider estimating the 'hot spot' (i.e. the region with the highest rate) and the rate itself. Table 5.4 provides a number of estimates of these quantities: the elementwise posterior medians for the full (that is, including ϕ) Bayesian analysis (the column labelled median in Table 5.2), the hierarchical Bayes estimates (the columns labelled HB in Table 5.3), the constrained hierarchical Bayes estimates (the columns labelled CHB in Table 5.3), the posterior mode of the estimands' joint distribution (the estimate which is appropriate for 0–1 loss), and the Bayes estimate for the extreme-value loss function (5.22). For each estimate, we report the (sample) expected value of the extreme-value loss function (on the $\boldsymbol{\theta}$ scale) based on 2000 draws from the posterior distribution of the model parameters. These estimates of the posterior expected loss are provided for several different values of the underestimation penalty h. Naturally, the Bayes estimate for the extreme-value loss function of Section 5.5 has the smallest expected loss in each column. The HB estimate with $\phi = 0$, which ignores the spatial nature of the data and does not address the variability in the underlying distribution of rates, seems to perform poorly. Interestingly, the CHB estimate with $\phi = 0$ does quite well (better even than the CHB estimate with the more reasonable $\phi = 0.14$) as long as the underestimation penalty h is small, even though $\phi = 0$ implies no spatial association. Its performance deteriorates as h increases. The estimates derived with $\phi = 0.14$ appear to do better for larger h. The CHB estimate with $\phi = 0.14$ is nearly equal to the optimal point estimate for large values of the underestimation penalty. The posterior mode does reasonably well for all values of h. It should be emphasised that these estimates have been evaluated on only a single data set. Additional study is required to determine if results like these are typical. For example, there is a suggestion in Table 5.4 that CHB might be calibrated by an appropriate choice of ϕ to yield an approximately Bayes estimate with respect to the extreme-value loss function.

ACKNOWLEDGEMENTS

The authors are grateful to Hsin-Cheng Huang, Craig Liu, and Deanne Reber for their computing assistance. Cressie's research was supported by the Environmental Protection Agency (CR822919-01-0) and the Office of Naval Research (N00014-93-1-0001). Stern's research was supported by the National Science Foundation (DMS-9404479). Computation was carried out using equipment purchased with NSF SCREMS grant, DMS-9707740.

APPENDIX: PROOF OF THEOREM 1

Let $t(\mathbf{Y})$ denote an estimator of $\boldsymbol{\theta}$. We show that the expected posterior loss is minimised by taking $\mathbf{t}(\mathbf{Y})$ as (5.16) in Section 5.4.2. Recall that we assume that the hyperparameters, $\boldsymbol{\beta}, \tau^2$, and ϕ are known (and that $\sigma^2 = 1$). The proof begins by expanding the expected posterior loss as follows:

$$
\begin{aligned}
&\mathrm{E}[(\boldsymbol{\theta} - \mathbf{t}(\mathbf{Y}))'\Phi^{-1}(\boldsymbol{\theta} - \mathbf{t}(\mathbf{Y})) \mid \mathbf{Y}] \\
&\quad = \mathrm{E}[(\boldsymbol{\theta} - \mathrm{E}(\boldsymbol{\theta} \mid \mathbf{Y}) + \mathrm{E}(\boldsymbol{\theta} \mid \mathbf{Y}) - \mathbf{t}(\mathbf{Y}))'\Phi^{-1}(\boldsymbol{\theta} - \mathrm{E}(\boldsymbol{\theta} \mid \mathbf{Y}) + \mathrm{E}(\boldsymbol{\theta} \mid \mathbf{Y}) - \mathbf{t}(\mathbf{Y})) \mid \mathbf{Y}] \\
&\quad = \mathrm{E}[(\boldsymbol{\theta} - \mathrm{E}(\boldsymbol{\theta} \mid \mathbf{Y}))'\Phi^{-1}(\boldsymbol{\theta} - \mathrm{E}(\boldsymbol{\theta} \mid \mathbf{Y})) \mid \mathbf{Y}] \\
&\qquad + \mathrm{E}[(\mathrm{E}(\boldsymbol{\theta} \mid \mathbf{Y}) - \mathbf{t}(\mathbf{Y}))'\Phi^{-1}(\mathrm{E}(\boldsymbol{\theta} \mid \mathbf{Y}) - \mathbf{t}(\mathbf{Y})) \mid \mathbf{Y}] \\
&\qquad + 2\mathrm{E}[(\boldsymbol{\theta} - \mathrm{E}(\boldsymbol{\theta} \mid \mathbf{Y}))'\Phi^{-1}(\mathrm{E}(\boldsymbol{\theta} \mid \mathbf{Y}) - \mathbf{t}(\mathbf{Y})) \mid \mathbf{Y}].
\end{aligned}
$$

The final term is zero because only the first factor is random when conditioning on $\mathbf{Y}$ and it has mean zero. The first term does not involve $\mathbf{t}$ so that we need only minimise $\mathrm{E}[(\mathbf{t}(\mathbf{Y}) - \mathrm{E}(\boldsymbol{\theta} \mid \mathbf{Y}))'\Phi^{-1}(\mathbf{t}(\mathbf{Y}) - \mathrm{E}(\boldsymbol{\theta} \mid \mathbf{Y})) \mid \mathbf{Y}]$ subject to the constraints. The outer expectation is not needed because each of the factors in the product is known or a function only of **Y**. We can expand this remaining quadratic form as follows:

$$\begin{aligned}
&(\boldsymbol{t}(\boldsymbol{Y}) - \mathrm{E}(\boldsymbol{\theta}|(\boldsymbol{Y}))'\Phi^{-1}(\boldsymbol{t}(\boldsymbol{Y}) - \mathrm{E}(\boldsymbol{\theta}|\boldsymbol{Y})) \\
&\quad = (\boldsymbol{t}(\boldsymbol{Y}) - P_\Phi \boldsymbol{t}(\boldsymbol{Y}) + P_\Phi \mathrm{E}(\boldsymbol{\theta}|\boldsymbol{Y}) - \mathrm{E}(\boldsymbol{\theta}|\boldsymbol{Y}))'\Phi^{-1} \\
&\quad\quad \times (\boldsymbol{t}(\boldsymbol{Y}) - P_\Phi \boldsymbol{t}(\boldsymbol{Y}) + P_\Phi \mathrm{E}(\boldsymbol{\theta}|\boldsymbol{Y}) - \mathrm{E}(\boldsymbol{\theta}|\boldsymbol{Y})) \\
&\quad = (\boldsymbol{t}(\boldsymbol{Y}) - P_\Phi \boldsymbol{t}(\boldsymbol{Y}))'\Phi^{-1}(\boldsymbol{t}(\boldsymbol{Y}) - P_\Phi \boldsymbol{t}(\boldsymbol{Y})) \\
&\quad\quad + (\mathrm{E}(\boldsymbol{\theta}|\boldsymbol{Y}) - P_\Phi \mathrm{E}(\boldsymbol{\theta}|\boldsymbol{Y}))'\Phi^{-1}(\mathrm{E}(\boldsymbol{\theta}|\boldsymbol{Y}) - P_\Phi \mathrm{E}(\boldsymbol{\theta}|\boldsymbol{Y})) \\
&\quad\quad - 2(\boldsymbol{t}(\boldsymbol{Y}) - P_\Phi \boldsymbol{t}(\boldsymbol{Y}))'\Phi^{-1}(\mathrm{E}(\boldsymbol{\theta}|\boldsymbol{Y}) - P_\Phi \mathrm{E}(\boldsymbol{\theta}|\boldsymbol{Y}))
\end{aligned}$$

where the first expression is a result of applying the first constraint, $P_\Phi \boldsymbol{t}(\boldsymbol{Y}) = P_\Phi \mathrm{E}(\boldsymbol{\theta}|\boldsymbol{Y})$. The first term in the last expression is fixed (i.e. does not depend on $\boldsymbol{t}(\boldsymbol{Y})$) by the second constraint and the second term does not depend on $\boldsymbol{t}(\boldsymbol{Y})$ at all. The final term is an inner product of $(\boldsymbol{t}(\boldsymbol{Y}) - P_\Phi \boldsymbol{t}(\boldsymbol{Y}))$ and $(\mathrm{E}(\boldsymbol{\theta}|\boldsymbol{Y}) - P_\Phi \mathrm{E}(\boldsymbol{\theta}|\boldsymbol{Y}))$ which, by the Cauchy–Schwarz inequality, is maximised (and the quadratic form minimised) if

$$(\boldsymbol{t}(\boldsymbol{Y}) - P_\Phi \boldsymbol{t}(\boldsymbol{Y})) = a(\mathrm{E}(\boldsymbol{\theta}|\boldsymbol{Y}) - P_\Phi \mathrm{E}(\boldsymbol{\theta}|\boldsymbol{Y})) \quad \text{w.p.1}, \quad a > 0. \tag{5.23}$$

Then, by taking the vector norms of both sides and applying the second constraint, we obtain

$$\begin{aligned}
&a^2[\mathrm{E}(\boldsymbol{\theta}|\boldsymbol{Y})'(I - P_\Phi)'\Phi^{-1}(I - P_\Phi)\mathrm{E}(\boldsymbol{\theta}|\boldsymbol{Y})] \\
&\quad = (\boldsymbol{t}(\boldsymbol{Y}) - P_\Phi \boldsymbol{t}(\boldsymbol{Y}))'\Phi^{-1}(\boldsymbol{t}(\boldsymbol{Y}) - P_\Phi \boldsymbol{t}(\boldsymbol{Y})) \\
&\quad = \mathrm{E}[\boldsymbol{\theta}'(I - P_\Phi)'\Phi^{-1}(I - P_\Phi)\boldsymbol{\theta}|\boldsymbol{Y}].
\end{aligned}$$

Thus,

$$a = \{\mathrm{E}[\boldsymbol{\theta}'(I - P_\Phi)'\Phi^{-1}(I - P_\Phi)\boldsymbol{\theta}|\boldsymbol{Y}]\lambda \mathrm{E}(\boldsymbol{\theta}|\boldsymbol{Y})'(I - P_\Phi)'\Phi^{-1}(I - P_\Phi)\mathrm{E}(\boldsymbol{\theta}|\boldsymbol{Y})\}^{1/2}. \tag{5.24}$$

Now by (5.23),

$$\begin{aligned}
\boldsymbol{t}(\boldsymbol{Y}) &= P_\Phi \boldsymbol{t}(\boldsymbol{Y}) + (\boldsymbol{t}(\boldsymbol{Y}) - P_\Phi \boldsymbol{t}(\boldsymbol{Y})) \\
&= P_\Phi \mathrm{E}(\boldsymbol{\theta}|\boldsymbol{Y}) + a(\mathrm{E}(\boldsymbol{\theta}|\boldsymbol{Y}) - P_\Phi \mathrm{E}(\boldsymbol{\theta}|\boldsymbol{Y})) \\
&= a\mathrm{E}(\boldsymbol{\theta}|\boldsymbol{Y}) + (1 - a)P_\Phi \mathrm{E}(\boldsymbol{\theta}|\boldsymbol{Y}),
\end{aligned}$$

where a is given in (5.24). We evaluate a by noting that

$$\begin{aligned}
\mathrm{E}[\boldsymbol{\theta}'(I - P_\Phi)'\Phi^{-1}(I - P_\Phi)\boldsymbol{\theta}|\boldsymbol{Y}] &= \mathrm{E}[\mathrm{tr}\{(I - P_\Phi)'\Phi^{-1}(I - P_\Phi)\boldsymbol{\theta}\boldsymbol{\theta}'\}|\boldsymbol{Y}] \\
&= \mathrm{tr}\{(I - P_\Phi)'\Phi^{-1}(I - P_\Phi)[\mathrm{E}(\boldsymbol{\theta}\boldsymbol{\theta}'|\boldsymbol{Y})]\} \\
&= \mathrm{tr}\{(I - P_\Phi)'\Phi^{-1}(I - P_\Phi)\mathrm{var}(\boldsymbol{\theta}|\boldsymbol{Y})\} \\
&\quad + \mathrm{E}(\boldsymbol{\theta}|\boldsymbol{Y})'(I - P_\Phi)'\Phi^{-1}(I - P_\Phi)\mathrm{E}(\boldsymbol{\theta}|\boldsymbol{Y}) \\
&\equiv H_1(\boldsymbol{Y}) + H_2(\boldsymbol{Y}).
\end{aligned}$$

Notice that $H_1(\boldsymbol{Y})$ can be simplified further:

$$\mathrm{tr}\{(I - P_\Phi)'\Phi^{-1}(I - P_\Phi)\mathrm{var}(\boldsymbol{\theta}|\boldsymbol{Y})\} = \mathrm{tr}\{\mathrm{var}[\Phi^{-1/2}(I - P_\Phi)\boldsymbol{\theta}|\boldsymbol{Y}]\}.$$

This completes the proof; however, we now address the remark made after the statement of Theorem 1. Consider using a Lagrange multiplier to perform the minimisation assuming only the second constraint. Then we find the constrained estimator by minimising the Lagrangian,

$$\begin{aligned} L &= \mathrm{E}[(\boldsymbol{\theta} - \mathbf{t}(\mathbf{Y}))'\Phi^{-1}(\boldsymbol{\theta} - \mathbf{t}(\mathbf{Y})) \,|\, \mathbf{Y}] \\ &\quad + \kappa(\mathrm{E}[\boldsymbol{\theta}'(I - P_{\Phi})'\Phi^{-1}(I - P_{\Phi})\boldsymbol{\theta} \,|\, \mathbf{Y}] - \mathbf{t}(\mathbf{Y})'(I - P_{\Phi})'\Phi^{-1}(I - P_{\Phi})\mathbf{t}(\mathbf{Y})). \end{aligned}$$

Upon setting $(\partial L/\partial \mathbf{t}) = \mathbf{0}$, we obtain

$$\begin{aligned} \mathbf{0} &= 2\Phi^{-1}(\mathbf{t}(\mathbf{Y}) - \mathrm{E}\,(\boldsymbol{\theta} \,|\, \mathbf{Y}) + 2\kappa(I - P_{\Phi})'\Phi^{-1}(I - P_{\Phi})\mathbf{t}(\mathbf{Y}) \\ &= 2\Phi^{-1}(\mathbf{t}(\mathbf{Y}) - \mathrm{E}\,(\boldsymbol{\theta} \,|\, \mathbf{Y}) + 2\kappa\Phi^{-1}(I - P_{\Phi})\mathbf{t}(\mathbf{Y}). \end{aligned}$$

Multiplying both sides by $X(X'\Phi^{-1}X)^{-1}X'$ gives $P_{\Phi}(\mathbf{t}(\mathbf{Y}) - \mathrm{E}(\boldsymbol{\theta} \,|\, \mathbf{Y})) = \mathbf{0}$, so that $\mathbf{t}(\mathbf{Y})$ from (5.16) automatically satisfies the constraint (5.14) whether it is stated as a condition of the theorem or not.

6

Edge Effects in Disease Mapping

Andrew B. Lawson
University of Aberdeen

A. Biggeri and E. Dreassi
Universita di Firenze

6.1 INTRODUCTION

In mapping geographical variation of risk or occurrence of disease in *tract count* data, information about the neighbouring areas is often incomplete. This is a problem found particularly in boundary areas. This lack of information could distort the estimates of their relative risks within the study region. Likewise, for *case event* data, the analysis of maps of disease incidence can be severely affected by the proximity of events to edges of the region.

The analysis of edge effects is a neglected area within spatial epidemiology. While there has been a considerable increase in research in the general area of spatial epidemiology within recent years, there has been little attention paid to edge effects. This is regrettable, since many analyses can be fundamentally altered by the inclusion of edge information in different forms. In this chapter we examine a number of basic edge effect problems within case event and tract count data. In addition we discuss possible schemes which make some allowance for, or compensate for, edge effects. We apply two such schemes to the example of mortality from gastric cancer in the municipalities of the Tuscany region of Italy, within a fixed time period. The results of this study suggest that the edge augmentation and weighting procedures used better reflect the underlying structure portrayed by the use of external guard municipality data.

Disease Mapping and Risk Assessment for Public Health. Edited by A.B. Lawson *et al.*
© 1999 John Wiley & Sons Ltd.

6.2 EDGE EFFECT PROBLEMS

6.2.1 Edge effects in case events

Define a study region W within which we observe m case event locations of a disease of interest. The locations usually are case address locations. We denote the locations as $\{x_i\}, i = 1, \ldots, m$. Also define an arbitrary region T which completely encloses W. For simplicity, we assume that the area of T outside W lies completely external to the study region (Figure 6.1). It is possible that for some study regions there may be areas internal to the main study region where no observations are possible. These external and internal areas can be regarded as areas where censoring of observations has occurred and we can apply appropriate methods to either type of area.

A variety of effects can arise due to the proximity of the external boundary of the region W to the observed data. First, if the case locations are spatially interdependent, then any measure that depends on this interdependence will be affected by the fact that observations are unavailable external to the study region. For example, if a measure of autocorrelation is to be applied over the study region, then the censoring of information at boundaries will affect this estimation process. Secondly, even when observations are independent, the estimation method used can induce edge effects in estimators. For example, a bias will be induced when a smoothing operation is applied to the event distribution. This is due to the unavailability (censoring) of information beyond the edge regions. A larger variance will also be found in edge areas due to the low proportion of small interevent distances found in that area. While edge effects may be minor when estimation of *global* parameters is considered, they may become severe when *local* estimates in regions close to the study boundary are to be made. Ripley (1981, 1988) discusses some aspects of the edge effect problem for point processes on the plain, and also notes the edge distortion with trend surface fitting to continuous data.

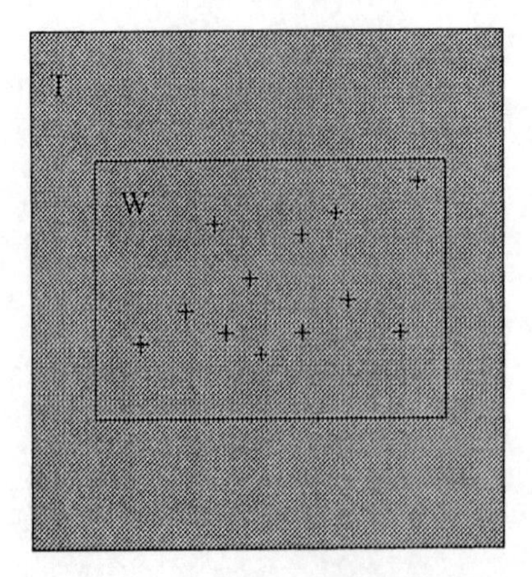

Figure 6.1 Idealised study region and associated areas. + = case event location

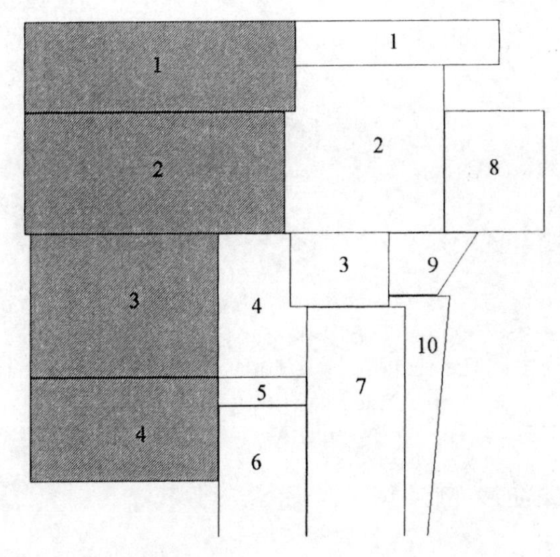

Figure 6.2 Idealised tract map. Tracts 1–4 (shaded area) are external to the study region and have counts $\{n_e\}$. Unshaded tracts are internal, and tracts (1,2,4,5,6) are in the set $\{n_i^*\}$

6.2.2 Edge effects in counts

In the case of counts within arbitrary tracts, similar considerations apply. Define the count of a disease within the ith tract as n_i. We assume there are m tracts within the study region. The inclusion criteria for tracts (i.e. which boundary tracts should be included or excluded from the study region) is an important issue and is discussed more fully by Lawson and Waller (1996).

We denote tracts within W which have a common boundary with the external region as $\{n_i^*\}$, whereas if we also can observe or otherwise estimate counts in external tracts, then we denote these counts as $\{n_e\}$ (Figure 6.2). Here the external region is defined to be any area *not* included within the study window. Usually this area lies adjacent to the window, but this is not a fundamental requirement. In addition, the external region may lie *within* the tracts where counts are observed. In that sense the external region may be regarded as having a missing observation. The comments above concerning global and local estimation apply here. The estimate of tract-relative risks at or near boundaries can be affected by the edge position, and by the requirement to use counts from neighbouring tracts in the estimation process, i.e. $\{n_e\}$ are censored. Even without the assumption of interdependence between events, any conventional smoothing operation applied, for example to the Standardised Mortality Ratio (SMR) (n_i/e_i) with e_i the expected number of cases in the ith tract, will also induce edge effects due to the use of neighbourhoods in such smoothing operations. Cressie (1993) has discussed this problem for lattice data, and an early reference to the problem was made by Griffith (1983).

6.3 EDGE EFFECT COMPENSATION METHODS

The two basic methods of dealing with edge effects are (i) the use of weighting/correction systems, which usually apply different weights to observations depending on their proximity to the study boundary, and (ii) the use of guard areas, which are areas outside the region which we analyse as our study region. The original study region could have as its guard area all the $\{n_i^*\}$ and so these areas are not reported, although they are used in the estimation of parameters relating to the internal tracts.

6.3.1 Weighting systems

Usually it is appropriate to set up weights which relate the position of the event or tract to the external boundary. These weights, $\{w_i\}$ say, can be included in subsequent estimation and inference. The weight for an observation is usually intended to act as a surrogate for the degree of missing information at that location and so may differ depending on the nature and purpose of the analysis. Some sensitivity to the specification of these weights will inevitably occur and should be assessed in any case study. Some basic weights are:

$$\text{for case events: } w_i = \begin{cases} 1, & \text{if } x_i \notin \{x_i^*\}, \\ m(d_i), & \text{if } x_i \in \{x_i^*\}; \end{cases}$$

$$\text{for tract counts: } w_i = \begin{cases} 1, & \text{if } n_i \notin \{n_i^*\}, \\ m(d_i), & \text{if } n_i \in \{n_i^*\}, \end{cases}$$

where $m(d_i)$ is a function of the distance (d_i) of the observation to the external boundary, and $\{x_i^*\}$ is the set of all events closer to the boundary than to any other event in the study region. The distance (d_i) could be the event–boundary distance for case events or the tract centroid–boundary distance for tract counts. Another possible surrogate for (d_i) in the case of tract counts is to use $m(l_{bi}/l_i)$, where l_i is the length of the tract perimeter and l_{bi} is the length of the perimeter of the tract which is in common with the external boundary. A simple choice would be $w_i = 1 - (l_{bi}/l_i), \forall_i$, which can be used for all tracts since non-boundary tracts will have $w_i = 1$.

Since the events are generated by a modulated heterogeneous Poisson process, weights could also be specified as functions not only of the distance from the boundary but also of the modulating population density. For example, defining an indicator for closeness to the boundary for each area, when in the tract count case some external standardised rates are available, it is possible to structure an expectation-dependent weight for a particular tract, e.g. based on the ratio of the sum of all adjacent area expectations to the sum of all such expectations within the study window. Other suitable weighting schemes could be based on the proportion of the number of observed neighbours.

6.3.2 Guard areas

An alternative approach is to employ guard areas. These areas are external to the main study window of interest and could be boundary tracts of the study window itself or could be added to the window to provide a guard area, in the case of tract counts. In the case event situation, the guard area could be some fixed distance from the external boundary (see, for example, Ripley, 1988). The areas are used in the estimation process but they are excluded from the reporting stage because they will be prone to edge effects themselves. If boundary tracts are used for this, then some loss of information must result. External guard areas have many advantages. First, they can be used *with* or *without* their related data to provide a guard area. Secondly, they can be used within data augmentation schemes in a Bayesian setting. These methods regard the relative risks or counts in external areas as a missing data problem (Tanner, 1996).

6.4 A HIERARCHICAL BAYESIAN MODEL FOR DISEASE MAPPING OF TRACT COUNT DATA

Data are usually represented by the observed n_i and expected e_i numbers of events in the ith area ($i = 1, \ldots, m$) of the region of interest. The expected number of cases is usually obtained by applying a standard reference set of sex–age-specific rates on the area population.

A simple estimate of the relative risk for a generic ith area is the SMR. It is obtained as

$$\mathrm{SMR}_i = \frac{n_i}{e_i},$$

which is the maximum likelihood estimate of the relative risk (θ_i) under a Poisson model with $\mathrm{E}(n_i) = e_i\theta_i$. This approach ignores the presence of unstructured extra-Poisson variation as well as the underlying 'spatial structure' of the relative risks. Besag *et al.* (1991) suggested a hierarchical Bayesian model where a Poisson model is defined for the observed number of events

$$n_i \sim \text{Poisson}\ (e_i\theta_i),$$

where the ith local estimate of the relative risk can be modelled in the Generalised Linear Mixed Model framework as

$$\log(\theta_i) = u_i + v_i,$$

where the logarithm of the relative risk is a linear function of u_i and v_i representing two random terms for unstructured and structured spatial components, usually referred to as *heterogeneity* and *clustering*, respectively (Breslow and Clayton, 1993). The heterogeneity random terms can be assumed to have prior distributions defined as

$$u_i \sim \mathrm{N}(0, (\lambda_u)^{-1}),$$

where λ_u is a constant precision parameter. The clustering components are modelled in a similar way except that the means and variances are dependent on the adjacencies $\bar{v}_i = \sum_{j \neq i} W_{ij}v_j/W_{i+}$, and $\mathrm{var}(v_i) = (W_i + \lambda_v)^{-1}$. The matrix W is a 0–1 adjacency matrix and the symbol + denotes summation over the appropriate subindex. An improper inverse exponential prior distribution for each parameter λ_u and λ_v is assumed, with default parameters df = 2 and scale parameter = 0. The posterior distribution, being intractable, is obtained via a Gibbs sampling approach (Besag *et al.*, 1991).

6.4.1 Corrections for the hierarchical Bayesian model

It is usually straightforward to adapt the hierarchical Bayesian model to accommodate edge-weighted data. Weighting each area can be done by introducing an offset term in the linear predictor of the regression model for the relative risks, in the same way as the expected cases. In the graphical representation of the Bayesian model this corresponds to adding a non-stochastic node directly to the node for the θ_i. When in the tract count case expected cases $\{e_e\}$ on external tracts are available, it is possible to structure an expectation-dependent weight for a particular tract, e.g. based on the ratio of the sum of all adjacent area expectations within the study window to the sum of all such expectations:

$$w_i = \frac{\sum_{j=1}^{m} W_{ij} e_j}{\sum_{j=1}^{(m+n)} W_{ij}^* e_j^*}, \tag{6.1}$$

with W^* the expanded adjacent matrix for the study region (m areas) and guard area (n areas) and e^* the expected cases for the $n + m$ areas. In practice this can be used as the weight for observed and expected cases before the use of the Bayesian model. This weighting system yields $w_i = 1$ for completely internal areas (i.e. with no censored neighbours) while it gives $w_i = 0$ for islands. This weighting also accounts for the number of uncensored neighbours that are adjacent to the area in question.

If guard areas are selected and observations are available within the guard area, then it is possible to proceed with inference by using the whole data but selectively reporting those areas not within the guard area. Note that this is not the same as setting $w_i = 0$ for all guard area observations in a weighting system. When external guard areas are available but no data is observed, then resort must usually be made to missing data methods. With limited external information for tract count data it is possible to proceed via the use of a data augmentation algorithm. In this approach it is possible to draw missing counts iteratively from the distribution of the counts given current relative risk estimates (imputation step) and to sample the full conditional distribution of relative risks (posterior step).

When the expected cases $\{e_e\}$ are unknown in the external regions it is simpler to regard the relative risks as the target parameters (without further evaluating the associated missing counts), and to employ the above algorithms as before on this smaller parameter hierarchy.

Let us start with a case where we know the population or the expected number of events for the out-of-the-border areas. The out-of-the-border areas are regarded as having missing values in the number of events. The Chained Data Augmentation Algorithm steps are:

(i) draw a sample $\{n_e\}$ from $p(n_e|\theta_e)$;
(ii) draw a sample $\{\theta\}$ from $p(\theta\,|\,n, \lambda)$,
(iii) draw a sample $\{\lambda\}$ from $p(\lambda\,|\,\theta)$,

where λ denotes the hyperparameters of the hierarchical Bayesian model, $\theta = (\theta_i, \theta_e)$ and $n = (n_i, n_e)$, and $p(\cdot)$ denotes the appropriate distribution. In step (i) we can take for n_e the corresponding expected number of cases (which implies equating the starting value for θ_e to 1.0). Step (ii) consists in sampling from the full conditionals used by the

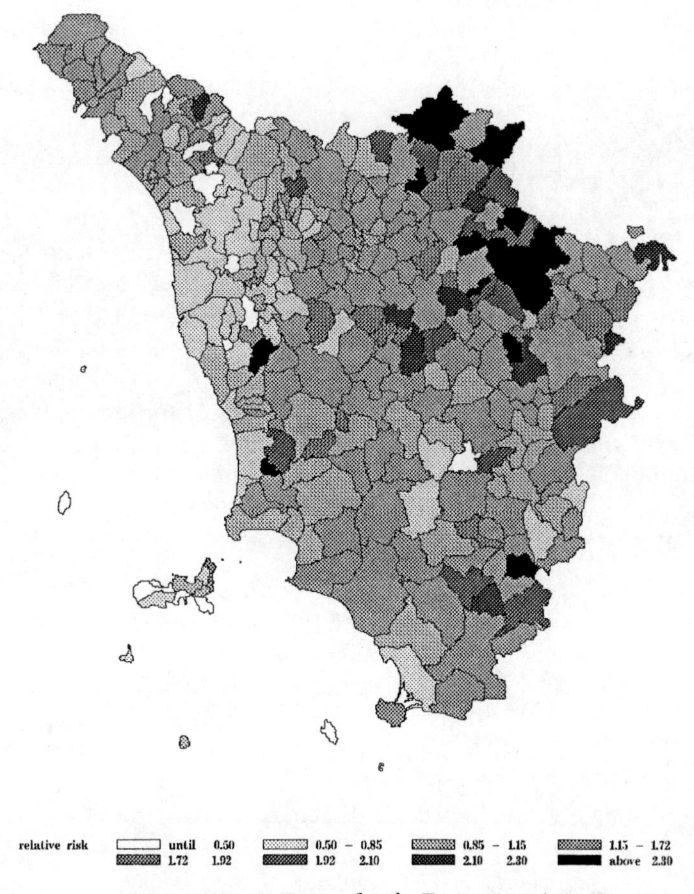

Figure 6.3 SMR map for the Tuscany region

Gibbs sampler (used also for the uncorrected method; see Besag and Green, 1993). Iterating steps until convergence, we obtain the final estimates of the relative risks. This approach is derived from the Data Augmentation Algorithm (Tanner, 1996).

If we do not know the expected number of events or the population for each out-of-border area, then we should consider as missing the parameters θ_e and not the number of events. Using the conditionals integrated over the distributions of the θ_e the sampling scheme becomes the following:

(i) draw a sample $\{\theta_i\}$ from $\int p(\theta_i | n_i, \theta_e, \lambda) p(\theta_e) \mathrm{d}\theta_e$,
(ii) draw a sample $\{\lambda\}$ from $\int p(\lambda | \theta_i, \theta_e) p(\theta_e) \mathrm{d}\theta_e$.

The integrals are obtained using Monte Carlo simulations with

$$p(\theta_e) = p(\theta_e | \theta_i, \lambda).$$

This approach can be viewed as an extension of Monte Carlo maximum likelihood methods to cope with missing data problems when the conditional distributions required by the algorithm are not available (Gilks *et al.*, 1996a)—here $p(n_e|\cdot)$ is undefined because we do not know the population or the expected number of events.

6.5 THE TUSCANY EXAMPLE

A selection of the above methods have been applied to a simple example. Municipality tract counts of gastric cancer mortality data in Tuscany (Italy) for males over 35 years were routinely collected at municipality levels (287 units) from 1980 to 1989. This choice was made because gastric cancer displays a high relative risk along the North-Eastern border of the region, so there may be great interest in the potential distortion due to edge effects when such a raised incidence is displayed. This distortion could appear in the estimation of 'true' relative risks within the study area. We have employed the weighting and augmentation corrections described above to the Bayesian hierarchical model of Besag *et al.* (1991) as implemented in BEAM (Clayton, 1994).

In what follows we examine four different scenarios for the data set:

(i) full Bayesian analysis of relative risk with structured and unstructured heterogeneity for the augmented region set using $\{n_e\}, \{e_e\}$ and $\{n_i\}, \{e_i\}$ (method C);
(ii) the same analysis applied to $\{n_i\}, \{e_i\}$ alone (method I);
(iii) edge weighting based on the data-dependent ratio of adjacent expected rates specified above (equation (6.1)) where a diagonal matrix of weights was introduced into the analysis and the weight is the proportion of observed adjacent area expectations over the sum of the total adjacent area expectations (method W) and
(iv) the edge-augmentation method discussed above (method R).

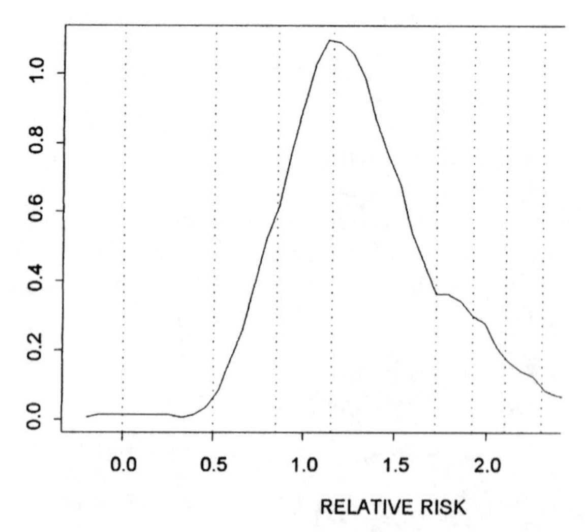

Figure 6.4 Cut points from the marginal empirical cdf function of the Bayes estimates for the augmented data

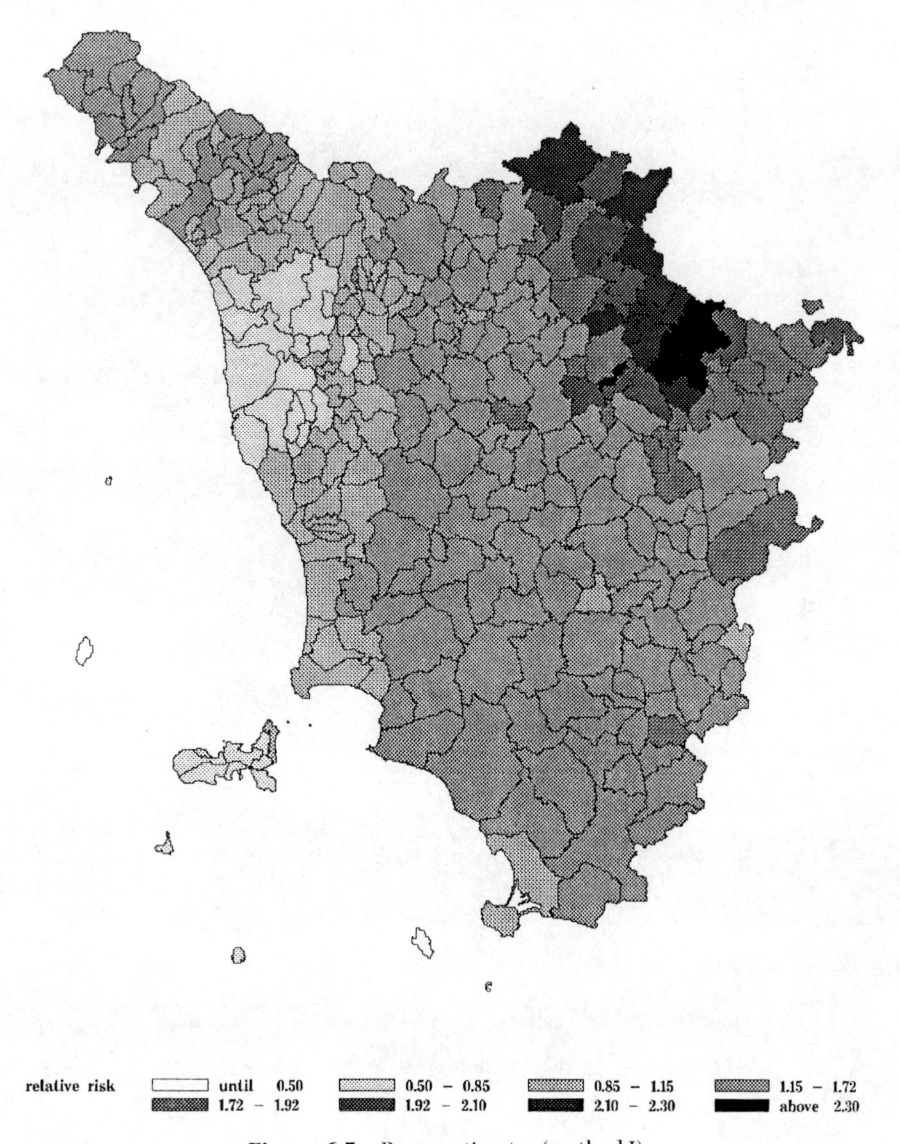

Figure 6.5 Bayes estimates (method I)

All sampling of conditional distributions were carried out by the BEAM program. The relative risks reported here are averages of 3500 samples after convergence of the Gibbs sampler, the convergence of the sampler being checked for heterogeneity and clustering parameters using CODA (Gilks *et al.*, 1996).

The map representing the SMRs (n_i/e_i) for the study region is shown in Figure 6.3. The maps of the different estimators are presented using absolute levels. These levels were chosen by inspection of the distribution of the full Bayesian estimates obtained from the observed augmented data and using them for each map (Figure 6.4). The Bayes

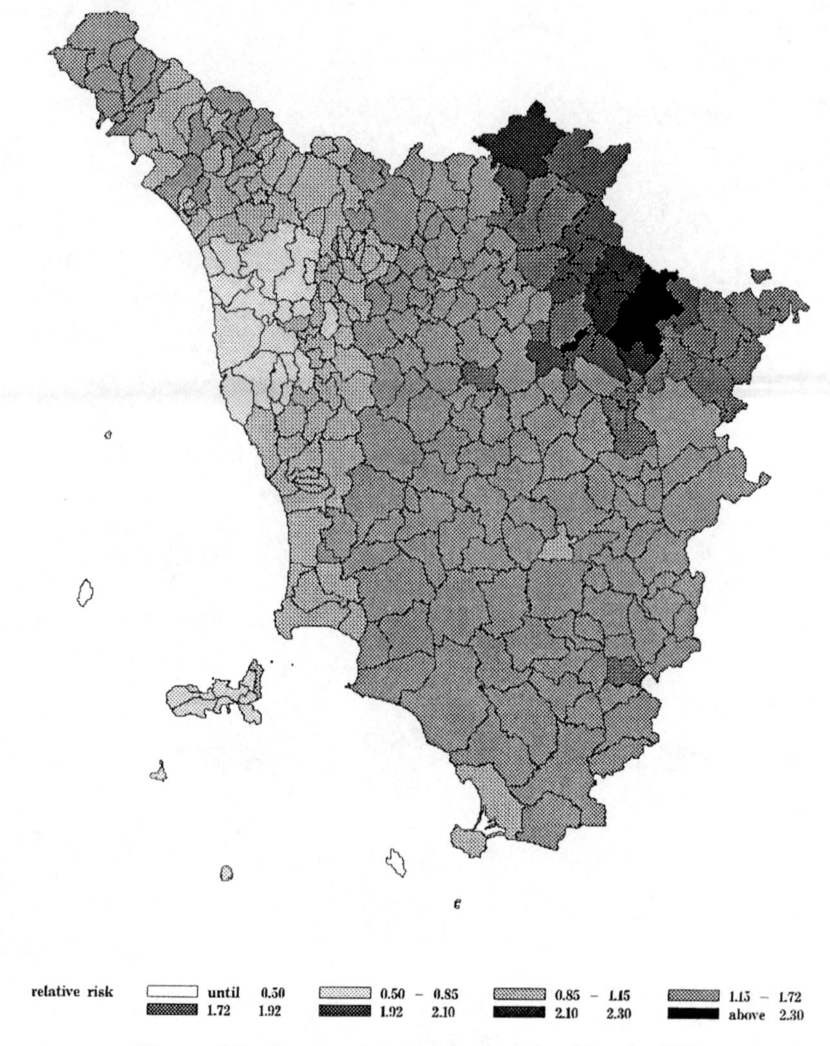

Figure 6.6 Bayes estimates: edge-weighted (method W)

estimates shown in Figures 6.5 and 6.6 display the *uncorrected* and *edge-weighted* approaches respectively (methods I and W). For the *edge-augmentation* method (method R) (Figure 6.7) with the observed number of cases as missing data, as initial value for $\{n_e\}$ we have used the expected numbers of deaths for each area. The relative risks obtained using the known gastric cancer mortality data for the external adjacent areas are shown in Figure 6.8 (method C).

Reported in Table 6.1 are the different estimators for the areas along the North-Eastern border of Tuscany (sorted from North to South). Three subregions are of particular

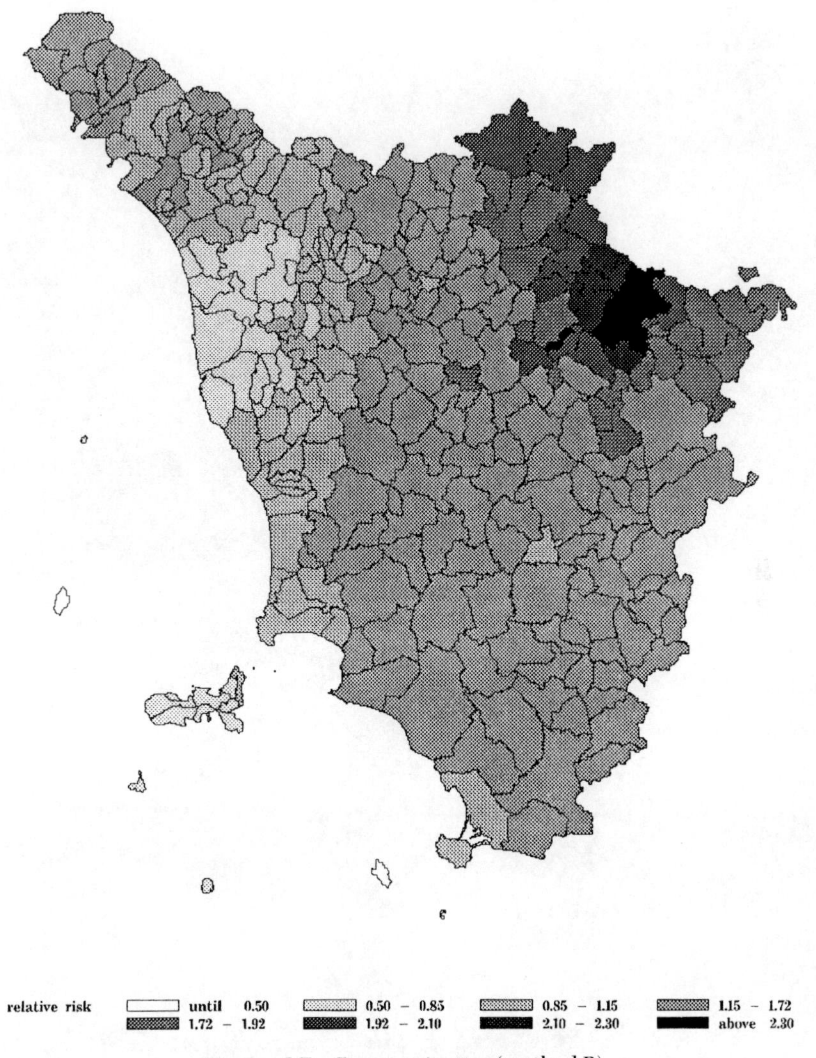

Figure 6.7 Bayes estimates (method R)

interest: the Tuscan Romagna (Rt), the Casentino valley (Ca) and the River Tiber valley (Ti). Gastric cancer mortality is particularly high in the Casentino valley. The Bayesian estimates based on the complete data (C) showed that the areas in the Casentino valley ranked higher together with the far north-east area of the Tuscan Romagna. The estimates based on the incomplete data (I) failed to highlight this pattern. The weighted (W) and the data-augmented (R) Bayesian estimates more closely approximated the full Bayesian analysis. The properties of the augmentation algorithm are evident in the spatial smoothing which affected the areas at the edges of the Casentino valley: the

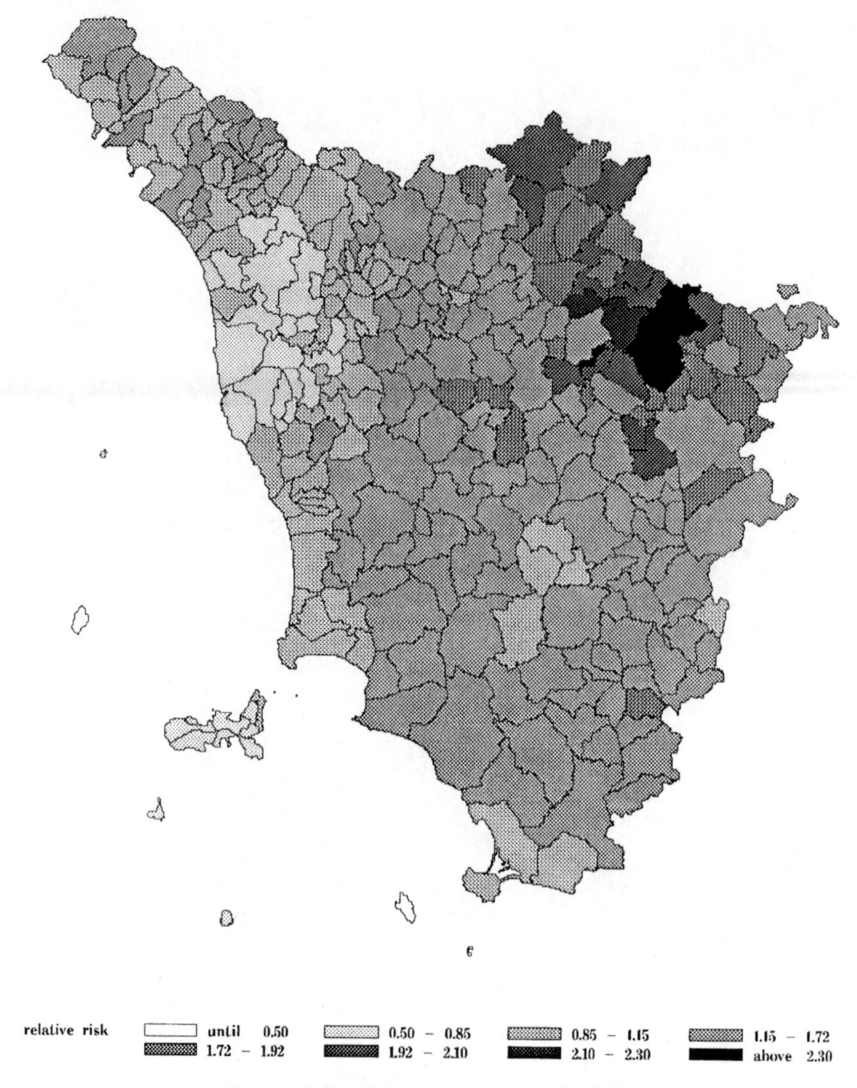

Figure 6.8 Bayes estimates (method C)

S. Godenzo municipality has been consistently estimated closer to the Casentino values while, at the southern border, the Chiusi Verna area has been underestimated due to its proximity to the lower risk areas of the River Tiber valley. It could be argued, with good reason, that a more accurate distance matrix would have provided more appropriate analyses, especially if geographical barriers like mountains are to be modelled. However, this is not a concern of the edge correction methods examined.

While this small example gives only an empirical snapshot of the edge effect problem displayed in a small data example, it does serve to highlight the importance of considering such effects in any mapping exercise.

Table 6.1 Comparison of different estimator for the area along the North-Eastern border of the Tuscany region (Italy). Gastric cancer, dealth certificate relative risk, 1980–89 males

Area name	SMR	I—Bayes	W—Bayes	R—Bayes	C—Bayes
Firenzuola (Rt)	2.73	2.26	2.11	2.09	1.97
Palazzuolo (Rt)	1.69	2.00	1.98	1.98	1.72
Marradi (Rt)	2.42	2.12	2.06	2.07	1.92
S.Godenzo (Rt)	2.01	2.11	2.07	2.07	1.83
Stia (Ca)	2.49	2.27	2.19	2.21	2.07
Pratovecchio (Ca)	1.99	2.17	2.15	2.14	2.04
Poppi (Ca)	3.08	2.62	2.55	2.48	2.59
Chiusi Verna (Ca)	1.60	2.01	2.04	2.01	1.97
Pieve S.Stefano (Ti)	1.71	1.79	1.79	1.83	1.75
Badia Tedalda (Ti)	1.70	1.82	1.80	1.83	1.64
Sestino (Ti)	2.14	1.99	1.91	1.85	1.58

6.6 CONCLUSIONS

We have presented a variety of possible edge effect problems that arise in disease mapping and also some possible solutions to these problems. In general, there is no one panacea for the incorporation of edge effects in models, and it is wise to evaluate the characteristics of the problem clearly before adopting a specific scheme.

In the situation where case events are studied, if censoring is present and could be important (i.e. when there is clustering or other correlated heterogeneity), then it is advisable to use an internal guard area, or an external guard area with augmentation via MCMC. In cases where only a small proportion of the study window is close to the boundaries and only general (overall) parameter estimation is concerned, then it may suffice to use edge-weighting schemes. If residuals are to be weighted, then it may suffice to label the residuals only for exploratory purposes.

In the situation where counts are examined, it is also advisable to use an internal guard area or external area with augmentation via MCMC. In some cases, an external guard area of *real* data may also be available. This may often be the case when routinely collected data are being examined. In this case, analysis can proceed using the external area *only to correct internal estimates*. Edge weighting can be used also, and the simplest approach would be to use the proportion of the region *not* on the external boundary. Residuals can be labelled for exploratory purposes.

The underlying assumptions in any correction method are that the model be correctly specified and that it could be extended to the not observed areas. In particular, it is questionable if an adjustment can really be obtained when ignoring the information on the outer areas. Edge-effect bias should be less prominent when an unstructured exchangeable model is chosen. Since the relative risk for each area would be regressed toward a grand mean, the information lacking for the unobserved external areas is very small compared with those from the observed areas. Of course such a simple model where common expectation is found is highly unlikely to be a good model in this area.

Extending the edge effect problem to the consideration of space–time data, the situation is more complex since spatial edge effects can interact with temporal edge effects. The use of sequential weighting, based on distance from time and space boundaries, may be appropriate (Lawson and Viel, 1995). For tract counts observed in distinct time periods only, the most appropriate method is likely to be based on distance from time and space boundaries, although it may be possible to provide an external spatial and/or temporal guard area either with real data or via augmentation and MCMC methods.

PART II

Clustering of Disease

7

A Review of Cluster Detection Methods

Andrew B. Lawson

University of Aberdeen

M. Kulldorff

University of Connecticut

7.1 INTRODUCTION

The analysis of disease clustering has generated considerable interest in the area of public health surveillance. Since the 1980s there has been increased interest in, and concerns about, adverse environmental effects on the health status of populations. For example, concerns about the influence of nuclear power installations on the health of surrounding populations has given rise to the development of methods that seek to evaluate clusters of disease. These clusters are regarded as representing local adverse health risk conditions, possibly ascribable to environmental causes. However, many diseases will display geographical clustering for other reasons, and possibly on a more global scale. The reasons for such clustering are various. First, it is possible that for some *apparently* non-infectious diseases there may be a viral agent, which could induce clustering. This has been hypothesised for childhood leukaemia (see, for example, Kinlen, 1995). Secondly, other common but unobserved factors/variables could lead to observed clustering in maps. For example, localised pollution sources could produce elevated incidence of disease (e.g. road junctions could yield high carbon monoxide levels and hence elevated respiratory disease incidence), or a common treatment of disease can lead to clustering of disease side-effects. The prescription of a drug by a particular medical practice could lead to elevated incidence of side-effects within that practice area.

Hence, there are many situations where diseases may be found to cluster, even when the aetiology does not suggest it should be observed. Because of this, it is important to be aware of the role of clustering methods, even when clustering *per se* is not the main

Disease Mapping and Risk Assessment for Public Health. Edited by A.B. Lawson *et al.*
© 1999 John Wiley & Sons Ltd.

focus of interest. In this case, it may be important to consider clustering as a background effect and to employ appropriate methods to detect such effects.

Here, we consider a number of aspects of the analysis of clustering. First, we review reasons for examining clusters. Basic definitions of clustering and their use in different studies are then considered. Secondly, we consider appropriate models based on these definitions. Thirdly, we examine the estimation of clustering as a background effect in studies where the prime focus is *not* clustering. Finally, we consider the use of testing for clusters and its application in different studies. A comparison of a restricted range of clustering methods has been made by Alexander and Boyle (1996) and a special issue of *Statistics of Medicine* (1996, Volume 15, 7–9) has focused on issues related to cluster studies.

Table 7.1 provides eleven examples of disease clusters with a known local cause. Except for leukaemia in Japan, the local excess of cases was first observed, then followed by an epidemiological investigation determining the cause of the cluster. In Japan, the suspected risk factor led to the detection of the local excess. The sizes of clusters range from a few city blocks, in the case of cholera, to several states with millions of inhabitants in the case of oral cancer among women. Both infectious and chronic diseases are on the list, and there is a wide range of different types of aetiology. In several cases, the detection of a cluster was an important step in establishing a previously unknown aetiology. With most, there were important public health benefits.

Table 7.1 Examples of disease clusters with known aetiology

Disease	Location	Aetiology	Reference
Cholera	Broad Street, London	Vibrio cholerae in drinking water	Snow (1854)
Nasal sinus adenocarcinoma	High Wycombe, England, UK	Occupational exposure in furniture industry	Macbeth (1965)
Leukaemia	Hiroshima and Nagasaki	Radiation from nuclear explosions	Ishimaru *et al.* (1971)
Malignant mesothelioma	Karain, Turkey	Erionite fibre exposure	Barris *et al.* (1978)
Lung cancer	Coastal Georgia, USA	Shipyard asbestos	Blot *et al.* (1978)
Pneumocystis pneumonia	Los Angeles	Human immuno-deficiency virus	Centre for Disease Control (1981)
Oral cancer among women	Southern USA	Snuff dipping	Winn *et al.* (1981)
Asthma	Barcelona harbour, Catalonia	Inhalation of soybean dust	Anto *et al.* (1989)
Down's syndrome	Rinyaszentkirály, Hungary	Trichlorfon in fish	Czeizel *et al.* (1993)
Kidney failure in children	Port-au-Prince, Haiti	Prescription drug with diethylene glycol	PCHR Group (1997)
Respiratory cancer	Armadale, Scotland, UK	Air Pollution	Lawson and Williams (1994)

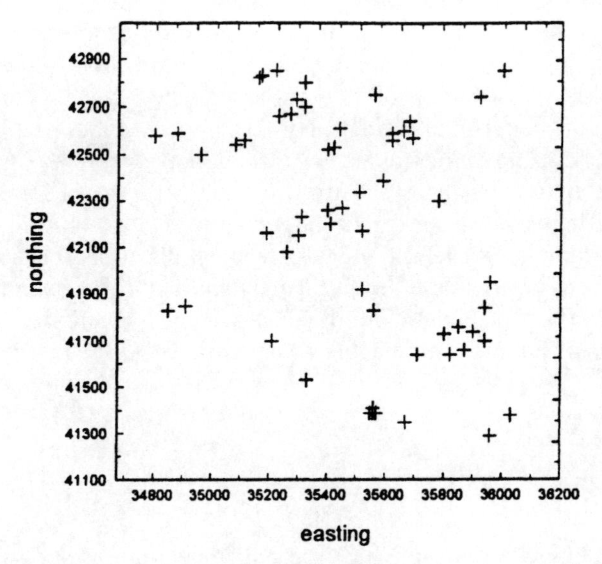

Figure 7.1 Larynx cancer case locations in Lancashire, UK (1974–83) (coordinates:map references)

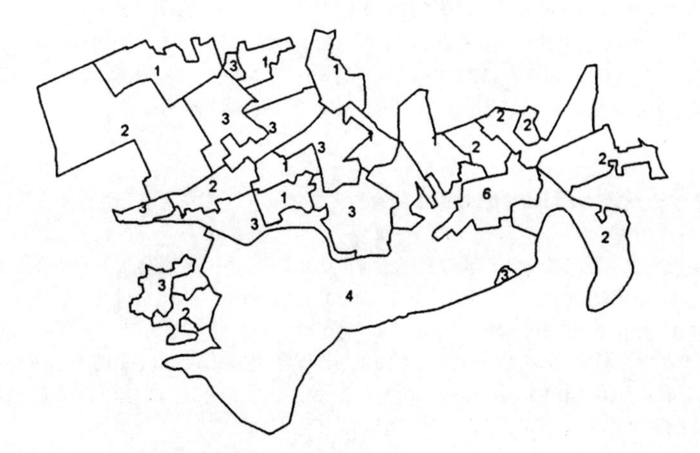

Figure 7.2 Counts of respiratory cancer cases in Falkirk, UK (1978–83)

Figures 7.1 and 7.2 display two examples of mapped small-area health data where clusters of disease are of interest. The two examples represent the two data formats commonly encountered in clustering studies: residential address locations (x, y coordinates), and the count of disease cases within arbitrary geographical regions, such as census tracts or postal districts. The first display shows the distribution of case residential addresses of cancer of the larynx in an area of Lancashire, UK, for a fixed time period (1974–1983). In this example, the spatial distribution of disease incidence can be examined to assess whether there is evidence of excess risk concentrated in areas of the map. Usually, the assessment of such risk must be made with reference to the spatial distribution of population that is 'at risk' of contracting the disease, and this requires the examination of the case map in relation to the map of 'at risk' population.

In the second example, a map of census tracts in the town of Falkirk, Scotland, UK is displayed (Figure 7.2). In the tracts are displayed the counts of respiratory cancer cases for the period 1978–83. Any areas of excess risk or clustering would appear as concentrations within small collections of tracts. It is also important in this case to assess such excess patterns in relation to the underlying distribution of 'at risk' population, and this is often achieved by comparing the count map with the distribution of the rates of disease incidence 'expected' in the tracts concerned.

In both these examples, no definition of clustering is discussed, because this aspect of the analysis is discussed in a later section. The purpose of these examples is to provide an introduction to the type of data found typically in such small-area studies. The definition of what constitutes a cluster or clustered pattern is of fundamental importance in any such study.

7.2 REASONS FOR STUDYING DISEASE CLUSTERING

There are three main situations in which the statistical analysis of disease clustering is important:

(i) in epidemiological research when trying to study the aetiology of a disease
(ii) in public health as part of geographical disease surveillance; and
(iii) in response to disease cluster alarms to evaluate whether thorough epidemiological investigations are warranted.

7.2.1 Finding the aetiology of a disease

There are many ways in which hypotheses about disease aetiology are generated. Differential disease rates observed in disease atlases have long been used for that purpose. A complementary approach is systematically to scan large areas for localised clusters without any prior idea of where they may be, hoping that detected clusters will give clues to previously unknown aetiologies. Different statistical methods for doing this are described in Chapters 8–12.

It is also of interest to search for local geographic clusters that are also localised in time. The Barcelona asthma cluster in Table 7.1 is such an example, where the excess was localised to the harbour area during those days when soybeans were unloaded from some ship (see Chapter 17).

Methods used to detect areas with high rates of a disease can also be used to find good places to conduct case–control or cohort studies. For cohort studies especially, it is important to enrol people with a reasonably high risk of getting the disease because it will either increase the power of the study, reduce the cost as fewer participants are needed, or both.

Whether purely spatial or space–time, a key factor in these methods is their ability to pinpoint the location of clusters. Sometimes, though, we are interested in the general nature of the spatial or space–time disease distribution without an interest in detecting the location of specific clusters. For example, we may simply want to know whether or not a disease is likely to be infectious. Or, we might want to know if the disease has related risk factors that vary geographically. Such information is unlikely to give us clues about any specific aetiology, but it can give us clues about directions in which to search. If a disease is infectious, then we would expect it to exhibit both space–time interaction and global spatial clustering. If a disease is related to risk factors that vary geographically but not over time, then it should exhibit global spatial clustering but no space–time interaction.

When involved in assessment of the relation of specific geographical variables with disease risk in an ecological analysis, it is important to adjust for any existing spatial autocorrelation/clustering not explained by the variables. General clustering methods could be used to test for the existence of clusters as well as for the estimation of the scale of such clustering.

In Chapter 8, Tango compares different tests for global spatial clustering, while Jacquez and Jacquez compare three different space–time interaction tests in Chapter 11.

Whether or not generated by geographical data, many aetiological hypotheses relate to risk factors that are of a local geographical nature. Examples include exposure to electromagnetic fields from power lines, exposure to radiation from nuclear power plants, contamination from toxic dump sites, occupational exposure in certain industries, and pollution from airports, roads or factories.

When individual exposure is unavailable (or too costly to obtain) for a preliminary study, we can use a focused cluster test in an ecological analysis as exemplified by Biggeri and Lagazio in Chapter 20 and by Dolk *et al.* in Chapter 29. The distance to the source is then used as a surrogate for exposure and a test is made to see whether individuals close to the source are more likely to have the disease than those farther away. Note that distance-from-source may not be a good surrogate for exposure in some cases, because exposure may not be monotonically decreasing with distance and may be highly dependent upon directional effects related to the prevailing wind regime, particularly when airborne pollution is thought to be important (Lawson, 1993). Other patterns of exposure can be related to, for example, the flow of ground water or percolation of hazards in the ground (Lawson and Cressie, 1999).

Focused cluster studies will not provide definite answers about aetiology, but as with other ecological studies, they provide an important link in the process from pure hypotheses to well-established risk factors. If a disease relationship is determined using a focused cluster test, the next step is to conduct a more ambitious case–control or cohort study using exposure measurements on individuals. For various papers on focused cluster tests, see Part IV of this book.

7.2.2 Evaluating disease cluster alarms

Disease cluster alarms are very common. In a survey of state health departments in the United States, Wartenberg and Greenberg (1997) found that altogether they received about 1500 requests for cancer cluster investigations in 1989 alone. Many cluster alarms are easily dealt with by providing information over the phone, but others require extensive investigations. It is not uncommon for media and politicians to take an interest in addition to the communities involved. Since the 1980s, a range of alarms have been investigated in different countries, from leukaemia incidence around Sellafield (Hills and Alexander, 1989) and Krümmel nuclear power stations (see Chapter 31 in this volume), to respiratory cancers around incinerators and dump sites (see, for example, Diggle 1990; Lawson, 1989, 1993b), to breast cancer on Long Island, New York (Kulldorff *et al.*, 1997).

In Chapter 10, Kulldorff reviews a number of statistical methods that are helpful when evaluating disease cluster alarms. Such methods can be used to confirm or reject alarms, as part of a decision on whether further resources should be spent on an investigation, to give a more solid scientific base when responding to public concerns, and to prioritise resources between competing needs. In Chapter 17, a review of the analysis of single clusters around putative health hazards is provided.

7.2.3 Public health surveillance

For many diseases there are well-established risk factors. If we detect a local disease cluster it is natural to look first for the presence of known risk factors. If found, we will learn nothing new about the aetiology of the disease, but public health surveillance can be informed of its presence and the risks can then either be removed or reduced in their effect on the population by appropriate public health intervention. Examples include environmental pollution and occupational exposures.

The opposite approach may also work: identify some known risk factor; search for clusters of the disease; and target those areas with preventive measures. For example, special anti-smoking campaigns could be launched in areas with exceptionally high lung cancer rates.

It is also possible to search for clusters with a high mortality rate, adjusting not for the total population at risk, but for the distribution of incidence. A cluster with high mortality compared with incidence may indicate substandard treatment of the disease, or lack of screening. If the latter is found for breast cancer, an improved mammography programme would be warranted.

Not only clusters of excess risk may be of interest. Areas of lower risk can indicate exceptional treatment or an exceptional screening programme that others should study to learn from. It could also be due to different drugs being popular in different areas, and cluster investigations could give an early hint of beneficial or detrimental effects from new drugs, both in terms of their intended purpose and their side-effects.

Local disease clusters may also be due to differential classification or reporting of disease incidence or mortality. As such, they may bring to attention problems that need to be resolved in order to maintain high-quality health statistics.

For these public health applications, we are interested in detecting the location of clusters. Several such methods are described in chapters in this Part of the volume.

7.3 DEFINITION OF CLUSTERS AND CLUSTERING

A wide variety of definitions can be put forward for clusters and clustering. However, it is convenient here to consider two extreme forms of clustering within which most definitions can be subsumed. These two extremes represent the spectrum of modelling from non-parametric to parametric forms and associated with these forms are appropriate statistical models and estimation procedures. First, because many epidemiologists may not wish to specify a priori the exact form/extent of clusters to be studied, a non-parametric definition is often the basis adopted. An example of such a definition is given by Knox (1989): 'a geographically bounded group of occurrences of sufficient size and concentration to be unlikely to have occurred by chance'. Without any assumptions about shape or form of the cluster, the most basic definition would be as follows:

- any area within the study region of *significant* elevated risk.

This definition is often referred to as *hot spot* clustering (see, for example, Marshall, 1991a). This is a simpler form of Knox's definition but summaries the essential ingredients. In essence, any area of elevated risk, regardless of shape or extent, could qualify as a cluster, provided the area meets some statistical criteria. Note that it is not usual to regard areas of significantly low risk to be of interest, although these may have some importance in further studies of the aetiology of a particular disease.

Secondly, at the other extreme, we can define a parametric cluster form as:

- The study region displays a prespecified (parameterised) cluster structure.

This definition describes a parameterised cluster form that would be thought to apply across the study region. Usually, this implies some stronger restriction on the cluster form and also some region-wide parameters that control the shapes and sizes of clusters. Both of these extremes can be modified by using modelling approaches that borrow from either extreme form. For example, it is possible to model cluster form parametrically, but also to include a non-parametric component in the cluster estimation part that allows a variety of cluster shapes across the study region (Lawson, 1995).

Besag and Newell (1991) first classified types of clustering studies and associated cluster definitions. We will extend their definitions here to include some extra classes. First of all, they defined *general* clustering as the analysis of the overall clustering tendency of the disease incidence in a study region. As such, the assessment of general clustering is closely akin to the assessment of autocorrelation. Hence, any model or test relating to general clustering will assess some overall/global aspect of the clustering tendency of the disease of interest. This could be summarised by a model parameter (e.g. an autocorrelation parameter in an appropriate model) or by a test that assesses the aggregation of cases of disease. For example, the correlated prior distributions used by Besag, Clayton, and Lawson and coworkers (Besag *et al.*, 1991; Clayton and Bernardinelli, 1992; Lawson *et al.*, 1996) incorporate a single parameter that describes the correlation of neighbouring locations on a map, whereas the methods of Cuzick, Diggle, and Anderson and coworkers (Anderson and Titterington, 1997; Cuzick and Edwards, 1990; Diggle and Chetwynd, 1991) for case events, and of Whittemore, Tango, Raubertas and Oden and coworkers (Whittemore *et al.*, 1987; Tango, 1995; Raubertas, 1988; Oden, 1995) for counts, consider testing for a clustered pattern within the study region, rather than model-parameter estimation.

It should be noted at this point that the general clustering methods discussed above can be regarded as *non-specific* in that they do not seek to estimate the spatial locations of clusters but simply to assess whether clustering is apparent in the study region. Any method that seeks to assess the locational structure of clusters is defined as *specific*.

An alternative *non-specific* effect has also been proposed in models for tract-count or case-event data. This effect is conventionally known as *uncorrelated* heterogeneity (or overdispersion/extra-Poisson variation in the Poisson likelihood case). This effect gives rise to extra variation in incidence and, for Poisson likelihoods, displays variability of observed counts that exceeds the mean of the observed counts. This marginal heterogeneity has traditionally been linked to 'clustering', as is evidenced by the use of the negative binomial distribution as a 'cluster' distribution (see, for example, Douglas, 1979). Often such effects can be considered to be modelled by the inclusion of *uncorrelated* heterogeneity within a parametric model. Hence, lognormal or gamma distributions are often used to model this component of the expected incidence. The result of using such as a *non-specific* effect is to mimic cluster intensity variation as a realisation of these distributions over the study region. This corresponds to a greater peakedness in intensity variation than that induced by *correlated* heterogeneity, and the earlier comments concerning the appropriateness of this approach for cluster structure also apply here.

Besag and Newell's second class of clustering methods is that of *specific* methods and these are divided into *focused* and *non-focused* methods. These methods are designed to examine one or more clusters and the locational structure of the clusters are to be assessed. Focused clustering is defined as the study of clusters where the location and number of the clusters is predefined. In this case, only the extent of clustering around the predefined locations is to be modelled. Examples of this approach mainly come from studies of putative sources of health hazard; for example, the analysis of disease incidence around prespecified *foci* that are thought to be possible sources of health hazard, such as nuclear power installations, waste dumps, incinerators, harbours, road intersections or steel foundries. In this chapter, we consider only the *non-focused* form of clustering, because focused clustering is discussed in Chapter 17, which is concerned with putative sources of hazard.

It is very important to consider, within any analysis of geographically distributed health data, the structure of hypotheses that could include cluster components. For example, many examples of published analyses within the areas of disease mapping and focused clustering consider the null hypothesis that the observed disease incidence arises as a realisation of events from the underlying *at risk* population distribution. The assumption is made that, once this *at risk* population is accurately estimated, it is possible to assess any differences between the observed disease incidence and that expected to have arisen from the *at risk* background population. However, if the disease of interest naturally clusters (beyond that explained by the estimated *at risk* background), then this form of clustering should be included also within the null hypothesis. Because this form of clustering often represents unobserved covariates or confounding variables, it is appropriate to include this as heterogeneity. This can be achieved in many cases via the inclusion of random effects in the analysis. Note that such random effects often do not attempt to model the exact form or location of clusters but seek to mimic the effect of clustering in the expected incidence of the disease. The correlated and uncorrelated heterogeneity first described by Clayton and Kaldor (1987), and Besag *et al.* (1991) come under this category. Note also that if clustering of disease incidence is to be studied under the alternative H_1, then not only would heterogeneity be needed under H_0 but some form of cluster structure must be estimable under H_1 as well. In Besag *et al.* (1991), an example is provided, in a disease mapping context, where a residual can be computed after fitting a model with different types of heterogeneity. This residual could contain uncorrelated error, trend, or cluster structure depending upon the application. Hence, such a residual could, in some cases, provide a simple non-parametric approach for the exploration of cluster form.

One disadvantage of the use of the *non-specific* random effects for background clustering so far advocated in the literature is that they do not exactly match the usual form of cluster variation in geographical studies. At least in rare diseases, clusters usually occur as isolated areas of elevated intensity separated by relatively large areas of low intensity. In this case, the use of a log Gaussian random effect model fitted to the whole region (as advocated by Besag *et al.*, 1991; Lawson *et al.*, 1996, and many others) will not closely mimic the disease clustering tendency, because the global extra variation described by such an effect will force a global cluster structure on a pattern that cannot be described by such a global form.

7.4 MODELLING ISSUES

The development of models for clusters and clustering has seen greater development in some areas than in others. For example, it is straightforward to formulate a non-specific Bayesian model for case events or tract counts that includes heterogeneity (Besag *et al.*, 1991; Clayton and Bernardinelli, 1992; Lawson, 1997; Lawson *et al.*, 1996). However, specific models are less often reported. Nevertheless, it is possible to formulate specific clustering models for the case-event and tract-count situation. If it is assumed that the intensity of case events, at location $\boldsymbol{x}$, as $\lambda(\boldsymbol{x})$, then by specifying a dependence in this intensity on the locations of cluster centres, it is possible to proceed. For example,

$$\lambda(\boldsymbol{x}) = \rho \cdot g(\boldsymbol{x}) \cdot m\left\{\sum_{j=1}^{k} h(\boldsymbol{x} - \boldsymbol{y}_j)\right\} \tag{7.1}$$

describes the intensity of events around k centres located at $\{\boldsymbol{y}_j : j = 1, \ldots, k\}$, where ρ is the overall rate, $g(\cdot)$ is a background modulating function describing the 'at risk' population distribution, and $m(\cdot)$ is a suitable link function. The distribution of events around a centre is defined by a cluster distribution function $h(\cdot)$. Conditional on the cluster centres, the events can be assumed to be governed by a heterogeneous Poisson process (HEPP), and hence a likelihood can be specified. As the number (k) and the locations of centres are unknown, then with a suitable prior distribution specified for these components it is possible to formulate this problem as a Bayesian posterior sampling problem, with a mixture of components of unknown number. This type of problem is well suited to reversible jump Markov chain Monte Carlo (MCMC) sampling (Green, 1995). The approach can be extended to count data straightforwardly, as

$$\mathrm{E}(n_i) = \rho \int_{W_i} g(\boldsymbol{x}) \cdot m\left\{\sum_{j=1}^{k} h(\boldsymbol{x} - \boldsymbol{y}_j)\right\} \mathrm{d}\boldsymbol{x}, \tag{7.2}$$

where the Poisson distributional model is assumed, W_i is the ith tract, and n_i is the count in the ith tract. Examples of the use of these models in case-event and tract-count data are provided in Lawson (1995, 1997) and Lawson and Clark (1999). Variants of this specification can be derived for specific purposes or under simplifying assumptions. The wide applicability of this formulation can be appreciated by the fact that $h(\cdot)$ could be non-parametrically estimated, in which case a data-dependent cluster form will prevail. This provides a modelling framework that can allow both vague prior beliefs about cluster form and also highly parametric forms.

7.5 HYPOTHESIS TESTS FOR CLUSTERING

The literature of spatial epidemiology has developed considerably in the area of hypothesis testing and, more specifically, in the sphere of hypothesis testing for clusters. Very early developments in the area arose from the application of statistical tests to spatio-temporal clustering, a particularly strong indicator of the importance of a *spatial* clustering phenomenon. The early seminal work of Mantel (1967) and Knox (1964) in the field of space-time cluster testing predates most of the development of *spatial* cluster testing. As noted above, distinction should be made between tests for general (non-specific) clustering, which assess the overall clustering pattern of the disease, and the specific clustering tests where cluster locations are estimated.

For case events, a few tests have been developed for non-specific clustering. Cuzick and Edwards (1990) developed a test that is based on a realisation of cases and a *sample* of a control realisation. Functions of the distance between case location and k 'nearest' cases (as opposed to controls) were proposed as test statistics. The null hypothesis of random labelling is tested against clustered alternatives, although not specifically of the form (7.1). Diggle and Chetwynd (1991) extended stationary-point-process descriptive measures (K-functions) to the case where a modulated population background is present. Their method uses a complete control disease realisation and provides a measure of scale of clustering also. Neither of these methods allows for first-order non-stationarity that may be present in many examples. Anderson and Titterington (1997) have proposed the use of a simple integrated squared distance (ISD) statistic for cluster assessment. This is closely related to the analysis of density ratios in exploratory analysis, and could be regarded as a type of non-parametric assessment of clustering. The advantage of this approach is that the assessment is not tied to a specific cluster model but detects departures from background. The major disadvantage, shared with all such statistics, is its low power against specific forms of clustering. Other simple forms of a global test can be proposed, where density estimates of cases are compared with intensity estimates of case events simulated from the control background. These could provide pointwise confidence intervals as well as global tests. There appears to have been little development of tests that detect uncorrelated heterogeneity in the intensity of the case-event process as a form of spatial clustering. It is unclear what aetoiological difference would be inferred when uncorrelated rather than correlated forms of heterogeneity were found.

General clustering tests based on tract counts can be classified into tests for correlated heterogeneity and tests for uncorrelated heterogeneity. Tests of the latter are not *spatial* in origin but are included here for completeness. We also consider the possibility of general cluster tests based on cluster sums. In the case of correlated heterogeneity, Whittemore *et al.* (1987) developed a quadratic form test statistic that compared observed counts and expected counts for all tracts weighted by a covariance matrix. This test was found to have reduced power in some situations by Turnbull *et al.* (1990).

Subsequently, Tango (1995) developed a modified general class of tests for general and focused clustering. An alternative procedure based on Moran's I statistic, modified to allow tract-specific expected rates, has been proposed by Cliff and Ord (1981) and Oden (1995). All these tests look for a divergence between count and expectation over the whole study region, and they are unlikely to perform well against specific localised clusters.

As mentioned above, some use has been made of tests for uncorrelated heterogeneity to assess clustering of tract counts. For example, the Euroclus project (Alexander *et al.*, 1996) tests for such heterogeneity across European states using the Pottohoff–Whittinghill (1966a,b) test and score tests for Poisson versus negative binomial distributions for the marginal count distribution (Collings and Margolin, 1985). The evidence of Euroclus suggests that for certain important forms of non-Poisson alternatives within the negative binomial family, these tests perform poorly (see Alexander *et al.*, 1996). Finally, as noted in Section 7.3, at least for rare diseases, it is easily possible that the marginal count distribution would not follow a negative binomial distribution and could even display multimodality.

Specific cluster tests address the issue of the location of putative clusters. These tests produce results in the form of locational probabilities of significances associated with specific groups of tract counts or cases. Openshaw *et al.* (1987) first developed a general method that allowed the assessment of the location of clusters of cases within large disease maps. The method was based on repeated testing of counts of disease within circular regions of different sizes. Whenever a circle contains a significant excess of cases, it is drawn on the map. After a large number of iterations, the resulting map can contain areas where a concentration of overlapping circles suggests localised excesses of a disease. It has been pointed out that a large number of 'significant' clusters will always be found due to the multiple testing involved. A similar method was proposed by Besag and Newell (1991), which includes a test for the number of clusters found. Their method involves accumulating events (either cases or counts) around individual event locations. Accumulation proceeds up to a fixed number of events of tracts (k). The number of k is fixed in advance. The method can be carried out for a range of k values. While the local alternative for this test is increased intensity, there appears to be no specific clustering process under the alternative and, in that sense, the test procedure is non-parametric, except that a monotone cluster distance distribution is implicit. One advantage of the test is that it can be applied to focused clusters also, while a disadvantage is that an arbitrary choice of k must be made and the results of the test must depend on this choice. Recently, Le *et al.* (1996) have proposed a modification to the Besag and Newell method that overcomes the arbitrary choice of k.

An alternative, a spatial scan statistic, has been proposed by Kulldorff and Nagarwalla (1995) and Kulldorff (1997), who employ a likelihood ratio test for the comparison of the number of cases found in the study region population (the null hypothesis), to a model that has different disease risk depending on being inside or outside a circular zone. The test can be applied to both case events and tract counts. The advantage of the test is that it examines a potentially infinite range of zone sizes and that it accounts for the multiple testing inherent in such a procedure. It relies on a formal model of null and alternative hypotheses, and it is possible to adjust for any number of known covariates. A limitation of the method relates to the use of circular regions, which tends to emphasise compact clusters, and the method has low power against other alternatives such as a long and narrow cluster along a river, or against an alternative with a large number of very small clusters at very different locations. Note that tests for clusters around *known* locations are discussed in Part IV of this volume (see, for example, Chapters 17 and 19).

It is also possible to apply two extreme forms of test of *either* a non-parametric (hot spot) cluster-specific test or a fully parametric form. First, if we assume n_i is the tract count of disease, and e_i is the expected rate in the ith tract, then we can compare

$$(n_i - e_i) \quad \text{with} \quad (n^*_{ij} - e_i),$$

for each tract, where n^*_{ij} $j = 1, \ldots 99$ are simulated counts for each tract based on the given expectation for the tract. If any tract count exceeds the critical level within the rankings of the simulated residuals, then we accept the tract as 'significant'. The resulting map of 'significant' tracts displays clusters of different forms. This does not use contiguity information. In the case-event situation, pointwise comparison of $\hat{\lambda}^*_i - \hat{\lambda}_i$ where $\hat{\lambda}^*_i$ is a density estimate based on the case events only and $\hat{\lambda}_i$ is a density estimate based on the controls only (assuming a control realisation is available), can be made. This could be compared with density estimates of sets of events simulated from the density estimate of the controls. Note that a variant of this procedure, where simulation is based on random allocation of cases and controls, has been suggested by Kelsall and Diggle (1995).

At the other extreme it is possible to test for specific cluster locations via the assumption of the cluster sum term of the form (7.1) in either the intensity of case events or, in the case of tract counts, the specification of the expected rate in each tract, as in (7.2). As the cluster locations and number of locations are random quantities, it would be necessary either to employ approximations that involve fixed cluster numbers, or to include testing within MCMC algorithms (Besag and Clifford, 1989) that sample the joint posterior distribution of number and locations of centres.

ACKNOWLEDGEMENT

We wish to acknowledge Peter Diggle for making available the Lancahsire larynx cancer dataset.

8

Comparison of General Tests for Spatial Clustering

T. Tango

Institute of Public Health, Tokyo

8.1 INTRODUCTION

The question of whether incident cases are clustered in space has recieved considerable attention in the statistical and epidemiological literature. Marshall (1991a) provides a thorough review and an issue of the *Journal of the Royal Statistical Society, Series A* (1989), an issue of the *American Journal of Epidemiology* (1990) and three issues of *Statistics in Medicine* (1993, 1995, 1996) have been devoted to the topic.

According to Besag and Newell's (1991) classification, tests for disease clustering are classified into two families. The first family consists of *focused* tests which assess the clustering around a pre-fixed point like a nuclear installation. The second family consists of *general* tests which are aimed at investigating the question of whether clustering occurs over the study region. The second family can be subdivided into two groups; the first group provides methods for examining *a tendency to cluster*, i.e. cases are located close to each other no matter where they occur, and the second group contains techniques for exclusively searching for the *location of clusters*. Furthermore, with regard to the data to be used, there is a division between methods that utilise population counts of small administrative subregions and those that employ a sample of controls.

This chapter is concerned with *general* tests and is organised as follows. Section 8.2 points out several problems associated with methods published recently. Section 8.3 describes several general tests which are free from statistical inappropriateness. Section 8.4 compares some of these methods.

8.2 INAPPROPRIATE TESTS

Although many tests have been proposed and used in the literature, most of them are not recommendable in the sense that they may produce spurious results in practice,

Disease Mapping and Risk Assessment for Public Health. Edited by A.B. Lawson *et al.*
© 1999 John Wiley & Sons Ltd.

primarily because *they cannot properly adjust for a heterogeneous population density* (see, for example, Besag and Newell, 1991; Marshall, 1991a). Tests due to Moran's (1948) spatial autocorrelation I or some variation thereof and Whittemore *et al.*'s (1987) mean-distance method are typical examples of inappropriate tests based on population counts. The former group of tests were originally devised for biological problems which need not take heterogeneous populations into account. Recently, Oden (1995) proposed two tests, I_{pop} and I^*_{pop}, adjusting Moran's I for heterogeneous populations. However, Oden assumed, incorrectly, normal asymptotic distributions for these two test statistics and thus Oden *et al.* (1996) have shown erroneous results that I^*_{pop} has the highest power in their simulation studies. Tango (1998) indicated that I_{pop} is essentially identical to Tango's (1995) test statistic and that the comparison of powers depends on the alternative hypothesis of clustering.

On the other hand, the inappropriateness of Whittemore *et al.*'s test,

$$W = \boldsymbol{r}^{\mathrm{T}}\boldsymbol{D}\boldsymbol{r},$$

seems not to be widely known (Tango, 1995, 1997), in which $\boldsymbol{r}^{\mathrm{T}} = (n_1, \ldots, n_L)/\ n_+$ denotes a vector of observed relative frequencies and $\boldsymbol{D}$ denotes a distance matrix. As a matter of fact, Whittemore *et al.*'s test statistic itself is a spatial version of Tango's (1984) clustering index in time, which *can only be used for homogeneous populations.*

To see its inappropriateness, the example used in Tango (1995) is shown here. Let us consider the study area comprising three regions with $d_{12} = d_{13} = d_{23}$, where d_{ij} (an element of the matrix $\boldsymbol{D}$) denotes the spatial distance between the ith region and jth regions, and $\boldsymbol{p} = \mathrm{E}_{\mathrm{H}_0}(\boldsymbol{r}) = (0.2, 0.3, 0.5)$. Then consider the two observed cases: (i) $\boldsymbol{r} = (0.2, 0.3, 0.5)$ and (ii) $\boldsymbol{r} = (0.5, 0.3, 0.2)$. In case (i), $\boldsymbol{r} = \boldsymbol{p}$, so we can judge that there is no clustering. In case (ii), on the other hand, we can clearly observe that the first region has higher incidence compared with other regions. However, the statistic W produces the same value in both cases. This example shows that this test cannot properly adjust for confounders, although it is the primary purpose of the test. It is surprising that many biostatisticians apply Whittemore *et al.*'s test to real investigations or compare it with their newly proposed tests without knowing or examining its inappropriateness. Recently, Ranta (1996) generalised Whittemore *et al.*'s test to a test for space–time interaction without knowing its inappropriateness (Tango, 1997). Tango (1995) proposed a more appropriate spatial version.

Methods utilising a sample of controls assume, as a starting point, two independent inhomogeneous Poisson processes with spatially varying density: $\mu(\boldsymbol{x})$ for cases and $\lambda(\boldsymbol{x})$ for controls. The null hypothesis then can be reduced to the *random labelling hypothesis* that the observed case series is a random sample from the entire sample combining cases and controls. Diggle and Chetwynd (1991) proposed a test statistic $\hat{D}(s) = \hat{K}_{11}(s) - \hat{K}_{22}(s)$, where $\hat{K}_{11}(s)$ is the estimate of $K_{11}(s)$ =(expected number of further cases within distance s of an arbitrary case)/(expected number of cases per unit area) and $\hat{K}_{22}(s)$ is defined analogously in terms of the control realisation. Under the null hypothesis, $\mathrm{E}\{\hat{D}(s)\} = 0$ and a positive value might suggest a departure from the null. However, this test statistic also becomes a member of inappropriate tests since we can easily find a pattern of clustered cases satisfying $\hat{D}(s) = 0$ for all s. One such typical example is the situation where $\mu(\boldsymbol{x}) = \lambda(\boldsymbol{x} - \boldsymbol{x}_0)$ for some $\boldsymbol{x}_0$. A special case of this example has been given by Kulldorff (1998).

8.3 AVAILABLE TESTS

Although some methods were originally designed for population counts data within small administrative subregions and others were devised for a sample of case–control location data, all these methods can be used for both types of data, at least in a Monte Carlo setting.

8.3.1 Tests originally designed for population counts

These tests assume that (i) the study region is partitioned into administrative subregions called *cells* and (ii) for each of a set of m cells $i = 1, 2, \ldots, I$, we have the number of cases n_i of the disease under study, the corresponding number of individuals at risk (population) ξ_i, and the coordinates of its administrative population centroid, $\boldsymbol{x}_i$. As far as is known, there are only four general tests that are free from statistically inappropriateness among tests published so far: Turnbull *et al.*'s (1990) and Kulldorff and Nagarwalla's (1995) tests for detecting locations of clusters, and Besag and Newell's (1991) and Tango's (1995) for detecting a tendency to cluster. Tango's test can be used for detecting locations of clusters.

Turnbull et al. (1990): searching for the location of clusters

For each cell, a 'ball' is constructed by absorbing the nearest neighbouring cells such that each ball contains just a pre-fixed number of individuals at risk, R. If $\xi_i < R$, then cell i is included in the ith ball and the cell whose centroid is nearest to that of cell i, say cell j, is included in the ith ball if $n_i + n_j < R$. If $n_i + n_j > R$, then a fraction of the population of cell j is added so that the total population at risk is exactly R. If $n_i > R$, then the ith ball contains only a fraction of cell i. As for the fractional cell included in a ball, the cases are also allocated in the same proportion to the ball as that of the population of the cell. In total, I overlapping balls are created with a constant population size at risk R. Test statistic, as a function of R, is given by

$$T_{\mathrm{TU}} = \text{maximum number of cases in the ball.} \tag{8.1}$$

Monte Carlo simulation is needed to evaluate the significance of the observed value of test statistic.

Kulldorff and Nagarwalla (1995): searching for cluster locations

This procedure is a spatial version of the scan statistic with a variable window size and is a generalisation of a method of Turnbull *et al.* For each cell, a 'circle' $\mathbf{Z}$ is constructed by absorbing the nearest neighbouring cells whose centroids lie inside the circle and the radius varies continuously from zero upwards until some fixed percentage (say 10%) of the total population is covered. Thus, for cell i, we can make, as a rough average, $I/10$ circles, $\mathbf{Z}_{ij}, j = 1, 2, \ldots$. Of course, Z_{ij} might be equal to Z_{kl} even if $i \neq k$ and $j \neq l$. In total, K different but correlated circles are created. Kulldorff and Nagarwalla assume that there exists a *clustered circle* such that for all individuals within the circle, the probability of being a case is p, whereas for all individuals outside the circle this probability is $q(< p)$. Therefore, the test statistic can be constructed via the likelihood ratio test which is given by the simple form (Kulldorff, 1997):

$$T_{\mathrm{KN}} = \sup_{Z} \left(\frac{n(Z)}{\xi(Z)}\right)^{n(Z)} \left(\frac{n(Z^{c})}{\xi(Z^{c})}\right)^{n(Z^{c})} I\left(\frac{n(Z)}{\xi(Z)} > \frac{n(Z^{c})}{\xi(Z^{c})}\right), \tag{8.2}$$

where Z^{c} indicates all the circles except for Z, $n(\cdot)$ and $\xi(\cdot)$ denote the observed number of cases and the population, respectively, and $I(\cdot)$ is the indicator function. To find the distribution of the test statistic, Monte Carlo simulation is required since the ordinary χ^2 approximation cannot apply.

Besag and Newell (1991): detecting a tendency to cluster

First, the size of cluster must be fixed, say k cases. The computational property of this method clearly indicates that it is applicable only for quite rare diseases and thus a typical value of k might be 2, 4, 6, 8 or 10. Each cell with non-zero cases is considered in turn as the centre of a possible cluster. When considering a particular cell, we label it as cell 0 and order the remaining cells by their distance to cell 0. We label these cells $1, 2, \ldots, I-1$ and define $D_i = \sum_{j=0}^{i} n_j$ so that $D_0 \le D_1 \le \ldots$ are the accumulated numbers of cases in cells $0, 1, \ldots$ and $u_0 \le u_1 \le \ldots$ are the corresponding accumulated numbers of individuals at risk. Now let $M = \min\{i : D_i \ge k\}$ so that the nearest M cells contain the closest k cases. A small observed value of M indicates a cluster centred at cell 0. If m is the observed value of M, then the significance level of each potential cluster is $\Pr\{M \le m\} = 1 - \sum_{s=0}^{k-1} \exp(-u_m Q)(u_m Q)^s / s!$, where $Q = n_+/\xi_+$. Then, the test statistic of overall clustering within the entire region is

$$T_{\mathrm{BN}} = \text{the total number of individually significant } (p < 0.05, \text{ say}) \text{ clusters.} \tag{8.3}$$

The significance of the observed T_{BN} may be determined by Monte Carlo simulation.

Tango (1995): detecting a tendency to cluster and searching for cluster locations

Tango (1995) proposed the test statistic C:

$$C = (\boldsymbol{r} - \boldsymbol{p})^{\mathrm{T}} \boldsymbol{A} (\boldsymbol{r} - \boldsymbol{p}) = \sum_i \left\{ \sum_j a_{ij}(r_i - p_i)(r_j - p_j) \right\} = \sum_i U_i, \tag{8.4}$$

where $\boldsymbol{r}^{\mathrm{T}} = (n_1, \ldots, n_I)/n_+$ denotes a vector of observe relative frequencies in I cells, $\boldsymbol{p} = \mathrm{E}_{\mathrm{H}_0}(\boldsymbol{r})$, and $\boldsymbol{A} = (a_{ij})$ is a matrix of the measure of *closeness* with $a_{ii} = 1$ and $a_{ij} = a_{ji} \le 1$, not of a distance measure. Asymptotically under H_0:

$$T = \frac{C - \mathrm{E}(C)}{\sqrt{\mathrm{var}(C)}} \sim \chi_\nu^2 \text{ distribution,}$$

where

$$\mathrm{E}(nC) = \mathrm{tr}\,(AV_p), \qquad \mathrm{var}\,(nC) = 2\mathrm{tr}\,(AV_p)^2, \qquad \nu = 8/\{\sqrt{\beta_1(C)}\}^2,$$
$$\sqrt{\beta_1(C)} = 2\sqrt{2}\mathrm{tr}\,(AV_p)^3/\{\mathrm{tr}(AV_p)^2\}^{1.5}, \qquad V_p = \mathrm{diag}(\boldsymbol{p}) - \boldsymbol{p}\boldsymbol{p}^{\mathrm{T}}.$$

This χ^2 approximation is generally quite accurate even for small n. In practice, the selection of a_{ij} is important and needs careful consideration of the disease under study. However, a simple exponentially decreasing function, $a_{ij} = \exp(-d_{ij}/\lambda)$, may be a natural one in many cases, where d_{ij} denote a distance measure and an appropriate constant λ should be fixed in advance. If we are interested in 'hot spot' clusters with maximum distance 2λ between clustered cases, then set $a_{ij} = 1$ for $d_{ij} \leq \lambda$ and $a_{ij} = 0$ otherwise. If we are interested in 'clinal' clusters, then such a simple model may be $a_{ij} = \exp(-4(d_{ij}/\lambda)^2)$. Needless to say, there are a plenty of other choices for a_{ij} depending on the situation. If the test result is significant, then the possible centres of such clusters may be indicated by the cell with large U_i, although this is of a descriptive nature and has no tests associated with individual U_is.

8.3.2 Tests originally designed for a sample of cases and controls

These tests assume that (i) n_0 cases are observed in the study region and denote their location by $(\boldsymbol{x}_1, \ldots, \boldsymbol{x}_{n_0})$; (ii) from all individuals at risk in the study region select at random a set of n_1 controls and denote their location by $(\boldsymbol{x}_{n_0+1}, \ldots, \boldsymbol{x}_{n_0+n_1})$; and (iii) case series and control series constitute two independent inhomogeneous Poisson processes with spatially varying densities: $\mu(\boldsymbol{x})$ for cases and $\lambda(\boldsymbol{x})$ for controls. As general tests applicable to these settings, in addition to the previous four methods, there are two further methods: Cuzick and Edwards' (1990) test for detecting a tendency to cluster, and Anderson and Titterington's (1997) test for detecting a tendency to cluster, although the latter requires further improvement on reliable computations.

Cuzick and Edwards (1990): detecting a tendency to cluster

This test statistic is a k nearest neighbours type which can be written a more in general form as

$$T_k = \sum_i \sum_j a_{ij}\delta_i\delta_j = \delta^T \boldsymbol{A}\delta, \tag{8.5}$$

where $\delta_i = 1$ if x_i is a case and $\delta_i = 0$ otherwise, and $a_{ij} = 1$ if $\boldsymbol{x}_j$ is in the set of k nearest neighbours to $\boldsymbol{x}_i$ and $a_{ij} = 0$ otherwise. It should be noted that Cuzick and Edwards' test is the same test as proposed by Alt and Vack (1991) and a special case of a test proposed by Cliff and Ord (1981). Furthermore, if we set $a_{ij} = d_{ij}$ if $\boldsymbol{x}_j$ is in the set of k nearest neighbours to $\boldsymbol{x}_i$ and $a_{ij} = 0$ otherwise, then this statistic becomes a generalised version of Ross and Davis's (1990) test. Several other measures of closeness can be used. Since the null hypothesis of no clustering can be reduced to the *random labelling hypothesis*, a permutational approach can be used. Cuzick and Edwards calculate the moments of permutational distribution of T_k but a Monte Carlo simulation is also carried out easily.

Anderson and Titterington (1997): detecting a tendency to cluster

This test statistic, called ISD (integrated squared difference), is defined as

$$\text{ISD} = \int_{\boldsymbol{x}\in\Omega} (\hat{\mu}(\boldsymbol{x}) - \hat{\lambda}(\boldsymbol{x}))^2 \mathrm{d}\boldsymbol{x}, \tag{8.6}$$

where $\hat{\mu}(\boldsymbol{x})$ and $\hat{\lambda}(\boldsymbol{x})$ are kernel density estimates in the study region Ω. This statistic incidentally becomes a continuous version of Tango's C with special case of $a_{ii} = 1$ and $a_{ij} = 0$ for $i \neq j$, indicating that ISD is a location-invariant statistic and is expected to have low power. Therefore,the following extended version would be more powerful:

$$C_{\text{ISD}} = \int_{\boldsymbol{x}\in\Omega}\int_{\boldsymbol{y}\in\Omega} (\hat{\mu}(\boldsymbol{x}) - \hat{\lambda}(\boldsymbol{x}))(\hat{\mu}(\boldsymbol{y}) - \hat{\lambda}(\boldsymbol{y}))a(\boldsymbol{x},\boldsymbol{y})\mathrm{d}\boldsymbol{x}\mathrm{d}\boldsymbol{y}, \tag{8.7}$$

where $a\,(\boldsymbol{x},\boldsymbol{y})$ denotes a continuous version of the measure of closeness between two locations $\boldsymbol{x}$ and $\boldsymbol{y}$. Statistical significance of these statistics requires a sophisticated Monte Carlo integration and simulation with the smoothed Bootstrap. One of the merits of this type of approach is that it allows for graphical comparison of the two surfaces.

8.4 DISCUSSION

Table 8.1 summarises the characteristics of tests available. It is impossible to compare their powers precisely since most of them are totally different and each test has its own unknown parameter except for the K&N method. However, estimating their powers under several realistic patterns of clusters is indispensable for understanding their strengths and weaknesses.

Table 8.2 is a trial example of simple comparative power study which is a reproduction of Table III of Kulldorff and Nagarwalla (1995), to which estimated powers of Tango's test are added. As expected, Kulldorff and Nagarwalla's approach is shown to be superior in this simulation where only one cluster is assumed. Although Kulldorff and Nagarwalla's approach has no pre-fixed parameters except for the shape of the 'circle' and seems to be very flexible, it is based on the maximum likelihood ratio test which tests the null hypothesis against the alternative hypothesis that there is one cluster in the whole study area. Therefore, the effect of a misspecified alternative on the performance should be investigated.

Table 8.1 Comparison of tests. All six methods can, in principle, be used for both kinds of data types: population counts and case–control data

Purpose[a]	Method	Unknown parameter	Test statistic	Null distribution
L	Turnbull	R = population	maximum number of cases in ball	Monte Carlo
L	K&N		LRT	Monte Carlo
T	B&N	k = cluster size	number of significant clusters	Monte Carlo
L,T	Tango	λ = radius	quadratic form	χ^2 approx.
T	C&E	k = size of nearest neighbours	number of cases in k nearest neighbours	Normal approx.
T	A&T	smoothing parameters	ISD	Monte Carlo

[a] L = searching for the location of clusters; T = detecting a tendency to cluster.

Table 8.2 Estimated power of three tests in the same Monte Carlo power study as Kulldorff and Nagarwalla's: On a square, we selected randomly the locations of 100 cells. A constant population of 100 is assigned to each cell and we placed another *square* with variable population size, n_{cl}, in the centre to constitute the true cluster. We then randomly assigned 1000 cases among the population in such a way that individuals within the true cluster had a relative risk that was *rr* times higher than those outside. Results of K&N's and of Turnbull *et al.*'s test are reproduced with permission from Table III of K&N (1995). A measure of closeness with 'clinal cluster' type, $a_{ij} = \exp\{-4(d_{ij}/\lambda)^2\}$ is used for Tango's test

Test	Cluster pop. n_{cl}: Relative risk *rr*:	100 3.0	200 2.5	400 2.0	700 1.7	1000 1.6	1400 1.5	2000 1.4	4000 1.35
Turnbull *et al.*	$R = 1000$	0.40	0.66	0.83	0.92	0.98	0.91	0.76	0.62
K&N		0.91	0.96	0.93	0.94	0.96	0.93	0.90	0.88
Tango	$\lambda = (\max d_{ij})/8$	0.46	0.71	0.77	0.89	0.85	0.81	0.77	0.74
	$\lambda = (\max d_{ij})/4$	0.36	0.64	0.73	0.85	0.88	0.86	0.81	0.77

To do further comparison we need reliable computer programs that can run on a PC or on a workstation since these tests are all computer-intensive. Recently Jacquez (1994a) developed STAT!, a statistical software for the clustering of health events, which includes many classical test statistics and also Cuzick and Edwards test. In terms of comparison of methods and also availability of these clustering methods to epidemiologists and statisticians, such software is very useful. Therefore, there is a great need for including the several recent important methods discussed in this chapter. Kulldorff *et al.* (1996) developed SaTScan for their spatial scan statistic, which is available via the Internet (`http://dcp.nci.nih.gov/BB/SATScan.html`). Tango's computer code in S-PLUS is available from the author upon request via e-mail (`tango@iph.go.jp`).

ACKNOWLEDGEMENT

I wish to thank Dr Kulldorff for his many constructive comments on an earlier manuscript.

9

Markov Chain Monte Carlo Methods for Putative Sources of Hazard and General Clustering

Andrew B. Lawson and A. Clark

University of Aberdeen

9.1 INTRODUCTION

Markov chain Monte Carlo (MCMC) methods have now become accepted as a general method for the exploration of posterior distribution or likelihood surfaces (see, for example, Gelman *et al.*, 1995). Within spatial epidemiology, only limited development of the use of these methods has been witnessed, mainly within disease mapping via Gibbs Sampling (e.g. the BUGS package). However, a wide range of models can be sampled via MCMC methods and a considerable degree of freedom exists with respect to areas of application. The earliest work on MCMC within statistical applications appeared in spatial statistics, particularly image processing, where the large numbers of parameters found encouraged the use of iterative simulation methods. In those applications, use is made of two main algorithms for posterior sampling: Metropolis–Hastings (MH) and Gibbs Sampling. The most general of these methods is the MH algorithm, while the Gibbs Sampler is a special case that requires the use of conditional distributions. The latter has been much favoured within the Bayesian community, partly because many basic non-spatial statistical models have tractable conditional parameter distributions. A statistical package designed for Gibbs Sampling such models has been developed (BUGS). As yet, no general purpose package for MH algorithms has been made available. In what follows, the use of MH algorithms for putative hazard assessment and general clustering will be discussed. The issues discussed here are found in greater detail in Lawson (1995, 1998a, 1999) and Lawson *et al.* (1996).

Disease Mapping and Risk Assessment for Public Health. Edited by A.B. Lawson *et al.*
© 1999 John Wiley & Sons Ltd.

9.2 DEFINITIONS

Within a study region, denoted as W, a number (m) of case addresses of the disease of interest are found. Denote these locations as $\{\boldsymbol{x}_i\}$, $i = 1, \ldots, m$. Often, the case-event addresses are not available and only the total count of cases within some small area are given. These small areas or *tracts* can be census enumeration districts, postal regions (sectors/areas) or other administrative areas. For this case, define the count of cases of disease within the ith tract in the study region as $\boldsymbol{n}_i$, $i = 1, \ldots, p$.

9.2.1 Case-event data

Usually the basic model for case-event data is derived from the following assumptions:

(i) individuals within the study populations behave independently with respect to disease propensity, after allowance is made for observed or unobserved confounding variables;
(ii) the underlying 'at risk' population, from which the cases arise, has a continuous spatial distribution;
(iii) the case events are unique, in that they occur as single spatially separate events.

Assumption (i) allows the events to be modelled via a likelihood approach, which is valid conditional on the outcomes of confounder variables. Furthermore, assumption (ii), if valid, allows the likelihood to be constructed with a background continuous modulating function, $\{g(\boldsymbol{x})\}$, representing the population 'at risk' from the disease. The overall intensity of the case events is defined as:

$$\lambda(\boldsymbol{x} \mid \theta) = \rho \cdot g(\boldsymbol{x}) \cdot f(\boldsymbol{x}; \theta), \tag{9.1}$$

where ρ is a constant overall rate for the disease, θ is a set of parameters, and $f(\cdot)$ represents a function of confounder variables as well as location. If these confounders are not included in the background $g(\boldsymbol{x})$ function specification, then they can be included in $f(\cdot)$ as regression design variables. The confounder variables can be widely defined, however. For example, a number of random effects could be included as well as observed covariates, as could functions of other locations. It can usually be assumed that the cases are governed by a heterogeneous Poisson process (see, for example, Diggle, 1993) with first-order intensity (9.1). Then, the likelihood associated with this is given by:

$$L = \frac{1}{m!} \prod_{i=1}^{m} [\lambda(\boldsymbol{x}_i|\theta)] \cdot \exp\left\{ - \int_W \lambda(\boldsymbol{u}|\theta) d\boldsymbol{u} \right\}. \tag{9.2}$$

By defining the parametric form of $f(\cdot)$, a variety of models can be derived. In the case of disease mapping, where only the background is to be removed without further model assumptions, a reasonable approach to intensity parameterisation is $\lambda(\boldsymbol{x}) = \rho \cdot g(\boldsymbol{x}) \cdot f(\boldsymbol{x})$. In this case, the $f(\boldsymbol{x})$ function acts as a relative risk function. It turns out that this formulation can also be extended to include the modelling of clusters of disease, as will be noted in a later section.

9.2.2 Count-event data

In the case of observed counts of disease within tracts, given the above Poisson process assumptions, it can be assumed that the counts are Poisson distributed with, for each tract, a different expectation: $\int_{W_i} \lambda(\boldsymbol{u}|\theta)\mathrm{d}\boldsymbol{u}$, where W_i denotes the area of the ith tract. The log likelihood based on a Poisson distribution, bar a constant only depending on the data, is given by:

$$l = \sum_{i=1}^{m} \left\{ \boldsymbol{n}_i \log \int_{W_i} \lambda(\boldsymbol{u}|\theta)\mathrm{d}\boldsymbol{u} - \int_{W_i} \lambda(\boldsymbol{u}|\theta)\mathrm{d}\boldsymbol{u} \right\} \tag{9.3}$$

Often, a parameterisation in (9.3) is assumed where, as in the case-event example, the intensity is defined as a simple multiplicative function of the background $g(\boldsymbol{x})$. An assumption is often made at this point that the integration over the ith tract area leads to a constant term ($\lambda(\boldsymbol{x}\,|\,\theta) = \lambda_i$ for all $\boldsymbol{x}$ within W_i), which is not spatially dependent, i.e. any conditioning on $\int_W \lambda(\boldsymbol{u}\,|\,\theta)\mathrm{d}\boldsymbol{u}$, the total integral over the study region, is disregarded. This assumption leads to considerable simplifications, but at a cost. Often, neither the spatial nature of the integral, nor the fact that any assumption of constancy must include the tract area within the integral approximation, is considered. The effect of such an approximation *should* be considered in any application example, but is seldom found in the existing literature (Diggle, 1993).

9.3 THE ANALYSIS OF HEALTH RISK RELATED TO POLLUTION SOURCES

The assessment of the impact of sources of pollution on the health status of communities is of considerable academic and public concern. The incidence of many respiratory, skin and genetic diseases is thought to be related to environmental pollution, and hence any localised source of such pollution could give rise to changes in the incidence of such diseases in the adjoining community.

In recent years, there has been growing interest in the development of statistical methods useful in the detection of patterns of health events associated with pollution sources. In this review, we consider the statistical methodology for the assessment of putative health effects of sources of air pollution or ionising radiation. Here, we consider the role of MCMC in modelling such problems and concentrate primarily on the analysis of observed point patterns of events rather than specific features of a particular disease or outcome.

There are two basic components in the model described by (9.1): $g(\boldsymbol{x})$ and $f(\boldsymbol{x};\theta)$. The first of these represents the background 'at risk' population and some consideration must be given to how this is accommodated or estimated within the analysis. Usually, an external reference distribution is used for the purpose of estimation of $g(\boldsymbol{x})$. This can take the form of a set standardised disease rates for the study area (which may be available *only* at an aggregate scale above the case-event scale, e.g. in census tracts), or the event locations of a 'control' disease. This disease should be little affected by the phenomenon of interest. For example, the incidence of respiratory disease around a putative air pollution source may require the use of control diseases such as cardiovascular disease or lower body cancers.

The function $f(\boldsymbol{x}; \theta)$ can be specified to include dependence on observed explanatory covariates. These can include measures relating to the source itself, such as functions of distance and direction, and also other measured covariates thought to provide explanation of the disease distribution. Pollution measurements made at fixed sample sties can be interpolated (with associated error) to the case-event locations as covariates.

It is also possible that population or environmental heterogeneity may be unobserved in the data set. This could be either because the population background hazard is not directly available or because the disease displays a tendency to cluster (perhaps due to *unmeasured* covariates). The heterogeneity could be spatially correlated or lack correlation, in which case it could be regarded as a type of 'overdispersion'. One can include such unobserved heterogeneity within the framework of conventional models as a random effect. In Part IV of this volume, a review of modelling approaches for putative sources of hazard is given.

9.4 THE ANALYSIS OF NON-FOCUSED DISEASE CLUSTERING

The analysis of disease clustering has generated considerable interest in the area of public health surveillance. Since the 1980s there has been increased interest in, and concerns about, adverse environmental effects on the health status of populations. For example, concerns about the influence of nuclear power installations on the health of surrounding populations have given rise to the development of methods that seek to evaluate clusters of disease. These clusters are regarded as representing local adverse health risk conditions, possibly ascribable to environmental causes. However, it is also true that for many diseases the geographical incidence of disease will naturally display clustering at some spatial scale, even after the 'at risk' population effects are taken into account. The reasons for such clustering of disease are various. First, it is possible that for some *apparently* non-infectious diseases there may be a viral agent, which could induce clustering. This has been hypothesised for childhood leukaemia (see, for example, Kinlen, 1995). Secondly, other common but unobserved factors/variables could lead to observed clustering in maps. For example, localised pollution sources could produce elevated incidence of disease (e.g. road junctions could yield high carbon monoxide levels and hence elevated respiratory disease incidence). Alternatively, the common treatment of diseases can lead to clustering of disease side effects. The prescription of a drug by a medical practice could lead to elevated incidence of disease within that practice area.

9.4.1 Definition of clusters and clustering

A wide variety of definitions can be put forward for clusters and clustering. Nonparametric definitions exist, e.g. Knox (1989): 'a geographically bounded group of occurrences of sufficient size and concentration to be unlikely to have occurred by chance'. This definition is often referred to as *hot spot* clustering. Here, we adopt a parametric definition:

- the study region has a prespecified cluster structure.

This assumption allows the formulations of likelihood models for the observed events, which is not available in a non-parametric formulation.

9.4.2 Parametric modelling issues

It is possible to formulate specific clustering models for the case-event and tract-count situation. If it is assumed that the intensity of case events, at location $\boldsymbol{x}$, is $\lambda(\boldsymbol{x} \mid \theta)$, then by specifying a dependence in this intensity on the locations of cluster centres, it is possible to proceed. For example:

$$\lambda(\boldsymbol{x} \mid \theta) = \rho \cdot g(\boldsymbol{x}) \cdot m\left\{ \sum_{j=1}^{k} h_1(\boldsymbol{x} - \boldsymbol{y}_j) \right\} \tag{9.4}$$

describes the intensity of events around k centres located at $\{\boldsymbol{y}_j\}$. Here, the function $f(\cdot)$ is replaced by a link function $m\{\cdot\}$ and sum of cluster functions. The distribution of events around a centre is defined by a cluster distribution function $h_1(\cdot)$. Note that it is possible to define different cluster distribution functions for each cluster (e.g. $h_{1_j}(\cdot)$), where different cluster variance parameters could describe each cluster separately. In some applications it may be advantageous to allow such variation in cluster form. In addition, variation in cluster form can be described by a cluster variance that has a spatial prior distribution (see, for example, Lawson, 1995). For simplicity, this has not been pursued in this example.

Conditional on the cluster centres, the events can be assumed to be governed by a heterogeneous Poisson process, and hence a likelihood can be specified. Because the number (k) and the locations of centres are unknown, then, with a suitable prior distribution specified for these components, it is possible to formulate this problem as a Bayesian posterior sampling problem with a mixture of components of unknown number. This type of problem is well suited to reversible jump MCMC sampling (Green, 1995). The approach can be extended to count data straightforwardly, as

$$\mathrm{E}(n_i) = \rho \int_{W_i} g(\boldsymbol{x}) \cdot m\left\{ \sum_{j=1}^{k} h_1(\boldsymbol{x} - \boldsymbol{y}_j) \right\} \mathrm{d}\boldsymbol{x} \tag{9.5}$$

under the equivalent Poisson distribution model, where W_i here signifies the integration is over the ith tract area, and n_i is the count in the ith tract.

9.5 A GENERAL MODEL FORMULATION FOR SPECIFIC CLUSTERING

Both in the analysis of putative health hazards (focused clustering) and in the analysis of non-focused clustering there is a common model framework that can be specified. Define a general intensity for events as:

$$\lambda(\boldsymbol{x} \mid \theta) = \rho \cdot g(\boldsymbol{x}) \cdot m(\boldsymbol{x}; \underline{\boldsymbol{y}}; \theta), \tag{9.6}$$

where $\underline{\boldsymbol{y}}$ denotes the vector of cluster centres, and

$$m(\boldsymbol{x}; \underline{\boldsymbol{y}}; \theta) = m_1(F(\boldsymbol{x})\alpha) \cdot m_2(\zeta(\boldsymbol{x})) \cdot m_3(\|\boldsymbol{x} - \underline{\boldsymbol{y}}\|). \tag{9.7}$$

This intensity encompasses trend or observed covariate effects (m_1), unobserved heterogeneity (m_2), and clustering effects (m_3), respectively. Note that m_1 could include variables measured to fixed focii such as putative sources of hazard, because the locations of these focii are fixed. Here, $F(\boldsymbol{x})\alpha$ is a linear predictor with a covariate design matrix $F(\boldsymbol{x})$ and a parameter vector α, $\zeta(\boldsymbol{x})$ is a spatially dependent random heterogeneity effect. Note that if $m_1 = m_2 = 1$, then a simple clustering model is derived, whereas if $m_3 = 1$, then, if our focus is some subset of putative source explanatory variables defined in the matrix F, we derive a model with unobserved background clustering suitable for use with, for example, childhood leukaemia around putative sources.

9.6 MARKOV CHAIN MONTE CARLO METHODS

The general approach to modelling putative hazards and to general clustering specified above, leads straightforwardly to a general approach to the use of Markov chain Monte Carlo (MCMC) methods in these applications. MCMC methods consist of a range of algorithms designed for the iterative simulation of joint posterior distributions found in Bayesian models (Gilks *et al.*, 1996a). As a general set of methods, they can be applied widely and are often the only methods available for posterior sampling in complex modelling found in spatial applications. The basic feature of the methods is that, from a current set of parameter values, new values are proposed and a comparison of the posterior probability for the new, as opposed to the current, values is made. The proposed values are generated from distributions and the new values are accepted based on a given probability criterion. If new values are accepted, then these become the current values. Two basic algorithms are now widely used: Gibbs sampling and the MH algorithm. Gibbs Sampling is based on conditional distributions and is a special case of the MH algorithm. When the number of parameters is unknown and must be sampled, special methods can be employed, such as reversible jump sampling (RJMCMC) (see, for example, Green, 1995).

Assume a general intensity for events as above in (9.6) and (9.7). For models where only the m_1 and m_2 components are present, conventional MCMC samplers can be employed. For example, with m_1 only, we have regression components that can be sampled from a joint posterior distribution (given suitable prior distributions specified for the components of α). In general, a MH algorithm can be employed as in Lawson (1995). Where random effects such as unobserved heterogeneity are felt to be present, m_2 must also be included. The inclusion of such effects can be sampled also via general MH algorithms akin to the formulations of Besag *et al.* (1991), who used special Gibbs sampling methods (see also Lawson *et al.*, 1996). If it is desired to estimate the locations of clusters by inclusion of m_3, then it is further required that the vector $\{\boldsymbol{y}_j\}$ be sampled from the joint posterior distribution with all other sampled parameters. Prior distributions for the cluster centre vector are available from the class of inhibition processes or, more simply, from a uniform distribution on the study region. The sampling algorithm required for this purpose is somewhat more complex than in other formulations. This is due to the inclusion of a mixture sum component in the intensity, where the number of components in the mixture is unknown. Reversible jump sampling (Green, 1995) can be employed for this case, using spatial birth–death diffusion transitions in a general sampler algorithm. In essence, this sampler explores the joint distribution of the k component centres and their locations $\{\boldsymbol{y}_j\}$, within a more conventional sampling algorithm for other parameters. Details of this approach are given in Lawson (1995, 1997). In the following section we have given an example of the use of such algorithms in a general clustering context.

9.7 PUTATIVE HAZARD EXAMPLE

In previous work in this area, two basic approaches have been adopted for this problem (for a recent comprehensive review, see Lawson and Waller, 1996). First, the background 'at risk' population has been estimated, and subsequent inference has been made conditional on this estimate. Secondly, for case-event data, where exact addresses of cases are known, and a 'control' disease is also employed, it is possible to 'condition out' the background from the analysis.

In the Bayesian method proposed here, we estimate the 'at risk' background, but incorporate the estimation in an iterative algorithm, and hence both provide an estimate of the background and an exploration of the posterior distribution surfaces for both the background estimation and putative source-related parameters.

In what follows, we examine case-event data only, but the methods can be extended to deal with other data formats. We consider a set of death certificate addresses, which are thought to be related to a single putative pollution source. Define a study region (W) within which the case disease locations, $\{\boldsymbol{x}_i\} : i = 1, \ldots, m$, are observed. The intensity of cases within underlying population is considered to be described by a two-component model such that

$$\lambda(\boldsymbol{x}\,|\,\theta) = \rho \cdot g(\boldsymbol{x}) \cdot f(\|\boldsymbol{x} - \boldsymbol{x}_0\|; \theta), \tag{9.8}$$

where $\boldsymbol{x}_0$ is a pollution source location, $g(\cdot)$ is a background 'at-risk' function, and $f(\cdot)$ is a parametrised function of $\boldsymbol{x}$ and $\boldsymbol{x}_0$.

Furthermore, we regard the cases as being independently distributed as a heterogeneous Poisson process with intensity given by (9.8). These are the standard model assumptions made in previous studies (see, for example, Diggle, 1990; Diggle and Rowlingson, 1994; Lawson and Williams, 1994).

Given (9.8), the likelihood for the data, conditional on m events in W, is derived from the conditional probability of an event at $\boldsymbol{x}$:

$$P(\boldsymbol{x}\,|\,\theta) = \frac{\lambda(\boldsymbol{x}\,|\,\theta)}{\int_W \lambda(\boldsymbol{u}\,|\,\theta)\,\mathrm{d}\boldsymbol{u}}$$

and the resulting likelihood is:

$$L(\boldsymbol{x}\,|\,\theta) = \prod_{i=1}^{m} P(\boldsymbol{x}_i\,|\,\theta), \tag{9.9}$$

where $\lambda(\boldsymbol{x}|\theta)$ represents the dependence of intensity on a parameter vector θ.

The dependence of $\lambda(\boldsymbol{x} \mid \theta)$ on $g(\boldsymbol{x})$ has led to a variety of proposals for the estimation of $g(\boldsymbol{x})$, and also the 'conditioning-out' of $g(\boldsymbol{x})$ from the analysis. The original proposals suggested that $g(\boldsymbol{x})$ should be estimated non-parametrically from a 'control disease' (Diggle, 1989, 1990; Lawson, 1989; Lawson and Williams, 1994). The 'control' disease was to be chosen to be matched closely to the 'at-risk' structure of the case disease, but should be unaffected by the pollution hazard of interest. A more conventional alternative to the use of a control disease is the use of standardisation (Inskip *et al.*, 1983) to provide small-area 'expected' rates for the case disease (Lawson and Williams, 1994). Usually, these 'expected rates' are only available in large spatial units (e.g. census tracts), and hence there is an aggregation difference between the case and expected rates. In the original approach to incorporation of $g(\boldsymbol{x})$, a two-dimensional kernel-smoothed estimate $\hat{g}(\boldsymbol{x})$, was substituted directly into (9.8), and all subsequent inference concerning $f(\cdot)$ was made conditional on $\hat{g}(\boldsymbol{x})$. This special form of profile likelihood ignored the variation inherent in the estimation of $\hat{g}(\boldsymbol{x})$.

In response to this problem, Diggle and Rowlingson (1994) proposed a conditional logistic model, which directly modelled the case locations and locations of a control disease as a bivariate point process where the probability of an event, within the joint realisation, being a case was taken conditionally on the observed locations. In this approach, $g(\boldsymbol{x})$ is 'conditioned out' of the model and need not be estimated. The attraction of this approach lies also in its lack of dependence on the definition of study window boundaries. However, there are a number of disadvantages to this conditioning method. First of all, the method relies on the availability of a point event map of a control disease, which may not be radially available. Secondly, matching of a control disease to a case disease can be very difficult. For example, some control diseases can be case diseases when targeted at specific air pollutants. A particular case of this is lower body cancers that can control for respiratory cancer, but not for the example of nickel pollution, which is known to target the kidney. Equally, respiratory cancer is often a case disease, but has been used as a control disease for larynx cancer (see Lawson *et al.*, 1996). In addition, it may be required that $g(\boldsymbol{x})$ be estimated so that the underlying risk surface is available.

In the approach advocated here, we do not restrict the possible data sources for the estimation of $g(\boldsymbol{x})$, while we incorporate $g(\boldsymbol{x})$ estimation in a Bayesian iterative algorithm. Both standardised rates or control diseases could be used to estimate $\hat{g}(\boldsymbol{x})$, and the algorithm provides an exploration of the posterior marginal distribution of possible $\hat{g}(\boldsymbol{x})$ surfaces.

9.7.1 Model development and the MCMC algorithm

In the original approach to this problem where $\hat{g}(\boldsymbol{x})$ was directly substituted in (9.8), the conditional likelihood can be written:

$$L(\boldsymbol{x}|\theta) = \frac{\prod_{i=1}^{m} \{\hat{g}(\boldsymbol{x}_i) \cdot f(\|\boldsymbol{x}_i - \boldsymbol{x}_0\|; \theta)\}}{\{\int_W \hat{g}(\boldsymbol{u}) \cdot f(\|\boldsymbol{u} - \boldsymbol{x}_0\|; \theta) \, \mathrm{d}\boldsymbol{u}\}^m} \tag{9.10}$$

This likelihood arises from the conditional distribution of the m events within W under the Poisson process assumption. In this likelihood, ρ does not appear and hence does not require estimation. With $\hat{g}(\boldsymbol{x})$ substituted into the intensity, the only requirement is to provide estimates of the θ vector. In what follows, we employ this conditional likelihood, because it gives a parsimonious parameterisation: a constant rate parameter is of little interest here as we focus on the spatial structure *within* the study region.

To include the estimation of $\hat{g}(\boldsymbol{x})$ within this likelihood formulation, we first assume that $\hat{g}(\boldsymbol{x})$ can be estimated by a two-dimensional kernel smoother representation and that this kernel smoother depends on smoothing constants h_x and h_y in the two directions. In what follows, we assume a common parameter h for both directions ($h = h_x = h_y$), because we assume there is no prior evidence of differences in density related to direction. We denote this dependence as $\hat{g}_h(\boldsymbol{x})$. We also assume that h can be regarded as the focus of estimation and not the function $\hat{g}_h(\cdot)$.

A Metropolis–Hastings (MH) MCMC algorithm

If we regard h as well as parameters in $f(\cdot)$ as conventional parameters, then within a Bayesian model we can ascribe prior distributions for the parameters, and proceed with exploration of the joint posterior distribution of these parameters. The novel feature of this approach is that we will include a prior structure for the smoothing constant h, and thereby avoid the singularity problems that arise when estimating the parameters of kernel smoothers (see, for example, Härdle, 1991 for a discussion of this problem). An alternative approach to the problem of estimation of the smoothing constant could be to employ separate likelihood ratios for smoothing constants and other parameters, where the smoothing ratios depend on 'leave one out' likelihoods as used in likelihood cross-validation. However, this alternation between different likelihood ratios could lead to poor mixing in iterative simulation.

The above formulation suggests the use of an iterative simulation method for the posterior exploration of parameter vectors h and θ. A method that can be used to explore posterior distributions for a wide a range of models is provided by the MH method (see, for example, Gilks *et al.*, 1996a; Tanner, 1996; or Gelman *et al.*, 1995, for an introduction to this method). One advantage of this method over the Gibbs Sampler is that it does not require the derivation of conditional distributions for parameters. In essence, the MH method directly evaluates the ratios of posterior distributions for different parameter combinations. These combinations are generated from a proposal distribution. Denote the proposal vector as $\boldsymbol{v}'$, and the current vector as $\boldsymbol{v}_c$. Define the posterior ratio criteria for parameters $\boldsymbol{v} : \{h, \theta\}$ as:

$$R(\boldsymbol{v}', \boldsymbol{v}_c) = \prod_{i=1}^{m} \left\{ \frac{\lambda(\boldsymbol{x}_i | \boldsymbol{v}')}{\lambda(\boldsymbol{x}_i | \boldsymbol{v}_c)} \right\} \cdot \left\{ \frac{\int_W \lambda(\boldsymbol{u} | \boldsymbol{v}_c)\, d\boldsymbol{u}}{\int_W \lambda(\boldsymbol{u} | \boldsymbol{v}')\, d\boldsymbol{u}} \right\}^m \cdot \frac{p(\boldsymbol{v}')}{p(\boldsymbol{v}_c)},$$

where $p(\boldsymbol{v})$ is the joint prior structure for $\boldsymbol{v}$. An arbitrary transition proposal function, $q(\boldsymbol{v}_c, \boldsymbol{v}')$ is also evaluated via an acceptance criterion:

$$\alpha(\boldsymbol{v}_c, \boldsymbol{v}') = \begin{cases} \min\{R(\boldsymbol{v}', \boldsymbol{v}_c) \frac{q(\boldsymbol{v}', \boldsymbol{v}_c)}{q(\boldsymbol{v}_c, \boldsymbol{v}')}, 1\} & \text{for } R(\boldsymbol{v}', \boldsymbol{v}_c) q(\boldsymbol{v}', \boldsymbol{v}_c) > 0 \\ 1 & \text{for } R(\boldsymbol{v}', \boldsymbol{v}_c) q(\boldsymbol{v}', \boldsymbol{v}_c) = 0 \end{cases}.$$

Here, the proposal $\boldsymbol{v}'$ is accepted with probability $\alpha(\boldsymbol{v}_c, \boldsymbol{v}')$. Otherwise, $\boldsymbol{v}_c$ is kept as the current parameter vector.

Specification of $f(\cdot)$ and prior distributions

To define the appropriate prior distributions for the parameters used, it is important to specify a parametric form for the $f(\cdot)$. To model the distance-related effects of proximity to a putative source, we define first the distance from source: $d_i = \|\boldsymbol{x}_i - \boldsymbol{x}_0\|$. We regard d_i as the main explanatory variable in our specification on $f(\cdot)$. Additional evidence of links with putative sources can be examined by the use of functions of the full polar coordinates of cases from the source location (see, for example, Lawson, 1993b; Lawson and Williams, 1994). For demonstration purposes here we confine our analysis to a simple function of distance. We define:

$$f(d_i) = 1 + \theta_1 \exp\{-\theta_2 d_i\} \tag{9.11}$$

as a suitable function for modelling the relation to a source. Following Besag *et al.* (1991), we do not wish to make strong assumptions concerning the prior structure of the $\{\theta_i\}$. Hence, we assume the following improper exponential joint prior distribution for these parameters:

$$p_1(\theta_1, \theta_2) \propto \exp\left\{-0.5\epsilon\left(\frac{1}{\theta_1} + \frac{1}{\theta_2}\right)\right\}.$$

We have set the value of ϵ equal to 0.01, after investigation of a variety of potential values. The value chosen provides a 'close-to' uniform distribution above ϵ, but penalises values close to zero, zero being an absorbing state of the chain used in the MCMC algorithm, i.e. close to zero the distribution has an asymptote, while across a range of positive values the distribution is relatively uniform. The alternative choice of using inverse Gamma distributions involves the specification of extra hyperparameters and is less parsimoneous.

In the case of the smoothing parameter h, the estimate of $g(\cdot)$ is provided by a standard kernel smoothing method

$$\hat{g}(\boldsymbol{x}) = \frac{1}{nh^2} \sum_{i=1}^{n} K\left(\frac{\boldsymbol{x} - \boldsymbol{x}_{ci}}{h}\right),$$

where $\{\boldsymbol{x}_{ci}\}$ are the locations of a control disease realisation within the study region, and $K(\boldsymbol{u})$ is a two-dimensional kernel function. We have assumed a Gaussian kernel in our example, denoted by

$$K(\boldsymbol{x}) = \frac{1}{2\pi} \exp\{-0.5\boldsymbol{x}'\boldsymbol{x}\}.$$

The exact specification of the kernel form is not as important as the estimation of the smoothing parameter h. For the prior distribution of h, we specify an inverse Gamma prior with parameters $\{\alpha, \beta\}$, and as these parameters are also strictly positive, these have an improper exponential hyperprior, i.e.

$$p_2(h) = \frac{\alpha^\beta}{\Gamma(\beta)} h^{-\beta-1} \exp\left(-\frac{\alpha}{h}\right),$$

$$p_3(\alpha, \beta) \propto \exp\left\{-0.5\epsilon\left(\frac{1}{\alpha} + \frac{1}{\beta}\right)\right\}.$$

It is important here to assume a prior for h that can penalise against extreme values of h, because these could be favoured in a smoothing operation, and hence the assumption of a uniform or 'close-to' uniform distribution for $p_2(\cdot)$ would not be appropriate.

Proposal distributions

A variety of possible proposal distributions could be used for h and θ. In our application, we do not have strong prior beliefs about the structure of $q(\mathbf{v}', \mathbf{v}_c)$ for the strictly positive parameters $\{h, \theta_1, \theta_2\}$, and we employ uniform proposals on the range $(0.5 * \mathbf{v}_c, 1.5 * \mathbf{v}_c)$ for these in the MH algorithm.

9.7.2 Data example

The incidence of respiratory cancer (International Classification of Disease (ICD) list A, 162) in Armadale, Scotland has been the subject of study since the retrospective analysis of the 'Armadale epidemic' (Lloyd, 1982; Williams and Lloyd, 1988). Armadale, a small industrial town in central Scotland (see Figure 9.1 for location), during the period 1968–1974, experienced a large increase in respiratory cancer deaths. Forty-nine cases were observed within a six-year period.

As a result of this apparent 'epidemic', a range of studies have been undertaken. It has been hypothesised that a centrally located steel foundry in the town may have been responsible for the adverse health effects experienced in the town. The foundry was regarded as a putative health hazard and a number of studies were executed to assess the effect of the foundry on the surrounding population. Most recently, an analysis of the spatial distribution of death certificate addresses has been made (Lawson and Williams 1994; Knox, 1989). In that study, the model considered in Section 9.7.1 was applied, with $g(\boldsymbol{x})$ estimated from the spatial distribution of a control disease (coronary

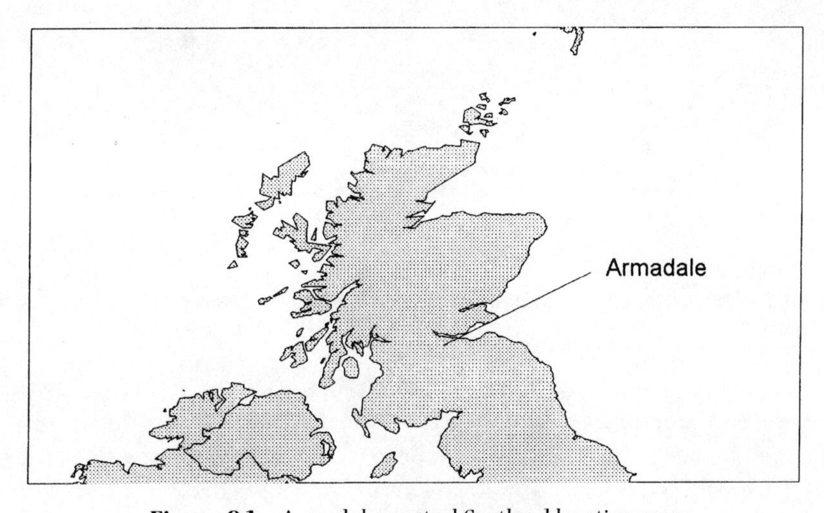

Figure 9.1 Armadale: central Scotland location map

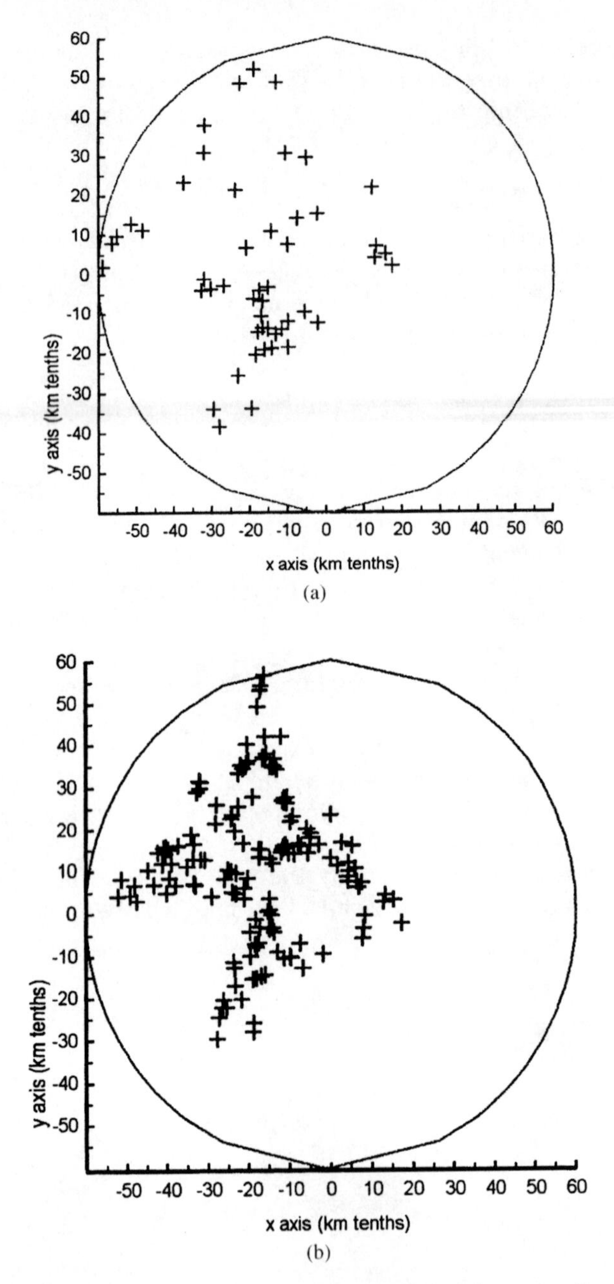

Figure 9.2 (a) Case-event realisation: Armadale. (b) Control disease realisation: Armadale

heart disease (CHD): ICD list A, 410–414) and also from standardised rates for respiratory cancer in enumeration districts. In addition, a hybrid model requiring no interpolation of population background was also proposed. the reader is referred to Lawson and Williams (1994) for further background and discussion of the dataset. Here, we propose to analyse the 49 respiratory cancer cases (Figure 9.2(a)) with a control disease of 153

CHD death certificates (Figure 9.2(b)) for the same period. The foundry is the centre of the circular study area in Figure 9.2(a) and 9.2(b)

The fitted model and results

For this example, we assume a simple distance-based model for $f(\cdot)$. The model specified in (9.11) has been used. Here, we have applied the above algorithm with the specified prior distributions of h and θ. Convergence of the sampler has been assessed by a variety of diagnostics, including summary measures on individual chains with multiple start values and by inspection of the joint log posterior surface (see, for example, Gelman

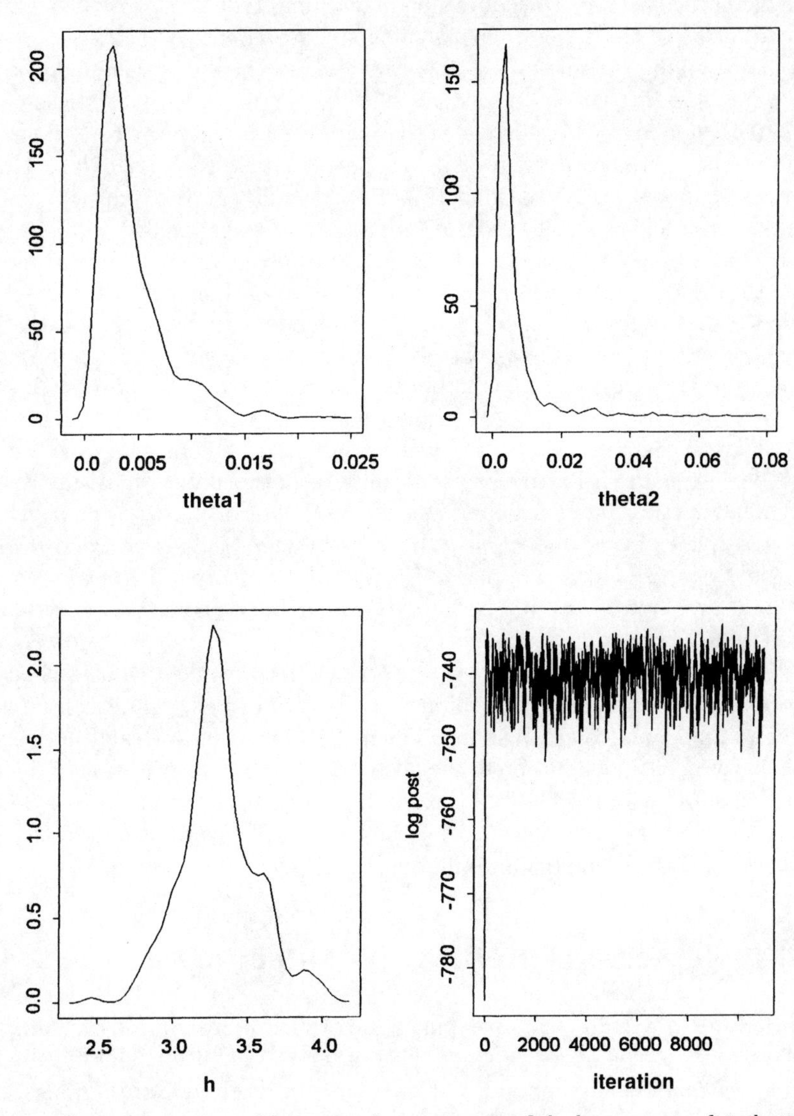

Figure 9.3 Plots of the marginal densities of parameters and the log posterior distribution versus iteration number

Table 9.1 Posterior analysis of model parameters

Parameter	mode	Central 95% interval	Marginal density smoother
θ_1	0.002	(0.001, 0.0108)	0.002
θ_2	0.005	(0.0016, 0.0505)	0.0098
h	3.287	(2.83, 4.01)	0.219

et al., 1995). Figure 9.3 represents the marginal distributions of these parameters, kernel-smoothed using the variance rule-of-thumb (Silverman, 1986).

The modal value of h is 3.287 and is associated with a marked peak in the marginal posterior distribution. This modal value is higher than the value *separately* estimated from CHD only (2.712) (Lawson and Williams, 1994). The main reason for this phenomenon, that of greater smoothing, is related to the incorporation of the conditioning on θ estimation and hence the use of case data, indirectly, in the estimation of $g(\boldsymbol{x})$. The parameters θ_1 and θ_2 appear to be reasonably well estimated. All the θ parameters display single well-defined density peaks and the joint posterior marginal distributions of scale and distance parameters versus smoothing (h) also show clear single peaks, although, for brevity, they are not displayed here. The application of this approach has produced some evidence for a slight radial decline in the incidence with distance from the source (mode of $\theta_2 = 0.005$) (Table 9.1). Of course, a fuller study of a variety of exposure models would be required to assess overall whether the source displays a relation with the surrounding area health status. As noted earlier and in Chapter 17 in this volume, models with only distance decline effects should be augmented by inclusion of directional effects to model exposure fully. In a previous analysis of this data, a marked directional effect was noted but a distance decline was not found. This disparity could be related to the use of a simple exposure model or to the greater smoothing. Here, we use a simple model for the purposes of illustration only. Formal tests for the parameters could also be employed, but have not been pursued here as our focus is on the posterior marginal distribution of the smoothing parameter.

In conclusion, the Bayesian method here proposed has a wide area of application in the analysis of non-parametric estimation of a background effect and can be applied when different data sources are available. The method provides a straightforward estimate of background risk in addition to the posterior marginal distribution of the kernel smoothing constant. The possibility of extending this approach to the smoothing of covariate fields (such as pollution fields), is also straightforward. Where a structured random effect field is to be included, similar approaches can be adopted.

9.8 NON-FOCUSED CLUSTERING EXAMPLE

Here, we briefly give an example of the application of the general model, including m_2, a random effect term, and m_3, a non-focused cluster term. We will consider the clustering tendency of a realisation of lymphoma and leukaemia in children from Humberside, UK, previously analysed by Cuzick and Edwards (1990) and Diggle and Chetwynd (1991). The data consist of the address locations of cases of lymphoma and leukaemia for a fixed

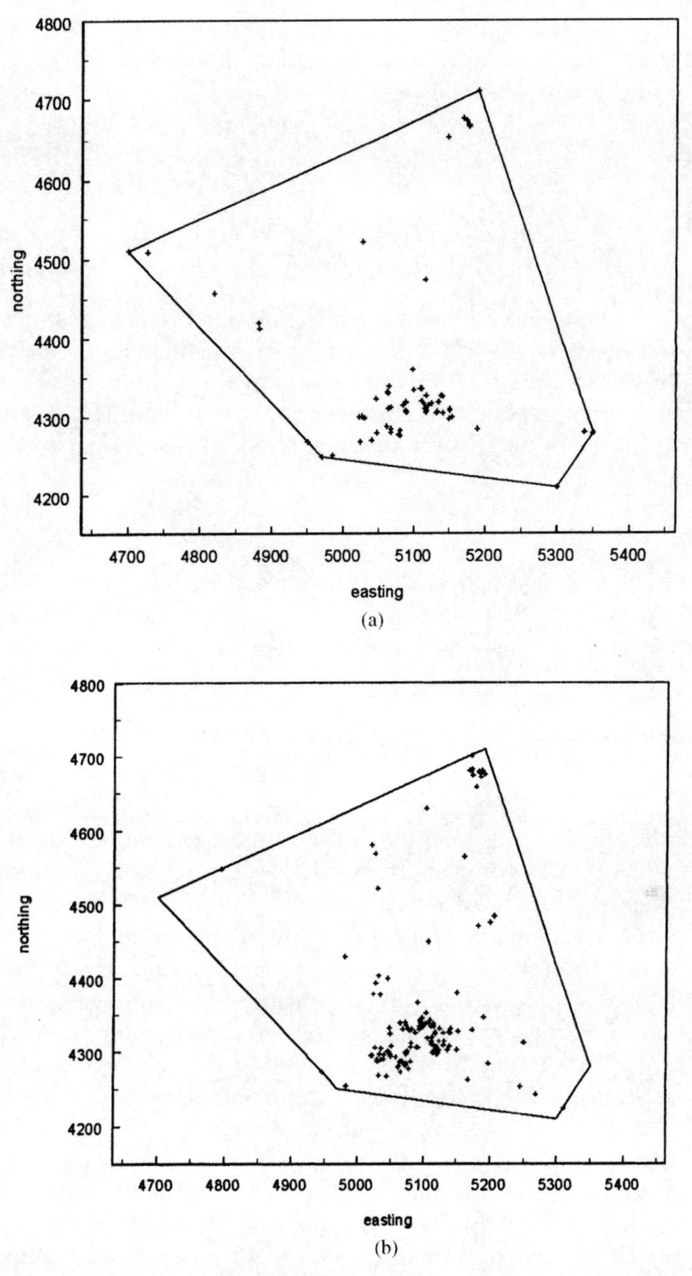

Figure 9.4 (a) Case-event locations: Humberside lymphoma and leukaemia (1974–86). (b) Control event realisation from birth register

period (1974–86) for the Humberside region of England. In addition, Cuzick and Edwards (1990) provide a control sample of birth locations from the birth register. The distribution of these data is given in Figures 9.4(a) and 9.4(b)

In the following analysis, we assume, as in previous analyses, that there are no trend effects and that:

$$\lambda(\boldsymbol{x}_i|\theta) = \rho \cdot g(\boldsymbol{x}_i) \cdot \zeta_i \cdot \left\{1 + \sum_{j=1}^{k} h_1(\boldsymbol{x}_i - \boldsymbol{y}_j)\right\},$$

where $g(\boldsymbol{x})$ is estimated from the control realisation (birth register sample) using non-parameteric density estimation. The smoothing parameter for this estimation (h) is incorporated in the MCMC algorithm and the prior distribution of this parameter is assumed to be inverse exponential. In addition, a random effect parameter (ζ_i) is assumed to describe any unobserved heterogeneity, and is defined as in Besag *et al.* (1991)

$$\zeta_i = \exp\{u_i + v_i\}.$$

The prior distribution for $\{v_i\}$ is standard normal with parameter σ. This can be regarded as a frailty effect that captures individual extra variation. The prior distribution for the correlated heterogeneity $\{u_i\}$ is an intrinsic autoregression with a distance weighting:

$$p_i(u_i|\ldots) \propto \exp\left\{-\sum_{k\in\partial_i} w_{ij}(u_i - u_j)^2\right\}, \tag{9.12}$$

where w_{ij} is a distance weight function, and ∂_i is a defined neighbourhood of the ith point. In the following, we assume $w_{ij} = \exp(-d_{ij})/\{2r\}$, where r is a range parameter. The neighbourhood ∂_i is assumed to be a fixed distance and in the example used here it is taken as half the maximum dimension of the window T. Note that the weights used here mimic the weights used in fully specified covariance priors. The two parameters of these distributions are r and σ. Both are assumed to have inverse exponential hyperpriors with $\epsilon = 0.01$, because we wish to provide a relatively uniform distribution with penalisation at zero. In addition, the prior distribution for the cluster centres and number of centres is Strauss inhibited, with fixed inhibition parameters ($\gamma = 0.95$, $R = 0.5 * \sqrt{|W|}$) and rate ρ_c, which has uniform prior distribution. The Strauss distribution describes the spatial distribution of a point process where the locations are inhibited. The amount of inhibition is controlled by the parameters (γ, R), while the number of centres is controlled by ρ_c. An inhibition prior is used to prohibit multiple response when recovering cluster centres within MCMC algorithms (see, for example, Baddeley and Lieshout, 1993; Lawson, 1996a). The cluster distribution function is assumed to be radial Gaussian:

$$h_1(u) = \frac{1}{\sqrt{2\pi\kappa}} \exp\left\{-\frac{u^2}{2\kappa}\right\}. \tag{9.13}$$

A MH sampler has been used for the conventional parameters (r, σ, h, κ) with uniform proposal distributions as described above. The joint sampling of $[k, \{\boldsymbol{y}_j\}]$ was achieved via a reversible jump MH sampler using spatial birth–death shifting (SBDS) transitions as described in Lawson (1995, 1996a, 1997 and Lawson and Clark (1999)). These transitions consist of randomly choosing a birth–death shift based on the current configuration of centres. A birth leads to the addition of a new centre, a death leads to the deletion of a centre, while a shift is a combination of both transitions. Proposal distributions for the centres are a function of $h_1(\cdot)$ and the current centre configuration, while k has a uniform proposal distribution. Convergence was assessed by the use of a variety of methods. Multiple start points have been used for both centres and conventional parameters and these confirmed the convergence reported here. Further summary measures were employed (Gelman *et al.*, 1995).

Using predictive inference methods, based on Bayesian information criterion (BIC) values, a variety of models were compared. These models formed a restrictive combination of correlated, uncorrelated, and cluster terms. It was found that the best BIC model was that which included only the heterogeneity term ζ_i. The full model results, i.e. for a model with all components included, are shown in Figure 9.5 and 9.6. The posterior sample spatial distributions for the best BIC model for u and v are displayed in Figure 9.7.

These results confirm that there is little support for a positive number of cluster centres (modal posterior marginal number of centres $= 0$). This result is similar to that found in the earlier studies of Cuzick and Edwards and Diggle and Chetwynd. Finally, for the best BIC model, the posterior marginal distributions of $\boldsymbol{u}$ and $\boldsymbol{v}$ are displayed in Figure 9.7. These show relatively small values and also spatial differentiation. Similar behaviour was noted by Besag *et al.* (1991).

Sensitivity to the specification of prior distributions and other model components can be an issue in complex Bayesian modelling such as demonstrated in the above example,

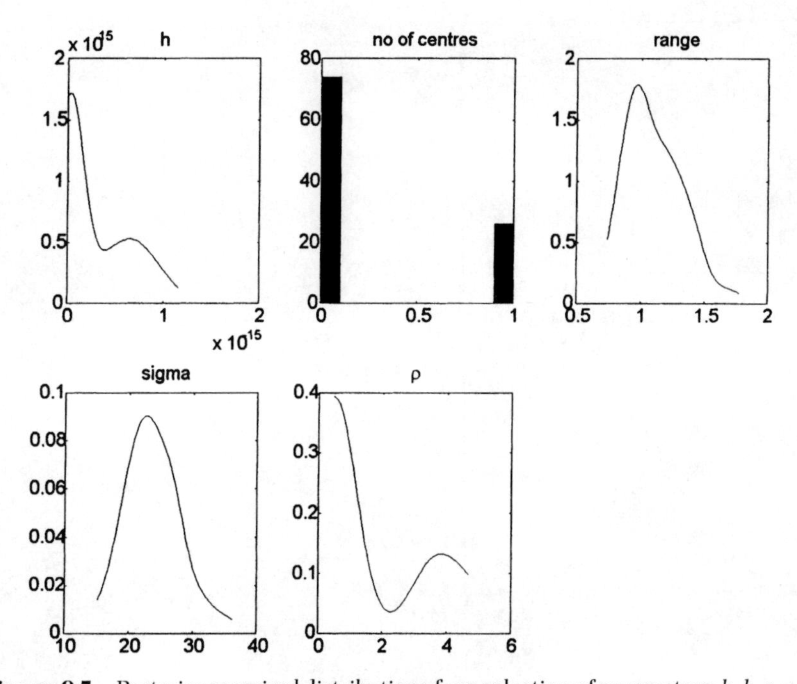

Figure 9.5 Posterior marginal distributions for a selection of parameters: h, k, r, σ, ρ

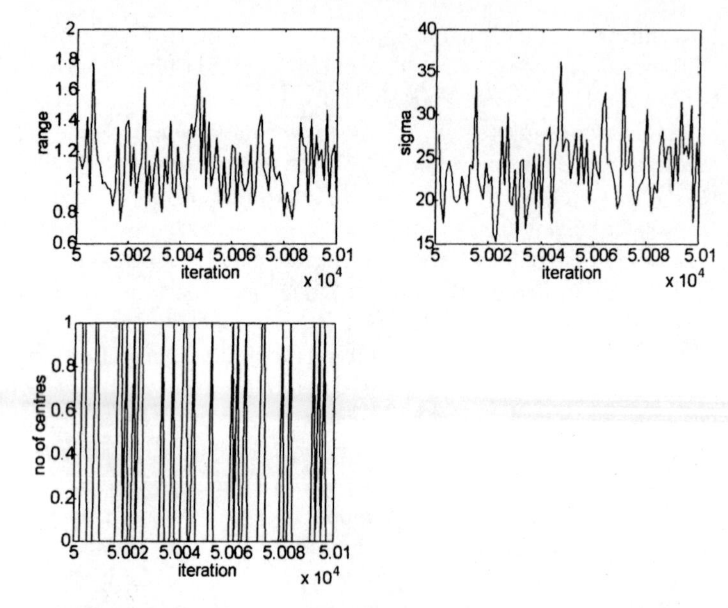

Figure 9.6 Final iterations: parameters versus iteration number

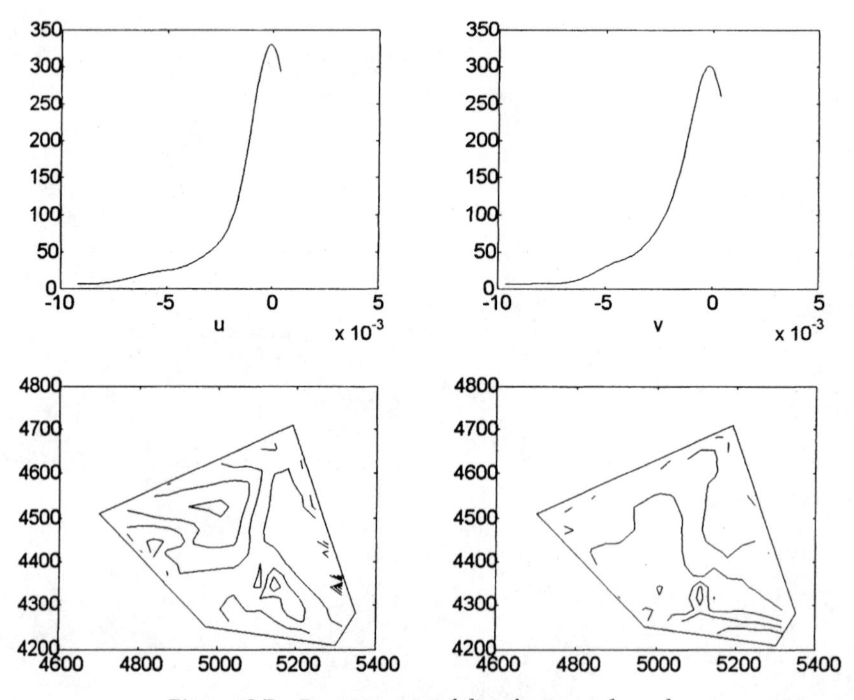

Figure 9.7 Posterior spatial distributions of u and v

and in most cases the use of distributions that are non-informative has been pursued to avoid the possible domination of the likelihood. We have not examined the effects of variation in the prior specification on the posterior samples, and this is an issue that will be addressed in the future. Variation in the likelihood model can also affect the posterior sampling in this analysis. The use of the Diggle and Rowlingson logistic likelihood formulation in this example has produced a modal number of the cluster centres of 1 (Lawson and Clark, 1999). Hence, there is a great need to examine the appropriateness of the model components as well as the prior distributional specification.

9.8.1 Extensions to count cluster modelling

It is possible to extend the basic point event methods to the case where only counts of the case disease are observed within tracts. This application is of considerable importance given the ready availability of such data and level of interest in its analysis.

We assume that the process is a heterogeneous Poisson process governed by $\lambda(\boldsymbol{x}\,|\,\theta)$ and that within the ith tract:

$$\mathrm{E}(n_i) = \int_{W_i} \lambda(\boldsymbol{u}|\theta)\mathrm{d}\boldsymbol{u} \equiv \Lambda(W_i), \tag{9.14}$$

where W_i denotes that ith tract, and n_i the disease count in the tract. Because disjoint regions are independent under the resulting Poisson count model, the $\{n_i\}$ are Poisson distributed with rates $\Lambda(W_i)$. Conditional on N, where $N = \sum_{i=1}^{p} n_i$, the likelihood for p regions is

$$L(\boldsymbol{n}|\theta) = \prod_{i=1}^{p}\left[\frac{\Lambda(W_i)}{\sum_{l=1}^{p}\Lambda(W_l)}\right]^{n_i}. \tag{9.15}$$

Now, in this case, we do not observe the point case events but only know their tract totals. At this point, it is possible to use conventional likelihood-based inference concerning parameters relating to *fixed* foci (such as putative source locations) or covariates, or to include conventional Bayesian methods incorporating prior distributions for random effects. However, for unknown foci locations $\{\boldsymbol{y}\}$ we can use directly the basic point process algorithms and replace the likelihood ratios with those based on (9.15). Note that if expected death variation is known for the ith tract, then the likelihood (9.15) can be written

$$L(\boldsymbol{n}|\theta) = \prod_{i=1}^{p}\left[\frac{\int_{W_i}\lambda(\boldsymbol{u}|\theta)\mathrm{d}\boldsymbol{u}}{\sum_{l=1}^{p}\int_{W_l}\lambda(\boldsymbol{u}|\theta)\mathrm{d}\boldsymbol{u}}\right]^{n_i}. \tag{9.16}$$

Often this can be reduced to

$$\prod_{i=1}^{p}\left[\frac{g_i \cdot \Lambda(W_i)}{\sum_{l=1}^{p} g_l \cdot \Lambda(W_l)}\right]^{n_i},$$

assuming that the background function $g(\boldsymbol{x})$ is constant over each tract. A further reduction to constant tract rate (λ_i) would allow the use of standard GLM software for fixed foci or covariate models. Note that the use of (9.16) requires integration over arbitrary tract regions. Hence, this approach utilises the counts directly in the algorithm.

Another approach extends this count algorithm by exploiting ideas based on data augmentation. Tanner (1996) discusses the different algorithmic approaches to data augmentation. The basic idea behind data augmentation arises from the need to deal with missing data. Augmentation refers to the idea that additional data can be provided within the analysis, which will improve the estimation of relevant parameters. In essence, the additional data required by the augmentation method is generated within an iterative algorithm. The algorithm generates the additional (missing) data at one iteration on the basis of current parameter estimates, and then conditioning on the augmented data set (i.e. the original data plus the additional data), the parameters are re-estimated. This alternation continues until a convergence is reached. One approach to augmentation is simply to consider any missing data as extra parameters within an MCMC algorithm. Hence, the above stages can be added to a conventional MCMC sampler.

In our approach, we use the ideas of data augmentation but implement them within conventional Gibbs or MH steps. In particular, we exploit the idea that censoring of observations leads to missing data and hence can be modelled by iterative augmentation of the missing portion. This can be applied in a variety of ways in point process modelling. For example, the use of an external border (U), which encloses the study region W, allows the iterative simulation of cluster centres outside the observation window. In addition, it could be possible to also simulate *data events* within U or, indeed, to simulate into internal areas of the window where events are censored (holes). In application to count modelling it is possible to regard the point process underlying the counts as a censored event set. In this way, we could conditionally simulate the point events within our count event model. This could lead to many realisations that were of comparable likelihood, to the inherent smoothness of the aggregated data. However, this does allow the reconstruction of the appropriate underlying point process intensity, which is not available when constant region rate models are fitted. In addition, one advantage of this approach is that spatially continuous covariates can be correctly incorporated in the model. In the following example we give a brief description of an algorithm that can be applied quite generally for count event modelling, which uses augmentation of the underlying point process. The approach is described here in the particular context of cluster modelling, but has wide applicability.

Define $\{\boldsymbol{z}_{ij}\}$, $j = 1, \ldots, n_i$, the point locations of case events within the ith tract. In each tract, the conditional distribution of any $\boldsymbol{z}$ given $\{\boldsymbol{n}, \theta\}$ is given by:

$$(\boldsymbol{z}|\boldsymbol{n}, \theta) \sim \frac{\lambda(\boldsymbol{z}|\theta)}{\sum_{i=1}^{p} \int_{W_i} \lambda(\boldsymbol{u}|\theta)\mathrm{d}\boldsymbol{u}}. \tag{9.17}$$

Hence, within the ith tract, the joint distribution of $\{\boldsymbol{z}_{ij}\}$ is

$$\frac{\prod_{j=1}^{n_i} \lambda(\boldsymbol{z}_{ij}|\theta)}{\{\int_{W_i} \lambda(\boldsymbol{u}|\theta)\mathrm{d}\boldsymbol{u}\}^{n_i}}. \tag{9.18}$$

Within an augmentation framework, this suggests the following iterative algorithm:

- initialise with $\theta^l, z^l, l=0$
- generate $\boldsymbol{z}_{ij}^{l+1}$ from $[\lambda(\boldsymbol{z}^l|\theta)/\lambda_{\max}(\boldsymbol{z}^l|\theta)]$ for each tract up to n_i, where $\max \equiv \max_{W_i}$
- either generate θ^{l+1} from the $(\theta|\{n_i\},\{\boldsymbol{z}_{ij}^{l+1}\})$ distribution or use a MH update, or maximise the likelihood

$$l = \frac{\prod_{j=1}^{n_i}\prod_{i=1}^{p}\lambda(\boldsymbol{z}_{ij}^{l+1}|\theta)}{\{\sum_{i=1}^{p}\int_{W_i}\lambda(\boldsymbol{u}|\theta)\mathrm{d}\boldsymbol{u}\}^N}. \tag{9.19}$$

After initialisation, the steps are repeated to convergence. Note that the method described is defined only for a likelihood-based model. This can be straightforwardly extended to accommodate prior distributions for components of the parameter vectors.

The important result of this algorithm is that the likelihood (or full posterior distribution) is now a function of the 'pseudo-data' and hence point process modelling can be used via augmentation to model count data. The extension of this algorithm to the cluster models described earlier is relatively straightforward. Assuming that it is required to estimate cluster centres $\{\boldsymbol{y}_j\}$ from the count data, as in the point process case, then suitable parameterisation of $\lambda(\boldsymbol{z}|\theta)$ with cluster terms and the inclusion of an inner MH iteration for $[k, \{\boldsymbol{y}_j\}]$, prior to the θ^{l+1} step, provides a cluster version of the algorithm.

Implementation of regional integration

The extension of the point process algorithms to count data requires the evaluation of integrals over irregular regions or tracts. It is possible to use finite element mesh generators to subdivide the tract into regular geometric areas: e.g. triangles. This shape is easy to generate (see, for example, George, 1991) particularly with straight line segment boundaries. In our method, we use the numerical approximation:

$$\int_{W_i}\lambda(\boldsymbol{u}|\theta)\mathrm{d}\boldsymbol{u} = \sum_{j=1}^{l} W_j\lambda_j, \tag{9.20}$$

where j denotes the jth mesh triangle for the ith tract, and W_j is the triangle area. The intensity, λ_j, is evaluated at the triangle incentre. The index l can be set initially to $l = v - 2$, where v is the number of region vertices. Once the initial triangular mesh is computed, it is straightforward to compute denser meshes on the basis of the addition of incentre points to the vertex set. The accuracy of the approximation in (9.20) depends on the mesh level chosen.

9.8.2 Data example: respiratory cancer in central scotland

A study of respiratory cancer incidence in central Scotland has been initiated. The purpose of the study is to examine the clustering tendency of a variety of diseases in Falkirk, a town formerly associated with a variety of heavy industries during the early to mid twentieth century.

For the purposes of this example, respiratory cancer (ICD code: 162) incidence in a subset of 26 contiguous Falkirk census enumeration districts (eds) has been recorded

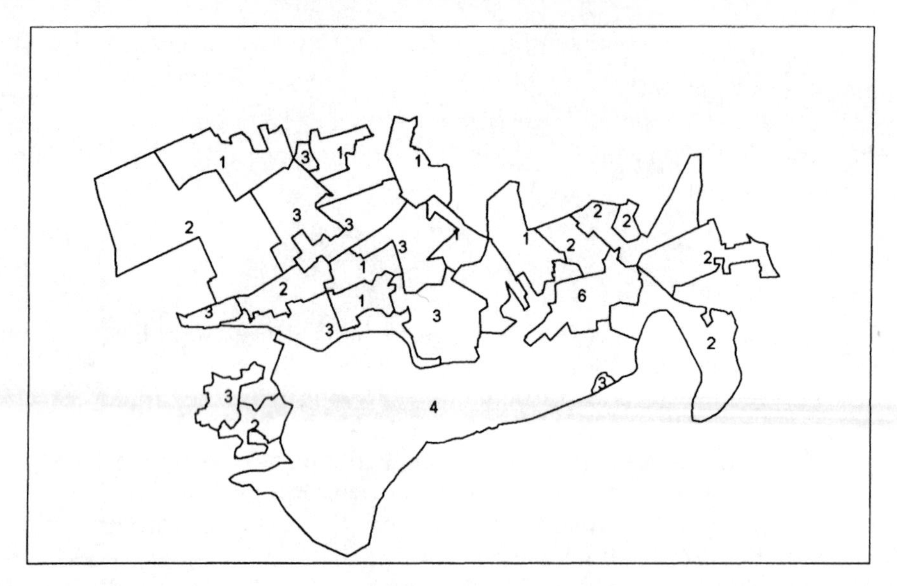

Figure 9.8 Counts of respiratory cancer in eds within central Falkirk

for a five-years period, 1978–83. The total cancer count, expected count based on 16 age $\times$ sex strata and external (Scottish) rates for the period, and digitised ed boundaries are available for this example. These boundaries were obtained from UKBORDERS output areas using MAPINFO. Figure 9.8 displays the outline ed map. Figure 9.1 shows the general location of the area (same as Armadale at this scale).

As part of a larger study, the clustering tendency of respiratory cancer is to be assessed. While such cancer is closely related to environmental health hazards such as air pollution, it is also related to lifestyle (e.g. smoking behaviours) (Lawson and Williams, 1994). At the large scale of this study it was not possible to obtain measurements of smoking behaviour. Deprivation indices (Carstairs, 1981) do not provide a perfect match of smoking lifestyle to deprivation status and in this case were not available. The intention in the following analysis is to demonstrate the application of the count data algorithm to the estimation of the cluster structure in this example.

We have applied the above augmentation algorithm of point events, with the following conditions. We initialise $\mathbf{z}$ with completely spatially random (CSR) events in the complete study region : $W_T = \sum_{i=1}^{p} W_i$. New values of $\mathbf{z}$ were rejection sampled from $\lambda(\mathbf{z}^l|\theta)$. These steps are based on the likelihood (9.19), with g_i assumed constant across regions and provided by the standardised rates mentioned above. The intensity for the case events was assumed to be $\lambda(\mathbf{z}_{ij}|\theta) = g_i \cdot C(\mathbf{z}_{ij})$, where

$$C(\mathbf{z}_{ij}) = 1 + \sum_{l=1}^{k} h_1(\mathbf{z}_{ij} - \mathbf{y}_l).$$

$C(\mathbf{z}_{ij})$ represents the cluster model terms. We have only included a cluster sum term for this example. The prior distributions used were as for the Humberside example, but no heterogeneity term was included. MH updates were used for the parameters κ and ρ_c. A

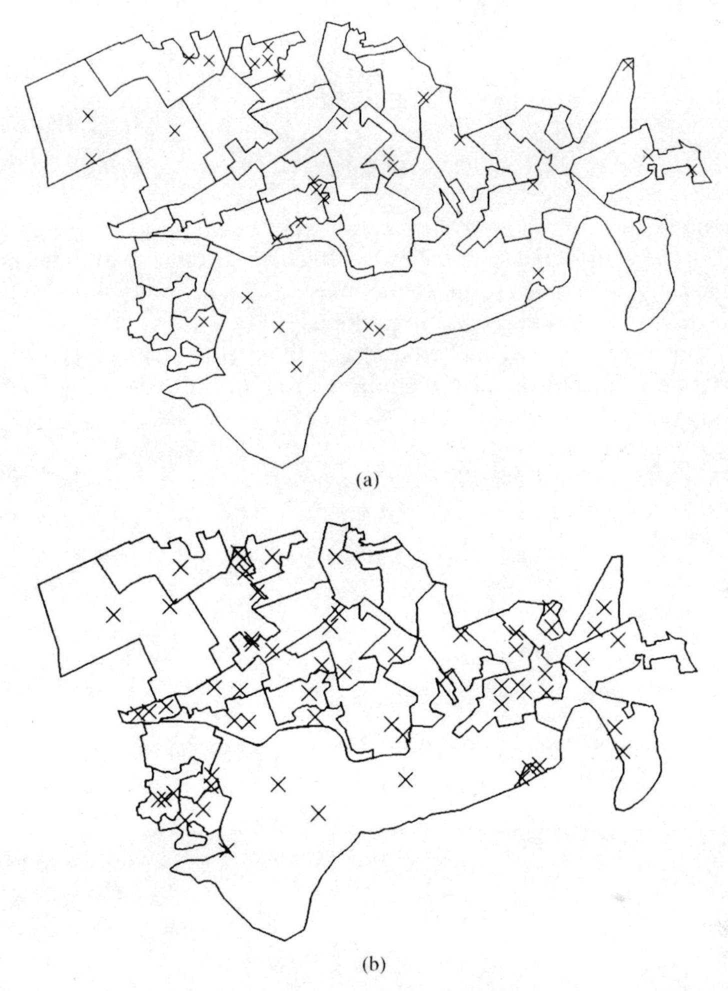

Figure 9.9 (a) Posterior marginal distribution of cluster centres. (b) Posterior sample distribution of $\boldsymbol{z}$

Markov (Strauss) prior has been included with parameters (γ, R) defined as for the Humberside example. Figure 9.9 displays the results of augmentation applied to this dataset.

Convergence occurred relatively quickly ($<80\,000$ iterations of the algorithm). Figure 9.9(a) displays the posterior marginal distribution of the cluster centres over the last 100 iterations. Figure 9.9(b) displays the augmented $\boldsymbol{z}$ realisation for the last iteration. There is some evidence that the number of centres lies in the range of one to three, although the parent rate mode is 1.12. The posterior marginal distribution of centres is relatively uniform.

9.9 CONCLUSIONS

The analysis above demonstrates a general approach to the modelling of putative source and non-focused clustering problems. The approach is parametric and relies heavily on iterative simulation methods in a Bayesian context. It also includes the use of *random object* effects (clusters), which are the spatial equivalent of conventional random effects. The results for the leukaemia/lymphoma data support the results of Diggle and Chetwynd (1991) who found little evidence of clustering in this data. The results of the Falkirk analysis support the conclusion of there being one centre, although its location has a number of site possibilities.

The advantage of our approach to parametric cluster modelling over previous approaches is that is allows the general modelling of a variety a spatial effects, this facility not being available in the descriptive methods used in previous cluster analyses, and avoids many of the restrictions apparent in the use of hypothesis testing, such as multiple comparisons, non-clustered background, etc. In addition, it extends the possibilities of *parametric* modelling in disease mapping to the area of cluster detection and so provides a unified approach to this area.

10

Statistical Evaluation of Disease Cluster Alarms

M. Kulldorff

National Cancer Institute

10.1 INTRODUCTION

During recent decades there have been frequent occurrences of disease cluster alarms, especially in Europe and North America but also in other areas of the world. Such alarms might be triggered by the observations of a local doctor or health official, by concerned citizens, or by members of the media. They are often accompanied by considerable worries in the communities affected. Some famous examples of cluster alarms include childhood leukaemia in Seascale, England (Gardner *et al.*, 1990a,b); leukaemia in Krümmel, Germany (see Chapter 31 in this volume); brain cancer in Los Alamos, New Mexico (Athas and Key, 1993); breast cancer on Long Island, New York (Jenks, 1994; Kulldorff *et al.* (1997)); and kidney failure in children in Port-au-Prince, Haiti (Public Citizen Health Research Group, 1997).

With most cluster alarms, a cause has not been found. A possible reason is that most clusters are simply reflections of random geographical variation in the disease rates, as just by chance, some areas are bound to have more cases than expected. There are also examples of cluster alarms where a cause has been found, leading to new aetiological knowledge or important public health benefits. Some such examples are listed in Table 7.1 of Chapter 7 in this volume. The aetiologies behind those cluster alarms range from viruses and bacteria, to occupational exposures and environmental pollution, to prescription drugs and personal tobacco consumption.

When there is a cluster alarm, the first thing to do is to establish the case definition. Then we need to check if the cluster area indeed has a rate higher than expected as compared with some larger region. If an excess rate does indeed exist, the next question is: Has the cluster occurred by chance alone or is the excess so great that it is probably due to some elevated risk factor of limited geographical extension? Only in the latter case would a more thorough epidemiological investigation be warranted, trying to identify that risk factor. Proper statistical evaluation of disease cluster alarms is

Disease Mapping and Risk Assessment for Public Health. Edited by A.B. Lawson *et al.*
© 1999 John Wiley & Sons Ltd.

important in order to minimise the time spent investigating random geographical noise.

To carry out a significance test by simply comparing the standardised incidence or mortality rate within the cluster area with what is observed in the larger geographic region is not a suitable statistical procedure. The spatial boundaries of the cluster are then defined from an observed set of cases, leading to pre-selection bias in the statistical analysis. This is the so-called Texas sharpshooter effect, so named after the Texan who first shot the gun and then painted the target around the bullet hole. If every cluster of cases that is 'statistically significant' using such a procedure was to be thoroughly investigated, then health officials would investigate a lot of random noise.

Although more seldom used, there are proper methods for how to statistically evaluate cluster alarms, eliminating the pre-selection bias. In this chapter we identify and describe three different approaches, discussing some of their pros and cons. The three are not mutually exclusive but complementary to each other. They will not lead directly to the aetiology causing a cluster, but they are important in determining whether or not to launch a thorough epidemiological investigation.

10.2 FOCUSED CLUSTER TESTS APPLIED AT OTHER SIMILAR LOCATIONS

If there is a localised source suspected of causing an excess number of disease cases, such as a harbour or a toxic dump site, then one can carry out a focused cluster analysis on a different area containing the same type of source. For example, after the leukaemia cluster alarm around the Sellafield nuclear power plant in England, Waller *et al.* (1995) looked at leukaemia around the nuclear power plants in Sweden, without finding any excess there. Sellafield is not only a nuclear power plant but also a nuclear waste reprocessing plant. So, Viel *et al.* (1995) looked at leukaemia around La Hague, a nuclear reprocessing plant in France, while Urquhart *et al.* (1991) studied leukaemia around Dounreay, a nuclear reprocessing plant in Scotland. Kinlen (1995), on the other hand, proposed the theory that the Sellafield cluster was caused by a virus that manifests itself more strongly in rural areas with recent immigration. He then looked at leukaemia in other areas with substantial immigration, but without a nuclear facility.

There are many proposed focused cluster tests, such as the Lawson–Waller score test (Lawson, 1993b; Waller *et al.*, 1992), Bithell's (1995) linear risk score test, Stone's (1988) test, and the focused version of Besag and Newell's (1991) test. They are discussed in detail in Part IV of this volume.

A drawback with this approach is that a negative result will not refute the alarm *per se*. The cluster could be real and due to something completely different than the suspected source, to some aspect of it that is not universal among such sources, or it could be a purely chance occurrence, and a negative result will not help to distinguish between the three. Also, there may not exist any other known places having the same type of suspected source.

It is important to point out that using a focused test on the cluster alarm area itself will not eliminate the preselection bias. It can be used in a preliminary stage to see if there indeed is an excess number of cases, and a negative result would indicate a chance occurrence. No conclusions can be drawn from a positive (significant) result though, as that could be due to the preselection bias, leading to the Texas sharpshooter effect.

10.3 POST-ALARM MONITORING

A second approach for evaluating disease cluster alarms is to forget about past and present data, and instead monitor future cases as they occur in the area of the alarm. This is a confirmatory type of analysis, and it avoids the preselection bias since the analysis is based only on cases diagnosed after the alarm occurred.

Chen *et al.* (1993) have proposed two different procedures based on the time interval preceding each of the first five cases subsequent to the alarm. A different number of cases can also be chosen. If these intervals are short, then the alarm is confirmed, while if they are long, then the alarm is rejected, and if they are somewhere in between, then the test is inconclusive. Parameters of the method can be chosen to obtain specific probabilities of falsely accepting the alarm under the null hypothesis of no excess risk, as well as the probability of falsely rejecting a true alarm under the alternative hypothesis of some specified excess risk.

One of the two methods, namely the median-based technique, uses the median of the five time intervals and accepts or rejects the alarm if this median is lower than some specified value or higher than some other specified value. The other is a mean-based technique that uses the mean instead of the median. If their is nothing in between the rejection and acceptance levels, then we have a traditional hypothesis test, although we may not necessarily want to use the traditional 0.05 or 0.01 rejection levels.

Of the two methods, the mean-based technique is preferable. The median-based technique only uses the information of whether a specific time interval is smaller or larger than the critical length, ignoring information on how much smaller or how much larger it is. With the mean-based technique, no information is lost. The advantage is confirmed by Chen *et al.* (1993) for one particular example, where the rejection level under the null hypothesis is higher for the mean-based technique, for a given confirmation level, and where the confirmation is higher under the alternative hypothesis, for a given rejection level.

It is not only the power of the test that is important, but also the time it takes until a conclusion is reached. Chen *et al.* (1993) pointed out that with the median technique it is sometimes possible to confirm the alarm immediately after the third or fourth case has occurred, while we always have to wait for all five cases with the mean-based technique. An opposite phenomenon is also true, and with the mean-based technique we can sometimes reject the alarm even before the first case occurs, while we always need at least two cases for rejection using the median-based technique, even if it takes several decades for the first two cases to occur. The key issue though is not how many cases we have to wait for, but how long we need to wait. For the mean-based technique, it is either at the time of the fifth case or at five times the critical mean value, whichever is smaller, and hence there is an upper limit. For the median-based technique, there is no upper limit on the time we need to wait, and hence the mean-based technique is preferable by this criterion as well.

A drawback of the post-alarm monitoring approach is that the cluster alarm cannot be evaluated immediately because we have to wait for new cases to occur. This is a problem especially if the disease is very rare. Also, if the cluster is limited not only in space but also in time, then we would not expect an excess of cases after the alarm has occurred.

10.4 THE SPATIAL SCAN STATISTIC

A third approach is to expand the study in space rather than in time. If we collect data for a larger region in which the cluster alarm is located, such as the whole country, then the spatial scan statistic (Kulldorff and Nagarwalla, 1995; Kulldorff, 1997) can be applied to the whole region to see if there is a significant cluster where we would expect it to be based on the alarm. The scan statistic gives us a measure of how unlikely it is to encounter the observed excess in the cluster alarm area in the larger region of our choice. Many cluster alarms could be quickly dismissed as a random occurrence were this technique to be used. The pre-selection bias is dealt with both in terms of the location and the size of the cluster.

The spatial scan statistic imposes a circular window on a map and lets its centre move across the study region. For any given position of the centre, the radius of the window is changed continuously to take any value between zero and some upper limit. In total the method uses a set $\mathscr{Z}$ containing a very large number of distinct circles, each with a different location and size, and each being a potential cluster. For each circle, the method calculates the likelihood of observing the observed number of cases inside and outside the circle, assuming either a Poisson or Bernoulli model for how the cases are generated.

Conditioning on the observed total number of cases, N, the definition of the scan statistic is the maximum likelihood ratio over all possible circles $Z \in \mathscr{Z}$:

$$S_{\mathscr{Z}} = \frac{\max_{Z \in \mathscr{Z}} L(Z)}{L_0} = \max_{Z \in \mathscr{Z}} \frac{L(Z)}{L_0}, \qquad (10.1)$$

where $L(Z)$ is the maximum likelihood for circle Z, expressing how likely the observed data are given a differential rate of events within and outside the zone, and where L_0 is the likelihood function under the null hypothesis.

Let n_Z be the number of cases in circle Z. For the Bernoulli model, let M be the total number of cases and controls, and let m_Z be the combined number of cases and controls in circle Z. Then

$$L(Z, p, q) = p^{n_Z}(1-p)^{m_Z - n_Z} q^{N - n_Z}(1-q)^{M - N - m_Z + n_Z}, \qquad (10.2)$$

where p is the probability that an individual within zone Z is a case and where q is the same probability for an individual outside the zone. Maximising the likelihood over p and q gives

$$\frac{L(Z)}{L_0} = \frac{\max_{p>q} L(Z)}{\max_{p=q} L(Z)}$$

$$= \frac{\left(\frac{n_Z}{m_Z}\right)^{n_Z}\left(1 - \frac{n_Z}{m_Z}\right)^{m_Z - n_Z}\left(\frac{N-n_Z}{M-n_Z}\right)^{N-n_z}\left(1 - \frac{N-n_Z}{M-m_Z}\right)^{M-N-m_Z+n_Z}}{\left(\frac{N}{M}\right)^{N}\left(1 - \frac{N}{M}\right)^{M-N}} \qquad (10.3)$$

if $n_Z/m_Z > (N - n_Z)/(M - m_Z)$, and one otherwise.

For the Poisson model, let $\mu(Z)$ be the expected number under the null hypothesis, so that $\mu(A) = N$ for A, the total region under study. It can then be shown that

$$\frac{L(Z)}{L_0} = \left(\frac{n_Z}{\mu(Z)}\right)^{n_Z} \left(\frac{N - n_Z}{N - \mu(Z))}\right)^{N-n_Z} \tag{10.4}$$

if $n_Z > \mu(Z)$, and one otherwise. For details and derivations as likelihood ratio tests, see Kulldorff (1997), who also proves some optimal properties for these test statistics.

Since this likelihood ratio is maximised over all the circles, it identifies the one that constitutes the most likely disease cluster. Its p-value is obtained through Monte Carlo hypothesis testing, adjusting for the multiplicity of circles used. Calculations can be done using the SaTScan software developed by Kulldorff *et al.* (1998b).

The spatial scan statistic has the following features, making it suitable as a screening tool for evaluating reported disease clusters:

(i) it adjusts for the inhomogeneous population density and for any number of confounding variables;
(ii) by searching for clusters without specifying their size or location the method resolves the problem of pre-selection bias;
(iii) the likelihood ratio based test statistic takes multiple testing into account, and delivers a single p-value; and
(iv) if the null hypothesis is rejected, then it specifies the location of the cluster that caused the rejection.

In some cases a cluster alarm is not only related to a specific area, but is also claimed to be present during a limited time period. We can then use a space–time scan statistic (Kulldorff, 1997; Kulldorff *et al.*, 1998a). Instead of a two-dimensional circle it uses a three-dimensional cylinder of variable size, where the circular base represents a particular geographical area and where the height represents a number of consecutive years or months. The cylinder is then moved through space and time in order to detect the most likely cluster, and a p-value is calculated, taking the multitude of cylinders into account. The size of the circular base and the length of the time interval vary independently of each other.

A drawback with the spatial and space–time scan statistics is that we need geocoded data for a large region, which is not always available. Another drawback is that we can only evaluate cluster alarms that are reasonably compact in shape, because the test has low power for other types of clusters such as along a long and narrow stretch of river.

10.5 A PROACTIVE APPROACH

An alternative to post-alarm analysis is a proactive approach, systematically screening a region for geographical clusters of a large number of different diseases, in a surveillance setting. The spatial and space–time scan statistics can be used for this as well, detecting significant clusters in the data. Multiple clusters can be detected by looking at the local maxima of equation (10.3) or (10.4), where the different zones are non-overlapping. Such a proactive approach is of course contingent on the existence of some type of disease registry with geocoded data.

If a cluster alarm does occur, then it can be quickly evaluated by looking up the result for that particular disease. If there is not a significant cluster at the location in question, then the alarm can be quickly dismissed as a probable chance occurrence, although some further investigation may still be warranted depending on the exact nature of the alarm. If there is a significant cluster at the location of the alarm, then at least the public health officials are not taken by surprise, and they may even have had a head start on the cluster investigation.

In a surveillance system with many diseases, the expected number of false positives may be too high unless we also adjust for the multiplicity of disease. This can easily be done by adjusting each individual significance level through a Bonferroni-type argument.

10.6 DISCUSSION

Assuming that the proactive approach is not in place, which of the three different methods do we choose? Considering the strengths and weaknesses of each approach, they are complementary to each other and in many situations it is recommended that all three be used. First use a spatial scan statistic to check if the observed excess is significant. Even if it is not, but if there is a highly suspected source, use a focused test to see if there is an excess of cases around other similar sources. Finally, and regardless of previous results, monitor the disease in the cluster area to see if the excess persists through time or not. If either or all methods indicate that the cluster is not a chance occurrence, then a thorough epidemiological or public health investigation may be warranted to try to find the risk factor responsible.

There is another class of cluster tests that is worth mentioning, namely general tests for global spatial clustering. They are designed to answer questions about whether there is clustering throughout the study region, and examples include Tango's (1995) test,

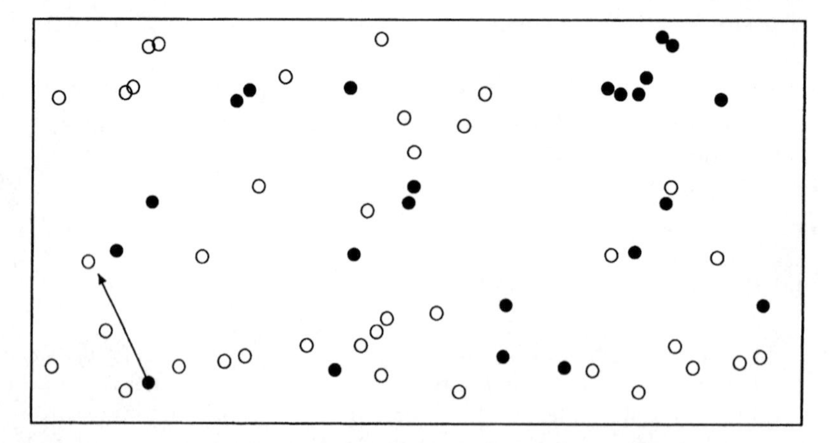

Figure 10.1 Keeping the population constant, but changing one case from one individual to another, as illustrated, will and should affect a general test for global clustering, but it should have no effect when determining whether or not there is a local cluster in the upper right hand corner (● = case, ○ = control)

Besag and Newell's (1991) test for the number of clusters, Walter's (1994) test, Grimson's (1991) test and Cuzick and Edward's (1990) special case of Cliff and Ord's (1981) test.

These are excellent for testing whether there is clustering globally throughout the study region, such as if a disease is infectious, but since they cannot pinpoint the location of clusters, they are not useful for evaluating disease cluster alarms. The same is true for the space–time interaction tests proposed by Knox (1964), Mantel (1967), Jacquez (1996a) and Baker (1996).

There is an important difference between a general test for global clustering on the one hand and the spatial scan statistic and focused cluster tests on the other. With a test of global clustering, the specific location of any case is important, no matter where it occurs, since moving it closer or farther away from other cases will determine the amount of evidence for global clustering. So, if one case is moved closer to another case, as in Figure 10.1, that would imply a stronger indication of global clustering. With local clusters, on the other hand, it is different. If two particular cases happen to be in Baltimore rather than one in Baltimore and the other in Washington, that should not be used as evidence for a local cluster in California at the other end of the country. Likewise, moving one case as indicated in Figure 10.1 does not give additional evidence concerning a potential local cluster in the upper right-hand corner of the map.

While the different types of tests will tend to have different power for different types of alternative hypotheses, that is not their main distinction. Rather, the main difference is in the type of analysis they perform, where the tests for global clustering makes no inference on the location of clusters while the spatial scan statistic does. For a formal mathematical theory on this, see Kulldorff (1997).

11

Disease Clustering for Uncertain Locations

Geoffrey M. Jacquez

BioMedware, Inc.

John A. Jacquez

The University of Michigan

11.1 INTRODUCTION

In general, disease cluster investigations are used in two contexts (Smith and Neutra, 1993): (i) to respond to alleged clusters brought forward by concerned citizens (a reactive investigation) and (ii) to survey health event data for significant excess in space, time and space–time (a proactive investigation). Many countries have in place procedures for reactive cluster investigation (see, for example, Centres for Disease Control, 1990), and the call for proactive surveillance programs is growing (Thacker *et al.*, 1996; Hertz-Picciotto, 1996). This has sparked research on innovative techniques for disease clustering, which typically use two kinds of data: rates within areas (such as postal code zones and census tracts) and points, such as the geographic coordinates of a place-of-residence where a case occurred. For rates, the spatial support is usually well-defined and knowledge of health event location is not required beyond 'the event occurred within this area'. Point data implicitly assume the locations of health events are known precisely, and the statistical machinery for analysing spatial point distributions is well-developed (consult reviews such as Marshall, 1991a; Jacquez *et al.*, 1996a, 1996b; Lawson and Waller 1996). Point-based methods include Knox (1964), Mantel (1967), and Cuzick and Edwards (1990), to name a few, and are often used when health events are rare and the number of observations is small. They are coming into increasing use as spatially referenced health data become commonly available (Openshaw, 1991), and models of spatial point processes (e.g. Diggle, 1993; Lawson and Waller, 1996), become increasingly sophisticated.

Disease Mapping and Risk Assessment for Public Health. Edited by A.B. Lawson *et al.*
© 1999 John Wiley & Sons Ltd.

Because of confidentiality issues, and lack of knowledge, exact health event locations may be unknown or unavailable. In these instances, one approach is to use centroid locations (e.g. centres of postal code zones, enumeration districts, census tracts), instead of the actual coordinates. The unit of clustering is then centroid rather than health event locations, and the question arises whether inferences based on centroids correspond to the inferences that would have been obtained using the actual locations.

Recently, Jacquez and Waller (1998) demonstrated that p-values calculated from centroids can differ markedly from those that would have been calculated from the actual locations, and that cluster detection capability decreased. That is not surprising since centroids are necessarily hyperdispersed (their spatial distribution is more uniform than expected under a spatial Poisson process) and thus can violate Poisson null models of many point-based cluster tests. The implications for public health policy are first, that statistical analyses using centroid locations should not be used as a quantitative basis for determining the future course of cluster investigations without first assessing the impact of centroids on the statistical results, and second, that the use of centroid locations can decrease the statistical power of cluster tests.

These observations prompted us to develop approaches to disease clustering that propagate location uncertainty through the statistical analysis. The first author's prior work presented a fuzzy algebraic approach (Jacquez, 1996a). This chapter deals with a new approach based on probabilistic location models of the geographic distribution of the at-risk population. Spatial randomisation methods assess the statistical outcomes (such as the value of the test statistic) possible given the uncertain locations and premised upon the location model. A new quantity, *credibility*, is defined which is based on the distribution of possible values for a test statistic that reflects uncertainty about the true locations of cases. In our simulations, credibility draws the correct inferences regarding the presence or absence of clustering more frequently than conventional tests based on centroid locations. In addition, because this approach models the possible locations from which samples are drawn, it provides an improved mechanism for evaluating statistical significance when actual locations are available. The approach is general, and applies to point-based methods that use randomisation distributions. It is best understood in the context of classical statistics and spatial randomisation tests.

11.2 CLASSICAL STATISTICS AND RANDOMISATION TESTS

Spatial data analysis and classical statistical inference have different theoretical backgrounds. Haining (1990) observed that classical statistics assume data from designed experiments that can be replicated, and samples that are drawn from a hypothetical universe defined by the sampling space. The inference framework is based on the comparison of a test statistic with its distribution under the null hypothesis (the reference distribution) for this sample space. A distribution of the test statistic can be obtained by replicating the experiment. Within this framework type I error (α) is the probability of rejecting the null hypothesis when it actually is true, and type II error (β) is the probability of accepting the null hypothesis when it actually is false. Statistical power—the probability of correctly rejecting the null hypothesis—is $1 - \beta$. Spatial data often violate the assumptions (e.g. iid random variables) of classical tests upon which their distribution theory is based. In addition, the asymptotic behaviour of classical statistics often cannot be used because sample sizes are small. Spatial randomisation methods generate reference distributions from the data, and are coming into increased use with the advent of desk-top computing. The remainder of this chapter is concerned with disease cluster statistics as randomisation tests.

Manly (1991) provides a succinct description of randomisation tests and their relationship to classical statistical theory. The data often are from only one sample; the concept of a 'designed experiment that can be replicated' does not apply, and classical statistical inference therefore is inappropriate. A commonly-used alternative is randomisation tests, which determine whether a pattern exists in a *sample*. The null hypothesis is that any pattern is a chance occurrence, and the alternative hypothesis is that a 'true' pattern exists. Some statistic, Γ, is selected that quantifies the pattern of interest. The value, Γ_A, from the observed data is then compared with a reference distribution obtained by repeatedly reordering the data at random, and by recalculating Γ for each repetition. The significance level is the proportion of the reference distribution which is as large or larger than Γ_A. Interpretation of this significance level is similar to conventional tests based on the classical model: if less than or equal to the α level (usually 5%) the null hypothesis of 'no pattern' is rejected. Manly further observed that randomisation tests have two principal strengths: they are valid even without random samples, and non-standard test statistics may be used. These advantages led to the wide use of randomisation tests for the analysis of spatial data. However, results pertain *only* to the sample, and this *lone sample is the sampling space upon which the reference distribution is based.*

11.3 CLUSTER STATISTICS AS RANDOMISATION TESTS

The gamma product of two $N \times N$ matrices, $\boldsymbol{A}$ and $\boldsymbol{B}$ is

$$\Gamma = \boldsymbol{A} \otimes \boldsymbol{B} = \sum_{i=1}^{N} \sum_{j=1}^{N} a_{ij} b_{ij}. \tag{11.1}$$

For spatial disease data we rewrite the gamma product as $\Gamma = \sum_{i=1}^{n} \sum_{j=1}^{n} \delta_{ij} c_{ij}$. Here n is the number of locations, δ is a proximity measure and c is calculated from the observations. The proximity metric may be geographic distances, spatial weights, adjacencies or nearest neighbour relationships. The observations may be on real, integer, or categorical data, and include case–control identity (e.g. Cuzick and Edward's test), time of diagnosis or death (e.g. Mantel's test), disease rates (e.g. Moran's I) and, for the multivariate version, exposure and confounder data (e.g. Mantel's multivariate extension). Several authors have shown that disease cluster tests are special cases of the Γ product (Haining, 1990, p. 230; Wartenberg and Greenberg, 1990; Marshall, 1991a; Getis, 1992; Jacquez, 1996a). In fact, most of the disease cluster tests in common use can be expressed as gamma products (Table 11.1).

Mantel's (1967) test for space–time interaction results when $\delta_{ij} = s_{ij}$ and $c_{ij} = t_{ij}$. Cuzick and Edwards' (1990) test results when $\delta_{ij} = n_{ijk}$ and $c_{ij} = d_i d_j$. Moran's I (1950) results when $c_{ij} = (z_i - \bar{z})(z_j - \bar{z})$ and δ_{ij} corresponds to elements of distance or adjacency matrices, as appropriate. The join-count statistic (Cliff and Ord, 1981) obtains when $c_{ij} = (x_i x_j)$ and δ_{ij} is the adjacency a_{ij}. The Pearson product-moment correlation

Table 11.1 Specification of seven cluster tests in gamma form (modified, with permission from BioMedware Press, from Jacquez, 1999)

Statistic	Data	δ	c	Reference/note
Mantel	(x, y, t) health event locations	distance between case pairs	waiting times between cases	(Mantel, 1967) space–time interaction test
Mantel extension	$(x, y, z_1, \ldots, z_p)$	distance between sample locations	multivariate distance calculated from $z_1, \ldots, z_p$	(Smouse 1986) test for spatial structure in multivarite data
Knox	(x, y, t)	spatial adjacency	temporal adjacency	(Knox, 1964) space–time interaction test
Jacquez	(x, y, t)	spatial nearest neighbour	time nearest neighbour	(Jacquez, 1996b) space–time interaction test
Cuzick and Edwards	(x, y, c); c is case idenfier	spatial nearest neighbour	$c_{ij} = 1$ if i and j are cases, 0 otherwise	(Cuzick and Edwards, 1990) spatial cluster test
Moran's I	(x, y, z) z is an attribute	spatial weight connecting locations i and j	$(z_i - \bar{z})(z_j - \bar{z})$	(Moran, 1950) spatial autocorrelation test
Join-count	(x, y, z), $z = 1$ if x, y is labelled	spatial adjacency	1 if both i and j are labelled	Cliff and Ord (1981), test spatial clustering of labelled locations

and multiple regression may be written in Γ form (Smouse *et al.*, 1986), as can the local autocorrelation statistics recently proposed by Anselin (1995) and Getis and Ord (1996).

One can use a normal approximation for the randomisation distribution of gamma (see Mantel, 1967, and Haining, 1990, for moments of this distribution) to assess the statistical significance of an observed value. This approach has been criticised (Mielke, 1978, Faust and Romney, 1985) and it is better to calculate the distribution under randomisation, and then compare the observed value with this distribution (Manly , 1991). This is accomplished under a statistical null hypothesis of independence (Cressie, 1993, terms this Complete Spatial Randomness or CSR) between the δ_{ij} and the c_{ij} using a randomisation approach equivalent to a relabelling of the locations so the observations are sprinkled at random across the locations. Note, however, that the limitations of randomisation tests apply: (i) inference applies only to the sample, and (ii) the spatial sampling space is assumed to consist solely and entirely of the sample locations.

11.4 SAMPLE-BASED RANDOMISATION TESTS ARE EPIDEMIOLOGICALLY UNREASONABLE

Spatial data usually are non-experimental and inference is undertaken within the framework of an exploratory data analysis whose purpose is to detect structure and pattern. Randomisation tests are widely used with spatial data because of difficulties in obtaining samples. These difficulties often preclude designed experiments, and randomisation tests, at first blush, seem particularly attractive. Given $z = (z_1, \ldots, z_n)$ values on a map, spatial randomisation tests permute the z-values over the sample locations. While randomisation tests may be appropriate when the experimental design justifies randomisation testing, they can be problematic for spatial data because they take the sampling space to be the locations at which the observations were made. *That is, spatial randomisation tests erroneously assume the universe of locations to consist entirely and solely of the sample locations.* In most situations we could have sampled other locations in the study area, but spatial randomisation tests, as currently implemented, ignore this fact. *This means that the sampling space is incorrectly specified, and the reference distribution pertains only to the sample, and not to the population at risk within the study area.* Epidemiologically, reference distributions from randomisation tests as currently implemented make little sense because they assume the at-risk population consists only of the sample.

Until now this issue has largely been ignored because data describing the geographic distribution of human populations have not been available, precluding specification of the universe of possible sample locations. This is no longer the case. Spatially referenced data are now available describing the global population density distribution within 5′ quadrilaterals (Tobler *et al.*, 1995), and address matching software can locate street addresses within an accuracy of 100 m (Rushton and Lolonis, 1996). Our approach uses location models and such spatial population data to correctly specify the spatial sampling space. We view this as a step towards the stronger inference model of classical statistics because it recognises that samples could have been taken at other locations in the study area (a sampling experiment). Our approach effectively increases the study's spatial sampling space and this, not surprisingly and as demonstrated in our research and as expected from theory, improves our ability to correctly detect disease clusters.

11.5 STATISTICAL INFERENCE

When exact locations are known, p-values are calculated as the probability of observing an outcome as or more extreme than that calculated using the actual locations under a hypothesis of no clustering (Figure 11.1 (a)). The reference distribution based on exact locations is denoted g_{A} and the test statistic based on exact locations is Γ_{A}. When locations are uncertain, centroid locations often are used instead, resulting in a reference distribution (g_{C}) and test statistic, Γ_{C}, calculated using the centroid locations (Figure 11.1 (b)). A centroid p-value is calculated using the same approach—by comparing the test statistic with the reference distribution—as for actual locations. However, p-values based on centroid data may differ considerably from p-values based on actual locations, and statistical power can be compromised (Jacquez and Waller, 1998). Instead, we use a new and entirely different approach; we model uncertain locations to obtain a reference distribution (g_{U}) of some statistic, Γ, and then obtain a distribution (g_{T}) of the test statistic given the uncertain locations (Figure 11.1 (c)).

One then determines the proportion of the g_{T} distribution which is as or more extreme than the $1 - \alpha$ critical value of the g_{U} reference distribution. This proportion is called *credibility*. When $\alpha = 0.05$ it is the proportion of g_{T} equal to or beyond the 95% critical value of g_{U}. We call this quantity credibility to distinguish it from statistical power, which is strictly appropriate only in the context of classical statistics where designed experiments can be replicated. The distribution of the test statistics g_{T} is not obtained by repeating a *designed* experiment, but rather from spatial randomisation

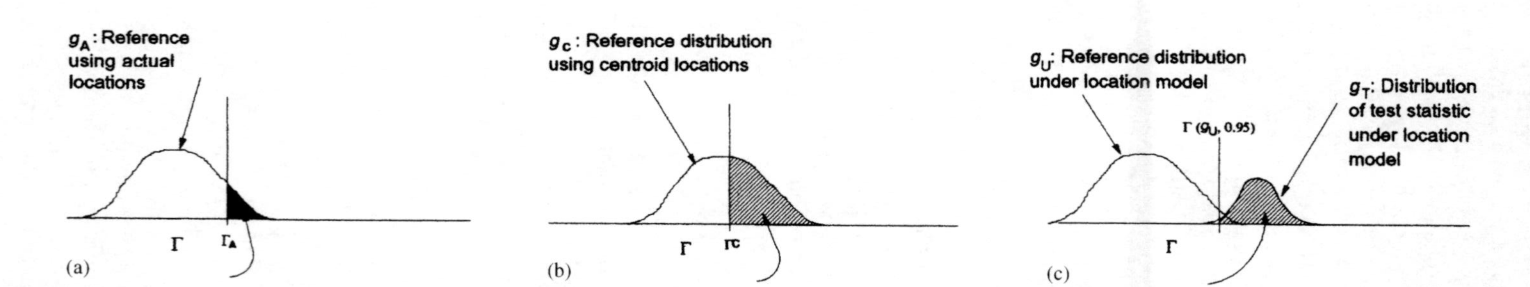

Figure 11.1 Probability of the cluster statistic using (a) actual locations, (b) centroid locations, and (c) modelled locations (a) $Pr(\Gamma \geq \Gamma_{\mathrm{A}})$: Probability under the actual distribution of observing an outcome as larger than Γ_{A}. (b) $Pr(\Gamma \geq \Gamma_{c})$: Probability under the centroid reference distribution of observing an outcome as large or larger than Γ_{c}. (c) $Pr(\Gamma \geq (\Gamma(g_{\mathrm{U}}, 0.95))$: Credibility of statistically significant clustering given uncertain locations

analogous to a *sampling* experiment. Thus credibility and power, while constructed in a similar fashion, have different theoretical underpinnings.

Now compare the three approaches to statistical inference — based on actual locations, centroid locations, and location models. The first two approaches (actual and centroid locations) assume that the locations in the sample *are the universe* of possible locations over which the z_i are distributed, while the approach using location models assumes that the universe is the locations of the at-risk population within the study area. *Both the actual and centroid approaches must be poor because they use the wrong universe.* From an epidemiological perspective, they incorrectly assume the at-risk population consists solely and entirely of the sample. Location models using g_{U} and g_{T} allow specification of a sampling space that correctly corresponds to the study area's at-risk population.

11.6 LOCATION MODELS

As noted earlier, there are sound statistical reasons why centroids are expected to be inadequate, and we present four alternatives that correspond to different levels of knowledge regarding spatial locations. The location models we have chosen are particularly appropriate in the health and environmental sciences, and their theoretical basis is now briefly described. Examples emphasise applications in epidemiology, but the models are general and broadly apply to all scientific fields working with spatial data. Our approach is to model location uncertainty as a probability density function whose specification depends on the amount of available knowledge (Jacquez, 1997).

General forms: Location models are based on discrete or continuous probability functions. *Discrete location models* are discrete probability functions for which the sample space, $\mathbf{s}(A)$, is the discrete set denoting all possible sample locations, $\mathbf{s}_j$, in region A:

$$\mathbf{s}(A) = \{\mathbf{s}_j : j = 1, \ldots, L(\mathbf{s}(A)); \forall \mathbf{s}_j \in A\}. \tag{11.2}$$

Here $L(\mathbf{s}(A))$ is the number of locations in $\mathbf{s}(A)$. The set of point probabilities for this sample space, $\mathbf{p}(\mathbf{s}(A))$, is

$$\mathbf{p}(\mathbf{s}(A)) = \{p_j : j = 1, \ldots, L(\mathbf{s}(A)); \forall \mathbf{s}_j \in \mathbf{s}(A)\}, \quad \text{where } p_j = p(\mathbf{s}_j). \tag{11.3}$$

The point probabilities are defined such that

$$\sum_{j=1}^{L(A)} p_j = 1. \tag{11.4}$$

The sample space and point probabilities define a *discrete location model*, represented graphically in Figure 11.2.

Continuous location models are probability density functions for which the sample space, $\mathbf{s}(A)$, is the infinite set denoting all possible locations in region A:

$$\mathbf{s}(A) = \{\mathbf{s} : s \in A\}. \tag{11.5}$$

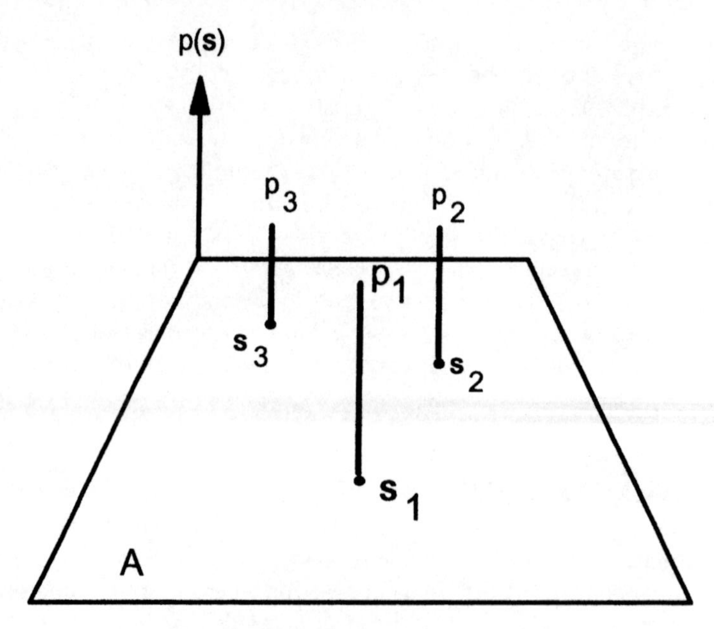

Figure 11.2 Discrete location model

The probability density function $f(\mathbf{s})$ is defined such that

$$\int_{|A|} f(\mathbf{s})\, d\mathbf{s} = 1. \tag{11.6}$$

Integration occurs over the entire area of region A, denoted $|A|$. The sample space and pdf define a *continuous location model*, represented graphically in Figure 11.3.

Variants of this model (epsilon, polygon, population) differ in pdf specification. Several components are needed to specify a location model for a particular application: the *sample space*, $\mathbf{s}(A)$; the point *probabilities*, $p(\mathbf{s}(A))$, or pdf, $f(\mathbf{s})$; and the *set of sampled points*, $\boldsymbol{x}'(A)$, which represent a single replicate of a sampling experiment based on that location model. We now specify four location models (point, polygon, population, and epsilon) useful for spatial studies in general and for epidemiology in particular.

A *point location model* is a discrete location model with sampling space defined in (11.2). Variants on the point model differ in specification of the point probabilities. In the general case (11.3) the point probabilities may or may not be equal to one another, and substantial knowledge is required to estimate them. In practice, knowledge may be limited, such that

$$p_1 = p_2 =, \ldots, = p_{L(\mathbf{s}(A))} = 1/L(\mathbf{s}(A)) \tag{11.7}$$

represents our best estimate of the point probabilities. Here $L(\mathbf{s}(A))$ is the size of sample space $\mathbf{s}(A)$; $L(\mathbf{s}(A)) = \text{card}(\mathbf{s}(A))$. Here 'card' is the cardinality of $\mathbf{s}$ and returns the number of locations in $\mathbf{s}$. For example, we may only know that a disease case occurred at one of the places of residence within region A, and we may reasonably assume, in the absence of more detailed information, that all locations in the sample space are

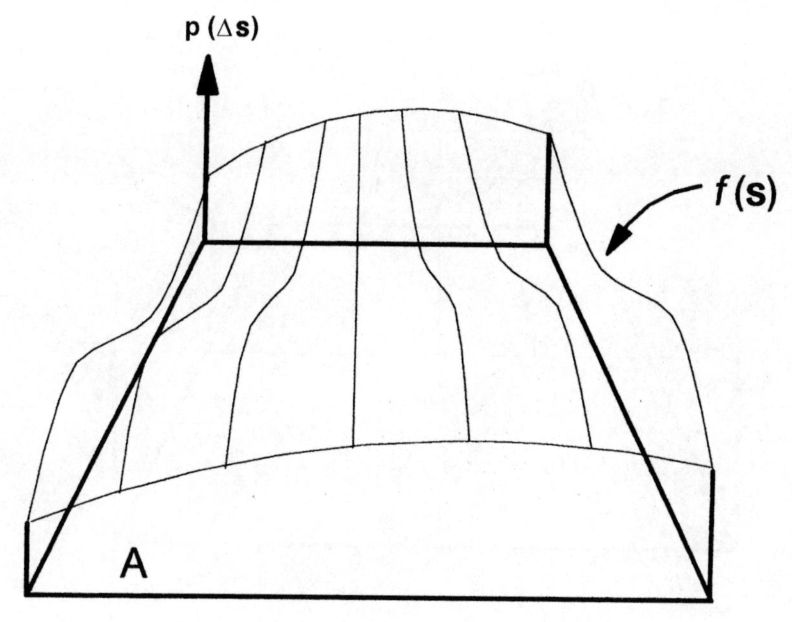

Figure 11.3 Continuous location model

equiprobable. In other situations we may know the number of residents at each place of residence in the sample space, $N(\mathbf{s}_j)$, and the estimated probabilities may be proportional to household size. The set of $n(A)$ points sampled from the sample space $\mathbf{s}(A)$ is

$$\{\boldsymbol{x}_i^{\mathrm{T}}: i = 1, \ldots, n(A); \boldsymbol{x}_i^{\mathrm{T}} \in \mathbf{s}(A)\}, \quad \text{where } \mathbf{s}(A) = \{\mathbf{s}_j : j = 1, \ldots, L(A); \forall \mathbf{s}_j \in A\}. \tag{11.8}$$

Here $\boldsymbol{x} = \{\boldsymbol{x}, \boldsymbol{y}\}$ is a location and $\boldsymbol{x}_i$ is the location of the ith point. This model is appropriate when the sample space can be enumerated as a finite set. Applications in epidemiology arise when locations of places of residence within areas (e.g. census tracts, enumeration districts, etc.) are known. Enumeration may be accomplished using technology such as address-matching software, digitising from maps that display places of residence, and through the use of Global Positioning Systems. This model is used when location uncertainty is such that an observation can occur at one of several locations, and we are uncertain which of those locations is the one.

A *polygon location model* is a continuous location model with sampling space defined in (11.5). The pdf $f(\mathbf{s})$ is uniform over A such that the probability of a health event occurring in the small area $|\Delta\mathbf{s}|$ is

$$p(\Delta\mathbf{s}) = \frac{|\Delta\mathbf{s}|}{|A|} \tag{11.9}$$

and all points within region A are equiprobable. The set of points sampled under this model is

$$\{\boldsymbol{x}_i^{\mathrm{T}} = 1, \ldots, n(A); \boldsymbol{x}_i^{\mathrm{T}} \in \mathbf{s}(A)\}, \quad \text{where } \mathbf{s}(A) = \{\mathbf{s} : s \in A\}. \tag{11.10}$$

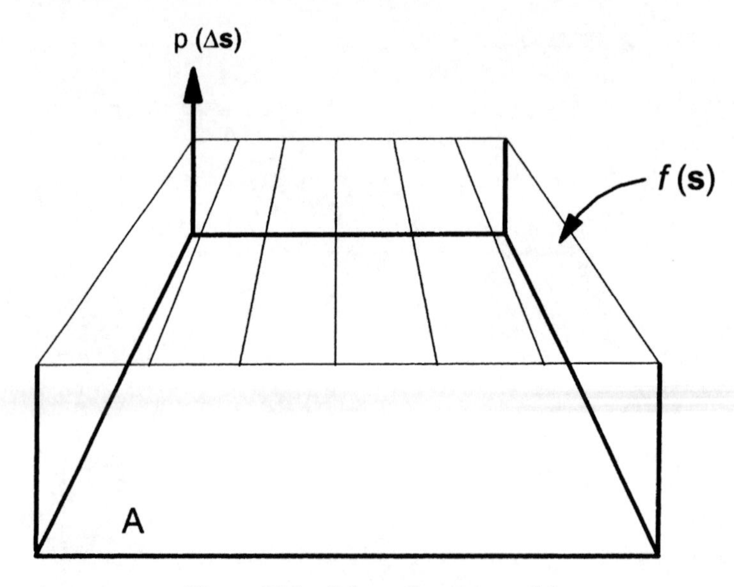

Figure 11.4 Polygon location model

This set is comprised of $n(A)$ locations sampled independently and with uniform probability from area A. Note that (11.10) and (11.8) differ in that the sample space is infinite in (11.10) and is finite in (11.8). This model is used when we know an observation occurred within an area, and our knowledge is limited such that the occurrence of the observation is deemed equiprobable within the area (Figure 11.4).

A *population location model* is a continuous location model with sampling space defined in (11.5). The pdf $f(\mathbf{s})$ is heterogeneous over A such that the probability of observing the uncertain location of the health event within the small area $|\Delta\mathbf{s}|$ is

$$p(\Delta\mathbf{s}) = \int_{|\Delta s|} f(\mathbf{s})\, d\mathbf{s}. \tag{11.11}$$

The function $f(\mathbf{s})$ depends on population density so the probability $p(\Delta\mathbf{s})$ is proportionate to the population density in area $|\Delta\mathbf{s}|$. A technique for estimating $f(\mathbf{s})$ from currently available population density data is described below. The sample space and pdf defining the population location model are represented graphically in Figure 11.3, with the distinction that $p(\Delta s)$ is as defined in (11.11). The set of $n(A)$ points sampled from the sample space $\mathbf{s}(A)$ is given in (11.10). This set is comprised of $n(A)$ locations sampled independently and with probabilities proportional to the local population densities in $|\Delta\mathbf{s}_i|$ of region A. The population model estimates the pdf proportional to population density, and is appropriate when the pdf can be estimated. For example, population density data are now available for any location on earth at the 1 km grid scale (Available on the CIESEN web site.) The problem then is to express probabilities in terms of the underlying population density grid. Let Q_j denote a population grid quadrat, with an associated population size $N(Q_j)$. The population size in area A based on the population grid data is

$$N(A) = \sum_{j=1}^{\nu} N(Q_j) \frac{|Q_j \cap A|}{|Q_j|}. \tag{11.12}$$

This is an allocation of the population equivalent to the proportion of the area of Q_j contained in A. Here ν is the number of quadrats that intersect (overlap) region A, and the indexing is assumed to be over these ν quadrats. $|Q_j \cap A|$ is the area of the intersection of the jth quadrat and region A. This model assumes a homogeneous population density within quadrats. Alternatively, one could smooth the population grid and integrate over the area of intersection.

The probability of sampling a point in quadrat Q_j is

$$p(Q_j) = \frac{N(Q_j)^{\frac{|Q_j \cap A|}{|Q_j|}}}{N(A)}. \tag{11.13}$$

Samples may be allocated under this model by constructing a number line between 0 and 1 divided into ν intervals whose widths are proportionate to the corresponding $p(Q_j)$. Observations on a uniform random variable $r \sim \mathrm{U}(0, 1)$ are then allocated to a quadrat, Q_j, when

$$\sum_{i=1}^{j-1} p(Q_i) < r \leq \sum_{i=1}^{j} p(Q_i). \tag{11.14}$$

The population model is used when we know only that an observation occurred within an area, and we can specify the corresponding pdf from population density data.

Epsilon location models are well described in the GIS literature (e.g. Mowerer *et al.*, 1997) and apply when the precision of the instrument used for measuring location is known and its error can be modelled as a bivariate distribution. The epsilon model is the special case of a continuous location model where the pdf is specified by the bivariate normal distribution, is useful for error modelling in GIS, and has been used in epidemiology as a 'mask' for protecting patient confidentiality (Rushton and Lolonis, 1996). Its theoretical basis is well known (see the above references) and will not be repeated here.

11.7 LOCATION MODEL APPLICATIONS

Point models occur when possible exact locations are available as a finite list. This situation arises when we know a case occurred within an area, but the exact place of residence is unknown. A list of possible locations is constructed as the coordinates of all places of residence within the area, and may be obtained from address-matching software which output latitude and longitude of street addresses. This model is preferred because it offers the greatest spatial resolution. Its greatest weakness is that a list of alternative locations may be difficult to construct or may not be available.

Population models are used when the underlying population density distribution is known. This distribution is then used to allocate possible case locations within each area. For example, locations with high population density are sampled most frequently. The resolution of this model depends on the available population density surface. We use 1 km population density rasters from CIESIN. The Nyquist sampling theorem states that a pattern can be detected only if it is greater than twice the resolution, and detectable spatial patterns therefore are > 2 km. High resolution data are increasingly available, and include British Census data reported by square kilometer (Rhind *et al.*, 1980), and Chinese population density in $1' 15''$ by $1' 52.5''$ quadrilaterals (Chenguri, 1987). The population model is appropriate for all of these.

Polygon models arise when the probability of sampling is assumed uniform within subareas. This model applies when possible places of residence (as for point models) are unknown and information on population density is lacking. Although the polygon model arises frequently due to data insufficiencies, it offers the least resolution because it assumes a homogeneous population density within subareas. Spatial uncertainty is modelled by sampling locations with uniform probability in each subarea.

Epsilon models appear to have few applications for modelling the locations of human health events. However, they are applicable to remotely sensed data, and undoubtedly will be used increasingly with the growth of Global Positioning Systems and remote sensing in public health.

To summarise, these models are appropriate for different levels and types of knowledge regarding the locations. Given an uncertain location, they allow us to specify a spatial sampling space containing the uncertain location, and the corresponding probabilities of having the health event occur at specific locations in that sample space.

11.8 SPATIAL RANDOMISATION

Once quantified as a model, uncertainty can be propagated through the gamma product using spatial randomisation procedures. We use a spatial randomisation approach to generate g_U and g_T from location models describing possible sample locations. Assume point data (x, y, z), where x, y is a geographic coordinate and z is an observation on a variable at x, y. Further suppose there are five observations $(z_1, z_2, z_3, z_4, z_5)$ in two areas (Figure 11.5 (a)). Our knowledge of exact locations is limited; we know z_1, z_2 occurred somewhere in area A and z_3, z_4, z_5 are in area B.

A location model in Step 1 (Figure 11.5(b)) generates possible sample locations constrained to the number of observations in each area. Thus two locations are generated for area A ad three for area B. Sample locations for one realisation of a location model are shown as X's in Figure 11.5 (a). In Step 2 the z_i are assigned to the sample locations generated in Step 1. For g_U the z_i are assigned at random over *all* sample locations (one z_i to each sample location). For example, z_3 is assigned with equal probability to any of the locations (X's), including those in area A. This corresponds to a statistical null hypothesis of no association between the z_i and their sampling locations. Under g_T the randomisation is spatially restricted *within areas*. Thus z_3 can be allotted only among the three sample locations (X's) in area B. This maintains any association between the z_i and their sampling areas (but not between the z_i and precise locations within the areas; within-area pattern is lost just as it is for centroid locations). In Step 3 the test statistic is calculated, and can be any statistic based on spatially referenced data. Steps

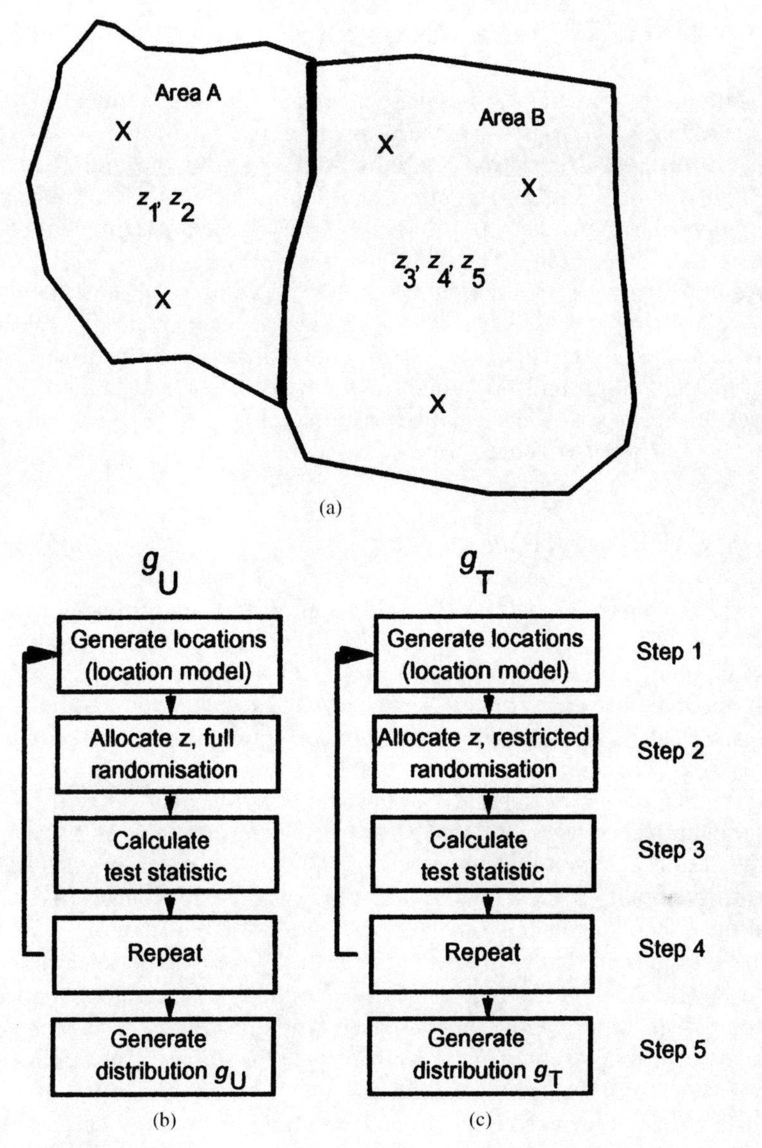

Figure 11.5 Generation of g_U and g_T distributions. $(z_1, z_2, z_3, z_4, z_5)$ occur in two areas, and the Xs are possible sample locations. Modified from Jacquez (1997) with permission from BioMedware Inc., USA

1 through 3 are repeated to generate g_U and g_T. g_U is the distribution of the statistic under the null hypothesis of no association between sample location and the variable z. g_T is the distribution of the test statistic under possible realisations of the uncertain locations. When the null hypothesis is true, g_T and g_U are similar. When the null hypothesis is false, the mean of g_T shifts right (assuming, for example, that nearby cases are clustered and have similar z_i) and credibility increases.

11.9 STATISTICAL INFERENCE FOR UNCERTAIN LOCATIONS

When locations are known exactly, p-values are calculated as shown in Figure 11.1 (a) and the arrangement of cases is deemed 'unusual' when the p-value is less than α. When locations are uncertain, the actual locations cannot be observed, and we instead calculate credibility as described in the previous section. Premised on the location model, credibility describes how likely an observation of statistically significant clustering is, given the uncertain locations. If credibility is greater than some *critical credibility*, then a putative cluster may warrant further study. The value used for critical credibility represents a trade-off between false positives and false negatives, and its selection therefore sets the context of the cluster investigation. For example, if the disease under scrutiny is highly contagious and fatal, then we might set the critical credibility low, so that any suggestion of clustering leads to further investigation. A higher critical credibility is chosen when false positives have severe consequences.

11.10 AN APPLICATION

Our research to date evaluated the credibility approach using simulated data for which the amount of clustering is known. We simulated clusters arising from an infectious disease in Michigan (refer to Jacquez and Waller, 1998, and Jacquez, 1996b, for simulation protocol) and compared the statistical power of the Knox, Mantel and k-NN (Jacquez, 1996b) tests. Health event location was assumed known only to the county level, and statistical power was compared for the actual locations, centroid locations, and credibility using the population model with pdf based on the 5 ft quadrilaterals of the Global Demography Project (Tobler *et al.*, 1995). Results are given in Table 11.2.

First, we found a substantial improvement in the proportion of correct inferences drawn using credibility relative to both centroid *and* actual locations. This improvement arises because location models include information on the underlying geographical population distribution that is not available from the sample. Because the asymptotic behaviour for the Knox and Mantel tests is questionable, it is common practice to calculate reference distributions using sample-based randomisation, as was accomplished for comparison purposes in this simulation study. Thus, the statistical power in Table 11.2 of both centroid *and* actual locations uses the sample to specify the sampling space. Location models allow us to correctly specify the study's sampling space, substantially increasing the proportion of correct statistical inferences. We thus have good reason to

Table 11.2 Proportion of correct inference drawn using sample locations, centroids and credibility. Cluster test (column 1); proportion of correct inferences using actual locations (column 2); centroid locations (column 3); and modelled locations (column 4) at critical credibility $= 0.05$ (column 4). $k = 3$ was used for evaluating the k-NN test

Test	Sample power	Centroid power	Credibility power
Mantel	0.59	0.46	0.57
Knox	0.40	0.32	0.68
k-NN	0.49	0.35	0.79

expect statistical power using credibility to be greater than that obtained using sample-based randomisation tests based on centroid or even actual sample locations.

Secondly, we found a favourable trade-off between type I and type II error rates so that, using credibility, a substantial decrease in type II error is obtained at a small increase in type I error. In contrast, tests using centroid locations have unacceptably high type II error rates resulting in low statistical power.

Thirdly, we found the spatial randomisation approach to be robust and to not depend critically on specification of the location model. Even the polygon model, which assumes uniform population density in subareas, is a substantial improvement over centroid locations.

11.11 DISCUSSION

It is convenient to classify location uncertainty into four sources: *measurement error, centroid assignment, surrogate assignment, and mobility assignment.*

Measurement error is a simple lack of sureness regarding a precise value, and occurs when precise locations are measured with an imprecise instrument, are transcribed incorrectly, and/or are represented in a fashion that introduces error. Examples include errors in digitising, Global Positioning Systems, and surveying instruments. Uncertainty also arises when data are gridded (as, for example, in raster-based GIS) and the coordinates of the nearest grid node are used instead of exact locations. All spatial data sets contain location uncertainty due to measurement error.

Location uncertainty due to *centroid assignment* occurs when a precise location is represented by the centroid location of the area that contains the precise location. It thus is due to a simple lack of sureness regarding the precise location. Uncertainty due to centroid assignment occurs frequently when analysing human data. An example is the centroid of a zipcode zone which contains a place of residence. Geographic Information Systems (GIS) assign centroids to areas with ease, and some authors advocate the use of centroid assignment (Croner *et al.*, 1996). Because of this case of centroid assignment, many statistical analyses of GIS data use centroids, although the impact of the resulting location uncertainty on the statistical results is only rarely addressed.

Surrogate assignment occurs when location is used as a proxy for the action of an unobserved, and, in some instances, even unknown variable. In disease cluster investigations places of residence are inherently uncertain whenever they are used as proxy measures of exposure. Increased knowledge of spatial location will not reduce this uncertainty, and surrogate assignment is thus inherently vague. An example is the use of proximity to a pollution source as a proxy measure of an individual's exposure. If more information were available, then we could explore dose–response relationships using controlled studies. When such detailed knowledge is lacking, cluster investigations often use the spatial relationships among cases, and their proximity to putative sources, as an uncertain measure of exposure.

Many of the events we represent by point locations are associated with objects (e.g. people) that actually are mobile, their representation as a point gives rise to location uncertainty, and this is termed *mobility assignment*. Tobler *et al.* (1995) observed that in modern society a person's daily activity space is approximately 15 km and varies widely. Health events and their causative exposures may occur anywhere within this activity space, and the use, for example, of place of residence, introduces uncertainty due to mobility assignment. Exact locations do not represent this kind of location uncertainty.

The problem of accounting for uncertainty in statistical inference dates back at least to Fisher (1949, p. 4) who observed

> We may at once admit that any inference from the particular to the general must be attended with some degree of uncertainty, but this is not the same as to admit that such inference cannot be absolutely rigorous, for the nature and degree of the uncertainty may itself be capable of rigorous expression. In the theory of probability, as developed in its application to games of chance, we have the classic example proving this possibility. If the gambler's apparatus are really *true* or unbiased, the probabilities of the different possible events, or combinations of events, can be inferred by a rigorous deductive argument, although the outcome of any particular game is recognised to be uncertain. The mere fact that inductive inferences are uncertain cannot, therefore, be accepted as precluding perfectly rigorous and unequivocal inference.

While concerned primarily with uncertainty associated with specific events under a stochastic process, Fisher's comments are germane to other sources of uncertainty as well, including location uncertainty. Rigorous inference is possible provided one is able to quantify the probabilities of the different possible events. We accomplish this using location models and spatial randomisation.

Risk assessment and analysis has a rich tradition of accounting for uncertainty (Vose, 1996), and uses terms such as 'credibility interval' as opposed to 'confidence interval' to distinguish variability attributable to uncertainty from variability due to sampling error. In risk assessment the emphasis usually is not on statistical inference, but rather on the impact of sampling error and uncertainty on Monte Carlo model estimates of exposure and risk. To our knowledge, the propagation of location uncertainty through spatial statistics for purposes of statistical inference is a new development.

Whether or not centroid locations may safely be used depends on many factors, including the statistical method, spatial resolution of the data, and size of the spatial support of the centroids. Until now techniques have not been available for assessing whether the use of centroids yields misleading results. The methods presented here provide a quantitative basis for answering this question.

While much work remains to be done, this initial research demonstrates the feasibility of incorporating knowledge of the spatial locations of the at-risk population into disease cluster statistics. Because of confidentiality and related issues our knowledge of the locations of health events will always be more or less uncertain. Technological advances in GPS, GIS and geocoding now support quantification of detailed location models describing the spatial locations of the at-risk population. In statistical terms this knowledge now allows us to specify more fully and correctly the spatial sampling space, resulting in increased statistical power.

The location models offer two advantages. First, they provide a ready mechanism for modelling the spatial distribution of the at-risk population. This effectively specifies the sample space and increases statistical power relative to sample-based randomisation tests. Secondly, they provide models of location uncertainty premised on readily available information. These models can be used to propagate location uncertainty within the context of disease cluster tests, as demonstrated in this chapter. They also have obvious utility for masking health event locations to project patient confidentiality.

Credibility offers greater statistical power and supports statistical inference when locations are uncertain. Note that the credibility approach is general and randomisation tests using only sample locations to define the sampling space are a special case of our new approach. This special case arises when we use the point model with the list of alternative locations set to be the actual locations in the sample. The distribution of the test statistic, g_T, condenses to a point mass at Γ_A, and the reference distribution g_U is the same as that calculated using just the sample locations ($g_U = g_A$).

The location models, proximity metrics, gamma product, and spatial randomisation methods are available in the Gamma software (contact the first author for availability). Developed with funding from the National Cancer Institute, Gamma is GIS-compatible and provides an intuitive environment for the visualisation and statistical analysis of spatially referenced human health event data (Figure 11.6). Its embedded spatial data structure integrates the location models, spatial queries, and statistical calculations to optimise performance. An Object Request Broker (ORB) mediates client/server interactions and provides unprecedented performance in distributed environments such as inter-and intranets.

Until very recently, population data for correctly specifying the spatial sampling space of disease cluster and other statistics simply have not been available. As a result,

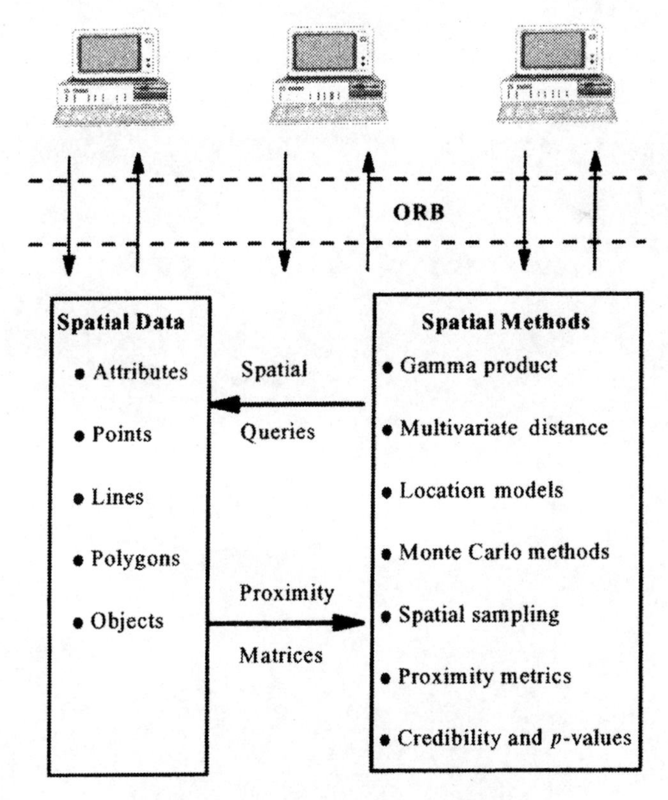

Figure 11.6 Gamma software design

cluster tests using randomisation distributions restricted the sampling space to consist solely of the locations in the sample. Clearly, the correct sampling space is the at-risk population in the study area. The theory, methods and software we are developing recognise this, and provide a ready means for calculating statistics using a properly defined sampling space. We recommend that research on means for incorporating detailed location knowledge from GPS, GIS, geocoding, etc. into statistical methods be fostered in order to improve the statistical power of disease cluster tests, and to better support public health decision making.

ACKNOWLEDGEMENTS

This research was funded by grants R43CA65366-01A1 and R44CA65366-02 from the National Cancer Institute (NCI). The perspectives are those of the authors and do not necessarily represent the views of the NCI. We thank John Chay for helpful discussions on the location models, and two anonymous reviewers for helpful comments.

12

Empirical Studies of Cluster Detection—Different Cluster Tests in Application to German Cancer Maps

Iris K. Zoellner

Landesgesundheitsamt Baden-Württemberg, Stuttgart

Irene M. Schmidtmann

Johannes Gutenberg-Universität Mainz

12.1 INTRODUCTION

Cancer mapping may provide clues to the underlying aetiological process. In many countries, the publication of a national cancer atlas has intensified the search for possible causes of spatial variations in the mortality rates of certain cancers. Disease cluster investigations have become an issue of public health policy.

One of the questions arising when incidence or mortality data—either as maps or simple lists—are screened concerns spatial correlation. Does an observed map display a non-random geographical pattern? Do adjacent regions show more similar rates than non-adjacent regions? To find data sets that deserve further investigation in a public health setting, objective tests may be helpful to distinguish non-random from random maps.

The purpose of the study was to investigate the application of a small selection of measures of spatial autocorrelation, including Moran's I (Moran, 1948, 1950) and Geary's C (Geary, 1954), to maps of the West German Cancer Atlas (see Table 12.1). The maps show age-standardised mortality rates of different cancer sites for 328 regions ('Kreise') as quintile maps. The four test statistics were chosen since they only made use of the information available in the maps. In other chapters of this volume reviews of some alternative cluster testing methods are given (see Chapters 7, 8, 10 and 12 in this volume).

Disease Mapping and Risk Assessment for Public Health. Edited by A.B. Lawson *et al.*
© 1999 John Wiley & Sons Ltd.

Table 12.1 Characterisation of the data included in the German Cancer Atlas

Area	**West Germany**
Number of regions	328
Median number of adjacencies per region	5
Range of person-years per region	170 000 to 9 585 000
Number of maps	44
Cancer site with highest mortality rate	50.19 [a] (lung cancer in men)
Cancer site with lowest mortality rate	0.16 [a] (laryngeal cancer in women)
Cancer mortality in men	183.3 [a]
Cancer mortality in women	117.4 [a]

[a] West Germany 1976–80, age-standardised rate per 100 000 person-years.

Since the distributions of Moran's *I*, Geary's *C* and the other spatial autocorrelation measures depend on the number of regions in the area, on the adjacencies between regions, and on the number of categories used for mapping, a simulation approach was chosen to obtain distributional information on these autocorrelation measures. In a first step, the different statistics were considered for random maps and for the 44 cancer maps. Relationships between the results of the four statistics are discussed. In a second step, an algorithm for the identification of 'extreme' clusters was applied to the maps, which combined spatial smoothing techniques with a method proposed by Grimson *et al.* (1981).

12.2 METHODS

In the German cancer atlas (Becker *et al.*, 1984) mortality rates are published for the area A of West Germany partitioned into $N = 328$ non-overlapping regions R_i ('Landkreise'). Two regions are considered adjacent if they share a common border of non-zero length. We define an adjacency matrix $W = [w_{ij}]_{i=1,\ldots,N;\, j=1,\ldots,N,}$ for the area $A = \cup_{i=1}^{N} R_i$ with

$$w_{ij} = \begin{cases} 1, & \text{if } R_i \text{ and } R_j \text{ are adjacent and } i \neq j, \\ 0, & \text{otherwise,} \end{cases}$$

and

$$w_{..} = \sum_{i=1}^{N} \sum_{j=1}^{N} w_{ij}.$$

Suppose the value of a random variate X has been observed in each of the regions. A set of N observations $\{x_i, i = 1, \ldots, N\}$ and a corresponding adjacency matrix represent a map.

12.2.1 Measures of spatial correlation

The following four cluster indices were considered as tests for spatial autocorrelation and applied to cancer maps of the atlas:

1. Moran's I;
2. Geary's contiguity ratio C;
3. cluster index T of Abel and Becker (1987);
4. test statistic χ_0^2 of Ohno and Aoki (1979, 1981).

Moran's I measures the covariation between rates in adjacent regions in relation to the total variance over all regions in the map and can be formulated as

$$I = \frac{N}{w_{..}} \frac{\sum_{i=1}^{N} \sum_{j=1}^{N} w_{ij}(X_i - \bar{X})(X_j - \bar{X})}{\sum_{i=1}^{N} (X_i - \bar{X})^2}.$$

The contiguity ratio C of Geary is formulated as the sum of squared differences between values in adjacent regions standardised by the total variance over all regions:

$$C = \frac{(N-1)}{2w_{..}} \frac{\sum_{i=1}^{N} \sum_{j=1}^{N} w_{ij}(X_i - X_j)^2}{\sum_{i=1}^{N} (X_i - \bar{X})^2}.$$

Abel and Becker proposed a cluster index T which is defined as the weighted sum of absolute differences between an observed value and its adjacent values:

$$T = \sum_{i=1}^{N} \sum_{j=1}^{N} w_{ij} \frac{|X_i - X_j|}{w_{i.}}$$

When the X_i are taken as ranks and W is symmetric, T is the same as the cluster index D suggested by Smans (1989) apart from weighting.

The approach described by Ohno and Aoki applies to categorised data which are normally displayed by different colours in choropleth maps. This test compares the observed number of adjacencies between regions of equal colour with the expected number of adjacencies of the same colour in random maps. If K denotes the number of categories and n_k, $k = 1, \ldots, K$, the number of regions in category k, then let

$$A_k = \frac{1}{2} \sum_{i=1}^{N} \sum_{j=1}^{N} w_{ij} \delta(x_i, x_j), \quad \text{with } 1 \le k \le K \text{ and } \delta(x_i, x_j) = \begin{cases} 1, & \text{if } x_i = x_j, \\ 0, & \text{otherwise.} \end{cases}$$

The test statistic of Ohno and Aoki is defined by

$$\chi_0^2 = \sum_{k=1}^{K} \frac{(A_k - E_k)^2}{E_k}, \quad \text{with } E_k = \frac{w_{..}}{2} \frac{n_k(n_k - 1)}{N(N-1)}.$$

To use the cluster indices in testing for spatial correlation, their null distribution was obtained by Monte Carlo simulation of 10 000 random quintile maps under the assumption of no clustering. The first part of the simulation results was obtained under the null hypothesis that any permutation of the observed $\{x_i\}$ is equally likely (permutation model) as described by Cliff and Ord (1981).

This null model ignores population heterogeneity. As shown by Alexander *et al.* (1988) and Besag and Newell (1991), regions with small population tend to exhibit extreme rates with higher probability than regions with larger population. Therefore Moran's I is considered to be biased.

For this reason we also used a multinomial model for distribution of cases under the null hypothesis. We kept the number of cases—and thereby the global rate p—fixed and allocated the set of cases $\{n_i\, i = 1, \ldots, N\}$, according to a multinomial distribution with parameter $\sum_i n_i$ and $\{p_i, i = 1, \ldots, N\}$, where $p_i = py_i / \sum_i y_i$, where y_i is the person-years in region R_i. This model reflects equal risks for all individuals in the study area. The second assumption is more reasonable than the permutation model, but it requires far more computing time, especially when many maps are examined as there has to be a separate simulation for each disease under investigation. So we chose to implement the uniform risk model for the maps of the most frequent cancer (lung cancer in men), of the least common site (laryngeal cancer in women) and of leukaemia in women as a disease of interest with a moderate mortality rate of 4.3 per 100 000. The intention of this step was to compare the distributions of the four test statistics under the different null models.

12.2.2 Identification of clusters of extreme rates

When global testing for spatial autocorrelation leads to the conclusion that a map displays a non-random pattern, a possible next step may be directed at detecting clusters of high rates. In our study we tested a combination of spatial smoothing procedures as suggested by Zoellner (1991) and an algorithm suggested by Grimson *et al.* (1981):

1. For each region compute a weighted average of the observed rate and its adjacent values (spatial smoothing).
2. Label those $k = 2, 3, \ldots, [N/10]$ regions exhibiting the highest values.
3. Compare the number of observed neighbourhoods (i.e. Moran's BB count) among labelled regions with the number expected given the random allocation of labels. Compute corresponding p-values.
4. Decide for the existence of a cluster if a sufficiently small local or global minimum of p-value is obtained.

The same procedure may be applied for the identification of clusters of low rates by labelling k regions with lowest rates, respectively (Schmidtmann and Zoellner, 1993).

12.3 RESULTS OF APPLICATION TO GERMAN CANCER MORTALITY DATA

In Table 12.2 descriptive parameters of sample distributions obtained by simulation of 10 000 randomised quintile maps are given. For Moran's I and Geary's C standardised values for I_s and C_s are tabulated.

In Figure 12.1 the distribution of Moran's I under different assumptions for random maps is shown: for the permutation model and uniform risk models based on the population figures per region and on the mean mortality rates of lung cancer, leukaemia, and laryngeal cancer, respectively. While the distribution of Moran's I tended to be robust under the different model assumptions, the moments and critical values of C, T and χ_0^2 differed substantially between the permutation model and the uniform risk models. Concerning robustness, Walter (1993b) and Oden *et al.* (1996) also found Moran's I to

Table 12.2 Moments, critical values, and extreme values of I_s, C_s, T_T, and χ_0^2, for the area of Western Germnay with 328 districts, Monte Carlo simulation (n =10 000 maps, permutation model)

	Mean	s.d.	$\alpha = 0.01$	$\alpha = 0.05$	Minimum	Maximum
I_s	– 0.005	1.003	2.426	1.676	– 3.920	3.978
C_s	0.004	1.005	– 2.35	– 1.645	– 3.546	3.783
T	523.1	14.0	489.13	500.07	465.773	580.007
χ_0^2	4.785	3.099	14.918	10.6405	0.025	27.961

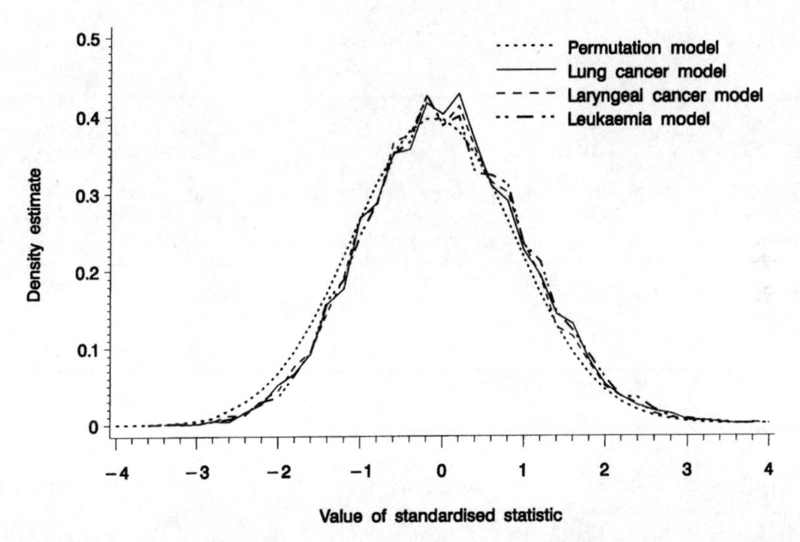

Figure 12.1 Monte Carlo simulation results on Moran's I under different model assumptions, (1) randomization model, (2) uniform risk model for lung cancer in men, (3) uniform risk model for leukaemia in women, (4) uniform risk model for laryngeal cancer in women; 10 000 random maps per model

behave well in the situation of no clustering. In the simulation by Oden *et al.* null models were based on permutations at the individual level similar to the uniform risk approach. Apart from the Lyme disease example, both Morans I and I_{pop} roughly keep the nominal level for moderate α. The results for the extreme tails look worse, but are difficult to judge since they are based on only 1000 replicates. For the alternatives considered there I_{pop} tends to be more powerful than I.

Tables 12.3 and 12.4 give the correlation coefficients of the four test statistics observed under the assumption of uniform risk and for the 44 cancer maps. While the values of Moran's I, Geary's C and the test of Abel and Becker show correlations between – 0.81 and – 0.88 even for random maps, the test of Ohno and Aoki seems to be of different type. The correlations of the test statistics depend very little on the frequency of the disease.

However, for the 44 cancer maps the correlation coefficients of χ_0^2 and I, C, and T took values of 0.88, –0.85, and – 0.88, respectively.

The values of the test statistics I_s, C_s T, and χ_0^2 for the 44 cancer maps are given in Table 12.5. Values in bold type relate to p-values less than 0.05. No spatial aggregation

Table 12.3 Correlation coefficients of the test statistics under the assumption of uniform risk: (1) lung cancer model, (2) leukaemia model, (3) laryngeal cancer model

Test statistic	Geary's *C*	Abel and Becker	Ohno and Aoki
	– 0.860 (1)	– 0.836 (1)	– 0.001 (1)
Moran's *I*	– 0.868 (2)	– 0.839 (2)	– 0.024 (2)
	– 0.877 (3)	– 0.841 (3)	– 0.003 (3)
		0.816 (1)	– 0.151 (1)
Geary's *C*	1	0.822 (2)	– 0.162 (2)
		0.815 (3)	– 0.114 (3)
			– 0.148 (1)
Abel and Becker		1	– 0.172 (2)
			– 0.155 (3)

Table 12.4 Correlation coefficients of the test statistics for the 44 cancer maps

Test statistic	Geary's *C*	Abel and Becker	Ohno and Aoki
Moran's *I*	– 0.990	– 0.991	0.883
Geary's *C*	1	0.989	– 0.854
Abel and Becker	1	– 0.880	0.880

was found by any of the test criteria for bone cancer and non-Hodgkin's lymphoma in woman. For Hodgkin's disease and ovarian cancer in women, melanoma in men, and leukaemia for either sex, only one of the four tests found spatial autocorrelation. Geary's *C* and *T* of Abel and Becker indicated spatial clustering for female kidney cancer and Hodgkin's disease in men, while the other tests did not reject the null hypothesis of a random spatial distribution of rates for these sites. The remaining sites did exhibit spatial aggregation according to at least three tests. For 13 maps i.e. stomach, bladder, and all sites for either sex, lung and larynx in men, female gall-bladder, colon, liver, thyroid and breast cancer, all four cluster indices had values that were not covered by the range of simulation data. The most distinct patterns were found in maps of stomach and lung cancer.

Comparing the results for all cancer maps, the test of Ohno and Aoki tended to be less sensitive than Moran's *I*, Geary's *C*, and *T* of Abel and Becker. I_S, C_S, and T tend to detect any deviation from spatial independence. When testing for more specific alternatives ('uni-coloured clusters') the test of Ohno and Aoki may be more appropriate. In consideration of the differences between the results of the permutation and the uniform risk approach we suggest using the uniform risk model in simulations when the corresponding parameters are available. In our study, Moran's I_S tended to be more robust against model changes than the other cluster indices considered, since moments and critical values of Moran's I_S changed only marginally under the different model assumptions. This corresponds to the results of Oden *et al.*

Clusters of high mortality rates were identified for stomach and liver cancer for both sexes. For stomach cancer a cluster of high rates was found in Bavaria in the south-east

Table 12.5 Test values for 44 German cancer maps (mortality data 1976–80)

Site	Sex	I_s	C_s	T	χ_0^2
Lung	M	**22.35**	**– 17.96**	**242.5**	**430.3**
Lung	F	**8.81**	**– 8.95**	**408.7**	**26.0**
Stomach	M	**16.15**	**– 14.63**	**309.4**	**244.0**
Stomach	F	**17.16**	**– 14.86**	**295.4**	**296.7**
Gall-bladder	M	**5.11**	**– 5.71**	**450.4**	**20.0**
Gall-bladder	F	**12.75**	**– 10.96**	**358.9**	**101.4**
Bladder	M	**12.12**	**– 10.72**	**367.3**	**79.9**
Bladder	F	**8.82**	**– 8.65**	**415.5**	**41.6**
Thyroid	M	**5.90**	**– 5.48**	**439.9**	**30.0**
Thyroid	F	**9.64**	**– 9.31**	**390.6**	**30.7**
Colon	M	**3.26**	**– 4.54**	**463.6**	**21.5**
Colon	F	**9.10**	**– 9.53**	**381.9**	**77.5**
Larynx	M	**8.98**	**– 8.50**	**405.3**	**43.4**
Larynx	F	**1.80**	**– 2.44**	**491.2**	**13.4**
Brain	M	**7.26**	**– 6.66**	**422.9**	**15.0**
Brain	F	**4.57**	**– 4.91**	**453.3**	**17.1**
Oesophagus	M	**6.84**	**– 6.98**	**431.7**	**20.9**
Oesophagus	F	**6.07**	**– 5.69**	**449.0**	**17.9**
Rectum	M	**6.75**	**– 6.82**	**418.4**	**29.4**
Rectum	F	**3.02**	**– 3.59**	**476.2**	6.5
Liver	M	**6.19**	**– 4.70**	**456.0**	**20.0**
Liver	F	**7.73**	**– 6.57**	**431.1**	**53.8**
Bone	M	**1.80**	**– 2.29**	**499.2**	7.3
Bone	F	1.10	– 1.28	509.7	7.6
Pancreas	M	**4.03**	**– 4.40**	**466.2**	10.3
Pancreas	F	**3.21**	**– 3.08**	**468.5**	**18.8**
Leukaemia	M	1.47	– 1.51	**495.0**	7.6
Leukaemia	F	1.64	**– 1.98**	506.4	2.8
Melanoma	M	1.43	**– 2.10**	503.6	5.3
Melanoma	F	**3.51**	**– 4.29**	**474.9**	**14.6**
Kidney	M	**2.88**	**– 3.88**	**494.2**	9.5
Kidney	F	1.66	**– 2.14**	**498.1**	2.7
M Hodgkin	M	1.15	**– 2.48**	**494.8**	2.9
M Hodgkin	F	0.96	**– 2.51**	506.1	6.7
NH Lymphoma	M	**1.99**	**– 2.40**	**491.7**	5.0
NH Lymphoma	F	0.21	– 0.88	524.8	3.8
Uterus	F	**5.91**	**– 6.28**	**440.6**	8.3
Cervix uteri	F	**6.04**	**– 6.37**	**448.6**	**13.8**
Breast	F	**7.83**	**– 7.01**	**440.0**	**38.9**
Ovary	F	1.38	– 1.14	**494.2**	3.4
Prostate	M	**3.42**	**– 4.15**	**477.1**	4.2
Testis	M	**2.78**	**– 3.98**	**468.5**	**14.7**
All sites	M	**15.05**	**– 13.92**	**332.8**	**160.5**
All sites	F	**8.81**	**– 8.47**	**408.1**	**80.0**

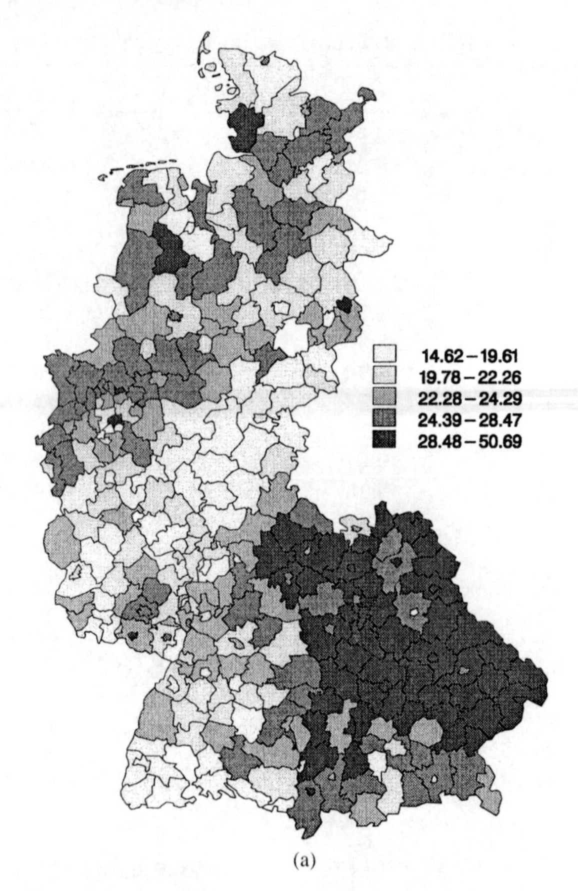

(a)

Figure 12.2 (a) Stomach cancer mortality in men 1976–80, West Germany (age standardised rates per 100 000 person-years); (b) laryngeal cancer mortality in women 1976–80, West Germany (age standardised rates per 100 000 person-years)

of Germany, while the highest liver cancer rates were concentrated in Rhineland–Palatinate. Spatial smoothing supported the identification.

Figure 12.2(a) displays stomach cancer mortality in men which exhibits the most prominent clustering, while Figure 12.2(b) displays laryngeal cancer in women which is the rarest disease in the atlas.

12.4 DISCUSSION

Alexander and Boyle (1996) have assembled an overview of applications of various methods for investigating localised clustering of disease to random data sets and to data sets generated by prespecified processes. This exercise did not yield a single most useful method. Instead, Alexander and Boyle suggest combining methods, e.g. start with a quadrat count method that could be followed by more complex analyses if appropriate. Oden's I_{pop} (Oden, 1995) was not considered in our comparison because the study was begun before the publication of his paper. Furthermore, the strong age dependency of

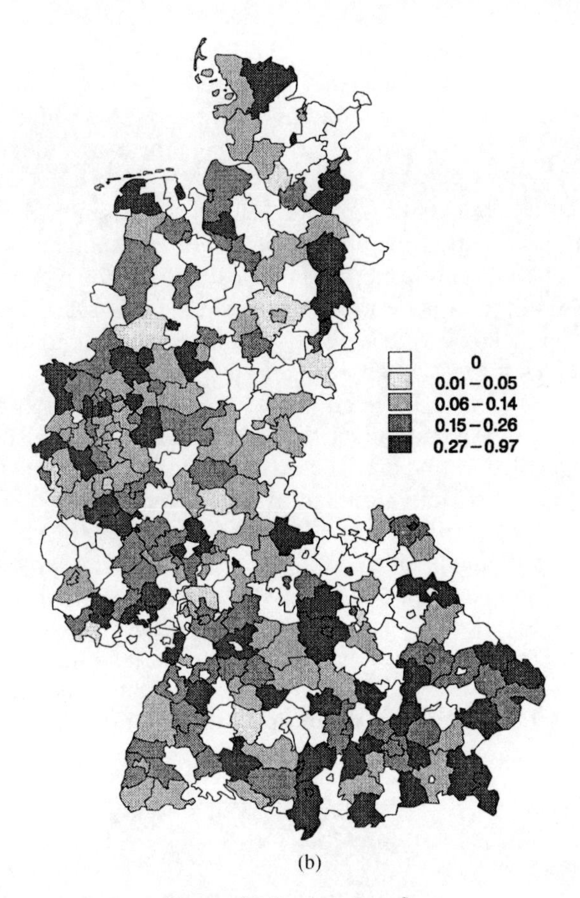

(b)

Figure 12.2 *(continued)*

cancer would require adjusting any measure for age-specific population distribution. This has also been pointed out by Gbary *et al.* (1995). However, such data were not readily available, so we took age dependency into account by basing our study on age-standardised rates.

If age dependency is not a problem, we suggest using I_{pop} which is also sensitive against alternatives of within-area clustering. Otherwise, if all necessary information is available, I_{pop} could be modified to take age-specific population distribution into account, or Tango's test (Tango, 1995) could be used.

The aggregation level of the data did not allow the application of point-based methods, e.g. the Cuzick–Edwards test (Cuzick and Edwards, 1990). The results of Oden *et al.* do not support the expectation that point-based methods necessarily perform better than area-based methods like I or I_{pop}. They stated that 'there are situations where Grimson's test or Moran's I are preferable. . . . their usefulness arises when the data at hand are less complete than our simulation provides'.

When we sorted all 44 maps by the value of Moran's I and screened them visually in this sequence we found that the visual impression of clustering and the values of I corresponded well. This would support the application of I or I_{pop} in a public health setting when many items could be mapped but one has to restrict attention to non-random patterns. Walter (1993a), in a more formal experiment, also found a correspondence between cluster index D and visual assessment. We agree with him that visual assessment may depend on shape, structure, and shading of the map and on the experience of the rater.

The Grimson method for cluster detection is a fairly coarse one; however, it is in line with the intention to use only the information displayed in the maps. If more information is available, then more subtle methods for cluster detection can be applied, e.g. local indicators of spatial autocorrelation (LISA) as defined by Anselin (1995) and Getis and Ord (1996). Similar indicators were suggested by Munasinghe and Morris (1996). Tango suggested using a focused test when a general test yielded a significant result (Tango, 1995). However, so far not much is known about the distribution of local statistics. Tango gives a normal approximation for the focused test, and Tiefelsdorf (1998) tackled the distribution of local Moran's I for the case of normally distributed random variables.

The cluster index and a method for identifying clusters should be chosen depending on the type and quality of available data, the disease rate, and the character of the clustering process and its regional variation. For rare diseases and large differences in the regional population, Tango's test, Oden's I_{pop} or point-based methods will be more appropriate, while for common diseases and regional populations of comparable size, Moran's I or Grimson's method will be applicable. On the basis of our simulation results we would not recommend the use of C_s, T, or χ_0^2 since their distributions differed substantially between the permutation model and the more reasonable uniform risk approach.

Part III

Ecological Analysis

13

Introduction to Spatial Models in Ecological Analysis

A. Biggeri and F. Divino

University of Florence

A. Frigessi

Norwegian Computing Centre

Andrew B. Lawson

University of Aberdeen

D. Böhning

Free University, Berlin

E. Lesaffre

Katholic University, Leuven

J.-F. Viel

University of Besançon

13.1 INTRODUCTION

Ecological analyses are epidemiological investigations examining associations between disease incidence and potential risk factors as measured on groups rather than individuals. Typically the groups involved are defined by geographical area such as country, region or smaller administrative area. In recent years there has been great effort on this kind of analysis, mainly because of the possibility of studying many variables and populations at low cost. Indeed, data from different sources can be linked more easily at the ecological level than at the individual one. On the other hand, there are some methodo-

Disease Mapping and Risk Assessment for Public Health. Edited by A.B. Lawson *et al.*
© 1999 John Wiley & Sons Ltd.

logical limitations. Ecological analyses could provide spurious results when there are individual-level predictors of the response which are associated with the aggregate variables of interest. This is due to the fact an ecological study might be subject to biases not present in corresponding individual-level study (for details about the problems arising when the level of analysis is different from the level of associations inference we refer to Morgenstern, 1998). Anyway an ecological analysis is valid when the ecologic effect is of primary interest or variables representing population features or environmental characteristics not available at the individual level (e.g. demographic density, type of welfare system, etc.) are involved. The aim of this chapter is to provide an introduction to ecological analysis problems and to review the statistical methods used to analyse geographical aggregate data. We want also to point out the importance of correctly specifing models in order to have valid inference on the association between disease and risk factor.

13.2 ECOLOGICAL FALLACY IN SPATIAL DATA

The regression analyses based on geographically collected data are subject to bias due to their aggregate nature and to the potential presence of spatial autocorrelation among the responses: these two aspects are related to each other, due to the fact that aggregation and scale change can lead to autocorrelation. Autocorrelation is also found due to unobserved confounder variables. In this way, making inference on the basis of ecologic associations to individual-level behaviour could have serious pitfalls. The problem, known as the *ecological fallacy* (also named *ecological bias*), was first pointed out by Robinson (1950), who demonstrated that the total correlation between two variables as measured at an ecologic level can be expressed as the sum of a within-group and a between-group component. Later Duncan *et al.* (1961) extended this result deriving the relationship between the regression coefficients in a linear model. The sources of ecological bias have been investigated by many authors (see, for example, Morgenstern, 1998; Richardson *et al.*, 1987; Piantadosi *et al.*, 1988; Greenland and Morgenstern, 1989; Greenland, 1992; Greenland and Robins, 1994; Steel and Holt, 1996). In addition to the individual-level sources (misspecification, within-group confounding, no additive effects, misclassification) special attention has been given to the bias due to grouping individuals (Brenner *et al.*, 1992; Greenland and Brenner, 1993). In particular, Greenland and Morgenstern (1989) analysed how grouping influences associations of exposure factors to disease, and they pointed out that ecological bias may also arise from confounding by group and effect modification by group. Now consider some ecological groups indexed by i and let p_i be the proportion of exposed subjects (a dichotomous variable), r_{0i} the individual rate in unexposed, and r_{1i} the individual rate in exposed at the site i. The crude rate in group i is given by

$$\begin{aligned} r_{+i} &= r_{0i}(1 - p_i) + r_{1i}p_i \\ &= r_{0i} + D_i p_i, \end{aligned}$$

where $D_i = r_{1i} - r_{0i}$ is the individual rate difference. Consider a population linear regression model of average disease level on the average exposure level in groups:

$$r_{+i} = \alpha + \beta p_i,$$

then $1 + \beta/\alpha$ is the ecological rate ratio. Greenland and Morgenstern demonstrated that the ecological regression coefficient β can be viewed as the expected rate difference at the individual level plus two bias terms. The mathematical relationship is given by

$$\beta = \mathrm{E}(D_i) + \frac{\mathrm{cov}(p_i; r_{0i})}{\mathrm{var}(p_i)} + \frac{\mathrm{cov}([p_i - \mathrm{E}(p_i)]p_i; D_i)}{\mathrm{var}(p_i)}.$$

The first bias component,

$$\frac{\mathrm{cov}(p_i; r_{0i})}{\mathrm{var}(p_i)},$$

is present when the unexposed rate is associated with the level of exposure in the group, and it may be viewed as a bias term due to confounding by group. It is plausible that such confounding acts because some external factor causing the disease is associated with groups having a higher level of exposure factor. The second bias component,

$$\frac{\mathrm{cov}([p_i - \mathrm{E}(p_i)]p_i; D_i)}{\mathrm{var}(p_i)},$$

is present when the risk difference in a group is associated with the level of exposure and it may be viewed as a bias term due to effect modification by the group. Such a remark commits ecological fallacy if we assume that the ecological rate ratio $1 + \beta/\alpha$ is only determined by the individual rate difference effect when, in fact, it may be also caused by the two bias components effect. Several strategies can be adopted to tackle the potential flaws in ecological modelling. First, we could try to estimate the joint distribution of outcome and explanatory variables within areas using a sample drawn from the populations investigated, and use the information collected to adjust the ecological regression coefficient and standard errors. This approach has been proposed by Plummer and Clayton (1996) and Prentice and Sheppard (1995). The reader should note that this derivation does not include spatial effects and it can also be viewed as an example of a mixed design with individual and ecological variables (see Chapters 14 and 16 in this volume; see also Lawson and Williams (1994), for an example of multiple level exposure risk modelling). When sampling within areas is not feasible, a second strategy could be to adjust for the correlation between area prevalence of the exposure variable and the baseline rate of disease, provided no effect modification occurs. If the level of aggregation is sufficiently thin, then a regression model for autocorrelated data would result in a sort of stratification by spatial closeness, where the baseline rates would be expected not to vary. Clayton *et al.* (1993) gave a justification of this approach in terms of a hidden spatially structured confounder. Indeed, where the spatial variation of the risk factor is similar to that of the disease, geographical location may act as a confounder.

For example, in Figure 13.1 hypothetical data on male suicide rates (externally standardised mortality ratio) and prevalence of unemployment in 30 areas are plotted. No relationship is apparent while a strong North–South gradient in suicide rates and unemployment is observed. The confounding by location is evident when reporting the predicted values from separate Poisson regressions for northern, central and southern areas. Finally, it should highlighted that spatially unstructured confounders are likely to be present in ecological data; these confounders are usually unknown due to the fact

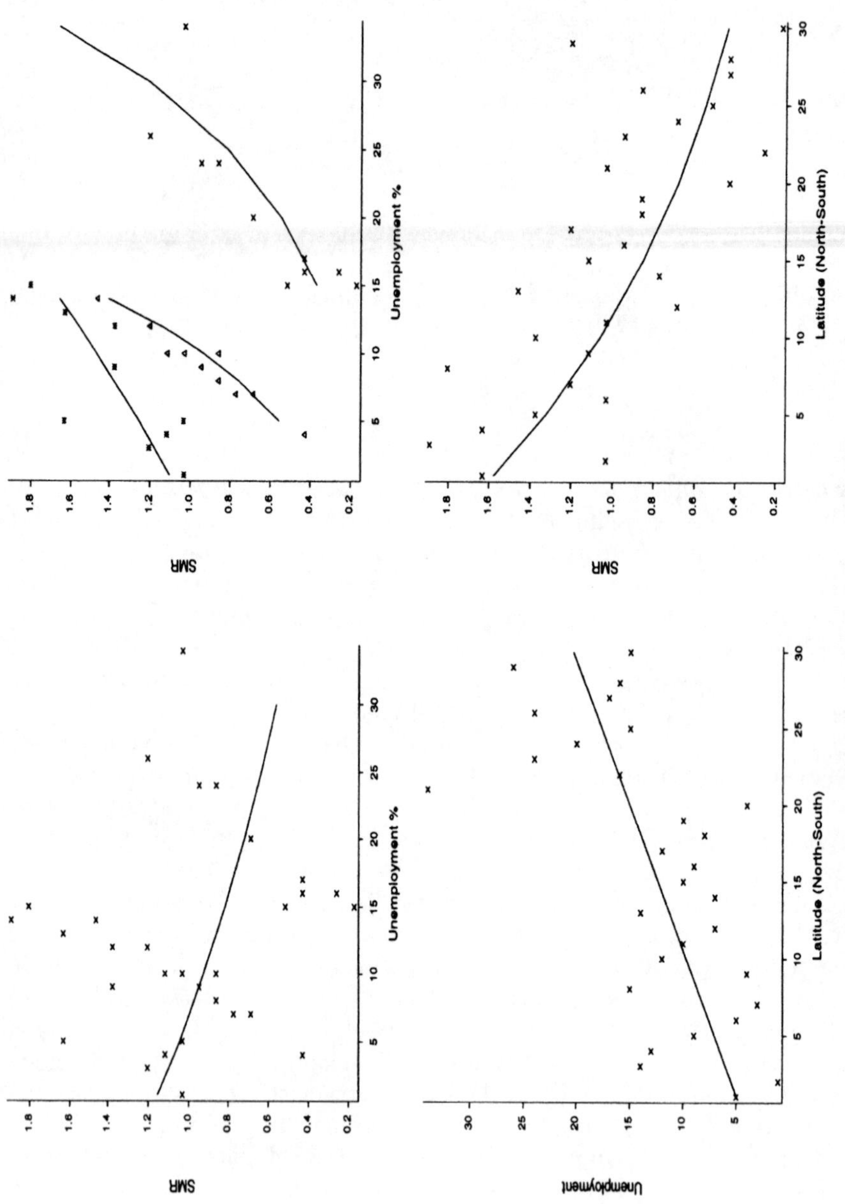

Figure 13.1 Hypothetical data on suicide and unemployment

that factors which are not confounders at the individual level can be confounders at the aggregate level (Greenland, 1992). In the data such hidden confounding will result in a certain degree of extra-variability (Clayton *et al.*, 1993).

In the remainder of this chapter statistical models that are able to cope with this sort of confounding will be reviewed. The control of small level spatial variation has many similarities to that developed in time series analysis, the main difference in geographical epidemiology is that the focus of the research is on the regression coefficient and not primarly in the interaction component, which in most applications is regarded as a nuisance.

13.3 STATISTICAL MODELS

Statistical methods proposed in the context of ecological regression, emphasising the role of autocorrelated models to ensure validity to this kind of analysis, are discussed below. These approaches do not cover all the points about ecological designs, leaving aside the problem of error-in-variables (for which see Chapter 14 in this volume and Bernardinelli *et al.*, 1997) and the strategies involving the use of individual data (Plummer and Clayton, 1996).

13.3.1 Poisson regression models

Before introducing spatial models we first consider the Poisson regression model that represents the starting point of statistical methods in ecological analysis. Let $\{Y_i, i = 1, \ldots, n\}$ be the set of the observed number of events of a certain disease and $\{E_i, i = 1, \ldots, n\}$ be the set of the expected number of events under a reference set of age-specific rates for n areas of the region of interest. Assume Y_i follows a Poisson distribution with expectation:

$$\mu_i = \theta_i E_i,$$

where θ_i is the relative risk for site i. The maximum likelihood estimates of θ_i, under a saturated model, are given by the standardised mortality ratios:

$$\mathrm{SMR}_i = \frac{Y_i}{E_i}.$$

This model can be extended to a set of explanatory variables $X_1, X_2, \ldots, X_H$ in a log-linear formulation:

$$\log \mu_i = \log E_i + \sum_{h=1}^{H} \beta_h x_{ih}.$$

Maximum likelihood estimates of the coefficients β_h can be obtained in a generalised linear models framework. A Poisson regression model can include the influences of many ecologic factors on a disease (the covariates X_h) but it does not control for the autocorrelation and for the extra-Poisson variability, which may arise due to, for example, unobserved confounding variables. Some authors argue that non-linear ecological models give biased estimates of the individual-level coefficients. Since this bias is negligible for moderately large risk ratios, we do not discuss this point any further (see, for example, Richardson *et al.*, 1987, and Greenland, 1992). Last, but not least, the reader should be aware of the potential bias introduced by regressing age-standardised effect measures (e.g. SMR) on non-standardised explanatory variables (Rosenbaum and Rubin, 1984).

13.3.2 Bayesian mixed models

Unstructured and structured extra-Poisson sources of variability can be taken into account by the following generalised linear mixed model:

$$\log \theta_i = t_i + u_i + v_i,$$

where

$$t_i = \sum_{h=1}^{H} \beta_h x_{ih}$$

denote the fixed regression component, u_i are the random unstructured terms (termed *heterogeneity*) and v_i the random spatial structured terms (termed *clustering*). Introducing the heterogeneity and clustering terms represents a way of controlling for unmeasured covariates. Defining appropriate prior distributions on the hyperparameters involved in the random part of the model, a Bayesian inference can be made using the posterior distribution of θ_i. In particular, estimates of the relative risks can be computed by running Markov chain Monte Carlo (MCMC) algorithms. This model was introduced by Clayton and Kaldor (1987) in disease mapping framework and then it was developed by Besag *et al.* (1991) and Clayton *et al.* (1993) in ecological analysis. (For details about ecological analysis in a Bayesian setting see Chapters 14 and 11 in this volume for an application of this model in specific clustering.)

13.3.3 Approximate hierarchical random effect models

Two approximate solutions have been proposed to fit the generalised linear mixed models for disease counts: the penalised quasi-likelihood approach (PQL) and the marginal quasi-likelihood approach (MQL) (Breslow and Clayton, 1993; Goldstein, 1995; see also Chapter 16 in this volume). These approaches are based on integrating the likelihood over the random effect terms to estimate the regression coefficients. The penalised quasi-likelihood approach comes from applying Laplace's method to an integral approximation:

$$\begin{aligned} IL &= \int l(\beta, e) \mathrm{d}e \\ &= c|\mathbf{D}(\sigma)|^{-1/2} \int \mathrm{e}^{-K(\beta,e)} \mathrm{d}e \\ &\approx c|\mathbf{D}(\sigma)|^{-1/2} \frac{\mathrm{e}^{-K(\beta,e)}}{|K''(\beta,e)|^{1/2}}, \end{aligned}$$

where β is the usual vector of fixed effects, e is the vector of random effects $e_i = u_i + v_i$, $\mathbf{D}(\sigma)$ is the covariance matrix of such random effects depending on an unknown variance vector σ, and $K''(\beta, e)$ denotes the matrix of second-order partial derivatives with respect to β of a function $K(\beta, e)$, including the link function and other specific components (see Breslow and Clayton, 1993, for details). Note that increasing domain asymptotics are assumed in the case, as opposed to infill asymptotics. The relevance of this assumption should be checked in any application. The above integral approximation gives on the log scale the following penalised quasi-likelihood function:

$$pql(\beta, e) = -\tfrac{1}{2}\log|\mathbf{D}(\sigma)| - \tfrac{1}{2}\log|K''(\beta, e)| - K(\beta, e).$$

The marginal quasi-likelihood approach approximates the populations mean by discarding the random terms in the linear predictor, i.e. given that the expected value of the relative risk θ is

$$\theta = h(\mathbf{X}\beta + \mathbf{Z}e),$$

under the PQL approach, it becomes:

$$\theta = h(\mathbf{X}\beta)$$

using the MQL approximation to the population mean. In the above equations $\mathbf{X}$ and $\mathbf{Z}$ denote the design matrices of the fixed and random effects, respectively, and h is the inverse link function. Computationally, the method iterates between the estimation of the fixed terms in the model by generalised least squares, and of the random terms, using the residuals and an appropriate design matrix. There are several proposals in the literature concerning how to introduce the spatially structured component v_i. Breslow and Clayton considered only one random term at a time, either the unstructured or the structered but not both. Langford *et al.* (see Chapter 16 in this volume) showed how to specify and estimate both random terms in a multilevel model using the generalised least squares algorithm (Goldstein, 1995). The MQL approach has been claimed to be more robust in practice but it may lead to biased estimates (Breslow and Clayton, 1993; Langford *et al.*, Chapter 16 in this volume). A generalised estimating equation approach (Liang and Zeger, 1986) to the inference on disease risk for mapping purposes has been proposed by Yasui and Lele (1997). This kind of solution treats the random structured terms as nuisance terms while providing correct covariances for the fixed part and can be described as a marginal approximation. All the approximate solutions are suitable for non-sparse data. Using the MQL approach one should be particularly cautious and MCMC or parametric bootstrap is recommended to check for biases (Goldstein, 1996b).

13.3.4 Truncated auto-Poisson models

A different way to address spatial interaction among areas is to introduce an autoregressive term into the systematic component of the Poisson regression model (Ferrándiz *et al.*, 1995). This equates to assuming that $Y_i|Y_{j\neq i}$ is Poisson distributed with expectation (on the log scale)

$$\log \mu_i = \log E_i + \sum_{h=1}^{H} \beta_h x_{ih} + \sum_{j \in S} \gamma_{ij} Y_j,$$

where S is the set of sites, and γ_{ij} is the spatial interaction term for the pair (i, j). This is the auto-Poisson model introduced by Besag (1974). It is shown that if $\gamma_{ij} < 0$, then the counts process $Y = (Y_1, Y_2, \ldots, Y_n)$ admits a joint probability distribution. Negativity on γ_{ij} can be avoided if the Poisson distribution is truncated. To estimate the parameters β_h and γ_{ij} by maximum likelihood, we need to compute the normalising constant of the joint distribution of the process Y, which is impossible to express analytically and cannot be computed numerically. Hence any appropriate algorithm must include some approximation of the integrals involved, for example using MCMC methods. Ferrándiz *et. al* (1995) were aware of this difficulty and proposed a maximum pseudo-likelihood approach. The pseudo-likelihood function, introduced by Besag (1974), is the product over all sites of the conditional probability distributions. As an approximation of the likelihood, the pesudo-likelihood approach allows for efficient parameter estimation at lower computational cost, because no untractable normalising constant is present. After having specified a threshold for the definition of a neighbourhood (for instance the set of nearest neighbours of each site), this model is usually parameterised in order to reduce the number of interaction parameters γ_{ij} to be estimated. If we denote by ∂_i the set of nearest neighbours of the site i, some popular choices are reported below:

(i) $\sum_{j \in \partial_i} \gamma_{ij} Y_j$, one for each pair of neighbours;
(ii) $\gamma \sum_{j \in \partial_i} Y_j$, one for all pairs of neighbours;
(iii) $\gamma \sum_{j \in \partial_i} a_{ij} Y_j$, where a_{ij} are known proximity indexes.

For example, Ferrándiz *et al.* (1995) used the third choice and proposed the following proximity index between pairs of sites:

$$a_{ij} = \frac{\sqrt{N_i N_j}}{d_{ij}},$$

where N_i and N_j stand for the population of areas i and j, and d_{ij} is the geographical distance between the pair (i, j).

13.3.5 Parametric interaction models

When appropriate, the degree of interaction between areas can be modelled explicitly as a function of covariates, e.g. the distance d. Several functions have been proposed in the literature, for example; denoting by ρ the auto correlation parameter, we can use:

(i) $1 - \rho d$, linear at the origin (triangular correlation);
(ii) $\exp(-\alpha d^{\beta})$, non-linear (exponential decay);
(iii) for a given T: threshold $\begin{cases} 1, & d \leq T \\ \rho(d), & d > T \end{cases}$, discrete.

The form of the interaction function had been selected for mathematical convenience and simplicity. Indeed, this is a crucial point in the analyses and deserves special considerations: for example Cook and Pocock (1983), and Pocock *et al.* (1982) described a thoughtful residual analysis to justify their choice of an exponential function (ii) with $\beta = 1$. Lawson (1997) and Lawson *et al.* (1996) utilised an exponential covariance model ((ii) above) within an autocorrelated Gaussian prior in a GLM framework with ecological applications. Their model can be applied to a range of likelihoods, and approximate maximum *a posteriori* estimates were derived and compared with Metropolis–Hasting MCMC sampling algorithms. (See also Chapter 16 in this volume.)

13.3.6 Non-parametric Poisson interaction models

Consider again the auto-Poisson model. If we substitute the parameter γ_{ij} with a function of covariates for the spatial interaction between sites, such as the geographical distance d_{ij} between pairs of area, then that model becomes:

$$\log\mu_i = \log E_i + \sum_{h=1}^{H} \beta_h x_{ih} + \sum_{j\in\partial_i}^{H} \varphi(d_{ij}) Y_j,$$

where φ could be any function to be estimated. In many applications it is important to suppose certain regularities for the functions involved in the model. We assume that φ is differentiable on a closed set D with an absolutely continuous first derivative and we denote the space of such functions by S_D. Following the *roughness penalising* approach (Green and Silverman, 1994), it is possible to define a penalised log likelihood and computationally simpler penalised log pseudo-likelihood (Divino *et al.*, 1998). We denote these functions by PLL(β, φ) and PLPL(β, φ), respectively. A classical roughness penalty is

$$\int_D [\varphi''(w)]^2 \mathrm{d}w.$$

Then PLL(β, φ) and PLPL(β, φ) are given by

$$\mathrm{PLL}(\beta, \varphi) = l(\beta, \varphi) - \lambda \int_D [\varphi''(w)]^2 \mathrm{d}w$$

and

$$\mathrm{PLPL}(\beta, \varphi) = Pl(\beta, \varphi) - \lambda \int_D [\varphi''(w)]^2 \mathrm{d}w,$$

where $l(\beta, \varphi)$ and $Pl(\beta, \varphi)$ are the logarithms of Poisson likelihood and pseudo-likelihood, respectively, and λ is a parameter governing the degree of smoothing that can be estimated by cross validation techniques. Under these assumptions, the estimators of β and φ maximising PLL(β, φ) or PLPL(β, φ) can be derived by applying standard results of classical cubic spline theory (Green and Silverman, 1994). Indeed, it can be shown (Divino *et al.*, 1998) that the optimal estimate of φ over the functional space S_D is the cubic spline with knots given by the ordered sequence (without repetitions) of spatial covariate values, e.g. the geographical distances d_{ij} between pairs of sites. We denote such a sequence by $(w_1, w_2, \ldots, w_{K_n})$, where the number of different elements K_n depends on the number of sites. To compute such an optimal spline we have to estimate the corresponding values $\varphi(w_k), k = 1, \ldots, K_n$. We can do this by applying the stochastic gradient algorithm to the penalised log likelihood (using MCMC methods to approximate the normalising constant) or by applying Newton's algorithm to the penalised log pseudo-likelihood, which avoids the use of any MCMC iterations. As for the parametric auto-Poisson model, in this case also the pseudo-likelihood approach allows for efficient joint estimates of β and φ at lower computational cost (for further details see the references in Divino *et al.*, 1998).

Table 13.1 Hypothetical data on suicide and unemployment

Area ID	Observed	Expected	Latitude 1-northest 30-southest	Unemployment (%)	Neighbours
1	12	11.633	2	1	3, 4, 20
2	140	116.333	7	3	5, 3, 4
3	13	11.633	9	4	1, 2, 4, 5, 20
4	12	11.633	6	5	1, 2, 3, 20
5	19	11.633	1	5	2, 3, 6
6	160	116.333	5	9	5, 7, 8
7	16	11.633	10	12	6, 8, 9, 10
8	19	11.633	4	13	6, 7, 9, 10
9	22	11.633	3	14	7, 8, 10
10	21	11.633	8	15	7, 8, 9
11	5	11.633	20	4	12, 13, 17, 19
12	9	11.633	14	7	11, 13, 15, 17, 19
13	80	116.333	12	7	11, 12
14	10	11.633	18	8	15, 16, 17, 27
15	11	11.633	16	9	12, 14, 16, 17, 27, 28
16	100	116.333	19	10	14, 15, 17, 19, 19
17	12	11.633	11	10	11, 12, 14, 15, 16, 18, 19
18	13	11.633	15	10	16, 17, 19, 20
19	14	11.633	17	12	11, 12, 16, 17, 18, 20
20	17	11.633	13	14	1, 3, 4, 18, 19
21	2	11.633	30	15	23, 25
22	6	11.633	25	15	23, 24, 25
23	30	116.333	22	16	21, 22, 24, 25
24	5	11.633	28	16	22, 23, 25, 26, 29, 30
25	5	11.633	27	17	21, 22, 23, 24, 26, 30
26	80	116.333	24	20	24, 25, 28, 30
27	11	11.633	23	24	14, 15, 28, 29, 30
28	10	11.633	26	24	15, 26, 27, 29, 30
29	14	11.633	29	26	24, 27, 28, 30
30	12	11.633	21	34	24, 25, 26, 27, 28, 29

13.4 EXAMPLE

Consider the hypothetical data on suicide and unemployment (see Figure 13.1 and Table 13.1). The estimated regression coefficients of suicide rates on the prevalence of unemployment under simple Poisson and Bayesian ecological regression models are reported in Table 13.2.

The simple Poisson regression analysis estimates a negative coefficient for unemployment. Including heterogeneity and clustering terms in the Bayesian model of Section of Table 13.2 results in a positive significant coefficient, since the confounding by location has been controlled for.

A second example is reported in Table 13.3. Here only unstructured extra-variability that is present in the data has been generated and again the statistical model which is designed for heterogeneity (a simple negative binomial regression, Clayton and Kaldor, 1987) is able to obtain the unbiased estimates (Section 13.3). Moreover, not taking into account the random terms in any models will produce incorrect standard errors.

13.5 CONCLUSIONS

The value of proof for a potentially causal relationship provided by an ecological regression has been considered very low, and these kinds of investigations have been labelled hypothesis generating, heuristic, explorative, descriptive and so forth. Here we have

Table 13.2 Coefficients and standard errors from Poisson regression and Bayesian ecologic regression

Variable	Poisson	Bayes
Unempolyment	– 0.0248	0.0302
SE	0.0056	0.0130
p-value	< 0.0001	0.0138
Heterogeneity	—	0.0427
SE	—	0.0153
Clustering	—	0.5556
SE	—	0.758

Table 13.3 Coefficients and standard errors from Poisson regression and Random effects Poisson regression

Variable	Poisson	Random effects
Unemployment	– 0.0234	0.0453
SE	0.0049	0.0201
p-Value	< 0.0001	0.0240
log(Heterogeneity)	—	– 1.0395
SE	—	0.3502
p-Value	—	0.0030

shown that new appropriate statistical models are available to take into account the bias induced by varying baseline rates among areas. These models can all be grouped under the common feature of being models for autocorrelated data. Epidemiologists should realise that the potential of ecological analyses has been greatly improved in terms of its validity by such new tools. Different models have been proposed and reviewed. The key point is the choice about the inclusion of a term(s) for spatial autocorrelation: while it is important to achieve validity this term(s) could mask the exposure–disease relationship when unnecessary (Breslow and Clayton, 1993). The Bayesian model is computationally intensive but it allows parameters for heterogeneity and spatially structured variability, leaving to the data an appropriate weighting of the two components. The truncated auto-Poisson models are interesting alternatives, especially when the simple pseudo-likelihood is used. However, these models could be more sensitive to the specification of the auto-correlation terms, which are fixed a priori. This argument applies also to the penalised and marginal approaches, which could also account for the heterogeneity among areas. Finally, some interesting developments are represented by non-parametric modelling of the spatial interaction, since it addresses directly the point of the possible overcorrection of the relationship under study. In addition to the purely ecological studies, mixed designs have now been proposed: among them sampling within areas can provide information to check and to control for ecological biases. We have recalled these studies just to mention the potential of multilevel or hierarchical models, which utilise both aggregate and individual data. As a final remark, some authors (see, for example, Cohen, 1990; Koopman and Longini, 1994; Schwartz, 1994; Susser, 1994 a, b) point out the different role of ecological analyses with regard to individual-level analyses. An ecologic study is not the mathematical equivalent of an individual-level one. Indeed, when the researcher ranges from individual level to ecological level, individuals are joined into groups, but the group — like a new study unit — acquires collective properties that are more than the simple sum of the individual properties of its members. In this way an ecologic variable is often referred to a different construct than its namesake at the individual level. Thus the difference between individual and ecological inferences may also be due to this different context measure and not only to an inherent fallacy of the ecological approach (Schwartz, 1994).

14
Bayesian Ecological Modelling

Nicola Best

Imperial College School of Medicine

14.1 INTRODUCTION

The term 'ecological study' is often used to describe epidemiological investigations in which associations between disease occurrence and potential risk factors (e.g. environmental agents or lifestyle-related characteristics) are studied over aggregated groups rather than at the individual level. Typically, the groups are defined by geographical regions, such as countries, states, counties, or smaller administrative districts like departments (France), electoral wards (UK) or census tracts (USA). Ecological studies are particularly useful when individual-level measurements of exposure are either difficult or impossible to obtain (for example, air pollution) or are measured imprecisely (for example, diet, sunlight exposure). (See English, 1992, for further discussion.)

Increasing media and public awareness of the health implications of environmental hazards has led to a growing number of ecological regression studies in the epidemiological literature, particularly in the context of hypothesised links between certain industrial pollutants and cancer. Ecological regression studies also provide a tool by which to address concerns in health services research, such as the extent to which socio-economic variations and lifestyle factors like smoking, alcohol and poor housing influence geographic variations in health needs; greater understanding of such interrelationships should better inform health services provision and resource allocation. When considering the results of such studies, however, it is crucial to recognise that their analysis and interpretation may be severely complicated by several inherent features of the data and study design (e.g. the spatial structure of the data, data quality, bias, confounding, presentation). This chapter describes some of these problems and discusses how Bayesian approaches to ecological regression modelling can attempt to address some of the issues raised.

Disease Mapping and Risk Assessment for Public Health. Edited by A.B. Lawson *et al.*
© 1999 John Wiley & Sons Ltd.

14.2 STATISTICAL ISSUES

14.2.1 Basic model formulation

Before considering the specific task of assessing the geographical association between a health indicator and some exposure variable, it is useful to address the problem of modelling the geographic distribution of the health variable alone. This topic is discussed in detail in Chapter 2 of this volume, so will receive only brief treatment here. Consider a geographic region that is divided into N sub-areas. The usual form for the data is a set of observed and expected disease counts for each area i, $i = 1, \ldots, N$ (denoted O_i and E_i, respectively), where the expected counts are calculated by multiplying the area-specific populations at risk (possibly stratified for known confounders, such as age and sex) by a common reference rate for the disease of interest. When the disease is rare and non-contagious, the observed count in each area is assumed to follow an independent Poisson distribution with mean $\mu_i = \lambda_i E_i$, where λ_i denotes the *relative risk* of disease in area i.

Various approaches are available for estimating the area-specific relative risks, λ_i. The choice will depend in part on the size of the areas and on the homogeneity of their sizes (and of the populations at risk in each area) across the study region. This paper focuses on *small-area* studies: here, the sparseness of cases within each area, particularly in those with low populations, typically results in the data exhibiting extra-Poisson heterogeneity or overdispersion (see McCullagh and Nelder, 1989, for a general discussion of this phenomenon). That is, the observed counts fluctuate about the mean within each area more than would be expected due to Poisson sampling variation alone. If ignored, this overdispersion can create the impression of spurious geographic variation in the disease rates. These considerations have led to the use of Bayesian statistical smoothing techniques whereby the relative risks λ_i are assumed to be *random effects* that arise from a probability distribution of risks (see, for example, Clayton and Kaldor, 1987; Clayton and Bernardinelli, 1992; Mollié, 1996). Typically, a normal distribution is specified for the logarithm of the random effects, i.e.

$$\log \lambda_i \sim \mathrm{N}(\mu_\lambda, \sigma_\lambda^2). \tag{14.1}$$

The unknown mean, μ_λ, and variance, σ_λ^2, of the random effects population distribution may be replaced by (say) their maximum likelihood estimates, leading to an empirical Bayes model (see, for example, Clayton and Kaldor, 1987). Alternatively, prior probability distributions may be specified for μ_λ and σ_λ^2. Common 'non-informative' choices are a normal prior with large variance for μ_λ, and a gamma prior with small shape and inverse scale parameters for $1/\sigma_\lambda^2$. This leads to a fully Bayesian analysis for estimating the *joint* posterior distribution of the area-specific relative risks λ_i and the population parameters μ_λ and σ_λ^2. Unlike empirical Bayes methods, this provides estimates for the area-specific relative risks which are fully adjusted for uncertainty in the population parameters μ_λ and σ_λ^2. However, sensitivity to the assumed prior distributions for the latter parameter should be checked, because this can often influence the posterior conclusions.

The above approach is an example of a Bayesian hierarchical or multi-level model whereby information on disease rates is pooled across areas, resulting in more reliable estimates of the area-specific relative risks.

14.2.2 Inclusion of covariate information

The model above extends naturally to include area-specific covariates, x_i, as follows:

$$\log\lambda_i = \alpha + \beta' x_i + \theta_i, \tag{14.2}$$

$$\theta_i \sim \mathrm{N}(0, \sigma_\theta^2). \tag{14.3}$$

'Non-informative' prior distributions (e.g. a normal density with large variance) are usually assumed for the intercept α and regression coefficients β. This formulation represents a standard log-linear mixed-effects model, whereby the covariate effects act multiplicatively on the overall relative risk e^{α}; e^{θ_i} represents the residual relative risk in area i after adjustment for covariates. It is often convenient to interpret e^{θ_i} as reflecting residual between-area variability due to unknown or unmeasurable risk factors such as the socio-economic characteristics of the study region, nutritional habits and genetic attributes of the population at risk, or an unmeasured environmental exposure. Such factors typically vary smoothly in space, which in turn induces spatial correlation between the observed disease counts in nearby areas. Spatial dependence between the O_i may also arise due to purely statistical features of the data, such as lesser variability of rates in neighbouring densely populated urban areas compared with more sparsely populated rural areas, or because the disease has an infectious aetiology. Failure to adjust for such spatial autocorrelation can lead to mis-specification of the variance structure of the model and misleading or false conclusions about the extent of geographical variation in health outcomes.

In ecological regression studies, the measured covariates themselves often exhibit positive spatial autocorrelation (i.e. nearby areas tend to have similar values). Ignoring these correlations during a regression analysis of such data tends to over-estimate the strength of any association, because the positive spatial dependence between the disease counts and between the covariate values across areas reduces the 'effective sample size' and the usual assumption of independent errors is violated.

14.2.3 Adjusting for spatial correlation

Recall that the residual random effects, e^{θ_i}, in (14.2) and (14.3) at least partially account for unmeasured covariates that may induce spatial correlation between the observed counts O_i. It follows that such dependence may be modelled by assuming a spatially structured prior distribution for these random effects. Various choices exist, the most popular of which is a special case of the conditional intrinsic Gaussian autoregressive (CIGAR) model (Besag *et al.*, 1991; Besag and Kooperberg, 1995):

$$\theta_i | \theta_{j \neq i} \sim \mathrm{N}(\bar{\mu}_{\theta_i}, \sigma_{\theta_i}^2), \tag{14.4}$$

$$\bar{\mu}_{\theta_i} = \frac{\sum_{j \neq i} w_{ij}\theta_j}{\sum_{j \neq i} w_{ij}}, \tag{14.5}$$

$$\sigma_{\theta_i}^2 = \frac{\gamma_\theta}{\sum_{j \neq i} w_{ij}}. \tag{14.6}$$

This models the log relative risk in area i, *conditional* on the risks in all other areas $j \neq i$, as being normally distributed about the weighted mean of the log relative risks in the remaining areas, with variance inversely proportional to the sum of the weights. The simplest choice for the weights w_{ij} is to regard any area sharing a boundary with area i to be its neighbour with weight 1; all other areas have weight 0. In this case, the conditional mean (14.5) reduces to $(1/n_i)\sum_{j\neq i}\theta_j$ and the conditional variance (14.6) reduces to $(1/n_i)\gamma_\theta$, where n_i is the cardinality (number of neighbours) of area i. Note that the CIGAR prior as defined above has an arbitrary level, and is therefore improper. The data O_i contain information on the location of each θ_i, leading to a proper posterior. However, if the model includes a separate intercept parameter, α, as in (14.2), then it becomes necessary to impose a constraint such as $\sum_i \theta_i = 0$ to ensure identifiability.

In practice, it is common to include both unstructured and spatially correlated heterogeneity in the same model by specifying two independent random effects according to a convolution Gaussian prior (Besag, 1989; Besag and Mollié, 1989). Equation (14.2) thus becomes:

$$\log \lambda_i = \alpha + \beta' x_i + \theta_{u_i} + \theta_{v_i}, \qquad (14.7)$$

where θ_{v_i} are modelled according to (14.3) and represent unstructured heterogeneity of the area-specific relative risks, whilst θ_{u_i} are modelled according to (14.4)–(14.6), with sum-to-zero constraint imposed, and represent spatially structured variation.

Examples of ecological regression studies that have implemented models of this form include Clayton and Bernardinelli (1992), who analyse the effects of urbanisation on breast cancer mortality in Sardinian communes, and Richardson *et al.* (1995), who analyse geographical variation of childhood leukaemia in UK districts in relation to natural radiation (gamma and radon). See also Best *et al.* (1998b), Knorr-Held and Besag (1998) and Chapters 2, 3, 5, 15 and 26 of this volume for further discussion and application of CIGAR and related Bayesian autoregressive models for spatially correlated disease data.

Spatially dependent covariates may be modelled in a similar manner. The observed area-specific covariate is assumed to be an imprecise measurement of the true value in each area; a CIGAR prior is then specified to model the exposure distribution for the unknown true covariate values. This approach is discussed in more detail in Section 14.3.2, and has been implemented by Bernardinelli *et al.* (1997) and Bernardinelli *et al.* in Chapter 26 of this volume in ecological regression analyses of the association between genetic susceptibility to malaria and the incidence of insulin-dependent diabetes mellitus in Sardinia.

14.3 DATA ISSUES

Inaccuracy and incompleteness of the measurements are major problems inherent in most routine sources of health and exposure data available for epidemiological study, and may cause severe bias if ignored during analysis. Bayesian hierarchical models provide the necessary general framework for handling missing data and measurement error problems. The unobserved data are treated in the same way as the other unknown model parameters by first assigning a prior distribution (in the case of missing response data, this is simply the likelihood function for the complete data; covariate measurement error models are described in Section 14.3.2). All unknown quantities (i.e. both the unobserved data and the parameters of interest, such as the regression coefficients) are then estimated *simultaneously* by conditioning on the observed data and applying Bayes theorem. This leads to posterior distributions for the regression coefficients that fully reflect the additional uncertainty associated with missing or inaccurate data. Recent work by Richardson and Gilks (1993a,b), Dellaportas and Stephens (1995) and Best *et al.* (1996) illustrate this approach for a variety of applications.

14.3.1 Numerator and denominator errors

A specific data quality issue arising in disease mapping and ecological regression studies relates to the completeness of the observed number of cases per area (the numerator) and to the accuracy of the small-area population estimates used to calculate expected counts (the denominator). Problems with the numerator data may arise due to, for example, diagnostic coding errors on death certificates or hospital notes, or failure of doctors/hospitals/health workers to notify the relevant national or regional disease registry of all incident cases of a disease. Often, the resulting pattern of missing data is not random, but is systematically concentrated in certain sub-areas of the study region.

Most published disease mapping and ecological regression studies regard the expected counts in each area as known quantities. Usually, these are calculated from census population counts. However, under-enumeration of census populations is a recognised problem, whilst migration during the inter-censual years can markedly alter the size and composition of small-area populations. Such problems, in turn, lead to inaccuracies in the expected number of cases. A further potential source of error involves the reference disease rate used in combination with the population data to calculate the expected counts; again, this is an unknown quantity that must be estimated from available data.

The danger of ignoring data quality issues such as the above is that high or low estimated risks may arise simply as artefacts due to missing cases or inaccurate baseline population data. Given suitable prior information, such as annual migration rates or a diagnostic coding calibration study, Bayesian missing data/measurement error models offer a promising approach by which to adjust for many of these problems. As yet, however, little has been published in the context of adjusting for numerator and denominator errors in ecological regression studies.

14.3.2 Covariate measurement error

Ecological regression studies are traditionally suited to situations where *within*-area measurement errors of the exposure of interest are small in comparison with between-area differences in mean exposure. Nonetheless, accurate measurement of many of the environmental and lifestyle-related covariates of interest in geographical epidemiology remains a major problem; ignoring such errors may result in biased estimates of the true ecological regression slope. Adjustment for such errors may be achieved via the following Bayesian 'errors-in-variables' model. Let z_i and x_i denote the observed and true covariate values in area i, respectively. For continuous variables, the classical measurement error model assumes that the observed value depends on the true value plus a normally distributed *iid* random error term with variance σ^2_{error}. That is,

$$z_i \sim \mathrm{N}(x_i, \sigma^2_{\text{error}}). \tag{14.8}$$

Since the data contain no information by which to estimate the error variance, either this must be specified a priori by setting σ^2_{error} to a constant value or assuming an informative prior distribution, or (preferably) σ^2_{error} should be estimated using a relevant calibration sample in which measurements of both z and x are available for some units. In addition, a prior distribution must be specified for x_i, representing the sampling distribution of the true exposure variable in the population. The dependent variable of interest is then regressed on the estimated true covariate value x_i, rather than on the observed value z_i.

Bernardinelli *et al.* (1997) describe a Bayesian ecological regression analysis with errors in covariates. Their application involves a regression of malaria prevalence on the incidence of insulin-dependent diabetes mellitus (IDDM) for 366 communes in Sardinia. Recent evidence supports a genetic link between resistance to malaria and the Human Leukocyte Antigen (HLA) system, whilst family studies suggest that genes in or near the HLA region are involved in susceptibility to IDDM. The observed covariate (z/m = malaria prevalence during the 1938–1940 epidemic) was thus used as an imprecisely measured surrogate for the true covariate of interest; namely, genetic adaptation of the HLA system to provide greater resistance to malaria. Bernardinelli *et al.* (1997) use a form of the classical measurement error model (14.8), specifically

$$z_i \sim \text{Binomial}(m_i, \pi_i), \tag{14.9}$$

$$\log(\pi_i/(1-\pi_i)) \sim \mathrm{N}(x_i, \omega), \tag{14.10}$$

where $\log(\pi_i/(1-\pi_i))$ is the true log odds of malaria in area i between 1938–1940, and x_i (the true covariate of interest) may be interpreted as a latent variable reflecting the average long-term 'endemicity' of malaria in commune i. The log odds of malaria at any given time thus fluctuate about the average endemicity with variance ω. Bernardinelli *et al.* (1997) specify an informative prior value for ω based on exploratory analysis, and, in addition, assume CIGAR prior distributions for both the residual area-specific relative risk parameters (see (14.4)–(14.6)) and the prior exposure distribution of the true covariate:

$$\begin{aligned} x_i | x_{j\neq i} &\sim \mathrm{N}(\bar{\mu}_{x_i}, \sigma^2_{x_i}), \\ \bar{\mu}_{x_i} &= \frac{1}{n_i}\sum_{j\neq i} x_j, \\ \sigma^2_{x_i} &= \frac{1}{n_i}\gamma_x. \end{aligned} \tag{14.11}$$

As noted in Section 14.2.3, this prior distribution is improper, but yields a proper posterior in the presence of any informative data z_i. An alternative formulation is to specify $x_i = \mu_x + \varepsilon_i$, where μ_x represents overall mean exposure and is assigned a diffuse constant prior, whilst ε_i is a zero-mean, spatially correlated error term for area i and is assigned a CIGAR prior with a sum-to-zero constraint imposed. Bernardinelli *et al.*'s full model thus allows for spatial correlation in both the response and explanatory variables. Such adjustments were found to increase considerably the magnitude of the estimated regression slope in comparison with a naive analysis using only the observed covariate values. A related model is described in Chapter 26.

14.4 PROBLEMS WITH THE INTERPRETATION OF ECOLOGICAL REGRESSION STUDIES

Even after adjusting for the statistical and data quality issues raised above, great care is required when interpreting the results of an ecological regression study. Of particular concern are the problems of socio-economic confounding and ecological bias.

14.4.1 Socio-economic confounding

Measures of social deprivation have been shown to be powerful predictors of the occurrence of many diseases (Jolley *et al.* 1992). Deprived areas do not occur randomly throughout a region, but tend to coincide with industrial sites and busy roads, and correlate with higher smoking rates. Failure to account for socio-economic deprivation could thus seriously bias ecological investigation of the impact of other lifestyle or environmental risk factors on ill health. Adjustment may be made by including, say, an area-specific deprivation score such as the Carstairs (Carstairs and Morris, 1991) index (based on small-area census statistics for the United Kingdom) as a covariate in the ecological regression analysis. Alternatively, indirect standardisation of the expected small-area disease counts can be done by stratifying on the socio-economic status of the areas as well as on age and sex. Modelling of spatial autocorrelation between small areas in an ecological regression study also provides some control for the effect of confounding due to location (recall that the autocorrelated random effects may be interpreted as accounting, at least partially, for residual variation due to unmeasured spatially dependent explanatory variables—see Section 14.2.3).

14.4.2 Ecological bias

The ecological regression slope estimates the relationship between area-level average response (e.g. prevalence of disease) and average exposure to the risk factor of interest. However, the quantity of real interest is the average of the *within*-area regressions of individual-level response on individual exposure. The discrepancy between these two quantities—namely, the group- and individual-level estimates of risk—is termed the ecological bias or fallacy and has been the subject of numerous publications in the epidemiological and statistical literature (e.g. Piantadosi *et al.* 1988; Greenland and Morgenstern, 1989; Cohen, 1990; Richardson and Hemon, 1990; Plummer and Clayton, 1996). Richardson (1992) identifies a number of situations in which ecological bias may arise, including when the individual-level exposure–response relationship is non-linear, or when the individual-level relationship is linear, but area-level confounders or effect modifiers lead to different intercepts (i.e. different baseline disease rates for the non-exposed) or different slopes (i.e. different exposure–response effects) in different areas.

Ecological bias may be reduced by improving the assessment of group-level exposure to the risk factor within each area. This is particularly important if the degree of heterogeneity of exposure is large, and/or the individual exposure–response relationship is non-linear. One approach is to combine routine area-level exposure data with more detailed individual-level survey data (where available) on a sub-sample of the population at risk in order to derive estimates of the full *marginal* exposure distribution within each area. In situations where an area-level confounder or effect-modifier is known to exist, Richardson (1992) suggests formulating hypotheses on the *joint* exposure distribution of the confounder and risk factor of interest. The modelling of such situations requires further investigation, but could be handled within the general framework of Bayesian hierarchical and measurement error models.

On a similar topic, Best *et al.* (1998a) have developed a Bayesian ecological regression model that allows response and covariate data to be modelled at different levels of spatial aggregation. This is achieved by relating all observable quantities to an underlying continuous random field model, and obviates the need to average covariate data such as point source exposures or modelled pollution surfaces to the same geographical scale as the response. This is particularly relevant for highly heterogeneous exposures, such as air pollution or water quality, where fine-scale patterns and local variations in the exposure gradient may be important in determining risk. These authors apply their model to an epidemiological study of the effects of nitrogen dioxide (NO_2) pollution on the prevalence of respiratory disorders in children. Here, the response data (cases of severe wheezing illness) are located by the grid reference of the child's home postcode, whilst population denominator data are only recorded at census enumeration district level (covering approximately 400–1000 households) and the exposure data are available as a regular grid of modelled NO_2 concentrations to a resolution of 250 m^2.

14.5 TECHNICAL IMPLEMENTATION

Implementation of Bayesian hierarchical models is technically demanding due to the high-dimensional integrations involved in computing the required posterior distributions. However, recent developments in computer-intensive Markov chain Monte Carlo (MCMC) methods (Smith and Roberts, 1993; Gilks *et al.* 1996a) and Bayesian graphical modelling (Spiegelhalter *et al.* 1996) have revolutionised the approach and opened the way for realistic modelling of complex data structures, such as those that arise in ecological regression studies. These computational methods are implemented in the BUGS software (Spiegelhalter *et al.* 1995), which offers an appropriate platform for real-time analysis of many of the models described in this chapter.

14.6 CONCLUSIONS

Ecological regression studies play an important role in epidemiology, and have led to some notable aetiological insights. In general, however, such studies have been concerned with hypothesis generation (i.e. qualitative identification of an association) rather than quantitative estimation of the strength of an exposure–response relationship. Such caution has rightly arisen due to major problems of bias and misinterpretation inherent in ecological studies. Nonetheless, there are many situations in which an ecological model offers the most suitable design for studying the impact of an environmental or lifestyle-related risk factor on ill health; in particular, when assessment of individual-level exposure is precluded due to the overwhelming imprecision of available measurements. Methodological developments should thus focus on extending and improving ecological study designs to handle appropriately such problems. The Bayesian modelling approach offers a natural framework within which to address this issue. In particular, it allows for the modelling of spatial correlation in the response and covariate data, and facilitates improved estimation of area-specific exposure distributions by adjusting for random measurement errors and enabling routine area-level data to be combined with individual-level survey data on exposures of interest. However, the Bayesian approach is not a panacea; considerable care is required concerning issues such as the sensitivity of the results to the assumed prior distributions, and the assessment of the convergence of the computer simulation algorithms used to estimate the posterior distributions of interest. Furthermore, methods for Bayesian model selection and model criticism are not well-established, although research on this topic is developing rapidly.

No statistical model, however sophisticated, can overcome basic deficiencies in the data. Where possible, efforts should be made to improve the quality and completeness of routine data sources used for epidemiological studies, and/or to collect additional individual-level calibration data to assess the scale of missing data, quantify measurement errors and assess exposure heterogeneity at the ecological level. Careful choice of the outcome and exposure variables; namely, well-defined disease groups with short time-lags between exposure and onset, should enhance the reliability and validity of the resulting ecological regression. In this context, recently available routine datasets such as the UK Hospital Episode Statistics offer a valuable source of information on acute as well as chronic health events. These should be exploited for ecological modelling of the short-term effects of environmental exposures on ill health, and for the purposes of health services research.

In summary, methodological advances, particularly in the field of Bayesian statistics, combined with efforts to improve the accuracy and specificity of epidemiological health and exposure data, should enhance the ability of ecological analyses to estimate the size of disease–exposure relationships, not merely to identify the possible existence of such associations. Nonetheless, problems of interpretation and bias remain, and wherever possible replication of the ecological studies in different areas or different time periods is recommended.

15

Spatial Regression Models in Epidemiological Studies

J. Ferrándiz and A. López

Universitat de València, Spain

P. Sanmartín

Universitat Jaume I, Spain

15.1 INTRODUCTION

Modelling disease risk in epidemiological studies often entails the regression analysis of mortality data on risk factors. In geographical studies, in order to avoid possible biases in the estimation of regression coefficients, the autocorrelation structure of data in neighbouring sites has to be considered.

Disease mapping has developed very quickly in recent years, offering a wide range of models, most of them based on different extensions of generalised linear models. Smoothing techniques have also been applied, but we are assuming here that our final goal is to understand the influence of risk factors on disease prevalence and, therefore, we will concentrate on regression models.

The geographical association of data from neighbouring sites may appear as a consequence of the direct influence of morbidity in contiguous locations (the case of infectious diseases), or because risk factors have a geographical structure, and contiguous areas will be exposed to similar levels of risk.

According to the first of these assumptions, we can model spatial dependence between observations directly. It is also a sensible procedure in the second possibility as well, because we can consider neighbouring mortality counts to be a surrogate for hidden risk factors with a geographical structure.

Besag's auto models are well suited to this end but, in spite of their elegant definition, the statistical analysis remains difficult. There is still much room for improvement. Only auto-Gaussian models have been considered with some extension in disease mapping so

Disease Mapping and Risk Assessment for Public Health. Edited by A.B. Lawson *et al.*
© 1999 John Wiley & Sons Ltd.

far (Richardson, 1992; Richardson *et al.*, 1992), although Ferrándiz *et al.* (1995) use a modified version of the auto-Poisson model.

The second assumption, that of contiguous areas sharing similar levels of risk factors, suggests a hierarchical model where hidden covariates are incorporated through latent variables in the second stage of the hierarchy. The spatial autocorrelation is considered in the variance–covariance matrix of these latent variables, usually modelled by means of a multivariate Gaussian distribution.

Clayton and Kaldor (1987) proposed the empirical Bayes approach of such models, while Besag *et al.* (1991) and Clayton and Bernardinelli (1992) developed full Bayesian analyses.

The frequentist (empirical-Bayes) approach just mentioned places disease mapping into the general framework of the Generalised Linear Mixed Models (GLMM) (Breslow and Clayton, 1993), and proposes the use of penalised and marginal quasi-likelihood functions to overcome its difficult statistical analysis. In the same vein, more recent works propose different generalised estimating equations (McShane *et al.*, 1997) or estimationg functions (Yasui and Lele, 1997).

Lee and Nelder (1996) extend the concept of GLMM to that of Hierarchical Generalised Linear Models, enlarging the use of non-Gaussian distributions for the latent variables. The tendency to use non-Gaussian distributions for latent variables is most likely to be reinforced in the future (a notable example is Ickstadt and Wolpert, 1996), improving GLMM relevance on disease mapping. This tendency will produce the need for more general likelihood-based methods of analysis.

The Bayesian analysis of GLMMs circumvents the difficulties of deriving exact posterior distributions by resorting to Markov chain Monte Carlo (MCMC) methods. This fruitful approach has led in different ways, to a spatio-temporal analysis of disease risk (Bernardinelli *et al.*, 1995b; Waller *et al.*, 1997b) and to a regression analysis with errors in covariates (Bernardinelli *et al.*, 1997, and Chapter 26 in this volume).

In either case, with Besag's spatial auto models or GLMMs, we face difficulties derived from an incomplete knowledge of the likelihood function or the posterior density.

The analysis of these complex models can benefit from MCMC optimisation techniques that have been receiving increasing attention in the literature. Geyer and Thompson (1992) achieve the maximum likelihood estimation of Gibbs distributions by optimising MCMC approximations of the likelihood function (see also Geyer, 1996). This is particularly useful with models defined via full conditional distributions, as in the case of Besag's spatial auto models. Gibbs sampling is well suited to perform the required simulation steps.

In a resembling line, Ferrándiz *et al.* (1995) use MCMC estimation of the derivatives of the likelihood function to perform Fisher's scoring maximum likelihood estimation for an auto-Poisson model. Their approach is based on some guidelines developed in Penttinen (1984) for point processes.

More recently, McCullogh (1997) has proposed a Monte Carlo Newton Raphson procedure to achieve the maximum likelihood estimation in GLMMs.

MCMC optimisation techniques can be applied in a very wide range of situations. Provided the likelihood function is smooth, the only requisite is a simulation procedure for the model. Markov chain simulation techniques have dramatically enlarged the scope of models susceptible to such treatment.

In this chapter we further explore the use of MCMC optimisation techniques proposed in Ferrándiz *et al.* (1995), and extend its use to the Bayesian framework, following a scheme close to that developed in Heikkinen and Penttinen (1995) for point processes.

In particular, we consider the truncated auto-Poisson model in the analysis of prostate cancer mortality in Valencia (Spain). We have performed the regression analysis of mortality data by considering age and nitrate contamination of drinking water as covariates. The main goal is to determine the possible influence of this contamination.

In Section 15.2 we present the truncated auto-Poisson distribution jointly with alternative models to be used in a subsequent comparative study. In Section 15.3 we describe the proposed MCMC optimisation procedure in general terms.

In Section 15.4 we describe the results obtained in the analysis of prostate cancer mortality in Valencia from a frequentist viewpoint. We compare maximum pseudo-likelihood with MCMC maximum likelihood, and discuss how the inclusion of spatial dependence in the model makes contamination of drinking water non-significant at the usual levels. Similar results are attained when we perform a Bayesian analysis of the same model.

To compare this last result with a widely accepted standard, we have adjusted GLMM to the same data from a Bayesian perspective along the lines of Mollié (1996). We then discuss the results obtained under both models.

15.2 TRUNCATED AUTO-POISSON VERSUS RANDOM EFFECTS POISSON REGRESSION

The Poisson distribution is widely accepted for modelling mortality counts in small regions. Thanks to the Generalised Linear Models (GLM) framework, Poisson regression is the first candidate to be considered in the analysis of mortality risk factors.

If we denote by $\boldsymbol{y}$ the random vector of mortality counts defined on a spatial irregular grid S of s fixed location, and $\boldsymbol{x}_i$ stands for the vector of values of covariates in location i, Poisson regression can be described as

$$[y_i|\boldsymbol{x}_i] \sim \text{Po}(\lambda_i), \quad i = 1, \ldots, s \text{ independent,} \tag{15.1}$$

$$\log\lambda_i = \boldsymbol{\beta}'\boldsymbol{x}_i. \tag{15.2}$$

In cases where spatial interdependence of mortality counts is suspected, two main generalisations of Poisson regression can be stated, depending on the way this interdependence is established.

15.2.1 Truncated auto-Poisson model

It we model spatial interdependence as a direct influence between neighbouring locations, then the full conditional distributions of the mortality count in each location given those in the remaining sites,

$$p(y_i|\boldsymbol{y}_{-i}, \boldsymbol{\theta}), \quad i \in S, \tag{15.3}$$

constitute an intuitive tool in order to build the joint distribution,

$$p(\boldsymbol{y}\,|\,\boldsymbol{\theta}) = \frac{\exp\{h(\boldsymbol{y}\,|\,\boldsymbol{\theta})\}}{c(\boldsymbol{\theta})} \tag{15.4}$$

Here $\boldsymbol{\theta}$ denotes all unknown parameters as well as covariates, while $\boldsymbol{y}_{-i}$ stands for the subvector $\{y_j : j \in S, j \neq i\}$ of all observed counts but that of location i. This subvector can be reduced to the smaller set of locations that are suspected to influence location i. Equation (15.4) expresses the joint density $p(\boldsymbol{y}\,|\,\boldsymbol{\theta})$ in terms of $h(\boldsymbol{y}\,|\,\boldsymbol{\theta})$, the logarithm of its kernel, and $c(\boldsymbol{\theta})$, the corresponding normalising constant.

Besag's (1974) auto models take $p(y_i\,|\,\boldsymbol{y}_{-i}, \boldsymbol{\theta})$ in the exponential family and limit the complexity of $h(\boldsymbol{y}\,|\,\boldsymbol{\theta})$. In particular, choosing the full conditionals as Poisson, we get the auto-Poisson model described as:

$$[y_i\,|\,\boldsymbol{y}_{-i}, \boldsymbol{\theta}] \sim \text{Po}(\lambda_i), \quad i \in S, \tag{15.5}$$

$$\log\lambda_i = \boldsymbol{\beta}'\boldsymbol{x}_i + \sum_{j:j\neq i} \gamma_{ij} y_j. \tag{15.6}$$

The autoregressive coefficient γ_{ij} is null whenever y_j does not contribute to the full conditional of y_i.

Equation (15.6) is the same as (15.2) but with autoregressive terms added. This characteristic makes the auto-Poisson distribution a natural autoregressive extension of the Poisson regression model. The condition of *independence* in (15.1) disappears in (15.5), making the statistical analysis much more difficult.

Full conditionals (15.3) are helpful when modelling the spatial dependence between neighbouring sites. The main caution is to verify the existence of the normalising constant $c(\boldsymbol{\theta})$, needed to make (15.4) a probability density function. For the auto-Poisson distribution, $c(\boldsymbol{\theta})$ is defined only if $\gamma_{ij} < 0$, thus limiting its use to model inhibition (see Besag, 1975).

We can recover the possibility of positive interactions if we truncate the Poisson distributions to a limited range. If we are working with mortality data, then we can restrict Poisson counts y_i to be smaller than the living populations n_i. Thus (15.5) would become an approximate binomial model with a large number of trials and a very small probability of success.

Therefore, we propose the truncated auto-Poisson model, substituting (15.5) with

$$p(y_i\,|\,\boldsymbol{y}_{-i}, \boldsymbol{\theta}) = \text{Po}(y_i\,|\,\lambda_i) I_{\{y_i \leq n_i\}}(y_i), \quad i \in S, \tag{15.7}$$

where $\text{Po}(\cdot\,|\,\lambda)$ stands for the corresponding Poisson density and $I_A(z)$ denotes the indicator function of the set $z \in A$.

In so far as our statistical procedures rely on simulation techniques, they remain unchanged when this truncation rule applies. All we have to do is to reject simulated values exceeding the established threshold. It is a pleasant characteristic of these methods. They are applied with the same ease as soon as we are able to simulate the model.

On the other hand, if we kept all autoregressive coefficients γ_{ij} distinct in (15.6), we would face an overparameterised model. To avoid this identification problem, we propose the following simplification:

$$\gamma_{ij} = \gamma q_{ij}. \tag{15.8}$$

Here q_{ij} is a known value depending on the characteristics of proximity between the pair of locations i and j. In our case study, to be presented later in the chapter, we have taken

$$q_{ij} = n_i n_j / d_{ij} \tag{15.9}$$

as a function of populations n_i and n_j and geographical distance d_{ij}. We have reduced the number of neighbours in each location by setting q_{ij} to zero if it is smaller than a previously specified threshold.

15.2.2 Random effects Poisson regression

If we model spatial interdependence through latent variables, then we are in the domain of hierarchical GLMs. Among the many possibilities available, we have followed the guidelines of Mollié (1996). After Besag *et al.* (1991) and Clayton and Bernardinelli, (1992), it has become a well accepted model. We can use it as a comparative reference to judge the performance of the previously described truncated auto-Poisson model.

We have chosen the model

$$[y_i | \boldsymbol{y}_{-i}, \boldsymbol{\theta}] \sim \text{Po}(\lambda_i), \quad i \in S, \tag{15.10}$$

$$\log\lambda_i = \boldsymbol{\beta}'\boldsymbol{x}_i + u_i + v_i, \tag{15.11}$$

$$\boldsymbol{u} \sim \text{N}(\boldsymbol{0}, \sigma^2 \boldsymbol{I}), \tag{15.12}$$

$$\boldsymbol{v} \sim \text{N}(\boldsymbol{0}, \tau^2 \boldsymbol{W}^{-1}). \tag{15.13}$$

Equations (15.12) and (15.13) state the distributions of the latent variables in the second stage of the hierarchy. The u_is are location-specific random effects allowing for Poisson extra-variance, while the v_is carry out the spatial interdependence through their variance–covariance matrix $\tau^2 \boldsymbol{W}^{-1}$. We have based the off-diagonal elements w_{ij}, of the precision matrix $\boldsymbol{W}$, on criteria similar to those in (15.8), to facilitate the comparison with the auto-Poisson model. Instead of choosing w_{ij} as the usual indicator function of the geographical contiguity of locations i and j, we have taken it to be one whenever the proximity index q_{ij} in (15.9) exceeds a predetermined threshold, and zero in the other case. The choice $w_{ii} = -\sum_{j:j \neq i} w_{ij}$ will make (15.13) an intrinsic autoregressive prior distribution (Besag and Kooperberg, 1995).

To perform a full Bayesian analysis of this model, we have specified prior distributions for all the unknown parameters $\boldsymbol{\beta}$, σ^2 and τ^2. We have followed the usual conjugate normal-gamma scheme,

$$\boldsymbol{\beta} \sim \text{N}(\boldsymbol{b}, \boldsymbol{B}), \quad \sigma^{-2} \sim \text{Ga}(a, d), \quad \tau^{-2} \sim \text{Ga}(e, f), \tag{15.14}$$

which will allow Gibbs sampling of the posterior distribution later in the chapter.

The hyperparameters $\boldsymbol{b}$, $\boldsymbol{B}$, a, d, e and f of these priors are fixed, and have to be assessed in the initial steps of the analysis. They are chosen to barely affect the likelihood function, but are needed to allow the implementation of the Bayesian learning process.

15.3 MODEL FITTING USING MONTE CARLO NEWTON–RAPHSON

The main problem in the statistical analysis of Besag's auto models relies on the inaccessibility of the normalising constant $c(\boldsymbol{\theta})$ of (15.4). It involves the parameters of the model $p(\boldsymbol{y}\,|\,\boldsymbol{\theta})$. Without $c(\boldsymbol{\theta})$ we cannot perform direct likelihood-based inference, or compute the Bayesian posterior distribution of these parameters.

Nevertheless, Monte Carlo techniques enable the possibility of circumventing this difficulty by estimating the likelihood function (or its main characteristics such as gradient and Hessian).

15.3.1 Monte Carlo Newton–Raphson

To simplify the notation, let the log likelihood and log posterior be expressed as

$$l(\boldsymbol{\theta}) = \log p(\boldsymbol{y}\,|\,\boldsymbol{\theta}) = h(\boldsymbol{y}\,|\,\boldsymbol{\theta}) - \log c(\boldsymbol{\theta}), \tag{15.15}$$

$$L(\boldsymbol{\theta}) = \log p(\boldsymbol{\theta}\,|\,\boldsymbol{y}) = \log p(\boldsymbol{\theta}) + h(\boldsymbol{y}\,|\,\boldsymbol{\theta}) - \log c(\boldsymbol{\theta}) - \log p(\boldsymbol{y}), \tag{15.16}$$

where $p(\boldsymbol{\theta})$ stands for the prior density of the parameter vector $\boldsymbol{\theta}$, and $p(\boldsymbol{y})$ denotes the predictive density $\int p(\boldsymbol{y}\,|\,\boldsymbol{\theta})p(\boldsymbol{\theta})\mathrm{d}\boldsymbol{\theta}$.

To find the maximum likelihood estimation $\hat{\boldsymbol{\theta}}$, Newton–Raphson is a widely used optimisation algorithm. Starting from a first guess $\boldsymbol{\theta}^{(0)}$, it proceeds in updating steps

$$\boldsymbol{\theta}^{(m+1)} = \boldsymbol{\theta}^{(m)} - [l''_{\theta}(\boldsymbol{\theta}^{(m)})]^{-1} l'_{\theta}(\boldsymbol{\theta}^{(m)}), \quad m = 0, 1, \ldots, \tag{15.17}$$

where, $l'_{\theta}(\boldsymbol{\theta}^{(m)})$ and $l''_{\theta}(\boldsymbol{\theta}^{(m)})$ are the gradient and Hessian of the log likelihood function evaluated at $\boldsymbol{\theta}^{(m)}$.

Provided the (quite general) conditions are met for exchanging the derivative and integral operations, it can be shown that

$$D_{\boldsymbol{\theta}} \log c(\boldsymbol{\theta}) = \mathrm{E}[h'_{\theta}(\boldsymbol{y}\,|\,\boldsymbol{\theta})\,|\,\boldsymbol{\theta}], \tag{15.18}$$

$$D^2_{\boldsymbol{\theta}} \log c(\boldsymbol{\theta}) = \mathrm{var}\,[h'_{\theta}(\boldsymbol{y}\,|\,\boldsymbol{\theta})\,|\,\boldsymbol{\theta}] + \mathrm{E}[h''_{\theta}(\boldsymbol{y}\,|\,\boldsymbol{\theta})\,|\,\boldsymbol{\theta}], \tag{15.19}$$

where D_{θ} and D^2_{θ} denote the gradient and Hessian of the function just to their right, as does the prime and double prime notation on $h(\boldsymbol{y}\,|\,\boldsymbol{\theta})$. The operators $\mathrm{E}[\cdot\,|\,\boldsymbol{\theta}]$ and $\mathrm{var}\,[\cdot\,|\,\boldsymbol{\theta}]$ denote the expectation and dispersion matrix of their arguments, given the value $\boldsymbol{\theta}$ of the parameter vector.

These last equations relate the derivatives of the log constant to the sampling moments of h'_{θ} and $h''_{\theta}(\boldsymbol{y}\,|\,\boldsymbol{\theta})$. Thus, although we could not derive them from the full conditionals specification (15.3), they can be estimated from any simulated sample $\{\boldsymbol{y}^{(i)} : i = 1, \ldots, r\}$ of our model.

These estimates can then replace their unknown theoretical values to provide the corresponding estimates,

$$\hat{l}'_{\theta}(\boldsymbol{\theta}^{(m)}) = h'_{\theta}(\boldsymbol{y}\,|\,\boldsymbol{\theta}^{(m)}) - \hat{\mathrm{E}}[h'_{\theta}(\boldsymbol{y}\,|\,\boldsymbol{\theta}^{(m)})], \tag{15.20}$$

$$\hat{l}''_{\theta}(\boldsymbol{\theta}^{(m)}) = h''_{\theta}(\boldsymbol{y}\,|\,\boldsymbol{\theta}^{(m)}) - \widehat{\mathrm{var}}\,[h'_{\theta}(\boldsymbol{y}\,|\,\boldsymbol{\theta}^{(m)})] - \hat{\mathrm{E}}[h''_{\theta}(\boldsymbol{y}\,|\,\boldsymbol{\theta}^{(m)})], \tag{15.21}$$

of the gradient and Hessian of the log likelihood function. These can be used in (15.17), in turn, to give an approximate updating step.

The same reasoning applies to the posterior mode $\bar{\boldsymbol{\theta}}$, if we consider the log posterior function $L(\boldsymbol{\theta})$ instead of the log likelihood $l(\boldsymbol{\theta})$. They differ essentially in the log prior term only, which we presume to be totally known. Then,

$$\hat{L}'_{\boldsymbol{\theta}}(\boldsymbol{\theta}^{(m)}) = \hat{l}'_{\boldsymbol{\theta}}(\boldsymbol{\theta}^{(m)}) - D_{\boldsymbol{\theta}} \log p(\boldsymbol{\theta}^{(m)}), \tag{15.22}$$

$$\hat{L}''_{\boldsymbol{\theta}}(\boldsymbol{\theta}^{(m)}) = \hat{l}''_{\boldsymbol{\theta}}(\boldsymbol{\theta}^{(m)}) - D^2_{\boldsymbol{\theta}} \log p(\boldsymbol{\theta}^{(m)}), \tag{15.23}$$

and we can proceed with the Newton–Raphson algorithm as before.

In the truncated auto-Poisson model of Section 15.2, it can be shown, following the steps in Cressie (1993, Section 6.5), that

$$h(\boldsymbol{y} \,|\, \boldsymbol{\theta}) = \boldsymbol{\beta}'\mathbf{X}'\boldsymbol{y} + 1/2\gamma\boldsymbol{y}'\boldsymbol{Q}\boldsymbol{y} - \sum_i \log(y_i!), \tag{15.24}$$

where $\boldsymbol{X}$ stands for the design matrix of covariates, the off-diagonal elements q_{ij} in $\boldsymbol{Q}$ are those of (15.8), and $q_{ii} = 0$.

Therefore, taking $\boldsymbol{\theta} = (\boldsymbol{\beta}, \gamma)'$ and $\boldsymbol{t} = (\mathbf{X}'\boldsymbol{y}, \boldsymbol{y}'\boldsymbol{Q}\boldsymbol{y})'$, (15.20) and (15.21) become

$$\hat{l}'_{\boldsymbol{\theta}}(\boldsymbol{\theta}^{(m)}) = \boldsymbol{t} - \hat{E}[\boldsymbol{t} \,|\, \boldsymbol{\beta}, \gamma], \tag{15.25}$$

$$\hat{l}''_{\boldsymbol{\theta}}(\boldsymbol{\theta}^{(m)}) = \widehat{\text{var}}[\boldsymbol{t} \,|\, \boldsymbol{\beta}, \gamma], \tag{15.26}$$

to be incorporated in the updating step (15.17).

15.3.2 Monitoring convergence

Given the specification (15.3) of our model via full conditionals, Gibbs sampling is especially well suited to produce simulated samples. Then, we have to supervise the convergence of each run to the stationary distribution to ensure that the generated sample is representative of our model. We have followed the proposal of Gelman and Rubin (1992b) of simultaneous multiple sequences.

In our case these Gibbs sampling runs are embedded in the sequence of updating steps of the Newton–Raphson algorithm, which has to be supervised for convergence as well. To this end, imitating the previous approach, we make simultaneous sequences of the optimisation process. Comparing their evolution, we can control the speed of the convergence, and verify when the random variability of the simulation mechanism impedes any further improvement in the precision of the estimates. Then we can either increase the sample size of the simulated sequences, or stop the process if we agree with the precision reached. For a more detailed discussion, see Ferrándiz and López (1996).

15.3.3 Approximate log likelihood and log posterior

It the maximum likelihood estimator is unique, then we can take advantage of the by-products of the optimisation process. Using Taylor expansion around this optimum, up to terms of second order, we obtain a quick approximation of the likelihood function (up to a multiplicative constant),

$$l(\boldsymbol{\theta}) = p(\boldsymbol{y}\,|\,\boldsymbol{\theta}) \approx K \exp\{-1/2(\boldsymbol{\theta} - \hat{\boldsymbol{\theta}})'[l''_{\boldsymbol{\theta}}(\hat{\boldsymbol{\theta}})](\boldsymbol{\theta} - \hat{\boldsymbol{\theta}})\}. \qquad (15.27)$$

Then we can obtain support regions, $\{\boldsymbol{\theta} : \ell(\boldsymbol{\theta}) > \alpha\}$, containing a prescribed amount of the total volume under $\ell(\boldsymbol{\theta})$, etc.

Although, under mild spatial interdependence of the observations, we can expect these support regions to be close to the corresponding confidence sets, this is not guaranteed. To estimate the sampling variance of $\hat{\boldsymbol{\theta}}$ we have to resort to bootstrap methods, simulating the estimation process for a sample of realisations $\boldsymbol{y}^{(r)}$ generated from $p(\boldsymbol{y}\,|\,\hat{\boldsymbol{\theta}})$.

In the Bayesian framework, a similar Taylor expansion around the mode $\bar{\boldsymbol{\theta}}$ leads to

$$[\boldsymbol{\theta}\,|\,\boldsymbol{y}] \approx \mathrm{N}(\bar{\boldsymbol{\theta}}, [L''_{\boldsymbol{\theta}}(\bar{\boldsymbol{\theta}})]^{-1}). \qquad (15.28)$$

Then we can build joint or marginal credible regions, etc.

15.3.4 Alternative fitting methods

To perform the statistical analysis of spatial auto models, avoiding their unattainable likelihood, Besag (1975) proposed the use of the pseudo-likelihood function obtained as the product of all full conditionals in (15.3). When the spatial interdependences between the observations are small, the situation approaches that of a random sample whose likelihood is the product of the likelihood of individual observations, and we can expect pseudo-likelihood methods to behave well.

The pseudo-likelihood function has a clear advantage in practice: the full conditionals in (15.3) are in the exponential family, and the pseudo-likelihood analysis can be performed using the common routines of GLMs. They are implemented in most of the modern statistical packages, and all we have to do is to let the program think that the autoregressive terms in (15.6) are new covariates. The GLM routines will perform maximum pseudo-likelihood, will compute pseudo-deviances for competing nested models, etc.

In the Bayesian context we are not aware of any general alternative to the proposed procedure for the auto-Poisson model. Methods based on Gibbs sampling from the posterior $p(\boldsymbol{\theta}\,|\,\boldsymbol{y})$ need the specification of the full conditionals $p(\theta_i\,|\,\boldsymbol{\theta}_{-i}, \boldsymbol{y})$ and it is not clear how to get them when the normalising constant $c(\boldsymbol{\theta})$ remains unknown.

In the random effects Poisson regression model, we can perform Gibbs sampling from the posterior because of the normal-gamma conjugation property between the second and third stages of the hierarchy. However, this possibility is heavily dependent on this condition.

15.4 PROSTATE CANCER IN VALENCIA, 1975–1980

The Public and Environmental Health unit at the University of Valencia detected a difference in mortality trend between Valencia and the other provinces in Spain for some types of cancer (see, for example, Morales *et al.*, 1993). They presumed that the high concentration of nitrate in drinking water, caused by the intensive agricultural activity of Valencia, could be a possible risk factor.

Here we apply the methods outlined in the preceding sections to the prostate cancer mortality in Valencia, in its 263 municipalities, for the period 1975–1980. The data were partially analysed in Ferrándiz *et al.* (1995). The data of nitrate concentration were taken from Llopis (1985).

We have stipulated mortality counts following (15.7) and (15.8) as a truncated auto-Poisson model, to allow for positive interactions. The proportion of the population older than 40, and the nitrate concentration in drinking water (mg/l) have been added as possible covariates to the logarithm of the population, which is always included as an offset. Then, the most complete model will be

$$\log \lambda_i = \log n_i + \beta_0 + \beta_1 \times \text{age} + \beta_2 \times \text{nitrate} + \gamma \sum_{j:j \neq i} q_{ij} y_j \tag{15.29}$$

The pseudo-likelihood approach via GLM produces the analysis of deviance results shown in Table 15.1. The row labelled Deviance shows this quantity for the model corresponding to each column (when the model includes the autoregressive term, the deviance becomes pseudo-deviance). The remaining rows show how this deviance decreases when the term corresponding to this row is incorporated into the model labelling the column.

We can appreciate how nitrate concentration is significant in the presence of age, although not as much as the autoregressive term. Once this last term has been incorporated, the nitrate concentration becomes non-significant.

Table 15.2 shows the estimates and 99% confidence intervals of the regression coefficient of the nitrate contamination. The first row corresponds to the maximum likelihood estimate in the Poisson regression model, with age and nitrate as covariates.

Table 15.1 Pseudo-likelihood analysis

Analysis Of deviance	**Age**	**Age + autoregression**	**Age + autoregression + nitrate**
Deviance	361.8	347.1	343.9
+ autoregression	14.7 *		
+ nitrate	7.9 *	3.2	

* Significant at the 0.01 level.

Table 15.2 Estimates of the Nitrate regression coefficient

Method	**Estimates** ($\times 10^{-3}$)	**99% C.I.** ($\times 10^{-3}$)
Poisson regression	2.09	[0.25, 3.90]
auto-Poisson (MPL)	1.41	[−0.56, 3.39]
auto-Poisson (MCML)	1.50	[−0.42, 3.42] (sup. reg.)

Maximum pseudo-likelihood (MPL) and Monte Carlo maximum likelihood (MCML) estimates for the auto-Poisson full model are shown in the two remaining rows. In the last case, the confidence interval is a support region.

We see how the inclusion of the autoregression term decreases the influence of the nitrate factor, making its relevance less plausible, although it still remains near the usual significance thresholds (chi-square p-value $= 0.074$). Being a case of regression variable selection, a possible confounding effect between the nitrate and the autoregressive terms could be partially responsible for this.

The disagreement between the maximum pseudo-likelihood and the MCML estimates is small for practical purposes, and we consider that, in this case, pseudo-likelihood analysis is reliable.

From a frequentist viewpoint, we could not compute the confidence intervals of the regression coefficients directly from the log likelihood approximation $\ell(\boldsymbol{\theta})$. We face a finite grid of locations and its is not clear how we can benefit from asymptotics in these kinds of situations. The estimation of standard errors by parametric bootstrap is still too demanding given the current technology.

To this end, things are much easier if we adopt the Bayesian paradigm. All relevant information is contained in the posterior density, and we have the approximate posterior distribution (15.27) at our disposal.

In fact, Bayesian analysis following (15.27) leads to a posterior mode of 1.50×10^{-3} of the nitrate coefficient, and $[-0.38 \times 10^{-3}, 3.38 \times 10^{-3}]$ as its posterior 99% credible interval. The use of a very flat prior has produced this agreement with the maximum likelihood results.

To judge the performance of the auto-Poisson model, we compare its results with those obtained with the random effects Poisson regression model of Section 16.2.2. The random effects Poisson regression have been analysed via Gibbs sampling from the posterior. We have used the same prior distributions for the regression coefficients of age and nitrate contamination as in the Bayesian analysis of the truncated auto-Poisson model in order to make them more comparable.

Figure 15.1 shows the posterior density of the regression coefficient of nitrate concentration (in the presence of age and spatial interaction) for both models. The histogram reports the sample obtained from the posterior of the random effects model, and the superimposed density (continuous line) corresponds to the auto-Poisson model.

We can see a good agreement in location between both distributions. The estimated mean is 1.50×10^{-3} for the auto-Poisson model and 1.69×10^{-3} for the random effects model. Nevertheless, the hierarchical model produces wider confidence intervals (variance $= 8.56 \times 10^{-7}$) than the auto-Poisson (variance $= 5.37 \times 10^{-7}$).

This fact could be partially explained by the local quality of the approximation based on Taylor expansion in the posterior of the auto-Poisson model. It can capture the central region of the posterior density better than their tails. Nevertheless, this behaviour was also to be expected because of the inclusion of extra-Poisson variability terms in the random effects model.

In particular, the interesting value $\beta_2 = 0$ lies outside the 95% credible interval following the auto-Poisson model ($P\{\beta_2 < 0 | \boldsymbol{y}$ auto-Poisson$\} = 0.020$), and the opposite is true if we consider the random effects Poisson regression model ($P\{\beta_2 < 0 | \boldsymbol{y}$ GLMM$\} = 0.036$). Nevertheless, both posterior distributions lead to 99% credible intervals containing the value $\beta_2 = 0$.

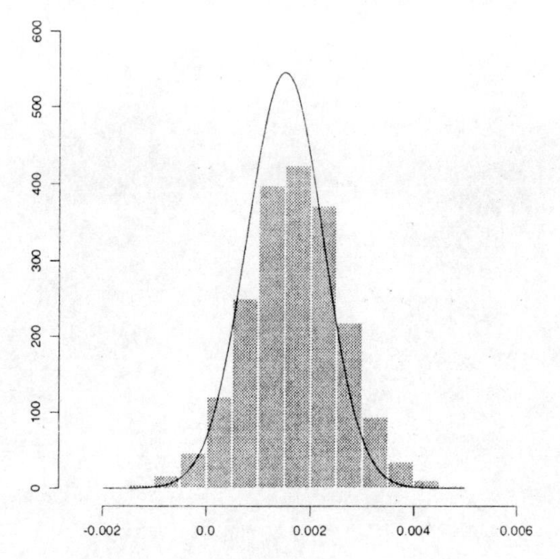

Figure 15.1 Posterior densities of the nitrate regression coefficient

Despite the quantitative differences in results reached for both models, they agree in essence with respect to the nitrate coefficient. They are saying that although there is no clear evidence of its influence, we cannot discard the revelation of a possible relationship through an improved model or a longer temporal series of new data.

To judge the whole relative merits of both models we have to compare them in a global sense. A usual Bayesian tool to compare non-nested competing models is the Bayes factor. Considered as the ratio of posterior to prior odds in favour of one model over the other, it reduces to the ratio of predictive distributions for the observed data, $p(\boldsymbol{y} \mid model1)/\text{p}(\boldsymbol{y} \mid model2)$.

How to estimate Bayes factors with MCMC techniques is still an open problem, and constitutes a very active area of research. Here we adhere to Gelfand's (1996) proposal of estimating the pseudo-Bayes (PBF) factor instead:

$$PBF = \prod_{i}^{n} \frac{p(y_i \mid \boldsymbol{y}_{-i}, \text{auto-Poisson})}{p(y_i \mid \boldsymbol{y}_{-i} \text{ GLMM})} \tag{15.30}$$

Each factor $p(y_i \mid \boldsymbol{y}_{-i}, \text{model}) = \int p(y_i \mid \boldsymbol{y}_{-i}, \boldsymbol{\theta}, \text{model}) p(\boldsymbol{\theta} \mid \boldsymbol{y}_{-i}, \text{model}) d\boldsymbol{\theta}$ in the numerator or denominator of (15.29) has to be estimated by Monte Carlo. For each location i we need a sequence $\boldsymbol{\theta}^{(r)}$ of simulated values of the parameters, taken from the distribution $p(\boldsymbol{\theta} \mid \boldsymbol{y}_{-i}, \text{model})$. Gelfand (1996) proposes an importance sampling technique in order to use only one sequence for each model, generated from the posterior distribution given all data $p(\boldsymbol{\theta} \mid \boldsymbol{y}, \text{model})$.

The Pseudo-Bayes factor modifies the Bayes factor in much the same way as pseudo-likelihood does with the likelihood function. It is easier to compute when the model has been specified via full conditionals, and its estimates, based on MCMC sampling of the posterior, are more stable than those for the Bayes factor.

In our case, MCMC sampling from the posteriors produces an estimate for the pseudo-Bayes factor (15.29) of 2.34 in favour of the truncated auto-Poisson model. We have to be

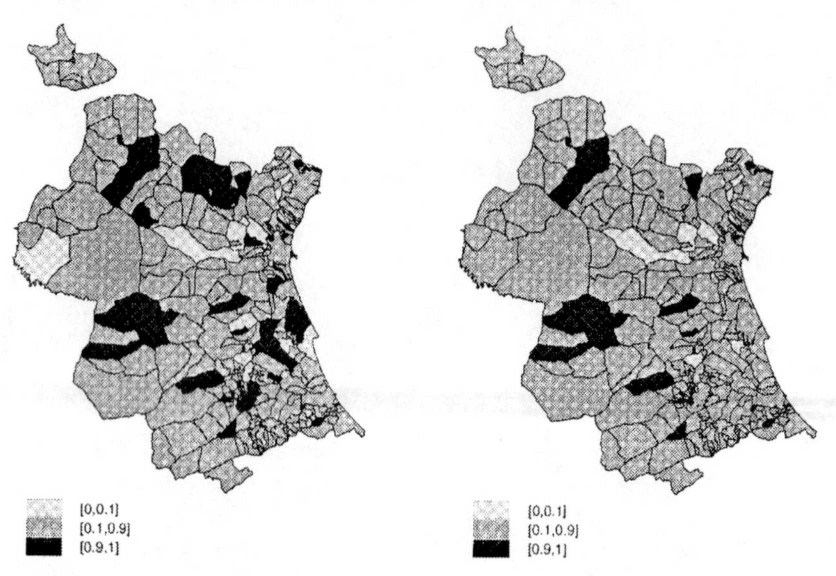

Figure 15.2 Estimated predictive posterior cumulative probabilities $\hat{F}(y_i|\boldsymbol{y}_{-i},\text{model})$. (a) Auto-Poisson; (b) random effects Poisson regression

cautious when interpreting this figure, given its stochastic nature. What it means to us is that both models are quite similar in capturing the spatial nature of our data. It seems that the more flexible structure of the GLMM, due to the Poisson extra-variance induced by its random effects, has been compensated for by a better proximity index (15.9) in the autoregressive model. In fact, we have tried the truncated auto-Poisson model using the weights w_{ij} of the precision matrix $\boldsymbol{W}$ in (15.13) instead of the proximity indexes (15.9) and we have obtained a much poorer fit. Compared with the previous models it gives a pseudo-Bayes factor of 527.03 in favour of the proximity index (15.9), and pseudo-Bayes factor of 224.80 in favour of the random effect model.

To detect locations that deviate from the model, the estimated values $\hat{F}(y_i|\boldsymbol{y}_{-i},$ model), of the predictive posterior cumulative probabilities, are very useful. They show how the fitted model predicts the observed mortality count of every location. In Figure 15.2 we can see the quantile maps of these posterior predictive distributions for both models. They are quite similar.

15.5 DISCUSSION

In the prostate cancer study developed in the previous section our main interest focused on the influence of a risk factor. Point estimation of its regression coefficient has proved to be quite stable with regard to modelling spatial interactions as autoregreesive or random effect terms. But confidence intervals differ substantially. Therefore we should consider the model to fit carefully in every case, and methods of comparing non-nested competing models deserve more attention. To this end, estimation of the pseudo-Bayes factor is a suggestive tool in the Bayesian context, although its properties have not yet been fully studied. In our case, both models have shown similar results. The truncated auto-Poisson model can be improved by including random effects allowing for extra-Poisson variability and the GLMM proposed in this chapter can be improved by trying a better precision matrix of the spatial random effects.

Autoregressive models can capture the effect of hidden risk factors with spatial structure through neighbouring mortality counts. Its use can be better justified when these spatial interactions between neighbouring sites obey a natural process, as in the case of infectious diseases. They are quite sensitive to the specification of the spatial autocorrelation structure.

If the spatial autocorrelation is not very strong, pseudo-likelihood analysis seems reliable, quick and easy. In other cases we have to resort to simulation methods like Monte Carlo Newton–Raphson, as proposed in this chapter.

Markov chain simulation techniques have revived Monte Carlo optimisation procedures, which have proven to be reliable and cover a wide range of practical situations. They produce approximate likelihood functions and posterior densities. It we adopt the frequentist paradigm of statistics, they need to be complemented with standard error estimators, which could be attained via the parametric bootstrap of the fitted models, a lengthy computing task. In the Bayesian framework, they are still useful when a unattainable normalising constant impedes the implementation of MCMC sampling from the posterior, as in the case of the truncated auto-Poisson model.

16

Multilevel Modelling of Area-Based Health Data

Ian H. Langford

University of East Anglia and University College, London

Alastair H. Leyland

University of Glasgow

Jon Rasbash

University of London

Harvey Goldstein

University of London

Rosemary J. Day

University of East Anglia and University College, London

Anné-Lise McDonald

University of East Anglia and University College, London

16.1 INTRODUCTION

Multilevel modelling is a form of random coefficient modelling which was first used in an educational setting (Goldstein, 1995). It is appropriate for data that have a natural hierarchy, such as educational data, where pupils are nested within classes within schools. However, it is also appropriate for geographically distributed data, where we may have individual cases of diseases nested within households, within postcode sectors, and so on. At larger scale, we may have ecological models where data are collected

Disease Mapping and Risk Assessment for Public Health. Edited by A.B. Lawson *et al.*
© 1999 John Wiley & Sons Ltd.

in larger administrative units, such as local authority districts, which are in turn nested within regions and nations. There are several examples of multilevel modelling being used in environmental and geographical epidemiology now appearing in the literature (see Langford *et al.*, 1998). Epidemiologists are increasingly using more complex methods of statistical analysis to investigate the distribution of diseases (Elliott *et al.*, 1992b, 1995), and the motivation behind this chapter is to introduce a multilevel modelling framework which also allows for the analysis of complex spatial processes such as:

(i) spatial autocorrelation in a Poisson generalised least squares model, and
(ii) simultaneous modelling of spatial effects that occur at different scales in a geographical hierarchy.

In the example given in this chapter we concentrate on investigating data that consist of observed and expected counts of disease occurring in discrete spatial units. Hence, for a sample of geographical areas we have a number of cases occurring within a distinct population at risk in each area. Whether we are embarking on an exploratory analysis, where we are simply interested in producing a map of the relative risks of disease, or an inferential analysis, where we are interested in investigating potential causal factors, it is useful to break down the likely effects on the distribution of a disease into three separate categories:

(i) Within-area effects, such as social characteristics of the population at risk. Since we are modelling aggregated data, we do not have information on individual cases, although we may have aggregate information on the mean and variance of social indicators such as income, employment and so on for each area. However, we can at least model these unmeasured variables by allowing for extra-Poisson variation in our model.
(ii) Hierarchical effects. These are due to the fact that small areas are grouped into larger areas, for administrative purposes, or for cultural and geographical reasons. For instance, in the example we present on mortality from prostate cancer in Scotland, local authority districts are grouped into Health Boards, which have different methods of treatment or classification of a disease, or different ascertainment rates. Again, if we have accurate information on these factors, then we could include them directly in the model, but we can allow for random variation between Health Boards even if we do not know the direct causes of this variation.
(iii) Neighbourhood effects. Areas that are close to each other in geographical space may share common environmental or demographic factors which influence the incidence or outcome of disease, but have a smoother distribution than that of the disease. For example, climatic factors such as temperature may vary between different part of a nation, but not at the smaller scale of a local authority district. In addition, as areas are usually formed from geopolitical boundaries that have nothing to do with the disease we are interested in, we may wish to spatially smooth the distribution or relative risks to remove any artifactual variation brought into the data by the method of aggregating the data.

The use of empirical Bayes and fully Bayesian techniques has allowed for alternative models of spatial and environmental processes affecting the distribution of a disease which rely on different underlying beliefs or assumptions about aetiology (Bernardinelli *et al.*, 1995a; Bernardinelli and Montomoli, 1992; Cisaghli *et al.*, 1995; Clayton and

Kaldor, 1987; Langford *et al.*, 1998; Langford, 1994; 1995; Lawson, 1994; Lawson and Williams, 1994; Mollié and Richardson, 1991; Schlattmann and Böhning, 1993). Two main statistical techniques have been used to model geographically distributed health data in this way. The first is Markov chain Monte Carlo (MCMC) methods, using Gibbs sampling (Gilks *et al.*, 1993) often fitted using the BUGS software (Spiegelhalter *et al.*, 1995), which can be used to fit fully Bayesian or emprical Bayesian models. The second set of methods is multilevel modelling techniques based on iterative generalised least squares procedures (IGLS) and are the focus of this chapter.

In the following section we discuss the basic multilevel Poisson model, and develop a computational method for modelling spatial processes within the software package MLn. MLn (for MS-DOS and Windows 3.1) and its successor MLwiN (for Windows 97 and NT) are widely used tools for multilevel modelling, and information about them can be found from a number of websites worldwide; for information, see `http://www.ioe.ac.uk/multilevel/`. We then present an example of how our model can be used using morbidity data for prostate cancer in Scottish local authority districts, and comment on how the results may be interpreted. The discussion section then focuses on methodological and substantive issues in a more general setting, and discusses work in progress to generalise the procedures we have developed.

16.2 DEVELOPING A POISSON SPATIAL MULTILEVEL MODEL

The basic model of fixed and random effects described by Goldstein (1995) and Breslow and Clayton (1993) is

$$Y = X\beta + Z\theta, \tag{16.1}$$

with a vector of observations Y being modelled by explanatory variables X and associated fixed parameters β, and explanatory variables Z with random coefficients θ. The fixed and random part design matrices X and Z need not be the same. θ is assumed to contain a set of random error terms in addition to other random effects. Goldstein (1995) describes a two-stage process for estimating the fixed and random parameters (the variances and covariances of the random coefficients) in successive iterations using IGLS. A summary of this process follows.

First, we estimate the fixed parameters in an initial ordinary least squares regression, assuming the variance at higher levels on the model to be zero. From the vector of residuals from this model we can construct initial values for the dispersion matrix V. Then, we iterate the following procedure, first estimating fixed parameters in a generalised least squares regression as

$$\hat{\beta} = (X^{\mathrm{T}}V^{-1}X)^{-1}X^{\mathrm{T}}V^{-1}Y \tag{16.2}$$

and again calculating residuals $\tilde{Y} = Y - X\hat{\beta}$. By forming the matrix product of these residuals, and stacking them into a vector, i.e. $Y^* = \mathrm{vec}\,(\tilde{Y}\tilde{Y}^{\mathrm{T}})$, we can estimate the variance of the random coefficients θ, $\gamma = \mathrm{cov}\,(\theta)$, as,

$$\hat{\gamma} = (Z^{*\mathrm{T}}V^{*-1}Z^*)^{-1}Z^{*\mathrm{T}}V^{*-1}Y^*, \tag{16.3}$$

where V^* is the Kronecker product of V, namely $V^* = V \otimes V$, noting that $V = \mathrm{E}(\tilde{Y}\tilde{Y}^{\mathrm{T}})$, Z^*

is the appropriate design matrix for the random parameters. Assuming multivariate Normality, the estimated covariance matrix for the fixed parameters is

$$\text{cov}(\hat{\beta}) = (X^{\mathrm{T}}V^{-1}X)^{-1}, \tag{16.4}$$

and for the random parameters, Goldstein and Rasbash (1992) show that

$$\text{cov}(\hat{\gamma}) = 2(Z^{*\mathrm{T}}V^{*-1}Z^{*})^{-1}. \tag{16.5}$$

We can therefore estimate random parameters, and their variances, in the same way as we estimate fixed parameters and their variances from the model. To compare what we are doing with ordinary least squares resgression, we are extend the process by modelling the random part of the model with respect to the structure of our data, estimating a set of parameters rather than simply having a residual error term.

However, we now need to develop a model for the relative risks of a disease. If we consider a population of areas with O_i observed cases and E_i expected cases and relative risk $\theta_i = O_i/E_i$, where E_i may be calculated from the incidence in the population N_i for each area as

$$E_i = N_i.\frac{\sum O_i}{\sum N_i} \tag{16.6}$$

and may be additionally divided into different age and sex bands, then we can write the basic Poisson model as

$$\begin{aligned} O_i &\sim \text{Poisson}(\mu_i), \\ \log(\mu_i) &= \log(E_i) + \alpha + x_i\beta + u_i + v_i, \end{aligned} \tag{16.7}$$

where $\log(E_i)$is treated as an offset, α is a constant and x_i is an explanatory variable with coefficient β (this may be generalised to a number of explanatory variables). We take account of the distribution of cases *within* each area by assuming that the cases have a Poisson distribution. In contrast, the μ_i represent heterogeneity effects between areas (Clayton and Kaldor, 1987; Langford, 1994), which may be viewed as constituting extra-Poisson variation caused by the variation among underlying populations at risk in the areas considered. The v_i are spatially dependent random effects, and may have any one of a number of structures describing adjacency or nearness in space (Besag *et al.*, 1991). Hence, we have a hierarchical model where within-area effects are modelled with a Poisson distribution (the first line of (16.7)) and relative risks between areas are considered as having a lognormal distribution (the second line of (16.7)). Other formulations for spatial effects are possible using normal approximations with covariance priors (see,for example, Besag *et al.*, 1991; Bailey and Gatrell, 1995; Lawson *et al.*, 1996).

Before discussing the structure of these spatial effects, we must first account for the fact that we have a non-linear (logarithmic) relationship between the outcome variable and the predictor part of the model. There are two options:

(i) If the cases in each area are sufficiently large, say $O_i > 10$, then it may be reasonable to model the logarithm of the relative risks directly (Clayton and Hills, 1993), assuming these follow a Normal distribution. In this case, heterogeneity effects can be accommodated by weighting the random part of the model by some function of the population at risk in each area.

(ii) When the Normal approximation is inappropriate, we can make a linearising approximation to estimate the random parameters. If we take the case of having heterogeneity effects only for the sake of simplicity, we can estimate the residuals $\hat{u}_i$ from the model using penalised quasi-likelihood (PQL) estimation with a second-order Taylor series approximation (Breslow and Clayton, 1993; Goldstein, 1995; Goldstein and Rasbash, 1992). After each iteration t we make predictions H_i from the model, where $H_t = X_i\hat{\beta}_t + \hat{u}_i$, and hence use these to calculate new predictions for iteration $t + 1$, so that

$$\begin{aligned} f(H_{t+1}) &= f(H_t) + x_i(\hat{\beta}_{t+1} - \hat{\beta}_t)f'(H_t) \\ &\quad + \hat{u}_i f'(H_t) + \hat{u}_i^2 f''(H_t)/2, \end{aligned} \tag{16.8}$$

where the first two terms on the right-hand side of (16.8) provide the updating function for the fixed part of the model, and $f(\cdot)$ is a link function. The third term comprises a linear random component created by multiplying the first differential of the predictions by the random part of the model, and the fourth term is the next term in the Taylor expansion about H_t. For the Poisson distribution:

$$f(H) = f'(H) = f''(H) = \exp(X_i\hat{\beta}_t + \hat{u}_i). \tag{16.9}$$

Hence, at each iteration we estimate about the fixed part of the model plus the residuals. A full description of this procedure can be found in Goldstein (1995) and Goldstein and Rasbash (1992). This can lead to problems with convergence, or with the model 'blowing up' if some of the residuals are particularly large. In these cases, the second-order term in (16.8) can be omitted, or, in extreme cases, estimates can be based on the fixed part of the model only. This latter case is called marginal quasi-likelihood (MQL: Breslow and Clayton, 1993; Goldstein, 1995), but may lead to biased parameter estimates. However, bootstrap procedures can potentially be used to correct for these biases (Goldstein, 1996a,b; Kuk, 1995). For (16.7) we substitute $\hat{u}_i + \hat{v}_i$ for $\hat{u}_i$, in (16.8) and (16.9).

There are several possibilities for specifying the structure of the random effects in the model (see, for example, Besag *et al.*, 1991, and Bailey and Gatrell, 1995). These models assume two components, namely a random effects or 'heterogeneity' term and a term representing the spatial contribution of neighbouring areas as in (16.7).

We adopt a somewhat different approach, which allows a more direct interpretation of the model parameters and can be fitted in a computationally efficient manner within a multilevel model. For the heterogeneity effects, this is not a problem, because we simply have a variance–covariance matrix with 1 or other specified values on the diagonal, and the model is analogous to fitting an iteratively weighted least squares model (McCullagh and Nelder, 1989). However, the case of the spatial effects is more complex, because we require off-diagonal terms in the variance–covariance matrix. This can be achieved through careful consideration of the structure of the spatial part. Our formulation of the spatial model is to consider the spatial effects v_i to be the weighted sum of a set of independent random effects v_i^* such that

$$v_i = \sum_{j \neq i} z_{ij} v_j^*. \tag{16.10}$$

The v_i^* can be considered to be the effect of area upon other areas, moderated by a measure of proximity of each pair of areas z_{ij}. The v_i^* can be estimated directly from the model—these are the residuals—due to their independence. Returning to the matrix notation used in (16.1), we can rewrite (16.7) as

$$Y = \{\log(E_i)\ 1\ x_i\} \begin{bmatrix} 1 \\ \alpha \\ \beta \end{bmatrix} + [Z_u\, Z_v^*] \begin{bmatrix} \theta_u \\ \theta_v^* \end{bmatrix}, \tag{16.11}$$

where Z_u is the identity matrix and $Z_v^* = \{z_{ij}\}$. With a variance structure such as

$$\text{var}\left(\begin{bmatrix} \theta_u \\ \theta_v^* \end{bmatrix}\right) = \begin{bmatrix} \sigma_u^2 I & \sigma_{uv} I \\ \sigma_{uv} I & \sigma_v^2 I \end{bmatrix}, \tag{16.12}$$

which is equivalent to

$$\text{var}\left(\begin{bmatrix} u_i \\ v_i^* \end{bmatrix}\right) = \begin{bmatrix} \sigma_u^2 & \sigma_{uv} \\ \sigma_{uv} & \sigma_v^2 \end{bmatrix},$$

the overall variance from (16.1), conditional on the fixed parameters, is given by

$$\text{var}\,(Y|X\beta) = Z \textstyle\sum_\theta Z^{\mathrm{T}}, \tag{16.13}$$

where $\sum_\theta$ is the variance of the random terms in θ. The structure of $\sum_\theta$ will often lead to simplifications; for example, in a random effects model when $\theta = \{u_i\}$ and $\text{var}(u_i) = \sigma_u^2$, $\text{cov}(u_i, u_j) = 0$ then $\sum_\theta = \sigma_u^2 I$ and so $\text{var}\,(Y|X\beta) = \sigma_u^2 ZZ^{\mathrm{T}}$. Similarly, in the spatial model defined by the partitions in θ and Z given by (16.11) and the variance structure of (16.12), we can see that

$$\text{var}(Y|X\beta) = \sigma_u^2 Z_u Z_u^{\mathrm{T}} + \sigma_{uv}(Z_u Z_v^{*\mathrm{T}} + Z_v^* Z_u^{\mathrm{T}}) + \sigma_v^2 Z_v^* Z_v^{*\mathrm{T}}. \tag{16.14}$$

There are many ways in which the z_{ij} can be formulated; in general we can write

$$z_{ij} = w_{ij}/w_{i+}. \tag{16.15}$$

The w_{ij} can either 1's and 0's representing an adjancency matrix, or be functions of the distance between areas (see Section 16.3). Common choices for the w_{i+} would be $w_{i+} = (\sum_{j \neq i} w_{ij})^{0.5}$, which ensures that the variance contribution is the same for all areas, or $w_{i+} = \sum_{j \neq i} w_{ij}$, in which case the variance of an area decreases as the information about that area (in terms of, for example, the number of neighbours in an adjacency model) increases.

Finally, there is the problem of specifying the random effects for heterogeneity and spatial effects within a generalised linear modelling framework, in this case using IGLS estimation within the MLn software. We do this by constructing weights matrices associated with the random effects and fit these directly into the model. The variance of the data conditional on the fixed part of the model, as given in (16.14), is expressed in terms of three matrices: $Z_u Z_u^{\mathrm{T}}(Z_u Z_v^{*\mathrm{T}} + Z_v^* Z_u^{\mathrm{T}})$, and $Z_v^* Z_v^{*\mathrm{T}}$. Expressing the model is terms of these design matrices overcomes the need to place multiple equality constraints upon the random parameters. This is generalisable to the non-linear model expressed in (16.7)–(16.9).

16.3 INCIDENCE OF PROSTATE CANCER IN SCOTTISH LOCAL AUTHORITY DISTRICTS

In this example we wish to investigate the hypothesis that the relative risk of prostate cancer is higher in rural than urban areas, as previous research has indicated an association between agricultural employment and the incidence of prostate cancer (Key, 1995). The data cover six years, from 1975 to 1980, of the incidence of prostate cancer in 56 districts in Scotland (Kemp *et al.*, 1985). Table 16.1 shows the observed and expected cases, plus the relative risks for incidence of prostate cancer.

To examine the effect of rural location, we use a variable measuring the percentage of the male workforce employed in agriculture, fishing and forestry industries as a surrogate measure of the rurality of an area. However, we have to look not only at the incidence of prostate cancer within districts, but account for a potential artifactual effect caused by differential diagnosis rates between Health Board areas in Scotland. Hence, we are modelling spatial effects caused by different processes at two different scales, namely:

(i) a spatial autocorrelation model at district scale, where we are accounting for the possibility that areas closer in geographical space have similar incidence of prostate cancer; and
(ii) a variance components model at Health Board scale, where we investigate the possibility that different Health Boards have different relative risks of prostate cancer, potentially because diagnostic criteria are variable.

Hence, we can extend (16.1) and (16.11) so that

$$\log(\mu) = \log(E) + X\beta + [Z_u Z_v^*] \begin{bmatrix} \theta_u \\ \theta_v^* \end{bmatrix} + Z_{\text{hb}}\theta_{\text{hb}}. \tag{16.16}$$

In this case the expected cases, E, have been calculated from national incidence rates for Scotland for discrete age bands. We use three explanatory variables in the fixed part of the model ($X\beta$) in addition to the intercept term (*CONS*), namely the proportion of the population in higher social classes (*SC12*); the estimated incidence to ultraviolet light at the earth's surface (*UVBI*); and the percentage of the male employment in agriculture, fishing and forestry (*AGRI*). Social class and ultraviolet light exposure have been included as these have previously been postulated as risk factors for prostate cancer.

The Z_v^* are calculated using distances between district centroids. The choice of distance decay function is largely user-dependent, and should ideally be based on some prior hypothesis about the data (Bailey and Gatrell, 1995). Here we have used a simple exponential decay model where we define the w_{ij} as

$$w_{ij} = \exp(-\lambda d_{ij}), \tag{16.17}$$

where d_{ij} are the Euclidian distances between the centroids of areas i and j, and λ is a constant to be estimated from the data. The estimation of λ is problematical because it is non-linear in the random part of the model. Goldstein *et al.* (1994) show that maximum likelihood estimates can be obtained using a Taylor series expansion for the Normal model. However, things become more complicated for a Poisson model, and in general

Table 16.1 Observed and expected cases, and relative risks for the incidence of prostate cancer in Scottish districts, 1975–1980

District	Health Board	Observed	Expected	SMR
Caithness	Highland	15	25.587	0.58625
Sutherland	Highland	18	12.319	1.46110
Ross–Cromarty	Highland	42	42.644	0.98489
Skye–Lochalsh	Highland	10	9.477	1.05520
Lochaber	Highland	22	18.005	1.22190
Inverness	Highland	51	51.173	0.99662
Badenoch	Highland	15	8.529	1.75870
Nairn	Highland	10	9.477	1.05520
Moray	Grampian	107	75.812	1.41140
Banff–Buchan	Grampian	95	74.920	1.25310
Gordon	Grampian	70	58.754	1.19140
Aberdeen	Grampian	249	189.530	1.31380
Kincardine	Grampian	52	38.854	1.33840
Angus	Tayside	104	86.236	1.20600
Dundee	Tayside	176	168.680	1.04340
Perth–Kinross	Tayside	148	108.030	1.37000
Kirkcaldy	Fife	145	135.510	1.07000
NE–Fife	Fife	91	56.859	1.60050
Dunfermline	Fife	117	115.610	1.01200
West-Lothian	Lothian	106	129.830	0.81646
Edinburgh	Lothian	538	402.750	1.33580
Midlothian	Lothian	87	77.707	1.11960
East-Lothian	Lothian	77	74.864	1.02850
Tweeddale	Borders	27	13.267	2.03510
Ettrick	Borders	38	29.377	1.29350
Roxburgh	Borders	44	33.168	1.32660
Berwickshire	Borders	23	17.058	1.34840
Clackmannan	Forth Valley	37	44.540	0.83072
Stirling	Forth Valley	100	72.969	1.37040
Falkirk	Forth Valley	149	136.460	1.09190
Argyll–Bute	Argyll & Clyde	56	60.650	0.92334
Dumbarton	Argyll & Clyde	80	72.969	1.09640
Renfrew	Argyll & Clyde	118	194.270	0.60741
Inverclyde	Argyll & Clyde	84	94.765	0.88640
Glasgow	Greater Glasgow	627	721.160	0.86943
Clydebank	Greater Glasgow	31	49.278	0.62909
Bersden	Greater Glasgow	31	36.958	0.83878
Strathkelvin	Greater Glasgow	57	82.446	0.69137
Eastwood	Greater Glasgow	43	50.225	0.85614
Cumbernauld	Lanarkshire	24	58.754	0.40848
Monklands	Lanarkshire	58	104.240	0.55640
Motherwell	Lanarkshire	100	141.200	0.70822
Hamilton	Lanarkshire	58	102.350	0.56670
East-Kilbride	Lanarkshire	40	78.655	0.50855
Clydesdale	Lanarkshire	47	54.0.16	0.87011
Cunninghame	Ayrshire & Arran	103	128.880	0.79919
Kilmarnock	Ayrshire & Arran	66	77.707	0.84934

Table 16.1 (*continued*)

District	Health Board	Observed	Expected	SMR
Kyle–Carrick	Ayrshire & Arran	108	106.140	1.01750
Cumnock–Doon	Ayrshire & Arran	29	42.644	0.68004
Wigtown	Dumfries & Galloway	28	28.430	0.98489
Stewartry	Dumfries & Galloway	40	20.848	1.91860
Nithsdale	Dumfries & Galloway	56	52.121	1.07440
Annandale	Dumfries & Galloway	48	33.168	1.44720
Orkney	Orkney	22	17.058	1.28970
Shetland	Shetland	17	21.796	0.77996
Western Isles	Western Isles	45	29.377	1.53180

an alternative is to fit a series of models with different values of λ_k and determine the residual deviance from each model, D_k. We can then regress the deviance against the distance decay parameter so that

$$D_k = a + b\lambda_k + c\lambda_k^2 + e_k. \tag{16.18}$$

Differentiating, the approximate solution will be where $\lambda = -b/2c$. Successive approximations then converge towards the estimated value. However, care must be taken when estimating λ, as the likelihood function may be multimodal (Ripley, 1988).

Z_{hb} is a vector of 1 which allows for a variance component for each Health Board to be estimated, and hence a measure of the variance at this scale, σ_{hb}^2. Table 16.2 presents the results for four different models, representing:

model A: a simple, single-level model with no spatial effects.
model B: a model with district scale spatial effects, but no Health Board effects.
model C: a model with only Health Board effects.
model D: a model with both district and Health Board effects as given in (16.16).

The simple model (model A) presented in Table 16.2 seems to indicate a strong and significant effect of rurality, as measured by percentage male agricultural employment (*AGRI*). However, this is weakened by fitting a spatial autocorrelation parameter in model B, which suggests that the effect of *AGRI* may be due to adjacent areas having similar mortality. The change in deviance between the two models is 14.89 on two degrees of freedom ($p < 0.001$: we have fitted a covariance parameter as well as a variance term). The third model (model C), using Health Boards as a level with no spatial autocorrelation between districts, shows how ignoring autocorrelation between residuals at a lower level of a multilevel model (in this case districts) could lead to misleading results at higher levels (in this case, Health Boards), as the parameter for the variance between Health Boards is statistically significant at $p < 0.05$, but the deviance statistic suggests that the model is not as good a fit to the data as model B.

Unexplained random variation at the district level can appear spuriously at the Health Board level, and the final model, with both Health Board effects and spatial effects between districts, suggests that this may be the case. The parameter estimate for *AGRI* becomes smaller in models B, C and D, although it is still significant at the

Table 16.2 Parameter estimates and standard errors for the prostate cancer models

	(A) Simple model		(B) Spatial effects		(C) Health Board effects		(D) Both effects	
	Estimate	**SE**	**Estimate**	**SE**	**Estimate**	**SE**	**Estimate**	**SE**
Fixed part								
Intercept	– 0.0257	0.584	– 0.513	0.605	– 0.0108	0.636	– 0.321	0.670
SC12	– 0.000645	0.00524	0.00145	0.00389	– 0.00339	0.00477	0.000825	– 0.000184
UVBI	– 0.0141	0.0635	0.0565	0.0704	– 0.00112	0.0705	0.0279	0.0764
AGRI	0.0272	0.00603	0.0163	0.00636	0.0180	0.00634	0.0167	0.00663
Random part								
σ^2_{hb}					0.327	0.0183	0.0225	0.00550
σ^2_u	0.0822	0.0155	0.0665	0.0141	0.0530	0.0117	0.0657	0.0126
σ_{uv}			0.000805	0.000414			0.000682	0.000130
σ^2_v			0.0000159	0.0000167			0.000	0.000
λ			7.23				7.00	
Residual deviance	18.98		4.09		10.93		– 4.97	

0.05 level. Hence, misspecification of the random part of a model can noticeably affect the fixed as well as the random parameters. Further work needs to be done on the analysis of residuals in these complex models: Langford and Lewis (1998) details some procedures for the general analysis of outliers in multilevel models.

However, we must be careful in drawing conclusions about the size of the parameter estimate for *AGRI*, because we are not postulating that there is some genuine spatial correlation between cases of prostate cancer, for example if the disease had an infectious aetiology and was transmitted between individuals by contact. The spatial effects are not, therefore, in this case an alternative causal factor, but merely a statistical manipulation to account for correlation amongst the residuals in our model. One problem is that the values of the variable *AGRI* are also spatially correlated, because rural districts tend to be adjacent to each other, as do urban ones. Hence, we must be cautious in making inferences from our models without corroborative evidence from elsewhere, although it is interesting to note that the parameter estimate for *AGRI* remains significant in all four model formulations.

16.4 DISCUSSION

We have attempted to demonstrate a general method for modelling geographical data which is distributed in hierarchical administrative units, but which also displays spatial autocorrelation. The models can be implemented within a widely available software package called MLn/MLwiN. However, there are several issues that still need to be addressed, both methodologically and substantively:

(i) We are extending the basic method to model multiple causes of disease simultaneously. Hence, we could model the joint distribution of prostate cancer and another cancer simultaneously. This is the equivalent of adding in another level to the model, so that we have diseases nested within districts within Health Board areas. A further extension to the model can be where areas, such as districts, are not discretely nested within higher level units, such as Health Boards. In this case, a multiple membership model (Goldstein, 1995) may be used, where weights are attached to allocate portions of districts to different Health Boards.

(ii) Space–time models are also possible, as time is simply an extra dimension that requires a variance parameter in the random part of the model, and covariance terms with any spatial parameters

(iii) The main problem in fitting the models is poor convergence properties, usually caused by a high correlation between the heterogeneity and spatial components of the model. One of the authors (AHL) is developing an orthogonalisation procedure to overcome this problem

(iv) The residuals from the model are measured with error, but the IGLS procedure used will tend to underestimate the variance of the residuals. To overcome this, MLwiN has the capability of using the IGLS convergence of the model as the starting point for either a Gibbs sampling or Metropolis–Hastings run of simulations which will provide for better estimates of, for example, the confidence intervals around the posterior relative risks of disease for each district or Health Board. These techniques could also be used to provide better estimates of the standard errors for fixed parameters in the model, rather than relying on those estimated from the model to judge statistical significance.

Substantively, multilevel spatial models suffer from similar problems of interpretation as single–level spatial models. It is often difficult to know whether one has modelled a genuine spatial pattern or merely accounted for unmeasured explanatory variables, and fitting a spatial smoothing parameter masks a genuine relationships with an explanatory variable which has its own distinct spatial distribution. However, we believe that the use of a multilevel model can shed light on different processes which may be operating at different spatial scales, and hence provide a valuable tool for the analysis of geographically distributed health data.

ACKNOWLEDGEMENTS

This work was supported by the Economic and Social Research Council, UK. The Public Health Research Unit is financially supported by the Chief Scientist Office of the Scottish Office Department of Health. The opinions expressed in this chapter are not necessarily those of the Chief Scientist Office.

Part IV

Risk Assessment for Putative Sources of Hazard

17

A Review of Modelling Approaches in Health Risk Assessment around Putative Sources

Andrew B. Lawson

University of Aberdeen

A. Biggeri

University of Florence

F.L.R. Williams

University of Dundee

17.1 INTRODUCTION

The assessment of the impact of sources of pollution on the health status of communities is of considerable academic and public concern. The incidence of many respiratory, skin, and genetic diseases is thought to be related to environmental pollution (Hills and Alexander, 1989), and hence any localised source of such pollution could give rise to changes in the incidence of such diseases in the adjoining community.

In recent years there has been growing interest in the development of statistical methods useful in the detection of patterns of health events associated with pollution sources. Here, we consider the statistical methodology for the assessment of putative health effects of sources of air pollution or ionising radiation. We consider inference and modelling problems and concentrate primarily on the generic problem of the statistical analysis of observed point patterns of case events or tract counts, rather than specific features of a particular disease or outcome.

Disease Mapping and Risk Assessment for Public Health. Edited by A.B. Lawson *et al.*
© 1999 John Wiley & Sons Ltd.

A number of studies utilise data based on the spatial distribution of such diseases to assess the strength of association with exposure to a pollution source. Raised incidence near the source, or directional preference related to a dominant wind direction may provide evidence of such a link. Hence, the aim of the analysis of such data is usually to assess the effect of specific spatial variables rather than general spatial statistical modelling. That is, the analyst is interested in detecting patterns of events near (or exposed to) the focus and less concerned about aggregation of events in other locations. The former type of analysis has been named 'focused clustering' by Besag and Newell (1991). The latter is often termed 'non-focused' clustering and is the subject of Part II of this volume, where a review of appropriate methods appears (Chapter 7). To date, most pollution-source studies concentrate on the incidence of a single disease (e.g. childhood leukaemia around nuclear power stations or respiratory cancers around waste-product incinerators).

The types of data observed can vary from disease-event locations (usually residence addresses of cases) to counts of disease (mortality or morbidity) within census tracts or other arbitrary spatial regions. An example of a data set consisting of residential locations is provided in Figure 17.1. Figure 17.1 displays the locations of respiratory cancer cases around a steel foundry (0,0) in Armadale, central Scotland for the period 1968–74. In this example, the distribution of cases around the central foundry is to be examined to assess whether there is evidence for a relation between the locations and the putative source (the foundry). In Figure 17.2 the counts of respiratory cancer for the period 1978–83 in Falkirk, central Scotland are displayed. A number of putative sources of health hazard are located in this area, most notably a metal processing plant (*).

The two different data types lead to different modelling approaches. Spatial point process models are appropriate for event–location data. In the case of count data, one may use properties of regionalised point processes. That is, an independent Poisson

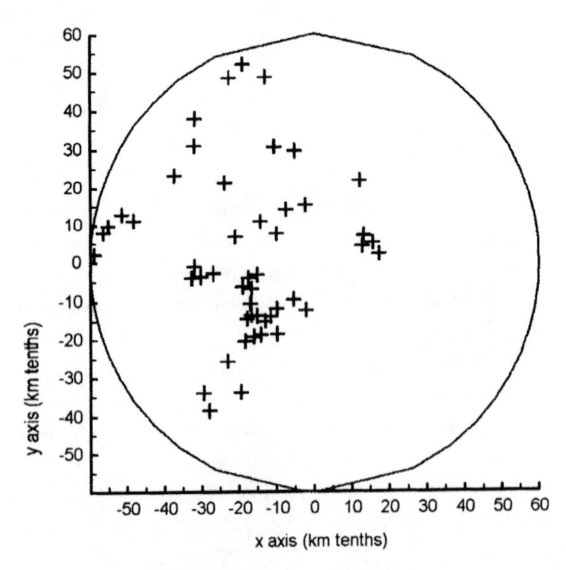

Figure 17.1 Armadale, central Scotland: residential address locations of respiratory cancer cases, 1968–74

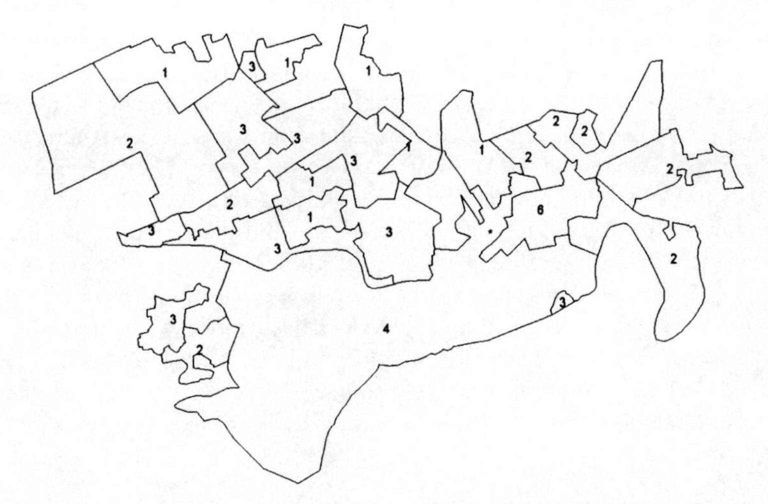

Figure 17.2 Counts of respiratory cancer for 1978–83 in Falkirk, central Scotland

model for regional counts is often assumed and one typically uses loglinear models and related tests (see, for example, Lawson, 1993b; Lawson and Waller, 1996).

The effects of pollution sources are often measured over large geographic areas containing heterogeneous population densities (usually both urban and rural areas). As a result, the intensity of the underlying point process of cases is heterogeneous. For an introduction to heterogeneous spatial point processes, and spatial point processes in general, see Diggle (1983). A review of spatial point process theory appears in Chapter 8 of Cressie (1993). The discussion below assumes some familiarity with terms and concepts related to spatial point processes.

A recent review of statistical methods for detecting spatial patterns of disease around putative sources of hazard appears in Lawson and Waller (1996). That review provides greater details of methods of analysis in such studies. A more general review appears in Lawson and Cressie (1999). In subsequent chapters in Part IV a number of different approaches to specific problems are discussed. In Chapter 18, exploratory methods for count data are discussed, while in Chapters 19 and 20 testing for effects and case–control methods are considered. In Chapter 21, a general discussion of the analysis of respiratory cancer around putative sources is provided.

Examples of analyses of data relating to putative source problems appear in Chapters 9, 20, 21, 29, and 30 of this volume.

17.2 PROBLEMS OF INFERENCE

The primary inferential problems arising in putative-source studies are (i) *post hoc* analyses, and (ii) multiple comparisons. The well-known problem of *post hoc* analysis arises when prior knowledge of reported disease incidence near a putative source leads an investigator to carry out statistical tests or to fit models to data to 'confirm' the evidence (Neutra, 1990; Rothman, 1990). Essentially, this problem concerns bias in data collection and prior knowledge of an apparent effect. In Hills and Alexander (1989) and Gardner

(1989) it is noted that both hypothesis tests and study-region definition can be biased by this problem. However, Lawson (1993b) noted that if a study *region* is thought a priori to be of interest because it includes a putative pollution source, one does not suffer from *post hoc* analysis problems if the internal spatial structure of disease incidence did not influence the choice of region.

The multiple-comparison problem has been addressed in several ways. Bonferonni's inequality may be used to adjust critical regions for multiple comparisons but the conservative nature of such an adjustment is well known. Thomas (1985) has discussed multiple-comparison problems and proposed the use of cumulative p-value plotting to assess the number of diseases yielding evidence of association with a particular source (see also Haybrittle *et al.*, 1995). An alternative approach is to specify a general model for the incidence of disease or diseases. Such an approach can often avoid multiple comparison problems. Modelling is discussed in Sections 17.4 and 17.5.

17.3 EXPLORATORY TECHNIQUES

The use of exploratory techniques is widespread in conventional statistical analysis. However, in point-source analysis one must exercise care about how the subsequent analysis is influenced by exploratory or diagnostic findings. For example, if an exploratory analysis isolates a cluster of events located near a pollution source, then this knowledge could lead to a *post hoc* analysis problem, namely inference based on a model specifically including such a cluster is suspect.

For case-event data, one can employ standard point process methods to explore data structure. For example, the intensity (i.e. points per unit area) of events can be mapped and viewed as a contoured surface, usually using non-parametric density estimation (Diggle, 1985). A natural model of spatial randomness is a heterogeneous Poisson process (HEPP) where the mapped surface represents the first-order intensity of the process (Diggle, 1983). Additionally, the Dirichlet tessellation or Delaunay triangulation of the points can demonstrate overall structure (Sibson, 1980).

If the intensity of controls is also mapped, then it is useful to assess whether the cases demonstrate an excess of events beyond that demonstrated by the controls (e.g. in areas of increased risk). Controls could consist of randomly selected individuals from the population at risk (perhaps matched on confounding factors), or a 'control disease' as mentioned above. A higher relative intensity of 'cases' to 'controls' near a pollution source, as compared with far away, may support a hypothesis of association.

Bithell (1990) has suggested that the ratio of density estimates of cases and controls be used to define a map displaying areas of increased risk. Lawson and Williams (1993) refined this method by using kernel smoothing and also provided crude standard-error surfaces for the resulting map. This type of 'extraction' of a control intensity is akin to the mapping of standardised mortality ratios for count data. Kelsall and Diggle (1995) further refined the original ratio estimator and described improved conditions for estimation of the ratio-surface smoothness. A case study of the application of a variety of such methods is given by Viel *et al.* (1995).

In the case of tract-count data, a variety of exploratory methods exist. One can use representation of counts as surfaces and incorporate expected count standardisation (e.g. through a standardised mortality/morbidity ratio (SMR)). While mapping regional

SMRs can help isolate excess incidence, estimates of SMRs from counts in small areas are notoriously variable, especially for areas with few persons at risk. Various methods have been proposed to stabilise these small area estimates. Two different approaches are based on non-parametric smoothing and empirical Bayes 'shrinkage' estimation.

Smoothing approaches include Breslow and Day's (1987) analysis of SMRs over time. Lawson (1993b) proposes a kernel smoothing approach using a single parameter to describe the surface smoothness. In Chapter 18 of this volume, Bithell discusses various issues relating to relative risk smoothing and exploratory analysis of tract count data. Some implementations of geostatistical prediction (Kriging) to obtain a risk surface have been proposed (e.g. Carrat and Valleron, 1992; Webster *et al.*, 1994), although some key characteristics of disease incidence may be violated by Kriging interpolation, such as an implicit assumption in variogram analysis of homogeneity of variances. Standard Kriging estimators can produce *negative* interpolant values, which are invalid for relative risk surfaces; this might be handled by Kriging the logarithm of relative risk and then back transforming. However, this may also fail due to singularities at zero relative risk. More complex smoothing approaches can be adopted (Diggle *et al.*, 1998; Lawson *et al.*, 1996).

In general, the use of non-parametric relative risk estimation, particularly combined with Monte Carlo evaluation of excess risk, is a powerful tool for the initial assessment of risk elevation. Care must be taken, however, not to prejudice further inference by the a posteriori focusing of analysis.

17.4 MODELS FOR POINT DATA

In this section we consider a variety of modelling approaches available when data are recorded as a point map of disease case events. Define A to be any planar region and $|A|$ as the area of A.

In analysing events around a pollution source, one usually defines a fixed window or geographical region A_1 with area $|A_1|$ and all events that occur within this region within a particular time period are recorded (mapped). Thus, the complete realisation of the point process (within A_1) is to be modelled. In the analysis of point events around pollution sources, the long-range or trend components of variation are often of primary concern. This leads one to consider heterogeneous (non-stationary) Poisson process (HEPP) models and to use the HEPP's first-order intensity to model the trend.

Event locations often represent residential addresses of cases and take place in a heterogeneous population that varies both in spatial density and in susceptibility to disease. Both Diggle (1989) and Lawson (1989) independently, suggested a method to accommodate such a population effect within a HEPP model.

Define the first-order intensity function of the process as $\lambda(\boldsymbol{x})$, which represents the mean number of events per unit area in the neighbourhood of location $\boldsymbol{x}$. This intensity may be parameterised as

$$\lambda(\boldsymbol{x}) = \rho \cdot g(\boldsymbol{x}) \cdot f(\boldsymbol{x}; \theta),$$

where $g(\boldsymbol{x})$ is the 'background' intensity of the population at risk at $\boldsymbol{x}$, and $f(\boldsymbol{x}; \theta)$ is a parameterised function of risk relative to the location of the pollution source. The focus of interest for assessing associations between events and the source is inference regarding parameters in $f(\boldsymbol{x};\theta)$, treating $g(\boldsymbol{x})$ as a nuisance function. The log likelihood of m events in A, conditional on m, is (bar a constant)

$$\sum_{i=1}^{m} \log f(\boldsymbol{x}_i;\theta) - m \log \int_A g(\boldsymbol{x}) f(\boldsymbol{x};\theta)\, \mathrm{d}\boldsymbol{x}. \tag{17.1}$$

Here, parameters in $f(\boldsymbol{x};\theta)$ must be estimated as well as $g(\boldsymbol{x})$. Diggle and Lawson proposed estimating $g(\boldsymbol{x})$ non-parametrically from the 'at risk' population. Diggle (1990) suggested using the locations of a 'control' disease to provide a kernel estimate, $\hat{g}(\boldsymbol{x})$, of the background at arbitrary location $\boldsymbol{x}$. Lawson and Williams (1994) illustrated an application where $g(\boldsymbol{x})$ is estimated from the expected death surface using the population's expected rates instead of a control disease.

Inferential problems arise when $g(\boldsymbol{x})$ is estimated as a function and then apparently regarded as constant in subsequent inference concerning $\lambda(\boldsymbol{x})$. One solution to this problem is to incorporate the estimation of the background smoothing constant in $\hat{g}(\boldsymbol{x})$ by the use of prior information in a Bayesian setting. Lawson and Clark (Chapter 9 in this volume) (also Lawson, 1998a) propose the inclusion of the smoothing constant within an MCMC algorithm. As an alternative, Diggle and Rowlingson (1994) proposed avoiding estimating $g(\boldsymbol{x})$ by regarding the control locations and case locations as a set of labels whose binary value is determined by a position-dependent probability:

$$\Pr(\boldsymbol{x}) = \rho f(\boldsymbol{x};\theta)/(1 + \rho f(\boldsymbol{x};\theta)).$$

This model avoids the use of $g(\boldsymbol{x})$ and hence avoids the inferential problems noted above. However, this model can *only* be applied when a point map of a control disease is available and when multiplicative relative risk is assumed.

An alternative model similar to the binary model above may be proposed. One conditions on the set of locations (cases and controls) and randomly assigns a binary label to each location indicating whether a particular location is a case or a control. Baddeley and Van Lieshout (1993) consider such a marked point process model. For example, if the points are a realisation of a Markov point process, then, conditional on the points, the marks form a binary Markov Random Field (MRF) (Baddeley and Møller, 1989). Note that a HEPP can be considered as a special case of a Markov point process. Variants to the above models are proposed where the case data and the data for estimation of $g(\boldsymbol{x})$ are at different resolution levels (see, for example, Lawson and Williams, 1994).

A further alternative is to incorporate the estimation of $g(\boldsymbol{x})$ into the procedure for the estimation of other parameters. Since $g(\boldsymbol{x})$ can be regarded as a stochastic process, we can ascribe prior distributions for its structure. One simple approach is to estimate $g(\boldsymbol{x})$ from a kernel density estimator where the smoothing parameter has a prior distribution. This smoothing parameter is estimated jointly with the other parameters, for example in a stochastic sampling algorithm such as Markov chain Monte Carlo (MCMC). In Chapter 9 of this volume Lawson and Clark provide an example of this approach. This approach has great generality, and can be the basis of a general approach to the incorporation of a stochastic process within a point process intensity. There are similarities between this approach and those discussed below, dealing with *unobserved* heterogeneity. The main difference is that an external data source (e.g. control disease or expected rate) is also available to focus the estimation. Since this type of approach to $g(\boldsymbol{x})$ estimation can be applied quite generally (e.g. to control diseases and expected rates), and avoids conditioning of the prior $g(\boldsymbol{x})$ estimate, it is likely to have wide adoption and application.

It is possible that population or environmental heterogeneity may be unobserved in the data set. This could be because either the population background hazard is not directly available or the disease displays a tendency to cluster (perhaps due to unmeasured covariates). The heterogeneity could be spatially correlated, or it could lack correlation, in which case it could be regarded as a type of 'overdispersion'.

One can include such unobserved heterogeneity within the framework of conventional models as a random effect. For example, a general definition of the first-order intensity could be

$$\lambda(\boldsymbol{x}) = g(\boldsymbol{x})\exp(F(\boldsymbol{x})\alpha + \zeta(\boldsymbol{x})),$$

where $F(\boldsymbol{x})$ is a design matrix that could include spatially-dependent covariates, α is a parameter vector, and $\zeta(\boldsymbol{x})$ is a random effect at location $\boldsymbol{x}$.

In this specification, $\zeta(\boldsymbol{x})$ represents a spatial process. If $\zeta(\boldsymbol{x})$ is a spatial Gaussian process, then conditional on the realisation of the process, any finite values of $\{\zeta(\boldsymbol{x}_i)\}$ will have a multivariate normal distribution. This distribution can include variance and covariance parameters representing uncorrelated and correlated heterogeneity, respectively. An alternative specification is to assume that the log intensity ($h(\boldsymbol{x})$, say) has a multivariate normal prior distribution, $\text{MVN}(F(\boldsymbol{x})\alpha, \Sigma)$, where Σ is the covariance matrix. Here, $\lambda(\boldsymbol{x}) \equiv \exp(h(\boldsymbol{x}))$, possibly with a modulating function $g(\boldsymbol{x})$ included. This is closer in spirit to the specification of a Cox process where the intensity itself is realised from a random process. This approach can lead to maximum aposteriori (MAP) estimators for α given Σ, similar to those found for universal Kriging in geostatistics, provided a likelihood approximation is made (Lawson, 1994). This approach can also be implemented in a fully Bayesian setting (see, for example, Lawson *et al.*, 1996).

17.4.1 The specification of $f(\boldsymbol{x}; \boldsymbol{\theta})$

It is important to consider the appropriate form for the function $f(\boldsymbol{x}; \theta)$, which usually describes the exposure model used in the analysis of the association of events to a source. Define the location of the source as $\boldsymbol{x}_0$. Usually the spatial relationship between the source and disease events is based on the polar coordinates of events from the source: $\{r, \phi\}$, where $r = ||\boldsymbol{x} - \boldsymbol{x}_0||$, and ϕ is the angle measured from the source. It is important to consider how these polar coordinates can be used in models describing pollution effects on surrounding populations. In many studies, only the distance measure, r, has been used as evidence for association between a source and surrounding populations (e.g. Diggle, 1989; Diggle *et al.*, 1997; Elliott, 1995; Elliott *et al.*, 1992a). However, it is dangerous to pursue distance-only analyses when considerable directional effects are present. The reason for this is based on elementary exposure modelling ideas, which are confirmed by more formal theoretical and empirical exposure studies (Esman and Marsh, 1996; Panopsky and Dutton, 1984). It is clear that differential exposure may occur with a change in distance *and* direction, particularly around air pollution sources (such as incinerator stacks or foundry chimneys). Indeed, the wind regime , that is prevalent in the vicinity of a source can easily produce considerable differences in exposure in different directions. Such directional preference or anisotropy can lead to marked differences in exposure in different directions and hence to different distance exposure profiles. Hence the collapsing of exposure over the directional marginal of the distribution could lead to considerable misinterpretation, and in the extreme to *Simpson's* paradox. In the extreme case, a strong distance relationship with a source may be masked by the collapsing over directions, and this can lead to erroneous conclusions. Many published studies by, for example, the SAHSU (Small Area Health Statistics Unit) in the United Kingdom (Diggle *et al.*, 1997; Elliott, 1995; Elliott *et al.*, 1992a, 1996; Sans *et al.*, 1995) have, apparently, ignored directional components in the distribution, and therefore the conclusions of these studies should be viewed with caution. Furthermore, if the analysis of a large number of putative source sites is carried out by pooling between sites and ignoring local directional effects at each site, then these studies should also be regarded with caution.

The importance of the examination of a *range* of possible indicators of association between sources and health risk in their vicinity is clear. The first criterion for association is usually assumed to be evidence of a decline in disease incidence with increased distance from the source. Without this distance–decline effect, there is likely to be only weak support for an association. However, this does not imply that this effect should be examined in isolation. As noted above, other effects can provide evidence for association, or could be nuisance effects which should be taken into consideration so that correct inferences be made. In the former category are directional and directional–distance correlation effects, which can be marked with particular wind regimes. In the latter category are peaked incidence effects, which relate to *increases* in incidence with distance from the source. While a peak at some distance from a source can occur, it is also possible for this to be combined with an overall underlying decline in incidence, and hence is of importance in any modelling approach. This peaked effect is a nuisance effect, in terms of association, but it is clearly important to include such effects. If they were not included, then inference may be erroneously made that no distance–decline effect is present, when in fact a combination of distance–decline and peaked incidence is found. Diggle *et al.* (1997) display data on stomach cancer incidence around a putative source, where peaking of incidence occurs at some distance from the source. Peaks of incidence compounded with distance–decline are clearly found in the Lancashire larynx cancer data also (see, for example, Elliott *et al.*, 1992). Further nuisance effects which may be of concern are random effects related to individual *frailty*, where individual variation in susceptibility is directly modelled or where general heterogeneity is admitted (Lawson *et al.*, 1996). A recent review of these critical issues appears in Lawson (1996c).

A general approach to modelling exposure risk is to include an appropriate selection of the above measures in the specification of $f(\boldsymbol{x}; \cdot)$. First, it is appropriate to consider how exposure variables can be linked to the background intensity $g(\boldsymbol{x})$. We define $f(\boldsymbol{x}; \theta) = m\{f^*(\boldsymbol{x})'\alpha\}$, where $m\{\cdot\}$ is an appropriate link function, and $f^*(\boldsymbol{x})$ represents the design matrix of exposure variables which is evaluated at $\boldsymbol{x}$. The link function is usually defined as $m\{\cdot\} = 1 + \exp\{\cdot\}$, although a direct multiplicative link can also be used. Usually each row of $f^*(\boldsymbol{x})$ will consist of a selection of the variables

$$\{r, \log(r), \cos(\phi), \sin(\phi), r\cos(\phi), r\sin(\phi), \log(r)\cos(\phi), \log(r)\sin(\phi)\}.$$

The first four variables represent distance–decline, peakedness, and directional effects,

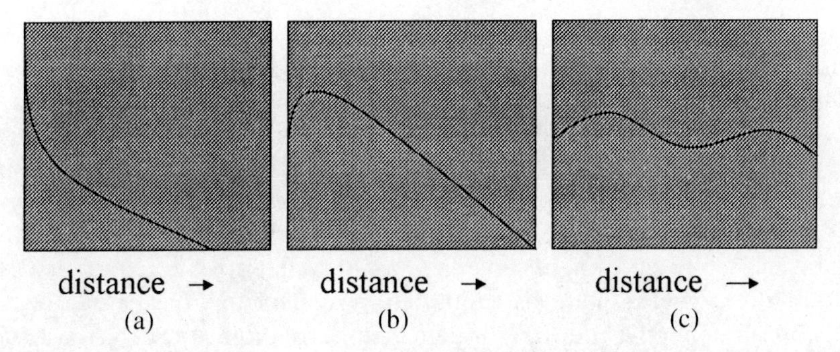

Figure 17.3 Possible distance–disease incidence relations: (a) monotonically decreasing; (b) peak-decline; (c) clustered decline

while the latter variables are directional–distance correlation effects (Lawson, 1993b). The directional components can be fitted separately and transformations of parameters can be made to yield corresponding directional concentration and mean angle (see, for example, Lawson, 1992a). Figure 17.3 displays different distance-related exposure models which could be used to specify $f(\mathbf{x};\theta)$. Note that in the figure nuisance effects of peakedness and heterogeneity appear in (b) and (c).

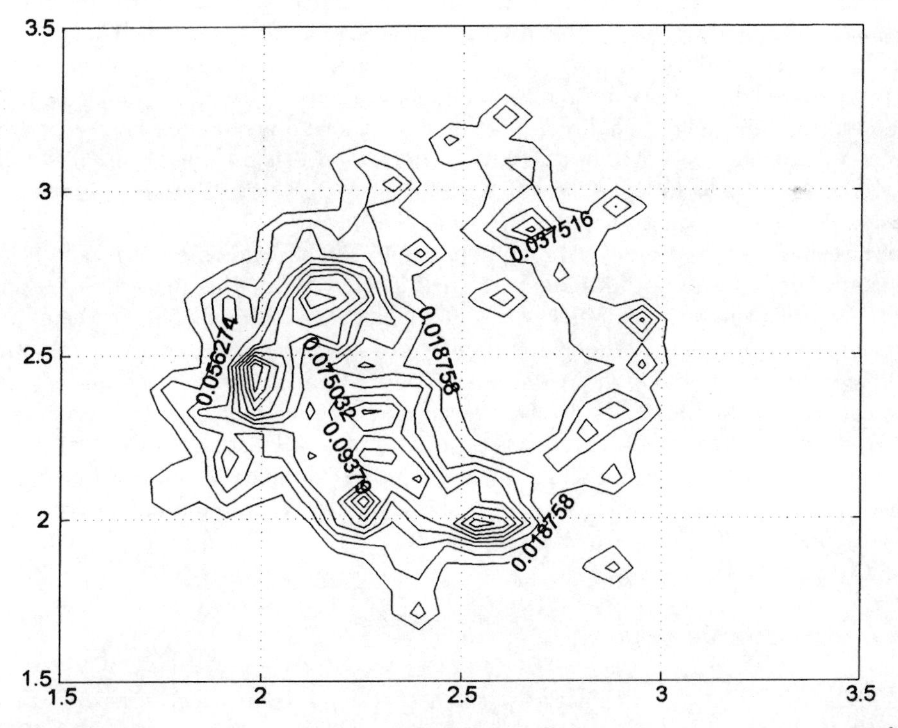

Figure 17.4 Simulation of a five-parameter dispersal model for a putative source, using a Weibull density for the distance marginal with scale and shape parameters and a von Mises with directional linear correlation for the directional marginal distribution. Source location: 2.5,2.5

Further examination of dispersal models for air pollution suggest that the spatial distribution of outfall around a source is likely to follow a convolution of Gaussian distributions, where in any particular direction there could be a separate mean level and lateral variance of concentration (dependent on r) (Esman and Marsh, 1996; Panopsky and Dutton, 1984). As a parsimonious representation of these effects it is possible to use a subset of the exposure variables listed above to describe this behaviour. Figure 17.4 displays the result of a simulation for a model which involves both peaked and distance–decline components and directional preference. Time-averaged exposure can be thought to lead to patterns similar to those depicted. Here a NW direction of concentration is apparent and the simulated exposure intensity surface was obtained from a five-parameter model for the distance and directional components. Note that averaging over the directional marginal of this distribution will lead to considerable attenuation of increased risk at distance from the source due to the anisotropic distance relations found.

17.4.2 Estimation

The parameters of the HEPP and modulated HEPP models discussed above can be estimated by maximum likelihood, conditional on $\hat{g}(\boldsymbol{x})$. In fact, it is possible to use GLIM or S-PLUS for such model fitting, if special integration weighting schemes are used (Berman and Turner, 1992; Lawson, 1992a). These schemes replace the normalising integral in (17.1) with $\sum_i W_i\lambda_i$: the dirichlet tile area and intensity at each location.

For the hybrid model of Lawson and Williams (1994), and the binary labelling model discussed above, direct maximum likelihood methods can be used. For the MRF model, pseudolikelihood can be used directly. In the case of spatially correlated heterogeneity, one may estimate the covariance components via restricted maximum likelihood (REML, cf. Searle *et al.*, 1992) and use an iterative algorithm for trend-parameter estimation (e.g. the expectation-maximization, or EM algorithm).

In the examples above, many estimation problems can be overcome by the use of Markov chain Monte Carlo (MCMC) methods such as Metropolis–Hastings or Gibbs sampling (see, for example, Gelman *et al.*, 1995; Gilks *et al.*, 1996a). The mathematical underpinnings of MCMC methods and their relationship to spatial statistics are found in Besag and Green (1993). The generalised Monte Carlo tests of Besag and Clifford (1989) show how MCMC methods can be used in hypothesis testing. Convergence of MCMC algorithms can be difficult to assess and there is still dispute on the best way to implement MCMC methods (Cowles and Carlin, 1996). Examples of some applications of MCMC methods in the analysis of putative hazard problems are found in Lawson (1995) and Lawson *et al.* (1996) and Chapter 9 of this volume.

17.4.3 Hypothesis tests

For standard HEPP models, Laplace's test can assess simple trend effects (Cox and Lewis, 1976). This is the score test for exponential trend and is uniformly most powerful (UMP) for monotone alternatives (provided a UMP test exists). In Lawson (1992b), tests for spatial effects in modulated HEPP models were presented. These include a variety of score tests for radial, directional, and directional–radial correlation. Note that both likelihood-ratio (LR) and score tests are available in statistical software package (such as GLIM, GENSTAT or S-PLUS) if one uses the special weighting schemes mentioned in Section 17.4.2.

Tests of monotonic radial decline assume that distance acts as a surrogate for exposure. Many proposed tests are based on radial decline models in point data and tract-count data. However, a wide variety of spatial effects could arise due to pollution from a fixed source, and overemphasis on radial decline can yield erroneous conclusions. For example, outfall from stack plumes tends to peak at some distance from the source. Hence, one would expect a peak-and-decline intensity to be present (Panopsky and Dutton, 1984). Simple radial decline tests can have low power when non-monotone effects, such as these, are present (Lawson, 1993b)

Figure 17.3 displays a variety of possible exposure types. If type (b) or (c) were realised, then simple radial decline tests (or model parameters) will have low power or unnecessarily high variance. Other exposure models have been proposed (Diggle and Elliott, 1995; Diggle *et al.*, 1997), which involve constant risk in a disc around the source. However, the justification for constant risk or exposure-path on epidemiological grounds seems scant (see also the comments in Lawson, 1996c).

The collection of data and spatial modelling of exposure levels should lead to increased power to detect pollution effects. Unobserved heterogeneity may be included as random effects following the generalised linear mixed models described by Breslow and Clayton (1993). Alternatively, the heterogeneity may be formulated in terms of nuisance parameters. Lawson and Harrington (1996) examined Monte Carlo tests, in a putative source setting, when spatial correlation is present and can be estimated as a nuisance effect under the null hypothesis.

Some recent examples of the application of tests to putative source problems have appeared (Le *et al.*, 1996; Viel *et al.*, 1995). In Chapter 30 of this volume a case study of the analysis of leukaemia incidence around a nuclear facility is presented.

17.4.4 Diagnostic techniques

The analysis of residual diagnostics for the assessment of goodness-of-fit of a model is common practice in statistical modelling. Usually such diagnostics are used to assess overall model goodness-of-fits as well as specific features in the data. If a spatial point process model fits well and all relevant covariates are included, then we expect spatially independent residuals.

Diagnostic techniques display 'outliers' or unusual features not accounted for by a model. If the underlying model assumes no clustering of events, then unusually strong clustering can be highlighted by clusters of positive residuals. Clustering may be reflected in positive spatial autocorrelation among residuals, or in isolated areas of positive residual clusters.

For case-event data, it is possible to use a 'transformation' residual (Diggle, 1990) to detect the effects described above. This residual relies on the transformation of distance from source and is often used in time-domain analysis (Ogata, 1988). Lawson (1993a) proposed a general deviance residual for heterogeneous Poisson process (HEPP) models and applied the method to putative pollution hazard data (Lawson and Williams, 1994).

17.5 MODELS FOR COUNT DATA

For a variety of reasons, outcome data may be available only as counts for small census regions rather than as precise event locations. As a result, a considerable literature has developed concerning the analysis of such data (e.g. Bithell, 1990; Clayton and Kaldor, 1987; Cressie and Chan, 1989; Devine and Louis, 1994; Hills and Alexander, 1989; Lawson, 1993b; Lawson and Harrington, 1996; Marshall, 1991a; Tango, 1984, 1995; Waller *et al.*, 1992; Whittemore *et al.*, 1987).

The usual model adopted for the analysis of region counts around putative pollution hazards assumes $\{n_i, i = 1, \ldots, p\}$ to be independent Poisson random variables with parameters $\{\lambda_i, i = 1 \ldots, p\}$. This model follows from an assumption that the location of individual events follows a HEPP. Any non-overlapping regionalisation of a HEPP leads to independently Poisson distributed regional event counts with means

$$\lambda_i = \int_{W_i} \lambda(\boldsymbol{x})\mathrm{d}\boldsymbol{x} \quad i = 1, \ldots, p,$$

where $\lambda(\boldsymbol{x})$ is the first-order intensity of the HEPP and W_i is the ith subregion.

The analysis and interpretation of models based on these assumptions is not without problems. First, many studies of count data assume that λ_i is constant within region W_i so that spatial variation across regions follows a step function (Diggle, 1993). When λ_i is parameterised as a loglinear function, one often treats explanatory variables (in particular exposure or radial distance or direction from a pollution source) as constants for the subregions or as occurring at region centres only. While such loglinear models can be useful in describing the global characteristics of a pattern, the differences between the W_is and any continuous variation in $\lambda(\boldsymbol{x})$ between and within regions is largely ignored. Secondly, the underlying process of events may not be a HEPP, in which case the independence assumption may not hold or the Poisson distributional assumption may be violated (Diggle and Milne, 1983). Assessments of model assumptions usually do not appear in studies of pollution sources (Elliott *et al.*, 1992a; Waller *et al.*, 1992), while they are often ignored in recommended epidemiological methodology (see, for example, Elliott, 1995; Elliott *et al.*, 1995). Analysis based on regional counts is ecological in nature and inference can suffer from the well-known 'ecologic fallacy' of attributing effects observed in aggregate to individuals. Finally, extreme sparseness in the data (i.e. large numbers of zero counts) can lead to a bimodal marginal distribution of counts or invalidate asymptotic sampling distributions (Zelterman, 1987).

While the above factors should be taken into consideration, the independent Poisson model may be a useful starting point from which to examine the effects of pollution sources (Bithell and Stone, 1989). Often, a loglinear model parameterisation is used, with a modulating value e_i, say, which acts as the contribution of the population of subregion i to the expected deaths in subregion $i, i = 1, \ldots, p$. Usually the expected count is modelled as

$$\mathrm{E}(n_i) = \lambda_i = e_i \cdot m(f_i'\alpha), \quad i = 1, \ldots, p.$$

Here, the e_i, acts as a background rate for the ith subregion. The function $m(\cdot)$ represents a link to spatial and other covariates in the $p \times q$ design matrix F, whose rows are $f_1', \ldots, f_p'$. The parameter vector α has dimension $q \times 1$. Define the polar coordinates of the subregion centre as (r_i, θ_i), relative to the pollution source. Often, the only variable to be included in F is r, the radial distance from the source. When this is used alone, an additive link such as $m(\cdot) = 1 + \exp(\cdot)$, is appropriate since (for radial distance decline) the background rate (e_i) is unaltered at great distances. However, directional variables (e.g. $\cos\theta$, $\sin\theta$, $r \cos\theta$, $\log(r) \cos\theta$, etc.) representing preferred direction and angular linear correlation can also be useful in detecting directional preference resulting from preferred directions of pollution outfall.

This model may be extended to include unobserved heterogeneity between regions by introducing a prior distribution for the log relative risks ($\log\lambda_i$, $i = 1, \ldots, p$). This could be defined to include spatially uncorrelated or correlated heterogeneity. The empirical and full Bayes methods described above often take this approach.

17.5.1 Estimation

One may estimate the parameters of the loglinear model above via maximum likelihood through standard GLM packages, such as GLIM or S-PLUS. Using GLIM or the GLM option on S-PLUS, the known log of the background hazard for each subregion, $\{\log(e_i), i = 1, \ldots, p\}$, is treated as an 'offset'. A multiplicative (log) link can be directly modelled in this way, while an additive link can be programmed via user-defined macros (see, for example, Breslow and Day, 1987).

Loglinear models are appropriate if due care is taken to examine whether model assumptions are met. For example, Lawson (1993b) suggests avoiding the violation of asymptotic sampling distributions by the use of Monte Carlo tests for change of deviance. If a model fits well, then the standardised model residuals should be approximately independently and identically distributed (i.i.d.). One may use autocorrelation tests, again via Monte Carlo, and make any required model adjustments. If such residuals are not available directly, then it is always possible to compare crude model residuals with a simulation envelope of m sets of residuals generated from the fitted model in a parametric bootstrap setting.

Bayesian models for count data can be posterior-sampled via MCMC methods, and a variety of approximations are also available to provide empirical Bayes estimates. Lawson (1994) gives examples of a likelihood approximation, while Lawson *et al.* (1996) compared approximate MAP estimates with Metropolis–Hastings modal estimates for a putative-source example.

17.5.2 Hypothesis tests

Most of the existing literature on regional counts of health effects of pollution sources is based on hypothesis testing. Lawson and Waller (1996) provide a discussion of the extensive literature in this area. Stone (1988) first outlined tests specifically designed for count data of events around a pollution source. These tests are based on the ratio of observed to expected counts cumulated over distance from a pollution source. The tests are based on the assumption of independent Poisson counts with monotonic distance ordering. A number of case studies have been based on these tests (Bithell, 1990; Bithell and Stone, 1989; Elliott *et al.*, 1992a; Turnbull *et al.*, 1990; Waller *et al.*, 1992 amongst others). Lumley (1995) has developed improved sampling distribution approximations for Stone's tests.

While Stone's test is based on traditional epidemiological estimates (i.e. SMRs), the test is not uniformly most powerful (UMP) for a monotonic trend. If a UMP test exists, then it is a score test for particular clustering alternative hypotheses (Bithell, 1995; Lawson, 1993b; Stone, 1988; Waller *et al.*, 1992). Waller (1996) and Waller and Lawson (1995) assessed the power of a range of such tests and found that all tests had low power against non-monotone or clustered alternatives. Unfortunately, these forms of alternative commonly arise in small-area epidemiological studies. Lawson (1993b) developed a distance–effect score test versus a non-monotone, peaked alternative, and also suggested tests for directional and directional–distance effects within a loglinear model framework. A procedure has been proposed by Besag and Newell (1991), which, though originally defined as a test for non-focused clustering, can be applied in a putative hazard application. The procedure involves the accumulation of observed counts for k tracts around the source and a comparison with the expected count. The choice of k is arbitrary, however. The procedure can also be adapted to case-event data applications.

A cautionary note should be sounded in relation to the use of tests for putative source locations. The results of recent power studies carried out on a range of distance–decline tests have shown that: '. . . many current tests of focussed clustering often have poor power for detecting the small increases in risk often associated with environmental exposures' (Waller, 1996). This supports the fundamental need to examine a range of approaches to putative sources analysis within one study as well as a range of association variables.

17.6 MODELLING VERSUS HYPOTHESIS TESTING

There are many examples of the application to putative health hazards of hypothesis tests as opposed to constructing general exposure models for the effects of the source (see, for example, Elliott *et al.*, 1992a, 1996; Gardner, 1989; Sans *et al.*, 1995). As for case-event data, the use of hypothesis testing as a general tool for the analysis of health hazards has a number of pitfalls. First, tests are often designed and used to assess single effects (e.g. distance decline (Stone's test or the Lawson–Waller score test) or directions (Lawson, 1993b) and hence can constrain the analysis by this focus. Often simple effects can yield misleading conclusions. As noted above, the distance effects could vary with direction and so collapsing over the directional marginal of the exposure path could lead to quite erroneous conclusions (i.e. Simpson's paradox). In addition, simple tests often cannot be modified easily to deal with more complex exposure scenarios (e.g. clustered background). If multiple testing is pursued, then that also has problems associated with, for example, multiple comparisons and the independence assumptions of tests (see, for example, Thomas, 1985). Many of the inadequacies of this approach arise from an overly simplistic view of what characterises exposure around sources. The main advantages of modelling exposure (rather than hypothesis testing) is that a variety of effects can be *jointly* assessed within a single model. In general, that model should include distance and directional effects as well as functions of distance (e.g. peaked effects at distances from the source) and should also include the possibility of including clustering in the background (i.e. if a disease naturally clusters, then this should be modelled under the background effect; Lawson *et al.*, 1996).

17.7 CONCLUSIONS

The analysis of small area health data around putative hazard sources has developed now to a stage where some basic issues are resolved and basic methods are in place. However, there is still considerable lack of agreement on a number of key issues relating to basic methods and also a number of underdeveloped areas worthy of further consideration. Perhaps the most contentious area of basic methodology is that of exposure modelling and how this should be carried out in the small area context. It is the firm belief of the authors that some degree of sophistication in exposure modelling should be attempted, since naive use of simple exposure models (e.g. distance-only models) can lead to erroneous conclusions. Both directional and distance-related effects should be included in any analysis, unless there are good reasons *not* to do so. The areas that remain underdeveloped lie mainly where space–time modelling is appropriate. As yet few attempts have been made to model the different types of space–time data that can arise naturally in this context.

18

Disease Mapping Using the Relative Risk Function Estimated from Areal Data

J. F. Bithell

University of Oxford

18.1 INTRODUCTION

An earlier paper (Bithell, 1990) introduced the Relative Risk Function (RRF), defined at each point of a geographical region $\mathscr{R}$, to provide an estimate of the risk of disease from samples of cases and controls. This chapter extends this methodology to data in the more conventional form of counts and expectations in small areas and illustrates how the RRF may be used to provide tests of the homogeneity of risk against various kinds of departure from it. Many methods have been used for the geographical mapping of disease (Bithell, 1998), though rather few of these are designed for case–control data. It is hard to overestimate the non-uniformity of population density even within towns. We argue that this non-uniformity is essential to the problem and should appear as a fundamental feature of any reliable method, rather than as entailing a 'correction' to methods appropriate to uniformity assumptions. Many analyses are model-based; recently the Bayesian approach has become particularly popular (Clayton and Kaldor, 1987; Lawson, 1994; Cressie, 1996). The methods are hard to evaluate, by the nature of Bayesian inference, which entails assumptions that are inherently unverifiable. Their value has consequently yet to be confirmed and they are often complicated to execute. Furthermore, there is a temptation to overparameterise Bayesian spatial models relative to the amount of information likely to be available in modest data sets, for example with regard to the structure of the auto-correlation; the effect of this on the reliability of the conclusions is hard to determine. Non-parametric methods, on the other hand, make fewer assumptions and may therefore give an estimate of an underlying risk surface which is relatively free of their influence. It is of course important to distinguish between exploratory analyses, in which one is merely observing an estimate of a risk function

Disease Mapping and Risk Assessment for Public Health. Edited by A.B. Lawson *et al.*
© 1999 John Wiley & Sons Ltd.

without prior conceptions of non-uniform risk, and analyses designed to test hypotheses formulated in advance. In practice, the epidemiological investigator often works between the two situations. It is therefore desirable to be able to test formally whether a 'cluster' of cases that appears to be close to some previously unidentified focus of risk is in fact more clustered than would be expected by chance. Methods based on testing the uniformity of a risk surface over a given region provide a partial answer to this problem, as we illustrate below.

18.2 DEFINITION OF THE RELATIVE RISK FUNCTION

The underlying idea of the RRF is based on that of the population density function. Instead of determining the incidence rate in a given geographical region $\mathscr{R}$, we assume that we can specify the (bivariate) probability density $\psi(x, y)$ of the location of residence of 'cases', i.e. persons affected by the disease; here (x, y) represent the geographical coordinates of the location of an individual, which may be the place of residence or of death, as appropriate. We then compare this distribution with that for the whole of the relevant population at risk, $\pi(x, y)$, and define the (conditional) RRF as their ratio:

$$\theta(x, y) = \psi(x, y)/\pi(x, y).$$

It is easy to see that $\theta(x, y)$ averages to one over $\mathscr{R}$ provided we use the underlying population density as a weight function, i.e.

$$\iint_{\mathscr{R}} \theta(x, y)\pi(x, y)\mathrm{d}x\mathrm{d}y = 1,$$

and that the Relative Risk so defined at any given point represents the risk relative to a population weighted average for the region as a whole. If we know the overall incidence for the region, then we can of course use it to scale the RRF to obtain the absolute risk function; or, for convenience, we may retain the concept of Relative Risk, but scale it so that national global rates form the basis of comparison. In other cases, for example where $\psi(x, y)$ and $\pi(x, y)$ are estimated from case and control data, the RRF itself can still be estimated. To the extent that different scalings produce essentially the same picture of variation of risk, they are relatively unimportant. For testing purposes, however, different sampling schemes may have to be distinguished, according to the underlying alternative hypotheses; we discuss this issue below. The use of case–control data involves an inversion in which the random variation is attributed to the place of residence of diseased and control individuals, rather than to the occurrence of disease at pre-defined locations. This operation may seem unnatural to those unfamiliar with the equivalent duality in classical epidemiology; it is precisely the inversion we make when we carry out a case–control (or retrospective) study rather than a cohort (or prospective) one. It would be normal in epidemiology to define a control as an individual known not to be affected by the disease. A sample of such controls is not of course quite the same as a random sample of members of the population. Using the different bases of comparison leads respectively to the odds ratio and the Relative Risk; the same distinction applies to the risk surface. The function $\theta(x, y)$ gives either the odds or the Relative Risk of getting the disease for an individual located at the point (x, y) relative to that for the region as a whole. The distinction hardly matters for rare diseases and, as is quite usual in epidemiology, we shall gloss over it when it is convenient to do so.

18.3 ESTIMATION OF THE RELATIVE RISK FUNCTION

There are numerous methods available for constructing estimates of population densities from areal data. Much the simplest is to adapt the idea of density estimation, which we can do by imagining that all the population in a given small area is located at its centroid. This approximation is not likely to matter provided we smooth sufficiently and, although this might not be a very good way of providing an accurate population density estimate, it will probably serve quite well at the degree of smoothing necessary for small samples of cases. The original paper (Bithell, 1990) used an adaptive kernel method for the density estimation of both cases and controls, following the method described in Silverman (1986). This methodology is comparatively computer intensive, however, requiring the construction of a pilot estimate in order to determine the degree of local smoothing through control of the bandwidth. This adaptivity is less important when the denominator of the RRF is derived from population data; moreover, there are more efficient methods of estimation now available. In particular, we have found that the average shifted histogram (ASH) method proposed by Scott (1992) is efficient and robust, giving very few problems in its implementation. We need the two-dimensional version and have modified the published routines so that they aggregate counts of cases occurring multiply at areal centroids for the numerator and corresponding population sizes or expectations for the denominator. The basis of this method is as follows. We suppose that the data consist of k small areas, for example electoral wards in UK analyses, in which Y_i cases are observed and may be compared with expectations e_i, computed, perhaps, using national rates, $i = 1, 2, \ldots, k$. We assume that we have a rectangular study region divided into $m \times n$ squares of side d. (If our geographical region of interest is non–rectangular, then we can simply enclose it in a rectangle and disregard the contributions from outside the geographical boundary.) This defines an $m \times n$ histogram into which we aggregate the observed and expected numbers of cases according to the positions of the centroids of their respective small areas. In effect we then reposition the grid by displacing the origin jh units East and kh units North, for $j, k = 0, 1, 2, \ldots, s - 1$, with $sh = d$, and reallocate the observed and expected numbers according to the positions of the area centroids in relation to each new grid. The resulting s^2 histograms are then simply averaged at each point to provide an overall estimate of density which, although strictly speaking discontinuous, will appear smooth for moderate or large s.

18.4 APPLICATION TO CHILDHOOD CANCER DATA

We illustrate the method by applying it to data on childhood cancer in parts of the UK counties of Oxfordshire and Berkshire. The dataset was originally constructed in connection with a collaborative study of the distribution of such cases in the vicinity of nuclear installations reported in the *British Medical Journal* (Bithell *et al.*, 1994). It consisted of the numbers of cases of malignant disease in electoral wards compared with expectations computed adjusting for various demographic factors related to socio-economic status (Bithell *et al.*, 1994). The data were divided into two sets; the first consisted of all children diagnosed as having leukaemia or non-Hodgkin lymphoma under the age of 15 in the years 1966–87. The results of the analyses of nuclear installations in England and Wales were published and were, broadly speaking, negative, with the clear exception of the well-known excess near Sellafield in Cumbria. The second, complementary, set consisted of all other childhood tumours. In the course of the analysis of the second set, it was observed that there is a slight excess in the general vicinity of various installations in Oxfordshire and Berkshire, notably the Atomic Energy Research Establishment at Harwell, though formal tests of the risk in relation to the sites themselves were inconclusive. The area is interesting for the presence of certain other sites with nuclear connections, namely Aldermaston, Burghfield, Culham and the American nuclear bomber base at Greenham Common, which has recently been the subject of newspaper and television publicity (Bithell and Draper, 1996). To examine the risk in the area as a whole we selected a rectangle approximately 50 km square in which there occurred 279 cases compared with 260.2 expected, an excess that is not statistically significant. They are depicted in Figure 18.1, in which the centres of the squares and circles indicate the population centroids of the 150 wards in which the cases are located. The sizes of the circles indicate the respective expectations, while the squares indicate the numbers of cases observed on the same areal scale.

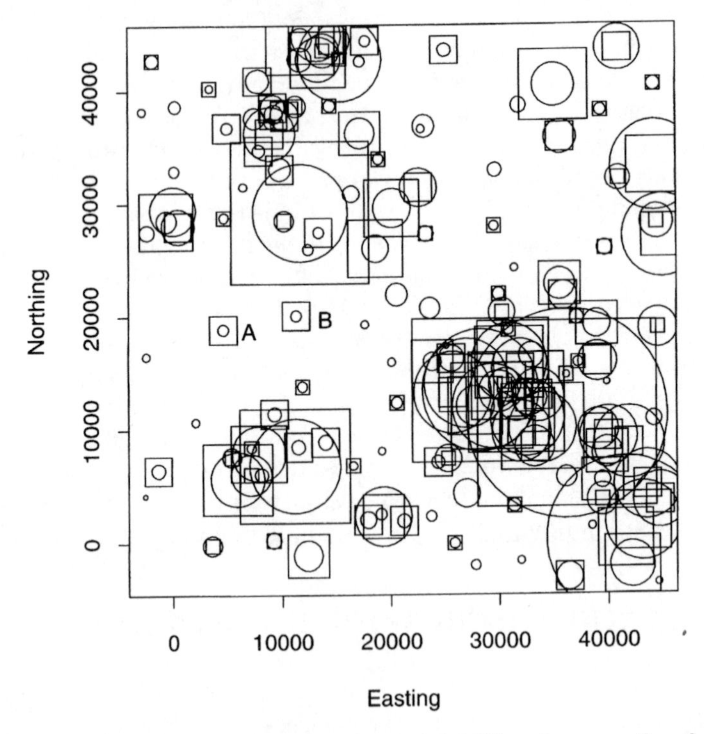

Figure 18.1 Observed and expected numbers of cases (of childhood cancer other than leukaemia or non-Hodgkin lymphoma under 15 years of age) in certain parts of Oxfordshire and Berkshire. The numbers are represented by the squares and circles, respectively, on the same areal scale and centred on the population centroids of the electoral wards. The axes in Figures 18.1 and 18.2 have been anonymised to meet ethical requirements

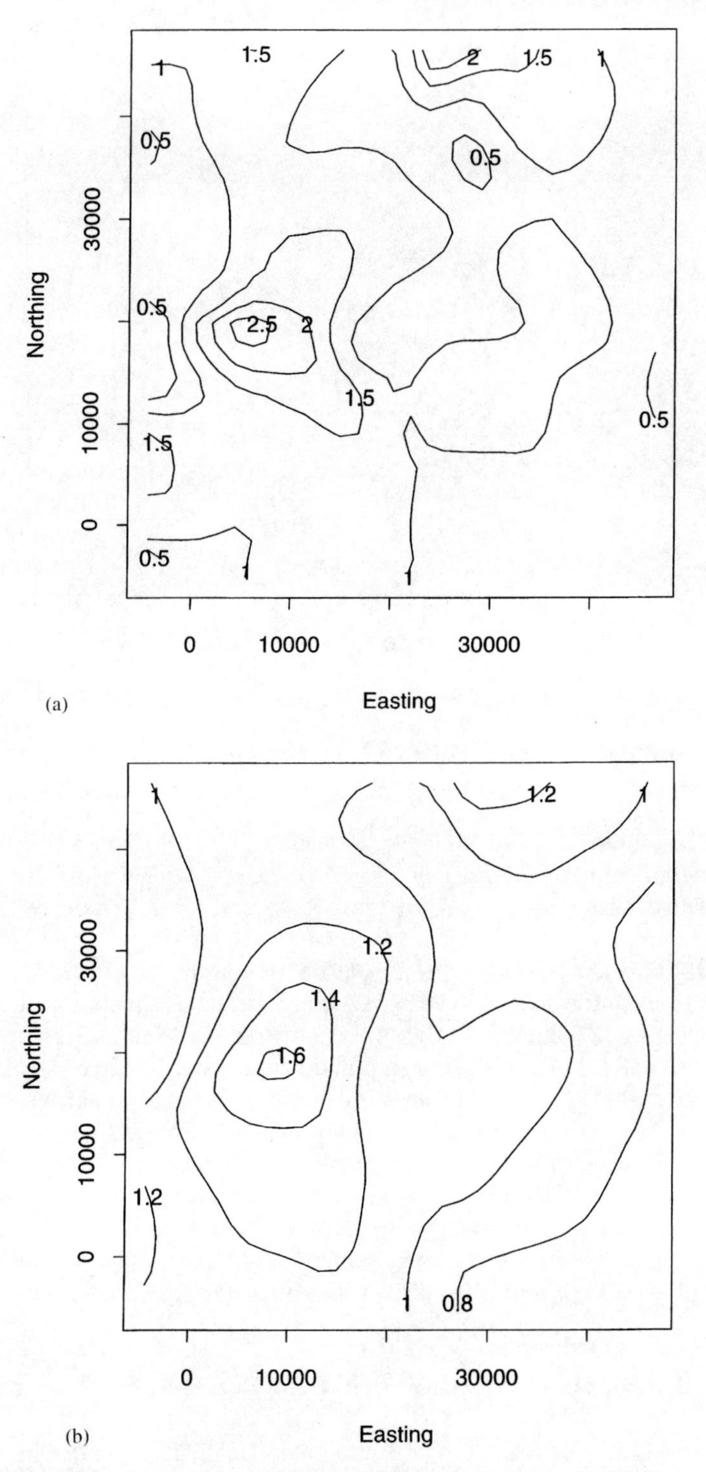

Figure 18.2 The RRF constructed as the ratio of density estimates calculated using ASH, with smoothing parameters $s = 5, 8, 12$ in (a), (b), and (c), respectively

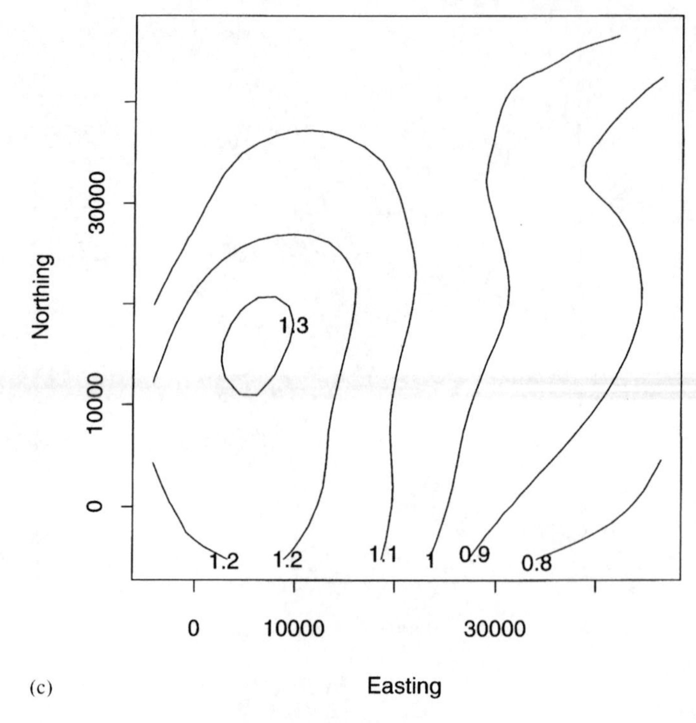

(c)

Figure 18.2 (*continued*)

Figure 18.2(a) shows the RRF constructed using the methods described above with the ASH default grid parameters of $m = n = 30$ and $s = 5$. Figures 18.2(b) and 18.2(c) show the result of changing the parameters to $s = 8$ and $s = 12$, respectively. The effect of these parameter changes is to increase the degree of smoothing progressively; how much smoothing is appropriate or in any sense optimal is discussed below. It will be seen at least from Figure 18.2(a) that there is a peak in the estimated Relative Risk and inspection of Figure 18.1 shows that this is due to the two isolated wards marked A and B in Figure 18.1. These are in areas of low population density and each contains two cases, with expectations of 0.63. It is clear therefore that the method is capable of at least pointing to an increase in risk resulting from conjunctions of excesses that are quite small and that would otherwise be undetectable. In fact the four cases concerned exhibit no particular similarity of tumour type or date of occurrence and it is likely that their proximity is due to chance. We should, however, aim to test formally whether the RRF with a given degree of smoothing departs significantly from what would be expected under the null hypothesis and we address this issue below.

18.5 GENERAL HETEROGENEITY OF RISK

Although a visual map of the risk is a useful epidemiological objective in itself, many investigations have a rather more specific purpose, namely to determine whether there is evidence of some variation of risk, particularly of a kind that can be associated with geographical features. Given areal data, a useful first step in this process is to see whether there is over–dispersion of the counts relative to their expectations. Many statistics have been proposed for this, but we suggest the following deviance statistic:

$$D = \sum \{e_i - y_i + y_i \ln y_i/e_i\},$$

where y_i is the observed count in the ith area. This could be referred to a chi-square distribution if the expectations were believed to be correct a priori and were of moderate size. In our case, the latter requirement is certainly not fulfilled, so we resorted to performing a Monte Carlo test, unconditional in the sense discussed below, i.e. by simulating sample datasets from independent Poisson distributions. The observed value of $D_{\text{obs}} = 155.4$ was exceeded in 66 out of 100 sample datasets, from which we conclude that there is no *prima facie* evidence of heterogeneity in the data. However, heterogeneity tests based on areal units are at best only able to detect spatial clustering at a scale comparable to the typical unit size, and do so only imperfectly; for example, an aggregation giving slightly increased expectations in several adjoining areas may not be detected. This is the rationale for constructing and testing a risk surface with different levels of smoothing. The RRF makes it simple to construct tests appropriate to a range of different alternative hypotheses. Two obvious possible hypotheses are that (i) there is a generalised and non-specific spatial variation of the true underlying RRF induced, for example, by unknown geographical factors, and (ii) there is an isolated peak due to some unsuspected focus of risk. A suitable statistic for alternatives of type (i) which measures the discrepancy in the RRF, with a weighting reflecting the population density, is given by

$$T_{\text{var}} = \int\int_{\mathscr{R}} \hat{\pi}(x, y)\{\hat{\theta}(x, y) - 1\}^2 \mathrm{d}x\,\mathrm{d}y.$$

An alternative, proposed by Anderson and Titterington (1997), measures the squared difference on the density scale, which has the effect of increasing the extent to which differences in Relative Risk are weighted according to population density:

$$\int\int_{\mathscr{R}} \{\hat{\pi}(x, y) - \hat{\psi}(x, y)\}^2 \mathrm{d}x\,\mathrm{d}y = \int\int_{\mathscr{R}} \hat{\pi}(x, y)^2\{\hat{\theta}(x, y) - 1\}^2 \mathrm{d}x\,\mathrm{d}y.$$

For an alternative of type (ii) it would seem more appropriate to consider a functional of the form $M = \max \hat{\theta}(x, y)$ and this is the statistic we employed for the childhood cancer data, again using a Monte Carlo test. It is important to appreciate that, as indeed with the Monte Carlo computation of the distribution of the deviance statistic, D, there are two ways of carrying out the necessary sampling, which we may conveniently term conditional and unconditional. In the former, the number of cases in the sample datasets is fixed to be equal to that in the observed sample and the observations are distributed among the k wards using the multinomial distribution with probabilities proportional to the expected values. In the unconditional case we sample from independent Poisson distributions with means equal to the e_i. The former approach is appropriate if we are doubtful about whether the expectations are really reliable for the study region, or if we are concerned exclusively with the spatial distribution of the cases observed. No contribution to the significance of the test would come from an overall excess in $\mathscr{R}$ in this case. If, on the other hand, we are confident that the expectations are correct and we wish to detect an increase in risk relative to the reference rates localised within $\mathscr{R}$, then it will be more appropriate to use the unconditional approach. These issues exactly parallel those arising in the context of tests for an excess risk near a putative source of risk (Bithell, 1995). We chose to carry out an unconditional test of the Oxfordshire data, since the expectations were felt to be trustworthy and the small overall excess incidence will contribute to the height of the peak in the RRF. The data value observed was $M_{\text{obs}} = 2.71$, which was exceeded by 62 out of 100 simulated values, from which we conclude that there is no evidence of clustering at this geographical scale.

18.6 SELECTING THE DEGREE OF SMOOTHING

The major difficulty with density estimation is that of choosing the degree of smoothing to use, though the same problem arises implicitly in most graphical methods, including the ordinary use of the histogram. For exploratory purposes it seems reasonable to use a subjective method. Too little smoothing will probably give a degree of variation which is quite implausible; too much will suppress potentially interesting features of the data. In our case it could be argued that, since there is no significant heterogeneity in the data, one should report the risk as being constant; however, it might still be worth investigating the excess in a small area such as that described above and to smooth too far will render this impossible. More objective, data-driven methods present some difficulties. For example, as pointed out by Kelsall and Diggle (1995), choosing the best bandwidth for one or both densities should not be expected to give the optimal result for their ratio, even if it can be agreed what are the most appropriate criteria. It is therefore not surprising that, with the highly multi-modal densities involved in population data, standard methods such as Least Squares Cross-Validation applied to the individual densities can lead to a very spiky RRF estimate (Rossiter, 1991). For the purpose of testing the degree of non-uniformity, this may not matter very much, though clearly the power of the test is likely to be greatest when the degree of smoothing is in some sense appropriate to the scale of the clustering implied by the alternative hypothesis. In the absence of an externally specified alternative, therefore, it would be reasonable to fix the smoothing parameter on the basis of previous analyses of similar data, though one should be careful to make the choice independent of the actual data in case one biases the analysis towards the degree of clustering observed. The first use of M for areal data was carried out by Hutchinson (1995), who also explored a number of methods of bandwidth selection. In our case, the default values in the ASH routines, producing the results shown in Figure 18.2(a), were used. There may be an element of data influence in this choice, since we might have been tempted to use other values if the resulting contours had been very erratic; this is a problem we would take more seriously if the result had been positive.

18.7 DISCUSSION

There are numerous methods for presenting maps of disease risk, most of which recognise the value of some kind of smoothing. Most such methods operate on estimates that are effectively population rates, quite frequently SMRs. The idea of smoothing case and population densities separately is distinctive and occurs less frequently in the literature, though the use of density estimation, as proposed by Bithell (1990), is gaining ground (Kelsall and Diggle, 1995). The original analysis used a deliberately small dataset with the intention of testing the methodology. The extension of this method to the use of areal data, in conjunction with a robust, efficient algorithm for density estimation, has led to much more stable estimates of the RRF. The method seems to have worked well for highlighting areas with an apparently elevated value of the RRF and lends itself to testing formally the extent to which any non-uniformity may not be ascribable to chance. It would be highly desirable to report the results of power comparisons, but we defer this to another paper. Rossiter (1991) studied the power of her test by comparison with the case–control clustering method of Cuzick and Edwards (1990) based on nearest neighbours and found it to be somewhat better. There are a number of reasons why one might expect this. In the first place, the RRF, with its extreme simplicity, is effectively interpolating counts of cases in a given area; in the limit, as the amount of data increases indefinitely, this would lead to a procedure based on a sufficient statistic for the expectation, whereas the nearest neighbour approach, attractive though it is, seems unlikely to be fully efficient. Secondly, the Cuzick-Edwards test is an aggregative test rather than an extremum test and would be expected to be better than M at detecting a generalised tendency to case aggregation as opposed to a single cluster. These arguments are admittedly rather speculative and a comparative analysis would be highly desirable. The number of disease mapping methods now available is quite large and each has, in principle, an associated possibility for testing. A comprehensive comparative analysis is therefore a formidable task.

ACKNOWLEDGEMENTS

The author thanks the referees of this chapter for numerous helpful comments. Permission to use the data is also gratefully acknowledged. The Childhood Cancer Research Group is supported by the Department of Health and the Scottish Home and Health Department.

19

The Power of Focused Score Tests Under Misspecified Cluster Models

Lance A. Waller

University of Minnesota

Catherine A. Poquette

St. Jude Children's Research Hospital, Memphis

19.1 INTRODUCTION

In recent years there has been considerable interest in the spatial distribution of disease cases and possible relationships to environmental exposures. Of particular interest are possible temporal or spatial aggregations of incident cases, i.e. 'disease clusters'. Knox (1989) defines a cluster as 'a bounded group of occurrences related to each other through some social or biological mechanism, or having a common relationship with some other event or circumstance'.

Hypothesis testing based on an underlying spatial point process of incident cases is a common tool used to assess disease clustering. Mathematical models of clustering differ between proposed tests, and often different models induce different underlying assumptions regarding mechanisms of disease incidence (Waller and Jacquez (1995)). A review of cluster tests is given in Chapter 7 of this volume, and of focused cluster methods in Chapter 17.

Besag and Newell (1991) categorise cluster tests as either 'general' or 'focused'. General tests determine whether or not cases are clustered anywhere in a study area, while focused tests assess whether cases are clustered around prespecified putative sources ('foci') of hazard. Foci may be point locations such as waste sites releasing carcinogens into the environment, or non-point locations such as agricultural fields and highways as (geographic) sources of exposure to pesticides and automobile emissions, respectively.

Disease Mapping and Risk Assessment for Public Health. Edited by A.B. Lawson *et al.*
© 1999 John Wiley & Sons Ltd.

Score tests of focused clustering seek to maximise statistical power versus local alternatives. A specified local alternative, corresponding to an a priori model of focused disease clustering, defines the particular form of the test. In this Chapter we explore the power of misspecified score tests, i.e. tests defined for one type of cluster model applied to clustering generated from a different model.

19.2 MODELS OF CLUSTERING AND SCORE TESTS

Typically, due to confidentiality requirements, disease incidence data arise as counts from small subregions of a larger study area. For each subregion we calculate expected counts, often standardised based on demographic factors. For this study we concentrate on chronic rather than infectious disease, and assume disease cases arise independently of one another. Suppose that the study area is partitioned into I subregions or cells, e.g. enumeration districts. Denote the population size of each cell as $n_i, i = 1, \ldots, I$, and the total population size $n_+ = \sum_{i=1}^{I} n_i$. The number of disease cases in cell i is a random variable C_i with observed value c_i. The total number of observed disease cases is $c_+ = \sum_{i=1}^{I} c_i$. For a rare disease under a null hypothesis of no clustering (H_0), we assume the C_i are independent Poisson random variables with $E(C_i) = \lambda n_i$, where λ denotes the risk of an individual contracting the disease, i.e. the baseline incidence rate. In the development below, we assume λ to be known. If λ is unknown, we use the maximum likelihood estimate (MLE) $\hat{\lambda} = c_+/n_+$, conditioning on c_+ and adjusting standard errors accordingly.

Focused tests utilise alternative hypotheses defining increased risk in areas exposed to a focus. Wartenberg and Greenberg (1990a) described two broad types of clustering models, 'clinal' and 'hot spot' clusters. In clinal clusters, disease incidence rates are elevated near a focus and decrease with increasing distance (decreasing exposure). In hot spot clusters, disease incidence rates are increased only in a small area near a focus but are lower and constant outside this area.

Waller and Lawson (1995) use a multiplicative model to address both hot spot and clinal clusters. Let g_i denote some measure of the exposure to the focus for each individual residing in cell i. Consider the alternative hypothesis

$$H_1 : E(C_i) = \lambda n_i(1 + g_i \varepsilon), \quad \text{for } i = 1, \ldots, I, \tag{19.1}$$

where the parameter $\varepsilon > 0$ controls the multiplicative increase in risk. A hot spot cluster assumes a dichotomous exposure, and divides the population into two groups: 'exposed' and 'unexposed'. We model hot spot clusters using equation (19.1) with

$$g_i = \begin{cases} 1, & \text{if cell } i \text{ is in the hot spot,} \\ 0, & \text{otherwise,} \end{cases}$$

for $i = 1, \ldots, I$. For hot spot clusters, the relative risk of disease for individuals residing in the cluster compared with those residing outside the cluster is $(1 + \varepsilon)$. For clinal clusters the relative risk for cell i, compared with the background disease risk, is $(1 + g_i\varepsilon)$. Note that we maintain independent Poisson cell counts.

For clinal clusters, we often use distance-based exposure surrogates in the absence of direct exposure data. Let d_i denote the distance from cell i to the focus. We consider three parametric distance exposure relationships, namely an *inverse distance* model,

$$g_i = d_i^{*(-1/\gamma)}, \quad i = 1, \ldots, I, \gamma > 0, \tag{19.2}$$

where $d_i^* = d_i/\min(d_i)$ is the rescaled distance from the focus, and two exponential functions of distance (Tango, 1995), namely an *exponential* model

$$g_i = \exp(-d_i/\tau), \quad i = 1, \ldots, I, \tau > 0, \tag{19.3}$$

and an *exponential-threshold* model

$$g_i = \exp(-4(d_i/L)^2), \quad i = 1, \ldots, I, L > 0. \tag{19.4}$$

Figure 19.1 compares inverse distance, exponential, and exponential-threshold exposures as a function of distance for $\gamma = 2, \tau = 5$, and $L = 10$. For comparative purposes, we rescale distance in (19.2) to remove the effects of the measurement scale of d_i (e.g. excess influence from areas less than one unit from a focus). The parameters γ and τ control the shape of the distance–exposure relationship, and the parameter L denotes a distance beyond which (essentially) baseline risk occurs. Note that the extent (or spatial range) of a cluster increases as the parameters γ, τ, and L increase. A Geographical Information System (GIS) easily calculates such functions of distance, but one must recall that proximity is often a poor surrogate for true exposures (Elliott *et al.*, 1995).

We explore the effects of these three distance-based exposure values on the power of two classes of focused clustering tests, namely those proposed by Lawson (1993b) and Waller *et al.* (1992, 1994), and those proposed by Bithell (1995). We outline each class of tests below.

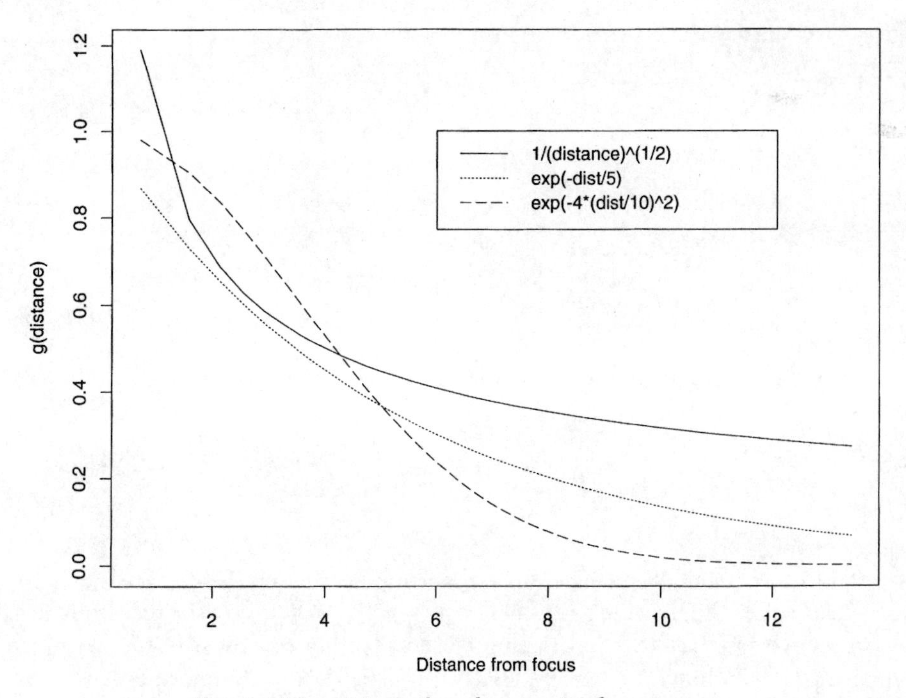

Figure 19.1 Distance-based surrogates for exposure

Lawson (1993b) and Waller *et al.* (1992, 1994) propose the score test statistic $U = \sum_{i=1}^{I} g_i(C_i - \mathrm{E}(C_i))$. Under $\mathrm{H}_0, \mathrm{E}(U) = 0, \mathrm{var}(U) = \sum_{i=1}^{I} g_i^2 \lambda n_i$, and $U/\sqrt{\mathrm{var}(U)}$ has an asymptotic standard normal distribution. The test is equivalent to tests of trend in Poisson random variables (Breslow and Day, 1987). Waller and Lawson (1995) note that the normal approximation is generally accurate for situations encountered in disease surveillance efforts. In cases where the normal approximation is questionable, exact tail probabilities may be obtained through numerical methods (Waller and Lawson, 1995).

Waller and Lawson (1995) derive the power of the score tests against the clustering alternative (19.1) using the normal approximation. Let u^* denote the α-level critical value of the standard normal distribution, and μ_{1_U} and $\sigma_{1_U}^2$ denote the mean and variance of U under H_1. Here, $\mu_{1_U} = \sum_{i=1}^{I} n_i \lambda g_i^2 \varepsilon$ and $\sigma_{1_U}^2 = \sum_{i=1}^{I} g_i^2 n_i \lambda(1 + g_i \varepsilon)$. For a one-sided test, the power of the score test with level α is

$$
\begin{aligned}
\Pr[U/\sigma_{0_U} > u^* | \mathrm{H}_1] &= \Pr\left[\frac{(U - \mu_{1_U})}{\sigma_{1_U}} > \frac{(u^* \sigma_{0_U} - \mu_{1_U})}{\sigma_{1_U}} | \mathrm{H}_1\right] \\
&= 1 - \Phi\left(\frac{u^* \sigma_{0_U} - \mu_{1_U}}{\sigma_{1_U}}\right),
\end{aligned}
\tag{19.5}
$$

where $\Phi(\cdot)$ denotes the cumulative distribution function of the standard normal distribution.

Bithell (1995) introduces a family of score test statistics including $B = \sum_{i=1}^{I} C_i \log(1 + g_i \varepsilon)$ for cluster alternative (19.1). Bithell notes that this test is equivalent to the Lawson and Waller score test as $\varepsilon \to 0$. The expectation of B under H_0 is $\mu_{0_B} = \sum_{i=1}^{I} n_i \lambda [\log(1 + g_i \varepsilon)]$. Under H_1, the expectation becomes $\mu_{1_B} = \sum_{i=1}^{I} [n_i \lambda(1 + g_i \varepsilon)] \log(1 + g_i \varepsilon)$. The associated variances are

$$
\sigma_{0_B}^2 = \sum_{i=1}^{I} [\log(1 + g_i \varepsilon)]^2 n_i \lambda
$$

under H_0, and

$$
\sigma_{1_B}^2 = \sum_{i=1}^{I} [\log(1 + g_i \varepsilon)]^2 n_i \lambda (1 + g_i \varepsilon)
$$

under H_1. We denote the standardised statistic by $B^* = (B - \mu_{0_B})/\sigma_{0_B}$. The power to detect alternative (19.1) is

$$
\Pr[B^* \geq u^* | \mathrm{H}_1] = 1 - \Phi\left(\frac{u^* \sigma_{0_B} + \mu_{0_B} - \mu_{1_B}}{\sigma_{1_B}}\right).
$$

In Section 19.3, we explore the effect on statistical power when the disease mechanism generating the data differs from the alternative hypothesis defining the test statistic. For this chapter we restrict attention to the misspecification of exposure (g_i) within the parametric cluster model defined by (19.1). In particular, we suppose a score test is defined with exposure g_i', when the true cluster is based on exposure $g_i, i = 1, \ldots, I$.

We obtain the power of the misspecified test using the results above with

$$\mu_{1_U} = \sum_{i=1}^{I} g_i g_i' \lambda n_i \varepsilon,$$

$$\sigma_{1_U} = \sum_{i=1}^{I} (g_i')^2 n_i \lambda (1 + g_i \varepsilon),$$

$$\mu_{1_B} = \sum_{i=1}^{I} [\log(1 + g_i' \varepsilon)] n_i \lambda (1 + g_i \varepsilon),$$

and

$$\sigma_{1_B} = \sum_{i=1}^{I} [\log(1 + g_i' \varepsilon)]^2 n_i \lambda (1 + g_i \varepsilon).$$

19.3 HOMOGENEOUS POPULATION RESULTS

For illustration, we arrange 400 cells into a 20×20 square grid, and assume all cells contain 3000 individuals at risk. We consider each cell to be a square with sides one distance unit in length. Our distance units are arbitrarily defined as 1/20 of the east–west and north–south dimensions of the study area. We assume the focus to be the centre of the grid, and assume the baseline incidence rate, λ, is known. We use three values of λ: 0.00017, 0.0005, and 0.001. The value $\lambda = 0.0005$ roughly corresponds to the ten-year childhood leukaemia rate in ages 0–15 (Doll, 1989). Population size $n_i = 3000$ roughly corresponds to the average for a US census tract.

We begin with a 'true' hot spot cluster in the 64 centre cells of the 400 (cells in the seventh through fourteenth rows and columns of the grid). We utilise cluster tests based on circular hot spots smaller and larger than the actual cluster. We choose ε in (19.1) so that a test based on the correct alternative has 99% power.

The power results based on (19.5) appear in Figure 19.2. The power of Bithell's score test is equivalent to four decimal places and is not shown. Figure 19.2 illustrates that the power of the test falls off more quickly when we underspecify rather than overspecify the extent of the hot spot. The power curve contains several 'plateaus' when an increase in the hot spot radius does not add any additional cells to the hot spot defining the test statistic. This effect particularly impacts underspecification of the cluster extent, and is exaggerated in our grid-based illustration due to many tied distances. Figure 19.2 also indicates higher power for rarer diseases (smaller λ), for any given hot spot size.

Next, we assume the true underlying cluster model is a clinal cluster with exposure measurements defined by (19.2)-(19.4). We begin by considering misspecifications only of the parameters γ, τ, and L, assuming that 'true' clusters are defined by (19.2)–(19.4) with $\gamma = 2, \tau = 5$, and $L = 10$, respectively. We calculate power functions for 40 parameter values, namely inverse distance clusters based on $\gamma = 0.2, 0.4, \ldots, 8$, exponential clusters based on $\tau = 0.5, 1, \ldots, 20$, and exponential-threshold clusters based on $L = 1, 2, \ldots, 40$. The specified parameter ranges define similar exposure–distance functions across models as illustrated in Figure 19.3.

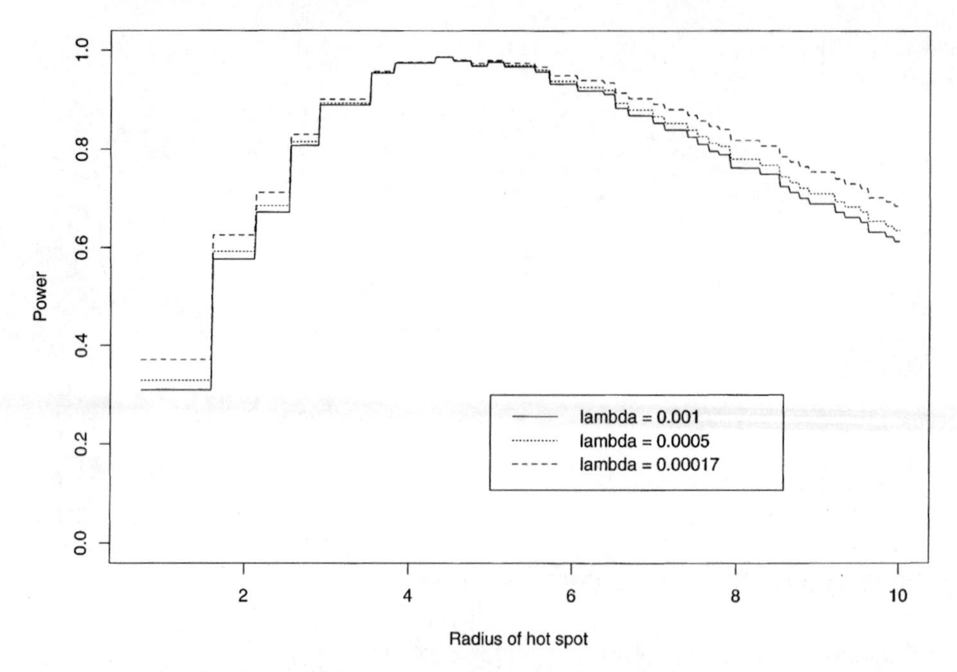

Figure 19.2 Plot of the power of the score test for circular hot spot clusters to detect a true 64-cell (square) hot spot. Correct specification of the hot spot results in 99% power

Figure 19.4 displays power curves for $\lambda = 0.001$. The three lines represent the choice of ε corresponding to 80%, 90%, and 99% power for the correctly specified test. Since all cells contain some excess risk, overspecification of the cluster extent results in increased power for all three definitions of g_i, contrary to the hot spot results. We see very similar behaviour for the three distance–exposure functions. The results suggest poor performance if the extent of the cluster is underspecified, even if the correct parametric family is used. In contrast, when all people have some causative exposure, the performance of the test does not decline under overspecification of the cluster extent.

Since the precise functional form of the distance–exposure relationship is likely unknown, we next consider the power of hot spot tests to detect clinal clusters. Figure 19.5 illustrates the power of score tests for circular hot spots of varying radii applied to data generated under clinal cluster models where the relative risk of disease is increased 20% in the cells nearest the focus. The top three panels illustrate power curves for the cluster models defined in (19.2)–(19.4). Each line represents power as a function of hot spot radius, for a true clinal cluster with a fixed value of γ, τ, or L in the left, middle, and right panels, respectively. The range of parameter values reflects those considered in Figure 19.3. The surfaces in Figure 19.5 present the same information in three dimensions.

Qualitatively, we see similar performance by the circular hot spot test versus each of the three clinal clustering models. Clinal models where disease risk declines more rapidly from peak exposures allow better approximation by hot spot tests. In fact, for the exponential-threshold model, we see an optimal hot spot radius indicated by a peak in each power curve in the top-right panel of Figure 19.5 and a 'ridge' in the corresponding three-dimensional representation.

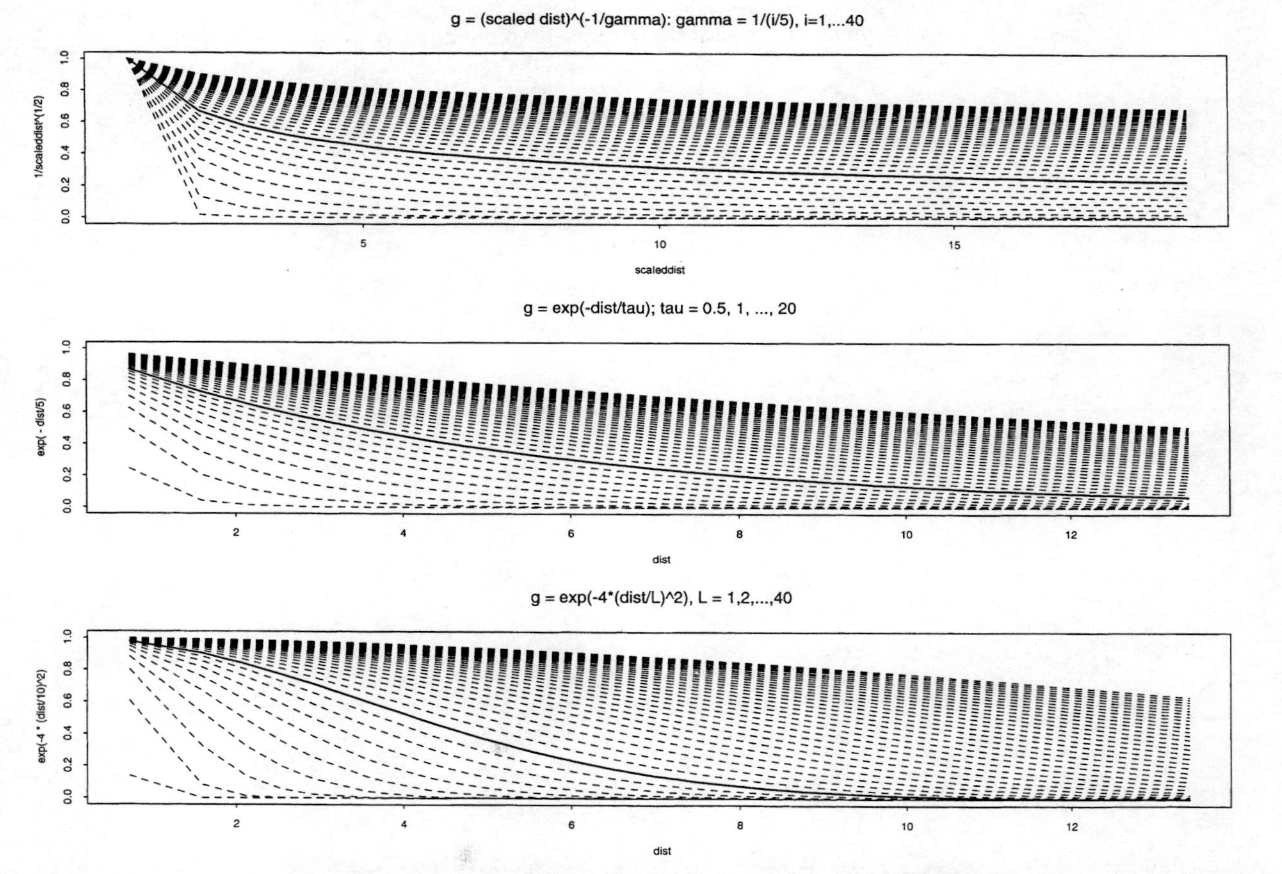

Figure 19.3 Clinal cluster models considered. The solid line represents the 'true' model generating data, and the dashed lines represent cluster models defining the tests under consideration. The cluster range increases with increasing parameter values (see text)

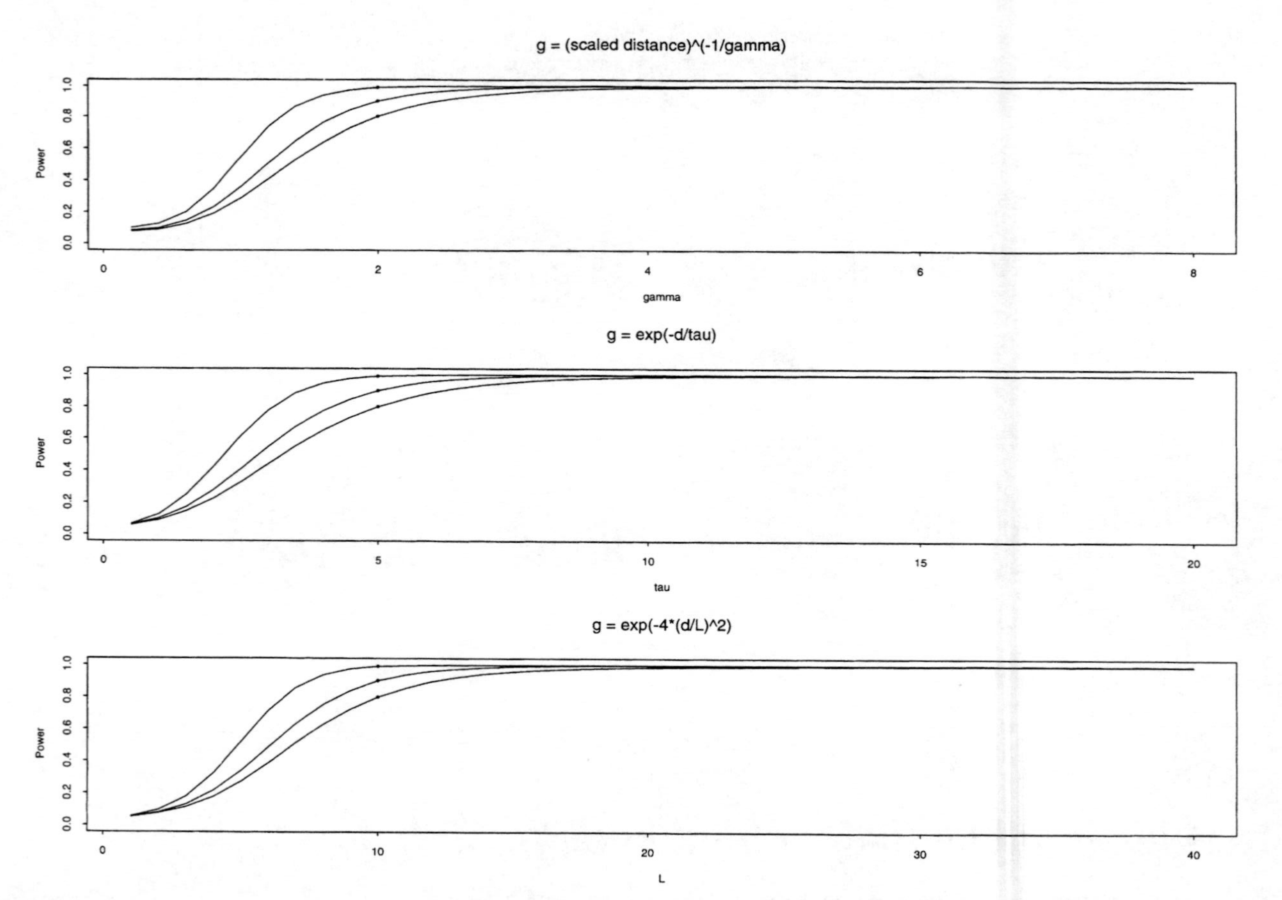

Figure 19.4 The statistical power of score tests for clinal alternatives with misspecified parameters. The background disease rate λ is 0.001. The three power curves in each subplot indicate the performance of tests with 80%, 90%, and 99% power, when correctly specified

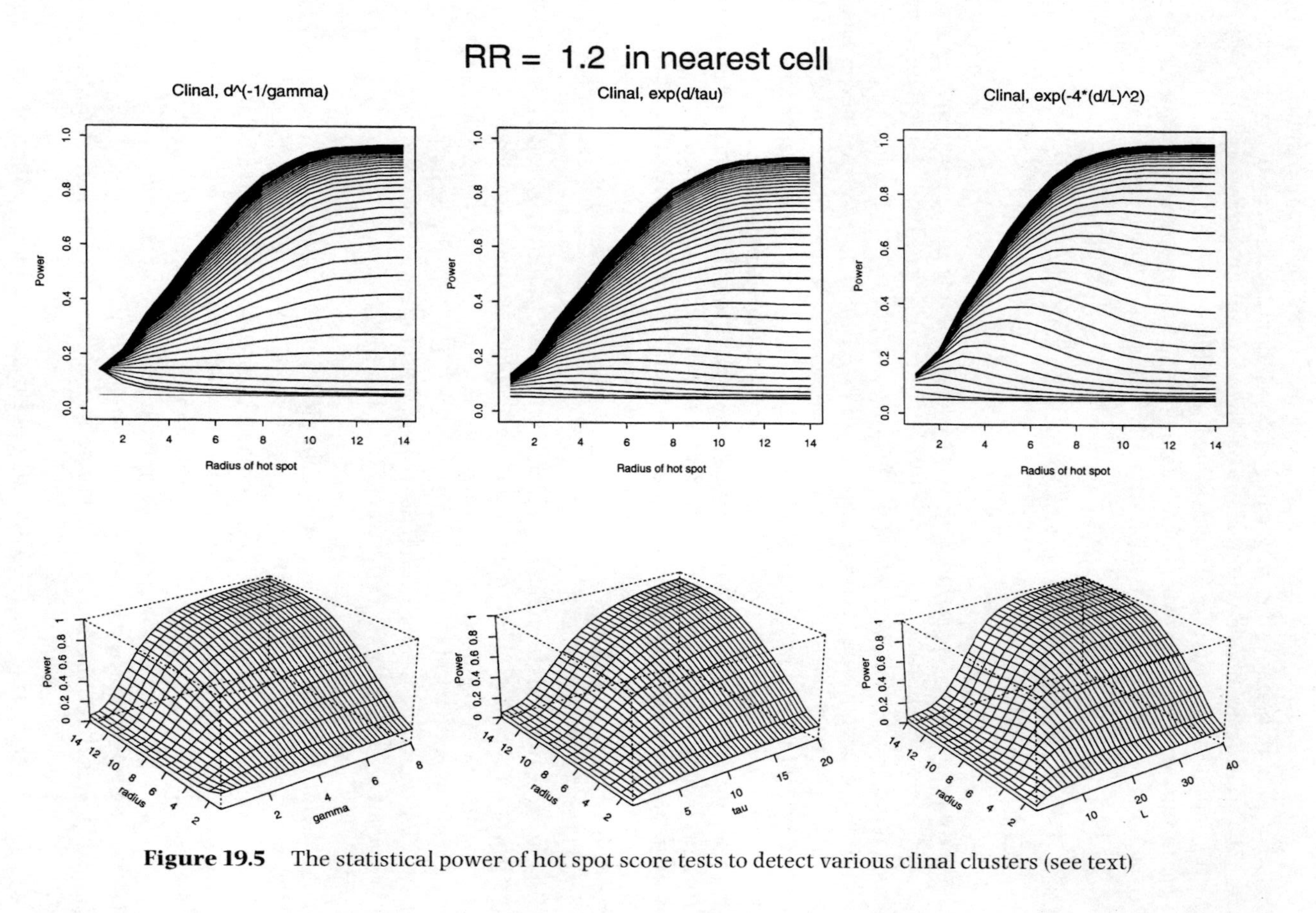

Figure 19.5 The statistical power of hot spot score tests to detect various clinal clusters (see text)

The examples above illustrate the performance of misspecified tests in an idealised environment, i.e. a regular grid with a homogeneous population density. Such data are rarely (if ever) observed in disease surveillance applications. More often, analysts encounter data in regions that vary widely in both population density and geographic area. As a result, power calculations rarely provide omnibus recommendations on test selection and study design for statistical focused cluster investigations. However, the power formulae above do provide helpful data-specific information regarding the ability of the chosen test to detect particular types of clustering as illustrated below.

19.4 *POST HOC* POWER ANALYSIS: NEW YORK LEUKAEMIA DATA

Since local vagaries in population density play a critical role in the performance of test statistics, we consider an alternative role for theoretical power calculations, namely *post hoc power analyses*, i.e. an assessment of the power available in a given dataset to detect clustering of predetermined type, strength, and extent. The 'true' cluster mechanism is never known, yet any particular score test presupposes a specific cluster model. By comparing the theoretical power of a particular score test (e.g. based on inverse distance) to detect a variety of possible 'true' cluster types (e.g. hot spot, exponential, or exponential-threshold clusters), we obtain a realistic assessment of the ability of the chosen test to detect any of a variety of clustering patterns. We use the term 'post hoc power analysis' to stress that the approach provides comparative results only in the context of the cells defining the particular study area in question. Such *post hoc* analyses quantify the degree to which a lack of statistical significance suggests a lack of different types of focused clustering in the data.

We illustrate the idea through an application to the leukaemia data reported and analysed by Waller *et al.* (1994). The data include incident leukaemia cases from 1978 to 1982 in an eight-country region of upstate New York. Foci include inactive hazardous waste sites documented as containing trichloroethylene (TCE). We illustrate *post hoc* power analysis for two foci. Site 1 is located near the town of Auburn, New York, and site 2 is in a rural area. The locations of the waste sites and the centroids of census regions are shown in Figure 19.6.

For illustration, consider the power of a clinal score test based on an inverse distance model with $\gamma = 1$. In the original analysis, Waller *et al.* (1994) report no significant clustering around either site based on the score test defined by such a model. Since the correctly specified score test is locally most powerful, we may feel confident that no inverse-distance clustering exists. However, can we assess our confidence that cases do not follow a 'true' (but unknown) hot spot cluster, given the results of the (now misspecified) inverse distance score test? Using *post hoc* analysis based on (19.5), we determine the power of the inverse-distance score test versus a hot spot cluster alternative for the New York leukaemia data.

We calculate the power of the inverse distance test based on two different 'true' clustering models. First, we find the power assuming the test was correctly specified, i.e. risk multiplicatively increases with exposure according to (19.1) with $g_i = 1/d_i$. To be consistent with the original analysis, we do not rescale exposure as in the preceding sections. Next, we consider the power of the inverse distance score test to detect circular

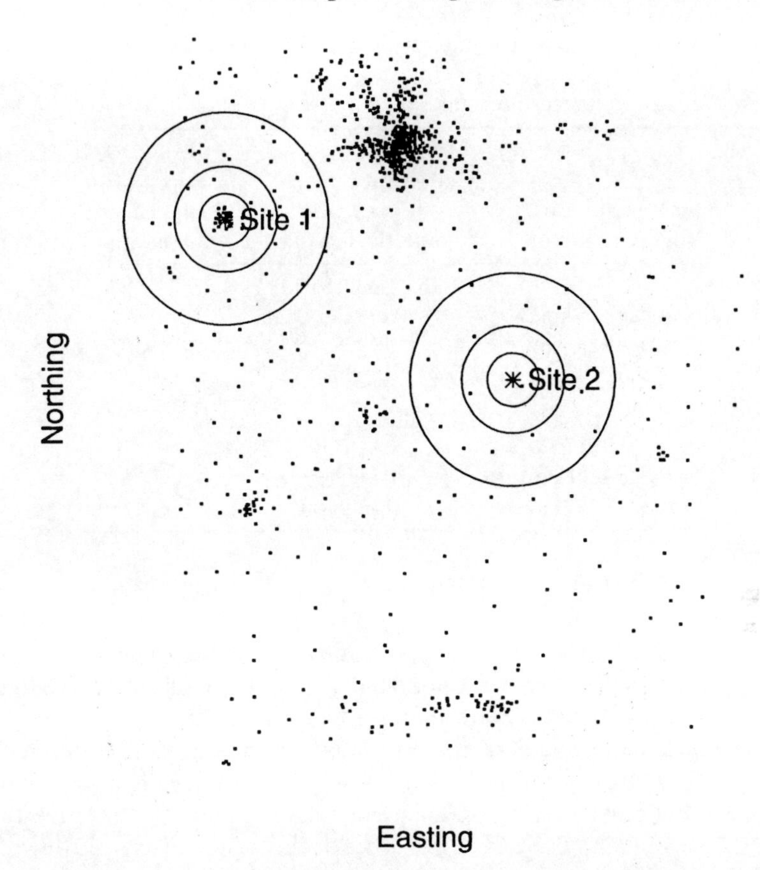

Figure 19.6 Census tract and block group centroids (1980 US, Census), and two hazardous waste sites in an eight-country region of upstate New York. Rings indicate distances of 5 km, 10 km, and 20 km, respectively

hot spot clusters with radii 5 km, 10 km, and 20 km, respectively. The definition of the test statistic provides σ_{0_U}, and the 'true' clustering model provides μ_{1_U} and σ_{1_U}. We define ε consistent with a doubling of the relative risk in the nearest census region. Finally, we assume a background disease rate of $\lambda = 0.0005$ (very near the observed disease rate for the entire study area).

Not surprisingly, if correctly specified (i.e. versus an inverse distance clustering alternative), the inverse distance score test has power near 1.0 for both sites. However, if the 'true' cluster is a hot spot cluster, then the inverse distance test is misspecified, and the power results change considerably. Table 19.1 shows power results and the size of the population at risk for hot spot cluster alternatives. The power of the misspecified inverse distance test versus hot spot clusters remains high for site 1, due to the relatively high population density in surrounding regions, but is reduced considerably for site 2, located in a rural area with much smaller population sizes.

Post hoc power analyses provide an assessment of the strength of evidence against clustering provided by a non-significant hypothesis test result. Non-significant results for a test with high power indicate a small probability of failing to reject the null hypoth-

Table 19.1 Power results for a misspecified inverse distance score test assuming true hot spot clusters for two foci in an eight-country region in upstate New York

Radius of true hot spot cluster	Power of inverse Distance score test vs. hot spot alternative	Population at risk in hot spot
	Site 1 (small town)	
Distance < 5 km	0.9943	34,407
Distance < 10 km	0.9999	39,103
Distance < 20 km	0.9999	79,910
	Site 2 (rural)	
Distance < 5 km	0.1263	879
Distance < 10 km	0.1419	1,649
Distance < 20 km	0.3406	14,845

esis when the specified alternative is true. In contrast, non-significant results for a test with low power allow a relatively large probability of failure to reject the null hypothesis (no clustering) when the specified alternative is true.

For the New York leukaemia data, the non-significant result for the inverse distance score test provides relatively strong evidence against an inverse distance clinal cluster centred at site 2, but much weaker evidence against a local hot spot increase in risk. As a result, we have confidently ruled out two types of clustering (inverse distance and hot spot) for site 1, but only one type (inverse distance) for site 2. Equation (19.5) allows us to consider a variety of clustering models, even those unrelated to the test under consideration.

19.5 DISCUSSION

In addition to the connection between the statistical power of a focused score test and the particular form of clustering assumed in the definition of the test, the results above reveal a dependence between the power properties of focused score tests and the structure of the population at risk. Other relevant issues include differential risk among individuals because of age, gender, and other confounding variables; differential exposure among individuals in the same cell; size and shape of the study region; and migration. The impact of such covariates works against strong performance by a particular test over a wide variety of possible situations.

In the homogeneous population density results, we find that, for the situations considered, it is often better to overspecify than underspecify the extent of clustering when defining a test. We also find no large differences in performance between the inverse (rescaled) distance, exponential, and exponential-threshold clinal models. Finally, we find that the adequacy of the hot spot approximation of clinal clusters largely depends on the 'tail' behaviour of the cluster, i.e. how quickly the relative risk drops off from the peak values. However, more complex relationships are likely to occur with more complicated clustering models.

It is difficult to extend the guidelines suggested from homogeneous grid-based data to the heterogeneous, enumeration-district-based data often encountered in disease surveillance. As we illustrate above, population density can impact the power of misspecified score tests, sometimes dramatically. Apart from the usual role of power in study design, we show how *post hoc* analysis provides a mechanism for interpreting and evaluating results, even allowing the assessment of clustering patterns not considered in the original analysis.

The sensitivity of hypothesis tests of clustering to features common in disease surveillance data (e.g. variable population density, misspecification of disease incidence models) encourages more comprehensive modelling of disease rates, accounting for confounders either through standardisation, comparison with 'control' (non-diseased) groups, or direct inclusion as covariates. The generalised linear mixed models addressed by Besag *et al.*, (1991) and Breslow and Clayton (1993) may provide an opportunity for a more accurate assessment of multiple covariate effects, adjustment for heterogeneous exposures, and quantification of residual spatial effects. Such models provide valuable opportunities for future work in disease surveillance.

ACKNOWLEDGEMENTS

The research work of Lance A. Waller was supported in part by National Institute of Environmental Health Sciences. NIH, grant number 1 R01 ES07750-01A1. The results reflect those of the authors and not necessarily those of NIEHS or NIH. The research work of Catherine A. Poquette was supported in part by Cancer Center Support CORE Grant CA 21765 and by the American Lebanese Syrian Associated Charities (ALSAC). The authors thank two anonymous referees for helpful suggestions.

20

Case–Control Analysis of Risk around Putative Sources

Annibale Biggeri and Corrado Lagazio

University of Florence

20.1 INTRODUCTION

Many epidemiological investigations on environmental hazards focus on a priori putative sources (see, for example, the studies on childhood leukaemia and nuclear reprocessing plant sites). The pattern of risk as a function of distance from the source is used as an approximation for the true distribution of risk by exposure levels and therefore such study design has a poor value as a proof of a direct causal relationship. This kind of geographical analysis may be used either to describe and evaluate the magnitude of the problem, or to surrogate the unknown exposures by distance measures.

To estimate this risk pattern aggregate data can be used, but the results could be affected by the ecological bias since individual risk factors are not taken into account (Diggle and Elliott, 1995). Theoretically the successful modelling of the risk pattern as a function of distance from putative sources depends on the availability of individual data and on information on major risk factors and predictors of the disease under study.

Individual information is usually of high quality but at greater cost than routinely collected aggregate data, and in epidemiological research special sampling designs are used to ensure an acceptable cost-effectiveness balance: case–control studies are one example of a highly efficient sampling strategy. In geographical analyses, the sample of controls is used in any situation where the distribution of the population by small areas is unknown or when it is too difficult or expensive to gather.

The present chapter focuses on the analysis of individual data generated by a case–control sampling design aimed to assess the risk gradient as a function of the distance from a putative source. The data are an example of a heterogeneous Poisson point process in the plane, in which the intensity of cases in a given location depends on the number of people at risk, on the distance from source and on individual covariates.

Disease Mapping and Risk Assessment for Public Health. Edited by A.B. Lawson *et al.*
© 1999 John Wiley & Sons Ltd.

20.1.1 Example

To investigate the relationship between four sources of air pollution (a shipyard, an iron foundry, an incinerator and car traffic in the city centre) and lung cancer, a case–control study has been conducted in Trieste (Italy) (Biggeri *et al.*, 1996). Seven hundred and fifty-five male cases of histologically verified lung cancer and 755 age-matched male controls were identified through the local autopsy register for the period 1979–86. Each subject next-of-kin was interviewed and information on demographic characteristics, smoking habits, occupational history and place of residence were collected using a structured questionnaire. Changes of residence in the last 10 years were recorded. The boundaries of the Province of Trieste were coded using geographical coordinates (Italian Army Geographical Institute, map 1:10 000) and the subject relevant residence was identified on the same map. The location of the four putative sources was identified similarly. For the analysis the distance and the angle from each subject residence location to each pollution source was calculated (north orientation). Historical data on air particulate deposition (g/m^2 per day) for the early 1970s were obtained from 28 stations that covered the city. Each subject residence was assigned the average value measured by the nearest station.

Alternatively one might have obtained the distribution of the male population by age class for each census tract during the study period 1979–86. In Italy this is almost impossible, since only for the census year (1981) is such information available. Furthermore, no information on smoking habits, and information on current job can only be derived from census data. Generally speaking a case–control study would be more efficient in term of costs even if census data had been available and in sufficient detail.

20.2 GENERAL DEFINITIONS

Let denote A the study area, R a region within A, and $\boldsymbol{x}$ a vector of generic coordinates (e.g. latitude and longitude) of a location in A. The spatial distribution of a set of points at locations $\{\boldsymbol{x}_i; i = 1, \ldots, n\}$ in A is the realisation of a two-dimensional point process, which can be described using a counting measure ϕ on A; $\phi(R)$ denotes the number of events in R (Cox and Isham, 1980). The first- and second-order intensity functions are defined, respectively, as

$$\lambda(\boldsymbol{x}) = \lim_{|\mathrm{d}\boldsymbol{x}|\to 0} \left\{ \frac{E\left[\phi(\mathrm{d}\boldsymbol{x})\right]}{|\mathrm{d}\boldsymbol{x}|} \right\},$$

$$\lambda_2(\boldsymbol{x}, \boldsymbol{y}) = \lim_{|\mathrm{d}\boldsymbol{x}||\mathrm{d}\boldsymbol{y}|\to 0} \left\{ \frac{E\left[\phi(\mathrm{d}\boldsymbol{x})\phi(\mathrm{d}\boldsymbol{y})\right]}{|\mathrm{d}\boldsymbol{x}||\mathrm{d}\boldsymbol{y}|} \right\},$$

in given infinitesimal regions $\mathrm{d}\boldsymbol{x}$ and $\mathrm{d}\boldsymbol{y}$ in A ($|R|$ being the area of R). Stationarity implies invariance under translation (the first- and second-order intensity functions are, respectively, $\lambda(\boldsymbol{x}) = \lambda$ and $\lambda_2(\boldsymbol{x}, \boldsymbol{y}) = \lambda_2(\boldsymbol{x} - \boldsymbol{y})$); isotropy implies invariance under rotation (the second-order intensity function depends only on the distance between the two locations, i.e. $\lambda_2(\boldsymbol{x}, \boldsymbol{y}) = \lambda_2(\|\boldsymbol{x} - \boldsymbol{y}\|)$) (Diggle, 1983). The simple homogeneous Poisson point process assumes the probability of two points occurring on the same location $\boldsymbol{x}$ be negligible and the counts of points on two disjoint regions R_1 and R_2 be independent. Then

$$\lambda(\boldsymbol{x}) = \lim_{|\mathrm{d}\boldsymbol{x}|\to 0}\left\{\frac{E[\phi(\mathrm{d}\boldsymbol{x})]}{|\mathrm{d}\boldsymbol{x}|}\right\} = \lim_{|\mathrm{d}\boldsymbol{x}|\to 0}\left\{\frac{\Pr[\phi(\mathrm{d}\boldsymbol{x})=1]}{|\mathrm{d}\boldsymbol{x}|}\right\},$$

$$\lambda_2(\boldsymbol{x},\boldsymbol{y}) = \lim_{|\mathrm{d}\boldsymbol{x}||\mathrm{d}\boldsymbol{y}|\to 0}\left\{\frac{E[\phi(\mathrm{d}\boldsymbol{x})\phi(\mathrm{d}\boldsymbol{y})]}{|\mathrm{d}\boldsymbol{x}||\mathrm{d}\boldsymbol{y}|}\right\} = \lim_{|\mathrm{d}\boldsymbol{x}||\mathrm{d}\boldsymbol{y}|\to 0}\left\{\frac{E[\phi(\mathrm{d}\boldsymbol{x})]E[\phi(\mathrm{d}\boldsymbol{y})]}{|\mathrm{d}\boldsymbol{x}||\mathrm{d}\boldsymbol{y}|}\right\} = \lambda^2,$$

and the number of points in a generic region R follows a Poisson law with expected value $\int_R \lambda(\boldsymbol{u})\mathrm{d}\boldsymbol{u}$, which is equal to $\lambda|R|$ under stationarity. Given $\phi(A) = n$, the set of points locations $\{\boldsymbol{x}_1, \boldsymbol{x}_2, \ldots, \boldsymbol{x}_n\}$ is distributed as a sample of independent identically distributed random variables with probability density (Cressie, 1993)

$$f_A(\boldsymbol{x}) = \frac{\lambda(\boldsymbol{x})}{\int_A \lambda(\boldsymbol{u})\mathrm{d}\boldsymbol{u}}.$$

The intensity of the population at location $\boldsymbol{x}$ under the Poisson assumptions is given by the probability of observing a subject resident at $\boldsymbol{x}$ divided by the unit area:

$$\lambda_{\mathrm{P}}(\boldsymbol{x}) \cong \frac{\Pr[\phi_{\mathrm{P}}(\mathrm{d}\boldsymbol{x}) = 1]}{|\mathrm{d}\boldsymbol{x}|}.$$

An incident case of disease can be observed, at a given time at location $\boldsymbol{x}$, only if she/he was healthy and present at the same location an instant before she/he got ill. Define $\mu(\boldsymbol{x})$ as the probability of being a case for a subject resident at $\boldsymbol{x}$, and recall that $\lambda_{\mathrm{P}}(\boldsymbol{x})$ is the intensity of the process for a subject at risk at $\boldsymbol{x}$. Then, the spatial point process that generates cases of disease is again a Poisson process with intensity function $\lambda_{\mathrm{CS}}(\boldsymbol{x})$ equal to $\mu(\boldsymbol{x})\lambda_{\mathrm{P}}(\boldsymbol{x})$ (Diggle, 1983; Cressie, 1993, p. 690). The expected number of cases in region R is

$$E[\phi_{\mathrm{CS}}(R)] = \int_R \mu(\boldsymbol{u})\lambda_{\mathrm{P}}(\boldsymbol{u})\mathrm{d}\boldsymbol{u}$$

and the probability density for a case, given the total number of cases in region A, is

$$f_{\mathrm{CS}}(\boldsymbol{x}) = \frac{\lambda_{\mathrm{CS}}(\boldsymbol{x})}{\int_A \lambda_{\mathrm{CS}}(\boldsymbol{u})\mathrm{d}\boldsymbol{u}} = \frac{\mu(\boldsymbol{x})\lambda_{\mathrm{P}}(\boldsymbol{x})}{\int_A \mu(\boldsymbol{u})\lambda_{\mathrm{P}}(\boldsymbol{u})\mathrm{d}\boldsymbol{u}}.$$

It should be noted that the intensity of the case disease process factorises into two terms: the first is the probability function of having the disease being at $\boldsymbol{x}$ and the second is the number at risk at $\boldsymbol{x}$. This resembles the factorisation used in survival analysis where the intensity process $\lambda(t) = \mu(t)Y(t)$ is defined as the product of the conditional probability (or hazard rate) $\mu(t)$ of failure at time t having survived at $t - 1$ and the process $Y(t)$ is the number at risk of failing at time t (Andersen *et al.*, 1992).

By the same arguments the intensity for a healthy subject (a non-case) is defined as

$$\lambda_{\mathrm{H}}(\boldsymbol{x}) = [1 - \mu(\boldsymbol{x})]\lambda_{\mathrm{P}}(\boldsymbol{x}).$$

The spatial intensity for the case process may then be expressed in terms of the *odds* of being a case of disease and of $\lambda_{\mathrm{H}}(\boldsymbol{x})$:

$$\lambda_{\mathrm{CS}}(\boldsymbol{x}) = \frac{\mu(\boldsymbol{x})}{1 - \mu(\boldsymbol{x})}\lambda_{\mathrm{H}}(\boldsymbol{x}).$$

20.3 ESTIMATION

In practice we have problems in estimating the spatial intensity of the cases when an accurate and precise estimate of $\lambda_{\mathrm{P}}(\boldsymbol{x})$ is difficult to obtain. Indeed, the census of the population is used to estimate the population intensities for aggregate data at the small-area level (e.g. post-code sectors or census tracts) but it is infeasible to determine the coordinates of the residence of each individual of the population.

Two alternative sampling strategies can be adopted.

1. Draw a sample from the population of healthy subjects with sampling fraction
$$\frac{1}{c} = \frac{\phi_{\mathrm{CN}}(A)}{\phi_{\mathrm{H}}(A)},$$
where $\phi_{\mathrm{CN}}(A)$ denotes the number of sampled subjects (controls) and $\phi_{\mathrm{H}}(A)$ denotes the total number of healthy subjects in region A.
2. Select a sample of healthy subjects (controls) proportional to the number of cases enrolled and in case matched to them for some characterics of interest (e.g. age; Rothman and Greenland, 1998).A consequence of this second strategy is that, being the total number of controls in region A chosen proportional to the number of cases in the study $\phi_{\mathrm{CN}}(A) = k\phi_{\mathrm{CS}}(A)$, the actual sampling fraction $1/c$ is not specified. This design is exemplified in Section 20.3.1 below.

In both cases the ratio of the number of cases, $\#\mathrm{CS}(\boldsymbol{x})$, to the number of controls, $\#\mathrm{CN}(\boldsymbol{x})$, at $\boldsymbol{x}$ estimates the ratio of the probability density of the cases and the healthy people at $\boldsymbol{x}$:

$$\frac{\#\mathrm{CS}(\boldsymbol{x})}{\#\mathrm{CN}(\boldsymbol{x})} \rightarrow \frac{f_{\mathrm{CS}}(\boldsymbol{x})}{f_{\mathrm{H}}(\boldsymbol{x})} = kc\frac{\mu(\boldsymbol{x})}{1 - \mu(\boldsymbol{x})}.$$

Thus, knowing the two sampling constants k and c, the *odds* of the probability of being a case for a subject resident at $\boldsymbol{x}$ is estimable.

Since under the second sampling strategy the proportionality constant c is not specified, the probability *odds* of being a case for a subject resident at $\boldsymbol{x}$ is not estimable. However, the probability *odds* at $\boldsymbol{x}$ relative to the *odds* at a reference area with coordinates $\boldsymbol{x}^*$ within A (i.e. the *odds ratio*) can be estimated:

$$\frac{\#\mathrm{CS}(\boldsymbol{x})/\#\mathrm{CN}(\boldsymbol{x})}{\#\mathrm{CS}(\boldsymbol{x}^*)/\#\mathrm{CN}(\boldsymbol{x}^*)} \rightarrow \frac{\mu(\boldsymbol{x})/[1 - \mu(\boldsymbol{x})]}{\mu(\boldsymbol{x}^*)/[1 - \mu(\boldsymbol{x}^*)]}.$$

This approach can be viewed as equivalent to conditioning on the total observed locations of cases and controls (see Diggle and Rowlingson, 1994).

The data consist of an indicator for each location $\boldsymbol{x}$ of an observed subject if she/he is a case or a control, and estimates of the *odds* of disease at $\boldsymbol{x}$ can be obtained by parameterising appropriately the function $\mu(\boldsymbol{x})$ or by estimating the intensities of the case and control processes non-parametrically (e.g. by density estimation techniques; see Bithell, 1990, and Lawson and Williams, 1994). This second choice is used to obtain a descriptive representation of the observed crude risk surface, independently of the point source location. The intensity of controls can be expressed as

$$\lambda_{\text{CN}}(\boldsymbol{x}) = \frac{1}{kc}\lambda_{\text{P}}(\boldsymbol{x})[1 - \mu(\boldsymbol{x})],$$

and the case intensity becomes

$$\lambda_{\text{CS}}(\boldsymbol{x}) = kc\lambda_{\text{CN}}(\boldsymbol{x})\frac{\mu(\boldsymbol{x})}{1 - \mu(\boldsymbol{x})},$$

i.e. a function of the odds of disease (the odds being the probability of being a case over the probability of not being a case for a subject resident at $\boldsymbol{x}$) (Biggeri *et al.*, 1996; for a similar approach contrasting two point processes, see Diggle and Rowlingson, 1994).

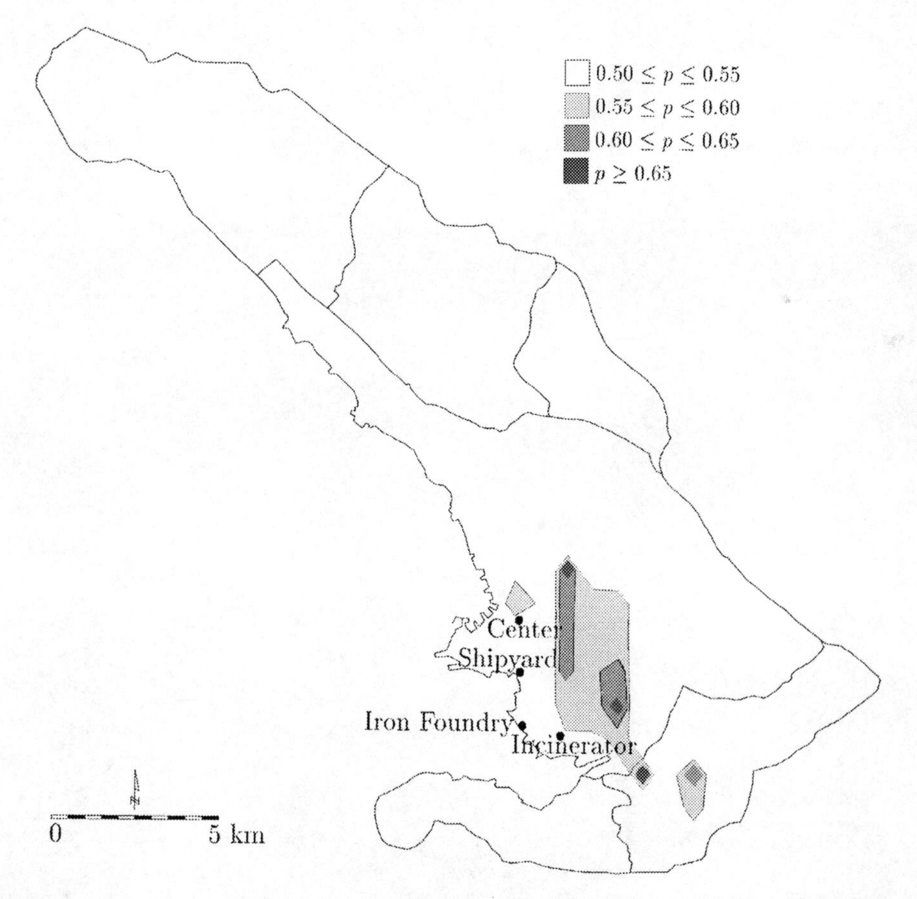

Figure 20.1 Location of pollution sources and contour plot of the probability of being a case

Both spatial intensities, for the case and the control series, can be estimated by

$$\hat{\lambda}(\boldsymbol{x}_i) = \sum_{j=1}^{n} h_i^{-2} G\left(\frac{\boldsymbol{x}_i - \boldsymbol{x}_j}{h_i}\right),$$

where the kernel function $G(\cdot)$ has the Epanechnikov functional form. The terms h_i are smoothing parameters that allow for local variation in the degree of smoothing. They are obtained as $h_i = \eta_i h$, where h is fixed in advance and η_i is a previous estimate obtained, for example, using the simple nearest neighbour technique (see Silverman, 1986, for a complete discussion of different choices of initial estimates). Apart from the proportionality constants kc, the ratio of the kernel estimates for cases and non-cases is the *odds* of being a case. To obtain easily interpretable contour plots we back-transform it: probability = odds/(1 + odds). However, since we do not know the sampling fraction $1/c$ this probability is not immediately interpretable as a disease probability and we can only interpret relative changes.

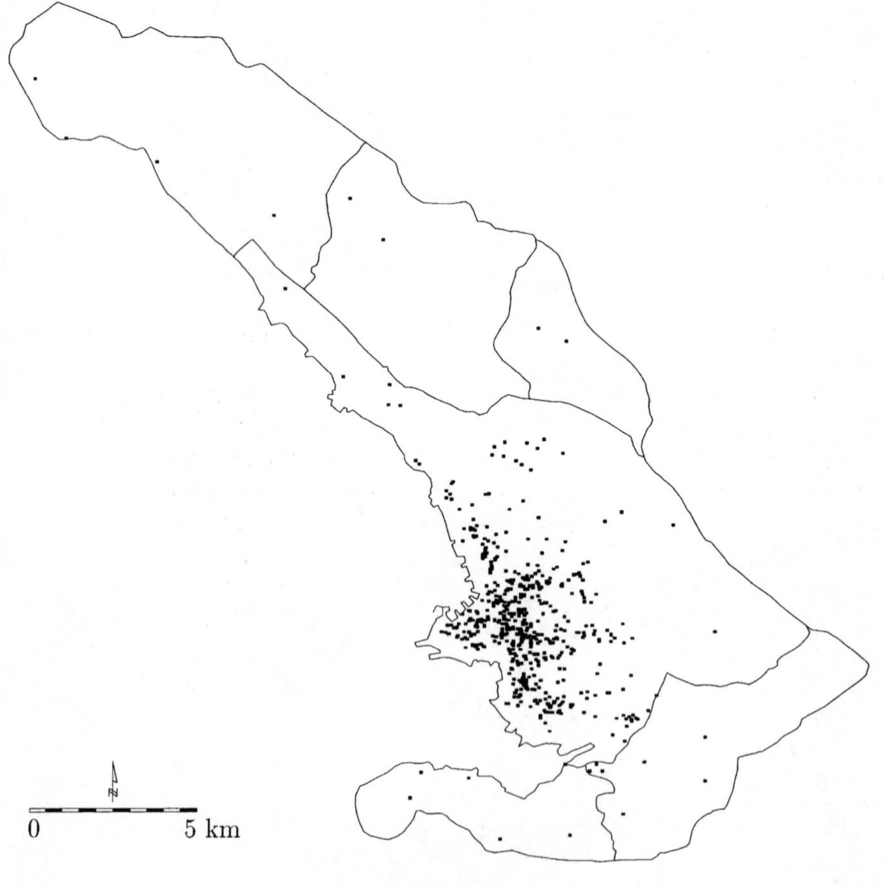

Figure 20.2 *(continued)*

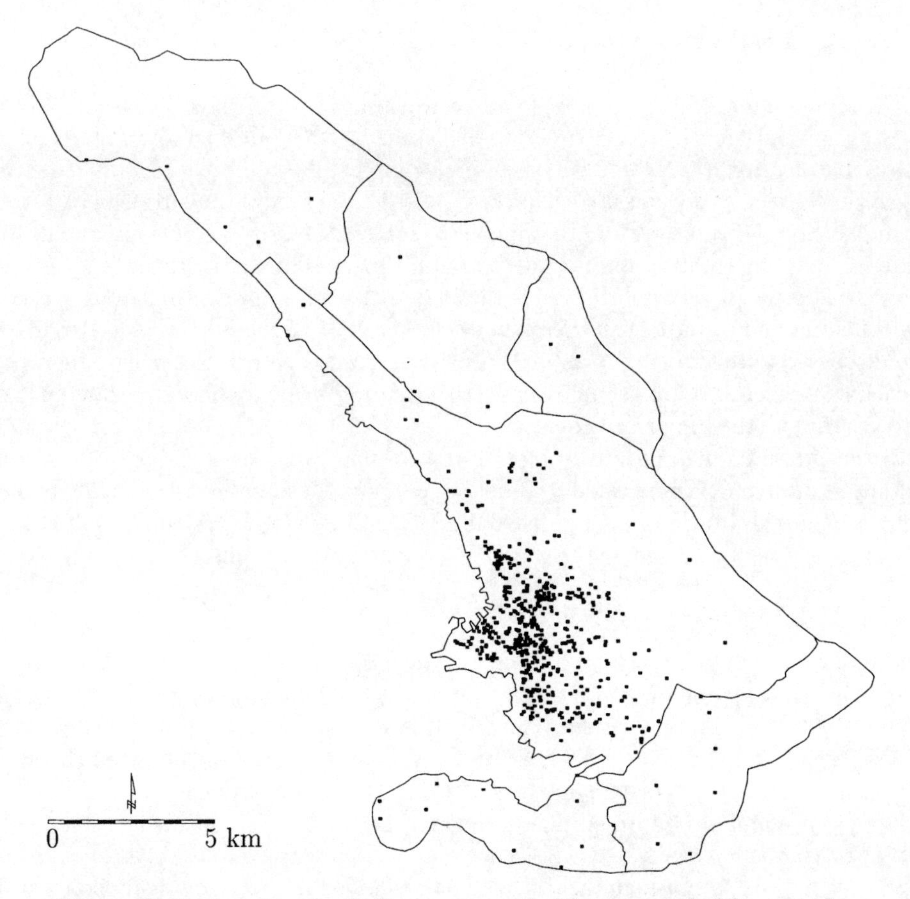

Figure 20.2 Location of pollution sources and of the cases and of the controls

20.3.1 Example

Figure 20.1 reports the location of the four pollution sources in the city of Trieste and the contour plot of the probability of being a case obtained using adaptive kernel estimators with a 500 m bandwidth and nearest neighbour initial estimates.

There appears to be a wide risk area with a spot near the city centre and two peaks east of the incinerator. The overall pattern suggests some directionality of sources effects due to the prevailing wind direction (west to east).

The reader should note that the spatial distribution of cases of disease cannot be used *per se* to test alternative hypotheses on higher risk in the neighbourhood of a putative source because of the spatially inhomogeneous distribution of the population. The sample of controls drawn from the population at risk is useful when that spatial distribution of the population is unknown. The locations of the cases and controls are then identified and geocoded. In Figure 20.2 the locations of the 755 cases and 755 controls enrolled in the Trieste study are reported. The concentration of the population and the sparseness or even absence in the proximities of the foundry and the shipyard are evident.

20.4 CRUDE TREND TESTS ON DISTANCE FROM SOURCE

In the remainder of the chapter we focus on the analysis of the risk of disease, parameterising appropriately $\mu(\boldsymbol{x})$ as a function of distance from a putative source of exposure. Using the case–control design we are interested in estimating the *odds ratio* of disease, taking as reference the odds at a location so far from the source that an effect of it would be implausible. First, we present the simpler trend tests of the *odds ratio* by distance from source. Since the existing methods (Stone and Cuzick–Edward tests) have low power, also two alternative possibilities are outlined here: the cumulative chi-squared test and the maximum chi-squared test (Hirotsu, 1983; Nair, 1987). Secondly, adjusted trend tests are discussed, either in the form of Mantel–Haenszel type or in a logistic regression model. These methods are useful to model the excess risk due to the source having taken into account all the important covariates.

In the presence of a putative point source of pollution the locations of cases are usually assumed to be generated by a heterogeneous Poisson process since the population is spatially inhomogeneously distributed with intensity $\lambda_{CS}(\boldsymbol{x})$ equal to $\mu(\boldsymbol{x})\lambda_P(\boldsymbol{x})$. The effect of the distance from the source can be modelled as follows:

$$\lambda_{CS}(\boldsymbol{x}) = \lambda_P(\boldsymbol{x})\mu(\boldsymbol{x} - \boldsymbol{x}_0; \theta),$$

where $\mu(\boldsymbol{x} - \boldsymbol{x}_0; \theta)$ is the risk, in its simplest formulation, as a function only of the distance from the location of the putative source ($\boldsymbol{x}_0$), modelled by the parameter vector θ (Diggle, 1990). The interest here is in making inference on the risk gradient. These kinds of tests are also called 'focused clustering tests'. Only two proposals exist in the literature that use case–control data for the statistical evaluation of the association between risk of disease and environmental pollution coming from a prespecified source (Stone, 1988; Cuzick and Edwards, 1990). Both of them can be considered non-parametric tests in the sense that no mathematical functional relationship between distance from the source and disease risk is specified and the alternative hypothesis states simply that the risk does not increase on increasing distance from the source.

Stone adapted the test he proposed to the case–control data. To apply the test, a complete ranking for the set of cases and controls on the basis of the distance of each from the point source (without ties) is needed. Given N as the total sample size (n cases and $N - n$ controls), let $Y_i = 1 (i = 1, \ldots, N)$ if the individual who is in the ith position, according to the distance ranking, is a case and $Y_i = 0$ if the individual is a control. The proposed test statistic is

$$T_p = \max_{1 \le i \le N} \sum_{j=1}^{i} Y_j / i.$$

Under the null hypothesis of absence of association between risk of disease and distance from the source, the distribution of the statistic could be analytically approximated, considering that, depending on the total number of cases (n) and on the total number of controls ($N - n$), the distribution of the $\{Y_i\}$ coincides with that of a random permutation of n successes and $N - n$ failures. The test turns out to have low power, unless the number of controls is particularly high and superior to that of the cases. In addition, the maximum value of the test statistics is often obtained for very low values of i (if $Y_1 = 1$

the maximum value of T_p is obtained for $i = 1$) and, therefore, all the further information gathered on the set of cases and controls becomes *de facto* unusable.

The second test is that proposed by Cuzick and Edwards (1990) and is based on the extension of their generalised clustering test. After having ordered the cases and the controls on the basis of their distance from the source, we calculate

$$T_{CE} = \sum_{i=1}^{K} Y_i,$$

where K is a sufficiently high fixed number (for example, equal to 5% or 10% of the sample size) and Y_i is defined as previously described. Under the null hypothesis of absence of association between source of pollution and disease, the test statistic has a hypergeometric distribution with parameters K, n and N, and, therefore, it is simple to determine its expected value and the level of significance. This test, like the previous one, presents problems of power. In addition, it must be emphasised that the determination of the value of K is arbitrary and that only the information regarding ordering by distance is used for the construction of the test statistic.

20.4.1 Cumulative chi-squared tests and their application to spatial analysis

We introduce here two alternative tests: the cumulative chi-squared test and the maximum chi-squared test (Hirotsu, 1983; Nair, 1987). Both tests have been proposed for the analysis of association in contingency tables in which a variable is naturally ordered. These tests are usually applied in controlled clinical trials and for statistical quality control (Nair, 1986).

It is assumed that it is possible both to subdivide the study area A into K circular bands, by means of concentric circles having as their centre the source of pollution, indicated by $k (k = 1, \ldots, K)$, and to be able to determine exactly the number of cases and controls in each circle, which we will indicate, respectively, by Y_{1k} and Y_{2k}. It is possible, therefore, to construct a $2 \times K$ contingency table in which the rows indicate, respectively, the cases and the controls and the columns indicate the circular bands. It is evident that the table presents a ranking in the columns, due to the distance from the source of pollution. The frequency observed in the (i, k)th cell is indicated by Y_{ik}. In what follows we indicate the marginal totals of the rows (the total of cases and controls, respectively) by $R_i = Y_{i1} + Y_{i2} + \ldots + Y_{iK}, i = 1, 2$ and the marginal totals of the columns, i.e. the total number of individuals observed in each circular band, by $C_k = Y_{1k} + Y_{2k}, k = 1, \ldots, K$. Finally, $Z_{ik} = \sum_{i=1}^{k} Y_{ij}$ indicates the cumulative frequencies of the rows (the number of cases and controls observed in the first k circular bands, respectively), $D_k = \sum_{k=1}^{K} C_k$ the cumulative frequencies of the columns totals, $r_i = R_i/N$ the marginal proportions of the rows, $c_k = C_k/N$ the marginal proportions of the columns and $d_k = D_k/N$ the cumulative proportions of the columns.

Assuming that the cases of disease are distributed according to a heterogeneous Poisson process of intensity $\lambda_{CS}(\boldsymbol{x})$, the number of events Y_{1k} within the kth circular band is Poisson distributed with expected value equal to $\int_{S_k} \lambda_{CS}(\boldsymbol{u}) d\boldsymbol{u}$, where S_k indicates the domain of the kth circular band. Since a n-tuple of random Poisson variables, conditional on their sum, is distributed according to a multinomial law (Johnson and Kotz, 1969), the K – tuple $\{Y_{1k}; k = 1, \ldots, K\}$ is distributed according to a multinomial distribution with probabilities

$$p_{1k} = \frac{\int_{S_k} \lambda_{CS}(\boldsymbol{u}) d\boldsymbol{u}}{\int_A \lambda_{CS}(\boldsymbol{v}) d\boldsymbol{v}}.$$

Analogous reasoning on the distribution of controls brings to the conclusion that $\{Y_{2k}; k = 1, \ldots, K\}$ is also distributed according to a multinomial law with probabilities

$$p_{2k} = \frac{\int_{S_k} \lambda_{CN}(\boldsymbol{u}) d\boldsymbol{u}}{\int_A \lambda_{CN}(\boldsymbol{v}) d\boldsymbol{v}}$$

and, thus, the K circular bands represent the categories of two multinomial populations, namely that of the cases and that of the controls, having sizes R_1 and R_2, respectively.

The cumulative probabilities of cases and controls are indicated, respectively, by $\tau_{ik} = \sum_{j=1}^{k} p_{ij}, i = 1, 2$. The null hypothesis of absence of association between risk of disease and source of pollution is tested against the alternative hypothesis:

$$H_1 : \tau_{2k} \leq \tau_{1k} \text{ or } \tau_{1k} \leq \tau_{2k}, \quad k = 1, \ldots, K-1,$$

in which, for at least one value of k, a strict inequality is valid. The first of the two positions corresponds to the hypothesis of an excess of cases of disease close to the source of pollution, the second to an excess of controls. The proposed test statistic is

$$T_L = \sum_{k=1}^{K-1} \bar{X}_k^2,$$

where $\bar{X}_k^2$ indicates the Pearson chi-squared statistic for the 2×2 contingency table obtained by grouping the circular bands from the first to the kth and from the $(k+1)$th to the Kth, that is:

$$\bar{X}_k^2 = \sum_{i=1}^{2} \frac{[Z_{ik} - R_i d_k]^2}{R_i d_k} + \sum_{i=1}^{2} \frac{[(R_i - Z_{ik}) - R_i(1 - d_k)]^2}{R_i(1 - d_k)}$$

See also Taguchi (1974) for a discussion of this test as part of a class of cumulative chi-squared tests. The p-value of the statistic T_L can be approximately determined by means of simulations or using the Satterthwaite (1946) approximation (Hirotsu, 1993), which in our experience tends to give probabilities of type I error larger than aimed at (Lagazio *et al.*, 1996).

20.4.2 The maximum chi-squared test

A possible alternative to testing the null hypothesis against the alternative hypothesis H_1 is the maximum chi-squared test, proposed by Hirotsu (1983). The test statistic is

$$T_M = \max_{1 \le k \le K-1} \bar{X}_k^2.$$

The derivation of the asymptotic distribution of T_M is somewhat complex, inasmuch as it is necessary to determine the distribution of $K - 1$ random correlated chi-square variables. For $K > 9$, as frequently happens in geographical epidemiology, it is necessary to use simulation methods in order to determine the level of significance of the test.

The maximum chi-squared test is interesting because it offers the possibility of determining, albeit approximately, the range of action of the source of pollution, identifying it by means of the distance at which $\bar{X}_k^2$ reaches its maximum (see Lagazio *et al.*, 1996). In a sense it is similar to the Stone test.

20.4.3 Example

The cumulative and maximum chi-squared tests were applied separately for each source of pollution on the Trieste data. The area under study was subdivided into circular bands with a span of 100 m, having as their centre, in turn, the examined source of pollution. The choice of the radius of the concentric circles was subjective and influenced the power of the two tests, but in this case it does not seem to have had any noticeable effects. The significance of the statistics T_L and T_M was determined on the basis of 10 000 Monte Carlo simulations. In Table 20.1 the results of the cumulative and maximum chi-squared tests are reported. The *p*-values obtained for the cumulative and maximum chi-squared tests suggest the existence of a significant association between sources of pollution and risk of disease. The two tests give discordant results as regards the possible effects of the incinerator and that the *p*-value of the statistic T_M relative to the foundry is at the limits of the rejection region. Such results therefore seem to indicate the existence of a significant excess of risk due, above all,to the polluting emissions of the shipyard and the city centre, while the relationship with the other two sources of pollution is more uncertain. However, relevant confounding is not taken into account: first, people working at the shipyard, foundry or incinerator could reside closer to those sources than people not working there; secondly, lower social class people with a higher

Table 20.1 Values of the statistics T_L and T_M for the four sources of pollution

	Cumulative chi-squared T_L		Maximum chi-squared T_M		
Source	**Obs. values**	**MC *p*-values**	**Obs. values**	**MC *p*-values**	**Range (km)**
City centre	494.62	0.001	19.43	0.000	3.7
Shipyard	503.27	0.000	17.19	0.000	2.6
Foundry	360.71	0.003	9.11	0.050	4.1
Incinerator	328.00	0.012	7.36	0.149	3.6

prevalence of smokers could be resident in more traffic-polluted areas; finally, the effect of one source could be confounded by other neighbouring sources, and this is the case for the shipyard and foundry with regard to each other and the city centre.

20.5 STRATIFIED ANALYSIS

A generalisation of the Cochran–Mantel–Haenszel test can be used to test the null hypothesis against the alternative hypothesis H_1 conditional on a confounder with M levels by the cumulative chi-square or the maximum chi-square tests. With aggregate data, a typical confounder in geographical analysis is the deprivation score which is a derived variable obtained by a combination of attained educational level, social class of appartenance and job. In general, using individual questionnaires, smoking behaviour should also be used as a major confounder for many environmental exposures.

The test statistic is derived simply by rewriting each Pearson chi-squared statistic for the 2×2 contingency table obtained by grouping the circular bands from the first to the kth and from the $(k+1)$th to the Kth, like the following chi-square from an $M \times 2 \times 2$ table:

$${}_{\mathrm{MH}}\bar{X}_k^2 = \frac{\left(\sum_m^M n_{11mk} - \sum_m^M n_{11mk}^*\right)^2}{\sum_m^M \mathrm{var}(n_{11mk})},$$

where $n_{11\mathrm{mk}} = Z_{\mathrm{mk}}, n_{11\mathrm{mk}}^* = R_{1\mathrm{mk}} d_{\mathrm{mk}}$ and $\mathrm{var}(n_{11\mathrm{mk}})$ is the variance of $n_{11\mathrm{mk}}$ conditional on the group totals (Cochran, 1954) or on the group and the response totals in the mth table (Mantel and Haenszel, 1959).

The stratified cumulative chi-square and the stratified maximum chi-square become, respectively:

$${}_{\mathrm{MH}}T_{\mathrm{L}} = \sum_k^{K-1} {}_{\mathrm{MH}}\bar{X}_k^2,$$

$${}_{\mathrm{MH}}T_{\mathrm{M}} = \max_{1<k<K-1} {}_{\mathrm{MH}}\bar{X}_k^2.$$

No adjusted test statistics have been proposed for the Stone or the Cuzick–Edwards tests.

20.6 LOGISTIC REGRESSION ANALYSIS

A different approach to evaluating the risk gradient from source consists in specifying a parametric function for $\mu(\boldsymbol{x} - \boldsymbol{x}_0; \theta)$. Given that using a case–control sampling design the *odds ratio* is estimable and that the case intensity is expressed by

$$\lambda_{\mathrm{CS}}(\boldsymbol{x}) = kc\lambda_{\mathrm{CN}}(\boldsymbol{x}) \frac{\mu(\boldsymbol{x} - \boldsymbol{x}_0; \theta)}{1 - \mu(\boldsymbol{x} - \boldsymbol{x}_0; \theta)},$$

a logistic regression model can be defined in terms of the odds of having the disease being resident at distance $d = (\boldsymbol{x} - \boldsymbol{x}_0)$ from the source:

$$\text{odds}[\Pr(Y_i = 1)] = \frac{\mu(\boldsymbol{x} - \boldsymbol{x}_0; \theta)}{1 - \mu(\boldsymbol{x} - \boldsymbol{x}_0; \theta)} = w[1 + f(\boldsymbol{x} - \boldsymbol{x}_0; \theta)],$$

where an additive scale for the relative risk is assumed (w is a proportionality factor and $f(\cdot)$ is a function that specifies the risk decay by distance). The additive scale is plausible because it provides that at infinite distance from the source the risk is unchanged, with choices of $f(\cdot)$ such that $\lim_{d \to +\infty} f(d) = 0$ (Diggle, 1990; Lawson, 1993b).

Individual risk factors can be introduced as multiplicative terms in the model

$$\text{odds}\left[\mu(\boldsymbol{x} - \boldsymbol{x}_0; \theta)\right] = w \prod_j \exp(z_j \gamma_j)[1 + f(\boldsymbol{x} - \boldsymbol{x}_0; \theta)]$$

and multiple sources could be accommodated in the following way:

$$\text{odds}\left[\mu(\boldsymbol{x} - \boldsymbol{x}_0; z; \theta, \gamma)\right] = w \prod_j \exp(z_j \gamma_j) \left[1 + \sum_s f(\boldsymbol{x} - \boldsymbol{x}_{0s}; \theta)\right],$$

where s denotes the sth source and γ_j the log odds ratio for the jth risk factor z_j. These models are known as mixed additive-multiplicative models for excess relative risk (Moolgavkar and Venzon, 1987).

Several parameterisations have been proposed for the function $f(\cdot)$ (see Diggle, 1990; Diggle *et al.*, 1997; Lawson, 1993b). The simplest choice is the exponential decay:

$$f(\boldsymbol{x} - \boldsymbol{x}_{0s}; \theta) = \alpha_s \exp(\beta_s d_s),$$

where the parameter α_s models the excess relative risk at the source location, d_s is the distance (in meters) from the sth source and the parameter β_s (being negative in sign) models the exponential decay of the excess relative risk for longer distances. For a given source, specific terms are added to the model to allow for directional effects:

$$f(\boldsymbol{x} - \boldsymbol{x}_0; \theta) = \alpha \exp(\beta_1 d + \beta_2 \sin(\vartheta) + \beta_3 \cos(\vartheta)),$$

where d is the distance and ϑ is the angle between the case or control location and the source location (Lawson, 1993b).

More sensible patterns of risk decay than the exponential decrease from the source have been suggested in the literature: Diggle and Rowlingson (1994) used d_s^2 and Lawson (1993b) proposed adding $\log(d_s)$ to the regressors. Moreover, a plateau of maximal risk at the source location and its proximity can be modelled introducing an additional parameter in the exponential function (Diggle *et al.*, 1997).

The logistic model described can be considered a semi-parametric model: the population intensity has not been estimated and only the locations of the controls are used to compute conditional probabilities to be a case of disease given a subject residence, while the risk gradient by location has been completely parameterised. The existence of an excess risk close to the source is assessed by means of score tests or likelihood ratio tests on the parameters of the risk gradient function. However, the parametric (on distance) additive-multiplicative models could present serious problems of convergence with some datasets and, moreover, the estimated standard errors can be unreliable (Diggle

Table 20.2 Excess risk of lung cancer as a function of distance from the city centre and from the shipyard, iron foundry and incinerator

Source	Models with one source at a time				Models with centre and one other source			
	α	β	Likelihood Ratio	p-value	α	β	Likelihood Ratio	p-value
City centre	2.2	– 0.015	7.4	0.02				
Shipyard	2.0	– 0.019	7.9	0.02	1.2	– 0.022	1.1	0.58
Foundry	1.7	– 0.017	5.3	0.07	5.9	– 0.161	4.9	0.09
Incinerator	1.5	– 0.015	4.7	0.09	6.7	– 0.176	9.2	0.01

et al., 1997, suggested using Monte Carlo standard errors). Therefore the analyst should be careful and should try to maintain the model as simple as possible, balancing over-parameterisation with sensibility.

Non-parametric modelling via generalised additive models (Hastie and Tibshirani, 1990) can be a useful alternative to describe the pattern of disease risk as a function of the distance from the source. The main problem here consists in the lack of the constraint that the excess risk be zero at infinite distance from the source, so that it could be estimated negative. A second problem could arise when analysing multiple correlated sources since the estimated non-parametric function for one source could absorb the effect of another, resulting in a very fuzzy picture. Currently no application of such methods to this context have appeared in the literature.

20.6.1 Example

Smoking habits, occupational exposure and air particulate deposition levels have been considered in a multiple logistic regression model on the Trieste data. The results are presented in Table 20.2. Among the confounders considered, the odds ratio estimates for particulate deposition levels are close to the null value and not statistically significative, since the distance from the sources captures all the information on the risk gradients by sources. Smoking and occupational exposures, however, are highly statistically significant. Among the sources, the most important contribution comes from the city centre, followed by the incinerator. The other sources are not statistically significant. The crude analysis has been only partially confirmed after adjustment for confounders and the inclusion of other sources. The sources appeared to be highly correlated and the geography of the city is heavily affected by its proximity to the coast and the uneven distribution of the population. For these reasons models with multiple sources are difficult to fit and we adopted a forward selection strategy. The selected model included terms for the city centre and the incinerator only. It is noteworthy that the effect of the incinerator emerged only after the city centre had been taken into account.

This geographical analysis supported and validated the results obtained using the historical measurement of air pollution and provided a more sensitive approach to risk modelling around putative sources. The case–control design allowed the adjustment for relevant individual risk factors (smoking habits and occupational exposures) as well as ecological variables (air particulate deposition) and provided efficient estimates of the excess risk gradients around putative sources. Finally, when interpreting the results from this kind of analysis, it is important not to forget the spatial definition of the determinant under study, which means that we have only a proxy of individual exposure, and the uncertainties and weaknesses in using multiple sources, which being close each other in a peculiar pattern could result in highly unstable estimated coefficients. The interested reader is referred to Biggeri *et al.* (1996) for a detailed discussion of this example.

20.7 CONCLUSIONS

We have reviewed the analysis of risk gradients as a function of distance from point sources by an individual case–control study. This kind of study can be viewed as a sort of mixed design, part based on individual information and part based on ecological measurement, distance of the residence from sources. In this sense it is superior to purely ecological studies based on aggregate data. On the other hand, this design is subject to all the biases which could affect the standard case–control study (Rothman and Greenland, 1998) plus those affecting the correctness of the location definition and assessment. Theoretically the relevant location should be that at which the exposure could be experienced by the subject. Examples of incorrect location definition are, for the cases, the residence at death (see the location of houses for old people in Gardner and Winter, 1984a); and for the controls, the more recent residence (since people could have moved away from a suspected source of pollution; see also Ross and Davis, 1990). Examples of imperfect assessment of location are the use of low definition geographical maps or incomplete recovery of residential histories (Diggle and Elliott, 1995).

The definite merit of this sort of investigation stands on the ability to model the risk gradients while adjusting for relevant individual risk factors in all the instances where historical data on exposure are difficult to obtain or unreliable or we want to develop an adequate valid environmental sampling strategy.

21

Lung Cancer Near Point Emission Sources

Göran Pershagen

Karolinska Institute

21.1 INTRODUCTION

Emissions from certain industries and power stations as well as from motor vehicles contain carcinogenic substances and other agents which may affect the cancer induction process, such as airway irritants (Pershagen, 1990). Heavy exposure to the same factors in the workplace have often resulted in increased risks of lung cancer. Although exposure levels are normally much lower in ambient air and extrapolation of risks from high to low doses is uncertain there has been considerable concern regarding cancer risks for the general population. Even low excess risks may be of importance from a public health point of view if large population groups are exposed.

Several descriptive studies have shown increased lung cancer rates in urban and industrialised areas (Katsouyanni and Pershagen, 1997). The design has often been ecological without detailed information on air pollution exposure and important risk factors. A number of epidemiological investigations have also been performed based on cohort and case–control methodology to assess lung cancer risks related to ambient air pollution. Information was generally included on tobacco smoking for each study subject as well as data on some other risk factors, such as occupational exposure. Unfortunately, the information on air pollution exposure was often poor, and based primarily on recent measurements, although earlier exposures may have had great relevance for lung cancer risk.

The purpose of this chapter is to assess the evidence regarding ambient air pollution and lung cancer, with particular emphasis on the methodology for estimating exposure and data analysis. Primarily case–control studies will be discussed which have often provided the most extensive and detailed schemes for assessing exposure. Lung cancer near point emission sources in focused, for example, near industries with high emissions of carcinogens, but studies on risks related to urban air pollution are also taken up since the methodological implications are similar. Finally, some recommendations

Disease Mapping and Risk Assessment for Public Health. Edited by A.B. Lawson *et al.*
© 1999 John Wiley & Sons Ltd.

are given for improving the quality of future epidemiological studies on ambient air pollution and lung cancer.

21.2 INDUSTRIAL AREAS

Ecologic studies on lung cancer have been performed in areas with industries of different types, including chemical, pesticide, petroleum, shipbuilding, steel, and transportation industries (Pershagen and Simonato, 1993). Most of the studies showed increased lung cancer risks, which did not seem to be fully explained by socio-economic factors. However, smoking habits were not controlled and neither was employment at the industries under study. For example, some investigations focused on lung cancer near iron and steel foundries and Scotland (Lloyd, 1978; Lloyd *et al*, 1985; Smith *et al*, 1987). The highest lung cancer mortality was seen in areas estimated to be most heavily exposed to emissions from the foundries. A decreasing pattern of lung cancer risk was observed with decreasing soil contamination of metals. This pattern become weaker when adjustment was made for socio-economic factors.

Several studies have been carried out in areas near copper, lead, or zinc smelters (Pershagen and Simonato, 1993). The emissions from the smelters are quite complex, but inorganic arsenic is often a major component. The studies come from four countries (Canada, China, Sweden and the United States) and are of ecologic or case–control design. Five ecological studies showed increased lung cancer rates among men living in areas near non-ferrous smelters with relative risks ranging from about 1.2 to over 2. Two case–control studies showed relative risks of 1.6 and 2.0 for men living near the smelters after adjustment for occupation and smoking (Brown *et al*, 1984, Pershagen; 1985). A subsequent study in one of these areas revealed that the excess risk was no longer present following a more than 98% reduction in the emissions of inorganic arsenic and other agents (Pershagen and Nyberg, 1995). This adds credibility to the hypothesis that the increased lung cancer risk observed during the early follow-up period was related to ambient air pollution.

A descriptive study in Lithuania showed a high lung cancer rate in a county with a mineral fertiliser plant where emissions of airway irritants had been substantial over several years, particularly of sulphuric acid (Gurevicius, 1987). A subsequent case–control study revealed that the excess risk was not related to residence in the vicinity of the plant or to occupational exposure at the plant (R. Petrauskaite, personal communication). Instead, the data indicated that the relationship between smoking and lung cancer was unexpectedly strong and that smoking was particularly prevalent in the county under study.

The assessment of exposure in the studies of lung cancer in industrial areas was often based on information regarding the last residence. Since emissions generally occurred from a point source, exposure estimation mostly used distance from this source. Unfortunately, this procedure was rarely validated against measured air concentrations. The paucity of data on levels of relevant air pollutants makes it difficult to estimate exposure–response relationships and to compare the results of different studies. Occupational exposure in the industries under study was often associated with an increased lung cancer risk and these subjects should be excluded or analysed separately if risks related to ambient air pollution are focused.

Overall, the studies suggest that emissions from some types of industries may increase lung cancer risk in the surrounding population. The evidence is strongest for non-ferrous smelters, where arsenic emissions may be of importance. In addition, increased lung cancer risks have generally been observed among persons employed in these industries, who are more heavily exposed to the same agents.

21.3 URBAN AREAS

The composition of ambient air in urban areas in quite variable and complex. Some examples of the environments under investigation in epidemiological studies include British towns and cities during the 1950s; and urban areas in Japan, China, Europe, and the United States from the 1960s to the 1980s. Emissions resulting from the use of coal and other fossil fuels for residential heating were dominating sources of pollution in some areas, while in others motor vehicles or industries were more important. The term 'urban' thus denotes a mixture of environments, which may show substantial differences, both in terms of actual exposures to various environmental pollutants and the influence of interacting or confounding factors.

The epidemiological evidence on air pollution and lung cancer has recently been reviewed by Katsouyanni and Pershagen (1997). Nine cohort studies on urban air pollution and lung cancer are available. All but one of the investigations contained information on smoking for all study subjects. The studies came from the United States (5), Sweden (2), Finland (1), and the United Kingdom (1). Smoking-adjusted relative risks for lung cancer in urban areas were generally of the order of 1.5 or lower in those cohort studies reporting increased risks. The findings pertain mainly to smokers. For non-smokers the number of cases was generally too small for a meaningful interpretation on urban–rural differences.

In general, the exposure to air pollution for the study subject was based on place of residence at the time of entry to the cohort. Detailed air pollution measurements were provided only in the more recent studies (Mills *et al.*, 1991; Dockery *et al.*, 1993; Pope *et al.*, 1995). However, the measurements mostly included criteria air pollutants only, such as O_3, SO_2 and suspended particulates. As a rule, data on carcinogens related to fossil fuel combustion were not provided.

In 13 case–control studies reviewed by Katsouyanni and Pershagen (1997), residential and smoking histories were obtained for the study subjects and sometimes also information on potential confounding factors, such as occupation. Increased relative risks for lung cancer were observed among men in urban areas in three British studies as well as in studies from Germany, Italy, Poland, China, and Japan. Two US studies found raised lung cancer risks in urban males, while another two failed to show an effect. One study including detailed information on histological types of lung cancer noted that the excess risk related to city centre residence seemed to be confined primarily to small cell and large cell carcinomas (Barbone *et al.*, 1995). The results for women are difficult to interpret because of small numbers, but at least three studies indicate a raised lung cancer risk for females in urban areas, also among non-smokers. The magnitude of the excess relative risks for lung cancer in urban areas reported in the case–control studies was similar to that in the cohort studies.

Most of the case–control studies included some information on air pollution levels in the areas under investigation. However, this was primarily based on recent measurements of criteria air pollutants at few stations and the relevance for cancer risk is uncertain. Exposure assessment was often based on the last residence, but some studies considered residential history extending over several decades (e.g. Xu *et al.*, 1989; Katsouyanni *et al.*, 1991; Barbone *et al.*, 1995). Then, the areas of residence were ordered according to air pollution levels and a weighted average was usually taken depending on the length of stay in each area.

In conclusion, the epidemiological evidence is consistent with a modestly increased lung cancer risk caused by ambient air pollution in urban area. However, the data are difficult to interpret. Many studies were not designed to study this relationship, which has implications for the detail and quality of the exposure information. It is clear that the environments under study show great differences in the types of exposures. Emissions resulting from the use of coal and other fossil fuels for residential heating were dominant sources of pollution in some areas, while in others, motor vehicles or industries were more important. It is not possible, from the data available, to separate effects of specific air pollution components.

21.4 METHODOLOGICAL IMPLICATIONS

Lung cancer has a long induction–latency period which may extend over several decades. This creates special problems for the acquisition of high quality information on exposure to ambient air pollution and potential confounders. Ideally, longitudinal exposure data should be available covering a relevant time period for disease induction. Some studies tried to estimate concentrations of indicator pollutants in various parts of the areas under investigation, mainly based on recent measurements. The agreement between these measurements and earlier, probably more relevant, concentrations is uncertain as is the representativeness of the indicator pollutants for cancer risk. Modern methodology for estimating ambient air concentrations based on emission data and dispersion modelling was rarely used.

Few studies provided detailed information on personal exposures, intra- and inter-person variability in exposure, and the correlation of personal exposures with levels measured at fixed-site monitors. Most people in industrialised countries spend about 80%–90% of their time indoors, and the major part at home (Nitta and Maeda, 1982; Ott, 1988). The time activity pattern is mainly determined by age, gender and work status (Schwab *et al.*, 1990). For example, time spent at work contributes significantly to the activity pattern for a major part of the population. It may be necessary to consider changes in individual time activity patterns over time if long-term exposure is to be estimated. Furthermore, it should be emphasised that microenvironments contributing little to the overall pattern might still be of great importance if exposures are high.

Retrospective assessment of exposure to air pollution generally involves identification of residential addresses during several decades. Validation studies show that such information may be obtained accurately via mailed questionnaires or interviews, also from next-of-kin (Pershagen, 1985; Schoenberg *et al.*, 1990). A high mobility reduces the possibility of achieving exposure contrasts between individuals, which has implications for the study power. Swedish studies indicate an average of about six residential addresses during a lifetime (Svensson *et al.*, 1989), but the number seems to be higher in North America (Letourneau *et al.*, 1994). It has not been investigated to what extent air pollution levels correlate between the places of residence in the same individual.

Confounding factors are of great concern in evaluating relative risks of the magnitude encountered in the studies of air pollution and lung cancer, i.e. of the order of 1.5 or lower. Validation studies indicate that high quality information on smoking habits and occupations can be obtained retrospectively, even from next-of-kin (Pershagen, 1985; Damber, 1986). Data on smoking habits were available in most of the cohort and case–control studies on air pollution and lung cancer, but there may still be residual confounding from smoking when different areas are compared. Occupational exposures may also be important confounders and this was often not controlled for in earlier studies. Other potential confounders include dietary habits and domestic radon exposure. For example, negative confounding may occur because residential radon levels tend to be higher in rural than in urban areas.

The available evidence on ambient air pollution and lung cancer suggests that there may be an interaction with smoking in excess of an additive effect, although the findings are not entirely consistent (Katsouyanni and Pershagen, 1997). The results have to be interpreted with caution, and there may be bias due to both crude exposure measures and uncontrolled confounding. However, the findings are consistent with data on occupational exposures to high doses of some of the agents present in ambient air pollution. In addition, it may be expected that there are interactions between various components of the pollutant mixture in urban and industrial areas, both of a synergistic and an antagonistic nature. It is not possible to assess the effects of such interactions in detail, but they may help to explain some of the variation in results between the epidemiological studies.

To obtain sufficient contrasts in exposure it may sometimes be useful to combine data on ambient air pollution and lung cancer from different locations. Such studies have been performed, and methods are available that combine ecologic-level contrasts of air pollution effects between areas with individual-level data on covariates (Prentice and Sheppard, 1995). Such studies can assess the effect of exposure to air pollution among different types of areas while controlling confounding by smoking, diet and other factors. This type of methodology should receive greater attention in air pollution epidemiology, although further development and testing of the designs and statistical methods is desirable. In particular, many potential confounders show spatial trends and thus covary with group-level exposure when this is related to geographical location.

Epidemiological studies near point emission sources are sometimes motivated by anxiety among the population and/or to assess clusters of disease. For example, studies were initiated in the United Kingdom because of an apparent cluster of respiratory tract cancers near a waste incinerator (Elliot *et al.*, 1992). Using cancer registration data near other similar facilities it was concluded that the incidence of cancer of the larynx or lung was not related to distance from incinerators for waste solvents and oils. Confounding may be of importance in these types of analyses as information is generally lacking at an individual, and even area, level regarding important risk factors for lung cancer. Consequently, this methodology has limited usefulness for the assessment of causal effects by air pollution but may well be justified for other reasons.

21.5 CONCLUSIONS

The occurrence of lung cancer shows substantial geographical variation, and ecological analyses have provided useful hints to explain these findings. However, detailed evaluation of the role of ambient air pollution requires individual data on important risk factors for lung cancer, particularly regarding smoking, to adequately assess confounding and interactions. A major drawback in the studies has been the poor characterisation of air pollution exposure. Ideally, measurements of air pollution in the study areas should span the time period relevant for disease aetiology and preferably include concentrations of suspected carcinogens. Exposure studies are needed to provide data on how individual exposure is related to the levels measured at fixed monitors, considering different activities and mobility. Furthermore, methods for retrospective exposure assessment covering periods of several decades should be developed. This could include the use of modern methodology for estimating ambient air concentration based on emission data and dispersion modelling. Epidemiological studies should also address the consequences of quality deficiencies in exposure assessment and implications for risk estimation.

Studies of lung cancer near point emission sources may sometimes be feasible for the assessment of risks related to ambient air pollution. However, small populations with excessive exposure resulting in limited statistical power argue in favour of combining data from several sites. International collaboration may be particularly useful and such initiatives should be encouraged. For example, the World Health Organisation has an important role in promoting international contacts and collaboration in environmental epidemiology.

Part V

Public Health Applications and Case Studies

22

Environmental Epidemiology, Public Health Advocacy and Policy

Jean-François Viel

Faculty of Medicine, Besançon

Steve Wing

University of North Carolina

Wofgang Hoffmann

Breman Institute of Preventian Research and Social Medicine (BIPS)

22.1 INTRODUCTION

The health consequences of industrial development and its attendant pollution are often of great interest to the public. Therefore environmental epidemiology is frequently interpreted in a political context of debate over the desirability of specific industries, products and processes (Wing, 1998). Threatened by a challenge to their appearance of neutrality, some epidemiologists have attempted to draw a careful distinction between research and policy or advocacy, and have been very cautious about the causal interpretation of epidemiological evidence (Lanes, 1985; Last, 1996; Wynder, 1996).

Other epidemiologists, in the company of many scientists outside the biomedical arena, doubt the possibility of a scientific practice that is free of social and cultural influences (Aronowitz, 1987; Brown, 1992; Dickson, 1984; Hardings 1991; Hubbard, 1990; Latour, 1987; Levins and Lewontin, 1985). While recognising the special character of science, they view epidemiology and science in general as inherently socially grounded, and believe that epidemiologists should behave as

Disease Mapping and Risk Assessment for Public Health. Edited by A.B. Lawson *et al.*
© 1999 John Wiley & Sons Ltd.

citizens by recognising more fully their participation in and responsibility to their society.

At the same time, public awareness about potential environmental hazards is growing. This leads to an increased demand for public health authorities to investigate adverse effects of environmental pollutants on human health. Moreover, in the mass society of the late twentieth century, the putative hazards of modern life often match the key criteria of newsworthiness, and the lay press is eager to publicise epidemiological findings. Industry and environmental groups are also interested in alerting the public to findings that support their political agendas.

These are new challenges to environmental epidemiologists whose training and experience are in design, measurement and analysis. Despite their caution, shyness or even conservativeness, they are now being asked to bring compelling evidence to the policy-makers, educate the public whenever data are informative, and behave as public health advocates, all of this with 'objectivity', in the sense of fairness, justice and intellectual honesty. Are we, as epidemiologists prepared to adopt such an approach? How can we broaden our scope to cope with this situation. What roles will we play in conflicts over the health consequences of industrial policy? Some new frontiers of environmental epidemiology are sketched below.

22.2 'WASHING WHITER': A PITFALL TO AVOID

22.2.1 Sources of funding

Declines in government research funds have increased dependence on private, and especially industrial funding. This does not *a priori* challenge the personal integrity and ability of epidemiologists working directly or under contract to industry. However, it represents three potential drawbacks. First, when working on a controversial topic there is a danger that researchers may overinterpret results, consciously or unconsciously taking sides in social or institutional struggles. Although theory and prior research are recognised positive influences on science, social and political aspects of assumptions and beliefs that influence the framing of questions, conduct of analyses and interpretation of evidence have not been widely recognised as they may affect the judgement of investigators in the process of causal inference (Wing *et al.*, 1997a,b; Wynder, 1996). The second concern is that industrial research funding is a financial investment, and is decided upon in terms of corporate goals and strategies. In this respect, positive (or incriminating) epidemiological results are rarely welcome by the industries under scrutiny or even, sometimes, by trade unions (when jobs and income are threatened), and can yield publishing constraints on investigators on the basis of legal or proprietary concerns (Wynder, 1997). Positive results are then more likely to be dismissed. Third, rightly or wrongly, public opinion considers financial independence to be a token of credibility for scientists. However, government funding may not be viewed as independent, especially when there is a revolving door between jobs in government and industry. Although government funding has been viewed at times as critical of industry (Proctor, 1995), in some cases it has been considered to be indistinguishable from industry funding (Geiger *et al.*, 1992).

22.2.2 Causal inference

The determination of whether or not an association is causal is a major policy issue in observational epidemiological studies. To ascertain causality in this non-experimental context, some general criteria (among which the Bradford Hill postulates are the most well-known) (Hill, 1965) have been proposed. It is commonly accepted that no single epidemiological study is persuasive by itself and that evidence (coming from different architectures, methodologies and subject groups) must accumulate before reaching a conclusion as to causation. However, this process may require decades before consistency starts to emerge, and cannot satisfy the expectation of politicians facing citizens concerned about new environmental hazards. Worse, more evidence may be accompanied by divergence in the scientific community, as in the case of radiation epidemiology, a field in which some investigators present evidence of health effects at ambient environmental or occupational levels, while others dismiss these findings claiming a no-effect threshold in the dose–response relationship or hypothesising even beneficial effects at low-level radiation (Bond *et al.*, 1996; Cronkite and Musolino, 1996; Joiner, 1994; Luckey, 1980). Meanwhile, new agents are introduced much faster than they could ever be evaluated scientifically, while many questions raised by previous research remain of interest.

Strict postulates such as the Bradford Hill criteria have already been partly relaxed. For example, specificity seems less relevant in environmental epidemiology due to recognition of multifactorial aetiologies and associations of many diseases with environmental agents (Traven *et al.*, 1995). Although the most important criteria are a very strong association and a highly plausible biological mechanism, some epidemiologists are satisfied with a relative risk of three or more, when some biological backup is lacking (Taubes, 1995). Often, the causal question is not about whether an agent is toxic, but about effects at low levels, an area of notorious difficulty for epidemiology. In this case, the null hypothesis must be able to be rejected even if the alternative hypothesis does not deal with commonly accepted theory. It is time for environmental epidemiologists to enter the field of risk assessment, which lies at the interface between science and policy. Bearing in mind that the appropriate criteria for regulating do not necessarily match the criteria for scientific consensus on causality, new guidelines need to be found to provide recommendations for public interventions that stress wider application of principles of precaution.

22.2.3 Absence of evidence is not evidence of absence

Several factors contribute to a lack of sensitivity of studies in environmental epidemiology. Health endpoints of interest to epidemiologists are sometimes either rare, associated with long latency periods, or both. Actual population exposures are generally difficult to quantify with precision. The magnitude of the expected association between low-level exposure and outcome is usually low and one can seldom demonstrably eliminate all potential sources of bias. Some models lack power when the genuine shape of the relationship is unknown (as in point source analyses looking for an effect declining with distance). Moreover, in many countries, either an underdeveloped public health surveillance system or tough legal constraints on data linkage preclude optimal epidemiological survey design.

Hence it is easy for studies to yield inconclusive, non-statistically significant results. Many people, still unfamiliar with the non-equivalence of statistical significance, conclude in this context that 'there is no risk' whereas the right conclusion is 'no risk has been found'. Epidemiologists must question whether the absence of evidence is a valid enough justification for inaction (Altman and Bland, 1995). Otherwise they could be accused of protecting the economic health of industry at the expense of the health of the population and of the ecosystem.

22.3 A SOCIETAL CONTEXT TO TAKE INTO ACCOUNT

22.3.1 Media coverage

The news industry is vital in mediating between researchers, policy-makers and the wider public. On the one hand, media are often regarded by scientists as heralding new results while leaving out the big picture, but on the other, they are inclined to publicise their own research since it represents one of the most efficient ways to get further support. In other words, media are asked by scientists to curb their appetite for such news, whereas scientists are asked by media to curb their craving for praise. Another institution which comes into play is the scientific journal that initially reports the study under consideration. Major medical journals issue press releases prior to publication, which brings publicity to the journal, sometimes overemphasizing (according to industry or environmentalists) the forthcoming studies. This leads to the so-called 'unholy alliance' between epidemiology, the journals and the lay press (Taubes, 1995). While epidemiological results deserve a subtle and cautious interpretation (taking the scientific and social context into account), at the end of the day, when the information reaches the public mind, an isolated finding can be interpreted as a universal truth. Some epidemiologists, more cautious than others, think that the pendulum of research, swinging back and forth as successive and apparently contradictory results come out, subjects the public to undue anxieties and fears (Gori, 1995; Janerich, 1991). Purposefully holding back evidence from the public, however, is a paternalistic attitude that can coincide with the 'evidence of absence' problem noted above, and in any case denies the fact that scientific research often follows a circuitous path while being transformed into knowledge, one that is profoundly influenced by the social and political context.

22.3.2 Judicial issues

Citizen may feel excluded from participation in decisions affecting health and the environment, which are monopolised by groups that are supposedly uniquely qualified to make decisions. Civil actions represent one of the ways that health issues come to the attention of the public and policy-makers (Wing *et al.*, 1997). Environmental epidemiologists will unavoidably be involved in the lawsuits on both sides. Are they prepared to be hired as consultants (sometimes with high fees) and to tear each other to pieces in a court? Self-awareness and careful reflection about the private and public interests that are represented in these conflicts can help epidemiologists avoid professional activities, including conducting studies that are insensitive or that side-step important policy questions, that are ultimately detrimental to public health interests.

Conversely, results regarded as 'unpleasant' by industrial owners can make them (or the government authorities) take legal action to obtain compensation for alleged damages to their public image or profits. In their opinion, researchers should be held accountable for these side-effects. The goal is obviously to put pressure on researchers to be even more cautious about the interpretation of results and to avoid publicity.

22.3.3 Policy decisions

Governments may be motivated to do little about public health issues, especially when coping with chronic diseases, which (almost by definition) can be put off until another day. But they tend to adopt new policies when a climate of public readiness is reached, using the principle that governments should not move far from what is perceived to be public opinion (Chapman, 1994). At this stage politicians need urgent answers to complex issues and turn towards scientists. They can afford neither delay nor uncertainty, whereas epidemiologists require time and funds to address issues, and then may offer only probabilistic conclusions. It is easier for politicians to call for short-term measures that do not threaten powerful constituencies, whereas the social responsibility of the professional is to recommend longer-term preventive interventions that benefit the general public and result in positive side-effects. Politicians and scientists run the great risk of an increasing mutual incomprehension. They should at least try to acknowledge, if not understand, their respective margins of action and frames of thoughts.

22.4 CONCLUSION

Academic epidemiology is said to have failed since the ecology of human health has not been addressed and the societal context in which disease occurs has been either disregarded or deliberately abstracted (Shy, 1997). Some public health workers feel that the *status quo* is no longer acceptable and that environmental epidemiologists should take lesson from the past. On the razor's edge between industry and community activists, they are operating in a rapidly evolving society. They should demonstrate a humility about the scientific research process and unrelenting commitment to playing a supportive role in larger efforts to improve public health. Citizenship and environmental equity are of primary concern (Wing, 1998). However, epidemiologists should not be left carrying alone this heavy burden at the expense of their professional and personal life. Public support is warranted by epidemiology's vital role in shaping public health policy and practice. It is now time to enter a new societal contract between citizens, politicians, and scientists, one that is based on principles of innovative science and social justice.

ACKNOWLEDGEMENTS

Steve Wing's effort was supported in part by grant R25 ES 08206-01 from the National Institute of Environmental Health Sciences under the program entitled 'Environmental Justice: Partnerships for Communication'.

23

The Character and the Public Health Implications of Ecological Analyses

Olav Axelson

Linköping University

23.1 INTRODUCTION

Many disorders and their determinants have been considered in a large number of epidemiological studies, especially during the past two decades. The character of the outcome is often used for classifying the studies as belonging to different fields of epidemiology, such as cancer-, cardiovascular- or neuro-epidemiology. An alternative view may be based on the kind of exposures or risk factors studied, for example life-style characteristics like smoking and drinking or dietary factors or long-term medication; other studies are concerned with occupational exposures. The health effects of widespread exposures to chemical or physical agents in the general environment have become an increasingly important sector of epidemiological research, providing results which, in principle at least, have far-reaching consequences for society.

Studies involving environmental exposures may be thought of as truly ecological in character as opposed to other fields of epidemiology. Commonly, however, an ecological study is taken as an epidemiological study design correlating environmental or other exposure data with disease rates on an aggregated level, that is, without information on the individuals. With such a view, 'ecological studies' or analyses may cover almost any determinants of occurrence of disease or mortality; some arbitrary examples of such designs can be drawn both from life-style and nutritional epidemiology (Colhoun *et al.* 1997, Sasaki, 1993; Norstrom, 1989). Other 'ecological studies' are merely descriptive as providing disease or mortality rates for circumscribed geographical areas, usually defined for administrative purposes. Some other term for this kind of study design would be preferable, however, since the word 'ecological' inherently and naturally relates to the character of human exposures rather than to a specific epidemiological study design.

Disease Mapping and Risk Assessment for Public Health. Edited by A.B. Lawson *et al.*
© 1999 John Wiley & Sons Ltd.

Ecological analyses or studies should perhaps also be thought of as a broader term than just corresponding to environmental epidemiology, and encompassing not only investigations of health outcomes in terms of registered disease or mortality rates but also to sometimes include only exposure aspects, that is, the occurrence and spreading of agents that are deleterious to human health. To the extent that the health consequences of a certain exposure are known, its distribution and spreading characteristics might be of greater interest to study than any final and often late-coming health effects. Nevertheless, in what follows the emphasis will be on the implications of studies directly concerned with the assessment of the occurrence of disease or deaths, essentially in relation to physical and chemical exposures in the general environment.

23.2 SOME REMARKS ON THE ASSESSMENT OF ECOLOGICAL EXPOSURES

As a background for considering the implications of ecologically oriented studies to public health, there may first be reason to consider some general characteristics of such investigations. A key issue in this context is whether or not any specific kind of exposure can be identified as likely affecting the disease or mortality rates of any studied area. To the extent that some disease mapping or cluster identification precedes the perception of any causative exposure, the value of such epidemiological efforts tends to be low from a public health point of view, especially when it comes to prevention. To the contrary, and as in epidemiology at large, a study leading to confirmation of a preformed hypothesis of a health risk from some particular exposure would attract the greater attention and possibly result in some preventive measures.

However, since the exposures usually involved in ecological analyses are more or less widespread, it is important to note the difficulties in achieving proper contrasts of exposure; most people, if not everybody, may have some degree of exposure. This is rather much in contrast to the study situation in, for example, occupational health risk studies or in pharmacoepidemiology, where totally unexposed comparison groups are available. Similar difficulties obtain in creating exposure contrasts in studies regarding nutritional factors. The geographical aspect involved in most, if not all, ecological studies implies some limitations regarding the specific issues that can be considered. It may be possible to study the health effects of air pollution and possibly also of water pollution by means of aggregated population data, whereas food items are widely transported and the possible effects of contaminated food can hardly be assessed without details on individual exposures. Furthermore, the mobility of population groups in modern Western societies tends to effectively attenuate any differences in disease rates that could be due to widespread exposures.

As a consequence of the difficulties in creating definite and strong exposure contrasts in most ecological studies, the effects tend to be relatively weak and difficult to distinguish clearly. This problem is reinforced by the fact that many, if not all, ecological exposures are of low grade in comparison with, for example, occupational exposures or life-style-related heavy exposures like smoking. However, quite large populations are often exposed to environmental risk factors, so the number of attributable cases may be considerable in spite of relatively low exposure levels and the correspondingly low risk estimates in epidemiological evaluations. Typically, the relative risks found in ecological epidemiology are of the order of 1.5 or lower.

A somewhat different situation, providing for sharper contrasts in exposure, may sometimes be at hand, however, for example when some exposure of the general population takes place around a polluting factory or a waste dump site. The risks may then be more pronounced and clearly distinguishable, but for a limited group of exposed people.

23.3 COMMENT ON STUDY DESIGNS

As implicitly indicated already, ecological analyses in a broader sense may involve not only studies based on aggregated data but rather the whole range of aetiologically oriented study designs that are available in epidemiology. Although not analytical in character the results of merely descriptive studies may sometimes be taken into account from the public health point of view, for example reports of changes in the annual incidence of some disease over a period of time or comparisons of disease rates between some administrative areas with certain characteristics. Many cancer registries in the world provide such descriptive information, but no specific guidance is obtained regarding what factors may determine any observed changes in incidence. Nevertheless, comparisons of disease rates between countries or regions and between native and emigrant groups suggest that environmental and also life-style factors are likely to play an important role for human health. The exposure patterns involved in such comparisons are too complex and diffuse, however, to provide any basis for preventive measures, but the reports may suggest a need for more detailed studies.

The effects of preventive measures taken against widespread risk factors, such as smoking and the risk of lung cancer, might be reflected in descriptive studies, but any findings of changing disease rates would nevertheless be too unspecific to permit any pertinent conclusions from the public health point of view. For example, since occupational exposures may play a great role in lung cancer, although in combination with smoking (Kjuus *et al.*, 1986; Kvåle *et al.*, 1986), any regional or national changes in lung cancer rates, both among men and women, may reflect not only shifting smoking habits but also disappearing (or sometimes perhaps even new) occupational exposures as well since women are now tending to take on work tasks that traditionally belonged to men.

The more explicit and useful information is obtained from specific and exposure-oriented studies. It is not uncommon that the starting point for ecologically oriented epidemiology is a report or some other indication that a more or less widespread exposure has taken place to an agent already identified as a health hazard, usually in an industrial context, or by animal experiments. A well conducted study based on individually assessed exposure would then be the preferred choice. A good example in this respect can be found in a study of the lung cancer risk in the general population due to arsenic emission from a copper smelter (Pershagen, 1985). In this study, it could be taken into account that highly exposed workers lived in the vicinity of the copper smelter, and also that there were miners in the area who had an increased risk of lung cancer due to radon exposure in the mines; that is, these occupational groups were separated into their own categories. Obviously, circumstances of this kind can hardly be properly handled in a study based on aggregated data. However, there may also be suspicions or observations regarding human health effects that are primarily of an ecological character, for example regarding the adverse impact of air pollution (US Environmental Protection Agency, 1996).

Predominantly, ecological epidemiology deals with open populations, i.e. populations with an turnover of individuals through migration, births and deaths (Miettinen, 1985). Consequently the study designs tend to be either of a correlation type, based on aggregated data (or 'ecological studies'—cf. above) considering incidence or prevalence rates in relation to (some degrees of) exposure, or a case–control (case–referent) approach is applied. There are many examples of the use of aggregated data in correlation studies regarding some truly ecological exposures, but confidence in the results from such studies has been limited through the years (cf. Hogan *et al.*, 1979).

Cohort studies have also been conducted to elucidate the role of exposures that may be seen as ecological in character. Such studies tend to be very expensive, however, since they require follow up of a large number of individuals over a long period of time, usually prospectively. The follow-up of Japanese A-bomb survivors with regard to the effects of ionising radiation (BEIR V, 1990) and the Seveso population accidentally exposed to polychlorinated dibenzodioxins (Bertazzi *et al.*, 1993) might be mentioned as well-known examples of such long-term prospective cohort studies of an ecological character. A cross-sectional approach with a comparison of the prevalence of some disorder among exposed and unexposed may sometimes be applicable. A study indicating a diabetogenic effect from arsenic in drinking water could be mentioned as an arbitrarily chosen example in this respect (Lai *et al.*, 1994).

Considering ecological analyses in a broader sense, the spectrum of outcome parameters, which deserves consideration from a public health point of view, is clearly more encompassing than usually included in classical epidemiology. Hence, studies that consider only exposure and not disease outcome, as well as surveys of the body burden of various agents, represent an important type of such ecological studies. Efforts to study and follow trends for lead (Anonymous, 1997; Baser, 1992) or cadmium in blood (Pocock *et al.*, 1998; Chia *et al.*, 1994) or organochlorine compounds in human breast milk (Albers *et al.*, 1996; Chikuni *et al.*, 1997; Schlaud *et al.*, 1995) may serve as examples in this respect, but also health concerns about ozone depletion and the greenhouse warming effect (Last, 1993) may be taken as stemming from exposure-related ecological analyses with public health implications.

In a more narrow perspective, however, there are also ecological analyses that include some subclinical effects as an outcome parameter. This aspect may be illustrated by mentioning studies of early effects on kidney function due to environmental exposure to cadmium (Staessen and Lauwerys, 1993; Järup *et al.*, 1995) or DNA-adducts of polyaromatic hydrocarbons due to heavy air pollution (Motykiewicz *et al.*, 1996).

23.4 CREDIBILITY OF ECOLOGICAL ANALYSES

The results obtained in ecologically oriented epidemiology as well as in other epidemiological studies are subject to consideration on several levels and in different contexts. First, there is the scientific level, including the possibility of getting results published, and involving essentially methodological scrutiny, but also critiques based on the scientific perceptions at the time when the study is presented. Then there are the reactions that may or may not follow from the regulating agencies in a country. Still other contexts are represented by discussions in the mass media and by the attitudes of the general public, which often strongly influence the health policy decisions finally taken by the authorities.

Whereas some of the scientific critique of a study may obtain in published form, there are also many rumours that stem from oral statements by scientifically involved persons. The latter kind of critique is not documented and therefore is difficult to describe in a proper way with pertinent references, buy may nevertheless strongly influence the final public health implications of a study. Similarly, the more or less political considerations by various agencies are usually even more difficult to account for because of the lack of any published material regarding why and how decisions were taken, as are the notions of the general public regarding health problems of an ecological character. Nevertheless, a variety of such poorly documented viewpoints might have a profound impact on the implications for public health of results from ecological analyses of different kinds. As a consequence, the comments given here can in essence only reflect the author's perception of how the results of ecological studies have been met in the past and are likely to be received in the future, coloured also by experiences of how some discussions have developed regarding other findings of an epidemiological character (Axelson, 1994).

23.5 SCIENTIFIC ASPECTS

The credibility of the results from any single epidemiological study tends to be relatively low, but so also are the results of any scientific observation. Repeatability is therefore an important and integral part of establishing a scientific concept and it is also a prerequisite for establishing causality, which in turn is the proper basis of preventive measures. Even so, there are often different views declared by different researchers regarding the interpretation of available data. In the ecological domain an illustrative example may be drawn from the disagreements around the acute respiratory effect of particulate air pollution (Dockery and Pope, 1994; Moolgavkar and Luebeck, 1996; Vedal, 1997).

Ecological studies with aggregated data often encompass large regions, or even nations. As a consequence there tend to be difficulties for any national repetition of the results from such a study. Corroborating results from several countries are therefore usually required before any more definite conclusions can be reached based on such studies. Should the results from different countries be inconsistent, the interpretation becomes problematic. Indications of a health risk in some of the studies may then be taken as a chance finding or suggested to be due to some confounding effect, for example from life-style and smoking, when considering particulate air pollution and mortality (Moolgavkar and Luebeck, 1996). Furthermore, but difficult to document, it seems common that studies showing no adverse health effects tend to be more readily accepted and attract little criticism even when clearly uninformative or biased (Ahlbom *et al.*, 1990).

Since studies involving aggregated data have an inherently poor assessment of exposure, they should be expected to not necessarily show any effect even when it is present in some subgroup of the population. Non-positive results in such studies can therefore hardly rule out an adverse role of an exposure if shown in properly conducted case–

control or cohort studies. Nor is it usually possible to evaluate the impact of modifying factors in studies based on aggregated data, although such factors may well explain the differences in results between studies; this aspect might be difficult to comprehend, however, even when individual data are at hand. Unless the exposure is properly quantified, there may also be considerable but not immediately apparent differences between studies with regard to the actual degree of exposure, especially if assessed in such terms as high, intermediate and low rather than described in measurable terms.

The complexity that can be involved in ecological analyses may be further illustrated by an example regarding residential radon exposure and lung cancer. Hence, judging from a large Swedish study (Pershagen *et al.*, 1994) the risk seems to be driven by a modifying effect from smoking, i.e. there was only a weak effect among the non-smokers over categories of exposure in terms of (time-weighted) radon gas concentrations, whereas the interaction effect from smoking was quite strong. At the same time, there was negative confounding from smoking. These features made it necessary to have individual exposure data to in any way be able to reveal a lung cancer risk from indoor radon. Furthermore, and as also further complicating the analyses, there was virtually no increase in risk among people who slept with a slightly open window.

The likely reason for this latter finding is that ventilation decreased the exposure, not only to radon gas in relation to what was measured at the time of the study, but especially to radon progeny as causing irradiation of the bronchi. The dosimetry in studies of indoor radon exposure and lung cancer is further complicated because ventilation also tends to reduce the particles in the air of a room. Thus, the relative fraction of so-called unattached progeny will increase, and this fraction is believed to be more efficient in causing lung cancer than the progeny attached to particles. The net effect of these phenomena could perhaps explain the discrepancies in results from the various studies that are available so far (Axelson, 1995; Lubin and Boice, 1997) and which might appear in the future.

The example regarding studies of indoor radon may be seen as complicated, but in contrast to many other study situations the assessment of exposure to indoor radon is fairly straight-forward, since exact measurements of current radon concentrations are possible. However, the current levels are not necessarily representative of those of the past. Most other exposures of interest in ecological analyses are much more difficult to characterise and assess over time than radon, however, a fact hampering the credibility of studies of any such exposures. The discrepant results regarding the health effects of exposures of this kind seem to imply that further studies, preferably in the own country, are awaited before preventive actions are taken and therefore the public health impact of any particular study tends to be low, especially when conducted in another country. A more rational attitude could be to only assess the exposure situation in a country or region and to rely on existing data on the health effects of the exposure at issue. Obviously the direct documentation of casualties in a country leads to more proper public health considerations and preventive actions.

Another aspect regarding the reproducibility and credibility of study results has to do with the lapse of time between early and later studies as potentially influencing exposure levels. Hence, early studies may have created some concern leading to precautions and, subsequently, also to decreased exposure levels even if not always officially admitted or otherwise recognised. Later studies may therefore be conducted under different exposure conditions than the first ones of its kind and therefore be unable to reproduce any pertinent increase in risk. Circumstances of this kind may easily lead to the conclusion that there might not be any risk present from some particular exposure. Although such a development already implies a kind of prevention, a particular exposure may no longer attract interest, even if widespread, and the long-term efforts to reduce a risk may come to an end.

Preventive measures can be expensive and a somewhat conservative attitude is clearly often justified to avoid expensive but, as it may turn out, unnecessary remedies. Sometimes no action is taken by responsible authorities because of the fear that information to the general public might lead to unnecessary anxiety. Such a concern certainly represents a most conservative attitude and reduces the public health impact of any epidemiological findings, ecological or otherwise. Adequate risk communication and perception may not be easily achieved because some people may react too strongly, whereas others do not care enough (Fischhoff *et al.*, 1993). Some measures, usually eliminative in character, are often quite simple and inexpensive, however, and should therefore be possible to implement even on a relatively vague indication of a health risk, especially as remedial actions often help to reduce anxiety among those people who are most concerned.

23.6 SOCIETAL ATTITUDES

The ultimate implications for public health from ecological studies depend on how the results are accepted by the community and the regulating or legislating authorities. The problem of credibility is a central issue in this respect and is strongly related to reproducible results. This is not specific to ecological analyses, however, although the results in this respect might be met with more reluctance from regulating agencies or industry than, for example, findings in epidemiological studies on life-style factors. The reason for this seems to be that society, or some industry or industrial branch, might have to take responsibility for environmental risk factors, whereas it might be more convenient to think of the individual as responsible for risk factors associated with life-style.

There is also another related complexity involved regarding the public health implications of epidemiological results regarding widespread exposures. Part of this is a tendency to apply a sort of comparison of risks as an excuse for not taking action. Often the risk of smoking is taken as a 'reference risk' in such comparisons and it is suggested that any particular risk is limited, if not negligible, in comparison with smoking. Such arguments may be heard regarding widespread exposures causing a relatively low individual risk, but rather many cases. On the other hand, there is sometimes little concern when it comes to high individual risks affecting small groups and therefore causing few cases, usually due to some occupational rather than an environmental exposure.

There are also contradictory attitudes involved in the appreciation of risks. Hence, the general public, or rather, particular subpopulations, tend to show greater concern about environmental risk factors than about those inherent in life-style. For example, it can be noted how people might be quite concerned about the health effects from pesticide residues in food but less so about their own smoking. The principle seems to be that health risks posed by others are seen as much more threatening than those relating to the individual's own behaviour. Similarly, a health risk, the elimination of which would lead to some costs to the individual, usually tends to be relatively ignored. Radiation risk is a good example in this respect, as there is much concern about radiation in general, e.g. around nuclear power plants, whereas residential radon is rather more neglected by home owners, perhaps with the exception of families with small children (Jansson *et al.*, 1989; Fischhoff *et al.*, 1993). The situation is somewhat similar regarding exposure to electromagnetic fields around power lines or at workplaces in comparison with the fields existing around all sorts of electrical equipment in homes. It may well be, however, that there are fairly consistent views within various subpopulations and that the overall impression of the debate is false as representing a mix of all attitudes taken by different interest groups. Nevertheless, this multitude of viewpoints tends to influence the overall public health implications of any epidemiological study results as to what actions are finally taken by society.

However, the seemingly irrational attitudes that sometimes can be seen with regard to various risk factors, ecological or otherwise, are not necessarily an undesirable phenomenon, but sometimes perhaps even the reverse, forcing prevention in some respects. Many health hazards of an ecological character can be reduced or eliminated through information, regulation or legislation, whereas hazardous life-style factors seem surprisingly difficult to eliminate. This view is certainly a counter-argument to the above notion that weaker risk factors may attract little attention as long as strong risk factors, like smoking, are not effectively eliminated. Some marginal preventive successes regarding rather weak ecological factors, are clearly better than little or no preventive achievement for some stronger, usually life-style-related, factors.

Another aspect relating to attitudes in society, and with some bearing on ecological analyses, has to do with the maintenance of already achieved preventive progress. Some regulations might be relaxed in times of economic problems because the preventive activities are no longer recognised as such and because the epidemiological (or other) foundations for the regulations tend to be forgotten. There are also problems in carrying out epidemiological studies when an adverse exposure has been reduced or has more or less disappeared. Positive study results from the latter part of this century may therefore fade with time, a development that might return later as a threat to human health; for the time being, this is not a major problem as the development of epidemiology and the provision of results is a fairly recent phenomenon. Water distribution might perhaps be seen as an example in this respect, because this is probably seen by many people in the developed countries as first and foremost a technical issue rather than a health preventive measure. The health consequences of relaxing standards may therefore not be immediately and clearly perceived; there are still hygiene problems with drinking water from time to time, even in developed countries.

23.7 CONCLUSIONS

There are different viewpoints from which to consider the public health implications of ecological analyses. Such analyses may include a variety of investigations on the occurrence of risk factors as well as clear-cut epidemiological studies associating disease with specific agents. The design issues are critical, as studies of widespread risk factors usually come up with fairly low risk estimates. The reason for the low risk estimates may be sought in both low grade exposure levels and in the problem of obtaining good exposure contrasts; this latter problem may increase when a very large part of a population is exposed to a widespread agent and to a fairly similar degree. As a consequence, also the influence of confounding could be relatively stronger when the risk estimates are low, causing credibility problems for the results obtained.

These inherent problems in ecological analyses may explain why the implications for public health often turnout to be limited, especially perhaps when the studies have been based on aggregated data. Most important for the ultimate implications of ecological studies are probably the more or less rational attitudes taken by society, sometimes influenced by the mass media, but also economic interests strongly influence what actions are finally taken. The low risk estimates usually found imply that the health concerns are more at the community level than representing a threat to the individual, which in turn reduces interest at the political level for taking actions and often leaves the responsible authorities with rather difficult decisions, especially if there is some inconsistency in the epidemiological studies.

24

Computer Geographic Analysis: A Commentary on its Use and Misuse in Public Health

Raymond Richard Neutra

California Department of Health Services

24.1 INTRODUCTION

The term 'geographic analysis' means different things to different people. For some the term denotes a statistical search for causes, which pays attention to geographic adjacency or connectivity. This was the primary way this term was used during the World Health Organisation meeting in Rome upon which this volume is based. The main focus of the meeting was on statistical tools for smoothing of maps to facilitate inferences about geographically located causes, tools to detect clusters of disease and tools for improving the quality of ecological studies to disease associations. As outlined below, I would include as well cartographic and statistical procedures with a descriptive intention as long as they pay attention to location, adjacency and connectivity of one place to another.

In particular, I believe that Geographic Information Systems (GIS) have many powerful and useful administrative and descriptive applications (Vine *et al.*, 1997). The developers of GIS and the developers of the statistical tools for map smoothing, cluster identification and enhanced ecological studies are optimistic about the utility of these tools for public health. Like any tools they have their appropriate and inappropriate uses (a hammer is good for nails, but not for screws). Below I lay out 13 generic analytical activities, which relate to location, adjacency or connectivity. I will argue that some will be common and useful in public health, while others will only rarely be helpful and some could be counter-productive. I group these analytical activities into the following categories:

Disease Mapping and Risk Assessment for Public Health. Edited by A.B. Lawson *et al.*
© 1999 John Wiley & Sons Ltd.

1. Hypothesis generating
2. Descriptive/administrative uses
3. Hypothesis testing

I assume that analytical activities are most frequently deployed with one or the other of these intensions, but some of the activities could occasionally be classified into more than one category. I predict that the descriptive/administrative and the hypothesis testing applications of geographic analysis will have the greatest utility, while the hypothesis generating techniques, which were the main focus of the Rome meeting, are more problematical.

Hypothesis generating

1. Mapping of smoothed or model adjusted rates and relative risks.
2. Detecting the clustering of disease rates or diseased cases at regional or neighbourhood levels of geography.
3. 'Ecological studies', i.e. studies that correlate the rate of disease in a series of populations with the prevalence of a risk factor in those populations while analytically controlling for the prevalence of other confounding factors and for spatial autocorrelation.

Descriptive/administrative uses of GIS

4. A convenient way to file geographically linked information.
5. A tool for gathering geographical facts off a map instead of in the field.
6. Mapping and counting populations adjacent to potential exposures.
7. Going beyond adjacency: estimating exposures from pollution sources.
8. Calculating optimum clinic locations and nursing routes.
9. Automatic address mapping.
10. Locating types of individuals or institutions requiring services.
11. Producing maps of areas with counts of subjects or institutions requiring service.

Hypothesis testing

12. Case–control or cohort studies that use geographic proximity to a pollution source or computer model estimates of exposure as a proxy for exposure (with or without adjustment for spatial autocorrelation).
13. Recognising a pattern of disease dispersal which, if found, would implicate a particular cause of disease (a sinuous distribution of hepatitis cases suggesting the existence of a hitherto unsuspected contaminated swale).

The plan of this chapter is to discuss the pros and cons of the various applications.

24.2 HYPOTHESIS GENERATION

In the early stages of epidemiological investigation, associations with variables of person, place and time are always examined. Place itself is powerfully associated with income, social class and life-style, but it may be associated with exposure to infectious agents, vectors of disease and chemical and physical exposures. Particularly when the disease has a short incubation period, when in- and out-migration is low, and relative risks vary greatly geographically, the consideration of place is a powerful hypothesis generating strategy.

24.2.1 Maps of smoothed and model adjusted rates and relative risks

When we want to demonstrate the effect of causal factors that cluster in adjacent areas we are tempted to adjust for other factors that might confound the pattern. Also, small-area variation may overemphasise differences, and smoothing procedures can increase the chance of seeing a true signal above the statistical noise. The eye may then see the effect of a geographically clustered pattern of life-style or physical or chemical agent.

While there are many situations in which smoothing is appropriate, we need to remember that it is inappropriate for administrative purposes where actual counts relate to workload. See the discussion below with regard to regional and neighbourhood clustering.

24.2.2 Cluster detection in regions, neighbourhoods or in parts of buildings

Cluster detection is challenging for statisticians and has taken on a life of its own. It has been used primarily to search registration districts, census tracts, counties or states to determine if rates of disease in adjacent areas are higher than expected. The same techniques have also been used to scan the distribution of individual cases and to compare this distribution with that of a sample of the underlying population. Some techniques merely detect the fact of clustering while others identify the cluster's location and assign a probability value to it. The techniques can be used at a neighbourhood or a national level of scale. In my view it is an interesting set of tools in search of a problem. What are the situations in which these tools would be helpful? First, there needs to be some cause that is operating so as to increase the incidence, duration, in-migration or prevalence of diseased cases in a particular location. Secondly, this increase must be small enough that it is not detectable by more obvious means. Thirdly, the detection of clustering would serve as a clue to the presence of a cause which we would have otherwise not suspected. If we suspected the cause, then we could simply compare rates of disease near and far from that cause. Fourthly, having detected an area where some cause is operating, we have a high likelihood of pinpointing the identity of that cause. Finally, having identified the cause we have resources and techniques to do something about it. Indeed, clustering techniques can be viewed as a kind of screening test which could lead to a definitive diagnosis (follow-up analytical epidemiology) and treatment (environmental or life-style remediation). As with any screening test we should employ it only if the overall campaign is cost effective, not just because the screening test part of the campaign is inexpensive. Below we will use this framework to examine the routine use of clustering techniques to determine the causes of cancer.

The techniques have been used widely with cancer, with few if any definitive discoveries as a result. In retrospect cancer, as opposed to infectious diseases, is the least promising of applications, because there are few general environmental exposures that would convey relative risks high enough to produce observable clustering and because the long incubation period and mobility of the population would obscure patterns left by the person-to-person spread of oncogenic viruses or the malign effect of pollution sources, even if they had been operating.

There have been some partial exceptions, however, and it is instructive to look at these. In China, simple inspection of Standard Mortality Rate disease maps (and not elaborate clustering or smoothing techniques) have identified dramatic elevated rates of nasopharyngeal cancer in one large region and female lung cancer in another. Note that the important confounders of these diseases, such as smoking, were not available for smoothing in any case. After follow-up case–control studies, the nasopharyngeal cancer has been attributed to the consumption of a kind of smoked fish (Yu, 1986), while the female lung cancer has been attributed to cooking with smoky coal in unvented huts (Mumford *et al.*, 1987). Both these practices represent widespread life-style factors in a country that used to have low mobility and great heterogeneity in these traditional practices. Such patterns can be expected in developing countries where only local foods are consumed and where traditional life-styles vary. These differences tend to be erased by the homogenising effect of modern mass markets. So searching for large-area clustering on any chronic disease in the very developed countries happen to have disease data available is like searching for a key under the lamplight where it is easy to see, even though one lost the key elsewhere in the shadows. Using these techniques for common short incubation diseases particularly in smaller areas could, on the other hand, be quite productive.

In any case, the general public tends to be more interested in neighbourhood-level cancer clusters and not in large regional variations. This is because the media has led them to believe that point sources will produce small-scale clusters of cancer or other dreaded chronic diseases. They assume that there are many such clusters just waiting to be found and, once detected, the cause will be pinpointed and dealt with.

This would be a reasonable expectation for infectious disease with short incubation periods, a good understanding of pathophysiology, ways to detect traces of the causal agent in bodily fluids or in the general environment and when almost all the cases in the cluster can be attributed to the one cause as judged by a high relative risk and a high population attributable risk percent.

These conditions almost never pertain to neighbourhood-level cancer clusters, so that even if a cluster were truly caused by one agent, it would be extremely hard to prove it *even if the agent were already known to produce cancer in other settings.* Asbestos and mesothelioma is a solitary and notable exception to this generalisation since it meets all of the above conditions (see the chapter in this book dealing with mesothelioma in Italy).

If a neighbourhood cancer cluster is due to a *hitherto unrecognised carcinogenic agent*, then the situation is even worse since we do not know where to start looking. I have argued elsewhere (Neutra, 1990) that of the hundreds of neighbourhood cancer clusters examined I could only find one that led to the discovery of a hitherto unknown carcinogen. This was the case of erionite, an asbestos-like mineral in Turkey. Previously known causes, with the exception of asbestos, have never explained an environmental cluster to my knowledge.

Should we use statistical tools to screen cancer registry data to discover regional or neighbourhood cancer clusters? *To answer this question we should answer the usual questions asked in evaluating any public health screening program:*

1. If you found a cluster, how good is your diagnostic ability to confirm that it is caused by an environmental or life-style factor?
   Answer: As we have seen above, our success rate is poor, and the few success stories involved decades of research.

2. If you confirmed the cause, what public health treatment have you to offer?
   Answer: For environmental factors there may be no practical remedy. Proving that the 9000-fold excess risk of mesothelioma in the small village of Kharain in Anatolia was due to naturally occurring erionite, simply left the villagers stigmatised, since the Turkish government could not afford to relocate them (Neutra, 1990). Knowledge without action was not helpful. This is a frequent dilema in environmental epidemiology and must always be considered.

3. Are there untoward side-effects of the screening program?
   Answer: Yes. Both for true positives like Kharain and for the inevitable false positives. One can anticipate decades of painful uncertainty, media stigmatisation of the community being studied and political pressure to devote scarce manpower to follow-up studies with a low probability of success.

4. What is the prior probability of a single cause for neighbourhood-level cancer clusters?
   Answer: Of course this is not ascertainable with certainty, but from environmental sampling in areas where clusters have been reported and in control areas, we know that the levels of carcinogenic pollution may be high enough to produce cancer of regulatory concern, but almost never reaches levels that would produce excess disease rates which would be epidemiologically detectable, particularly not in small populations. Thus, the prior probability of an environmentally caused and epidemiologically detectable cancer cluster is extremely small. Clusters of short incubation infectious diseases could be a different matter.

5. What is the probability of a false positive?
   Answer: That of course is determined by the alpha level chosen by the investigator. Since California cancer registries recognise 80 varieties of cancer which can be studied in some 5000 census tracts state wide, a probability value of 0.01 would generate 4000 false positive clusters per year. A probability value of 0.0001 would generate 40 per year. That would be nearly one per week! Each one would initiate a long painful wild goose chase.

6. What is the posterior probability of a true cluster, given that a statistical tool has determined it to be unlikely by chance?
   Answer: Once again, using Bayes Theorem, this depends on the prior probability, the true positive rate and the false positive rate. My best judgement is that the prior probability of environmentally caused cancer clusters in neighbourhoods is extremely low, maybe one or two per century. Environmental chemicals are probably causing cancer above the *de minimis* standard of one in a million life time risk, but we epidemiologists will not be able to detect this with the tools currently available to us. That means that even with a p-value of 0.0001, the vast majority of statistically significant neighbourhood cancer clusters will be false positives.

For all these reasons, and despite the fact that the statistical cluster-screening test itself is cheap, environmental epidemiologists have not dredged for cancer or chronic disease clusters as a fruitful strategy. Instead they have carried out occupational studies where doses were higher or focused in on exposures in large populations where a priori evidence suggested they had a chance of observing an effect.

What about applications to conditions of shorter incubation, such as low birth weight, foetal death or birth defects? Kharrazi *et al.* (Kharrazi, 1998) of our group recently applied a number of clustering techniques to a variety of such outcomes over a 5-year period around a hazardous waste site. Even though these outcomes were not associated with distance from the site, it was reasoned that clustering might have indicated an unanticipated route of exposure (for example, landfill gas travelling through a particular sewage line to a manhole outlet). After examining a number of outcomes some clusters were 'detected' but none related to unusual routes of exposure.

What disease outcomes would we *expect* to be caused by geographically located factors? Automobile accidents can be influenced by street design and traffic light timing. Homicides and assaults can be influenced by street lighting, the location of bars, brothels, automatic teller machines and gambling houses. Muggings may cluster near the residences of felons. Within schools and office buildings we might expect airborne infectious diseases to cluster. We could detect clusters and look for causes, or we could locate suspected causes on a map and test if cases were closer to them than expected. At the neighbourhood level we could imagine detecting several index cases of a malaria epidemic in an area where the vector was present, but the disease had been eradicated except for the occasional imported cases coming in from endemic areas. Here the techniques might help identify which secondary cases belonged to separate clusters and this would help cross reference to the location of travellers recently arrived from endemic areas of the world. Alternatively, clustering of Lyme disease might suggest areas where warnings should be posted or pesticide applied (Zeman, 1997). A clustering of the purchase of over-the-counter flu medications on a weekly basis might provide early warning of the time and location of an emerging influenza epidemic. We are using clustering techniques to determine if certain subareas convey extra risk of miscarriage in a county where a prospective study of miscarriage has shown that the consumption of tap water but not bottled water conveys excess risk (Swan *et al.*, 1998). There are undoubtedly interesting applications for these techniques but I doubt that they will ever play a major role in cancer epidemiology or the epidemiology of other diseases with long incubation periods.

24.3 ECOLOGICAL STUDIES

Next to techniques for detecting clusters, statisticians have focused effort on ways to improve the validity of ecological studies. This is probably because there is so much readily available health and environmental data and it is inexpensive to analyse. If only we could use such data for causal inference, it would be so convenient. Statisticians have devised methods to correct for the theoretical possibility of substantial spatial autocorrelation. When some unknown confounder of the association under study is associated with neighbourliness it can effect both the measure of association and the confidence limits around it. They have also tried ways to avoid the obviously false assumption that everyone in a census tract is exposed to the average of the individual exposures in that population. There was extensive discussion of these innovations in the Rome meeting which this volume documents. There is almost a tendency to equate the term 'Geographical analysis' with these 'improved' ecological study designs. But they are only one (and the least productive, in my opinion) of the study designs that have a geographical component. Epidemiologists recognise that a strong association at the individual level can often be reliably detected by ecological study designs, but the new 'improved' ecological designs will need to be validated on datasets where both individual and ecological analyses can be done. If silk purses can indeed be made from sows ears, then this will be an important and welcome development.

24.4 DESCRIPTIVE/ADMINISTRATIVE USES OF GIS AND GEOGRAPHIC ANALYSIS

24.4.1 A convenient way to file geographically linked information

Imagine the convenience of clicking on the residence of a lead-poisoned child and retrieving the age of the house, whether it was owned or rented, its assessed property value, the volume of traffic flowing on the adjacent street and the racial composition of that block of houses. GIS systems offer that kind of convenience which can make feasible, epidemiological studies previously prohibited by cost or time considerations.

24.4.2 A tool for gathering geographical facts off a map instead of in the field

Suppose we wanted to know the miles of four lane streets, or polluted river front beaches, or the square feet of grassy parks in a city. This could be found from the field, done by hand on a paper map, or easily computed with a GIS system. All these surrogates for exposure could be applied to an ecological analysis of census tracts relating disease rates to these surrogates and could be greatly facilitated with a GIS.

24.4.3 Mapping and counting populations adjacent to potential exposures

GIS makes it possible to classify each block face in a city according to traffic density. These could be colour coded on a map. We could use linked census data to add up, city wide, the number of people in each category of traffic density. By drawing a random sample of births we can estimate the population of children at risk of house paint lead poisoning in houses of different ages. We can locate the streets where water pipes and sewage pipes have not been replaced for 80 years and estimate the population at risk of hepatitis A from breakage and cross-contamination.

24.4.4 Going beyond adjacency: estimating exposures from pollution sources

This is a very powerful use of GIS technology which is very costly or impossible to achieve with other means. By linking the residential history of cases and controls we can reconstruct past exposures if we have records of things like air pollution, past traffic density, past electric current on adjacent power lines or past strength and frequency and antenna type of nearby radio frequency transmitters.

Note that such modelling can be related to present-day conditions which are open to validation or to the reconstruction of past conditions which, failing the availability of a time travel machine, could never be obtained by direct measurement.

There are a number of other powerful administrative uses of GIS in public health which I list below without further discussion:

- Calculating optimum clinic locations or nursing routes
- Automatically mapping addresses
- Locating types of individuals or institutions requiring service
- Producing maps of areas with counts of individuals or institutions requiring service

Note that 'smoothing' of such maps would be inappropriate and that counts, rather than rates, are what are wanted.

24.5 HYPOTHESIS TESTING

24.5.1 Case–control or cohort studies using adjacency, or modelled past exposures, with or without correction for spatial autocorrelation

Croen *et al.* (1997) recently published a study of several types of birth defects in which cases from all over the state were compared with a random sample of births as to being within one-quarter of a mile of any one of several hundred hazardous waste sites in the state of California. This was arbitrarily chosen as a 'close' distance from such a site. The sites were further classified as to the known contaminants there. Only 0.1% of the controls, corresponding to a few thousand at-risk births in the state, lived this close to a waste site. A handful of extra cases in this category represented a statistically significant doubling of the expected rate of one of the rare defects. The airborne exposure from these sites must have been low, but this adjacency analysis gives us no clue as to what it might be.

In San Diego, California, English, Reynolds and Scalf are proposing to compare asthma cases with randomly selected births as to the estimated traffic pollution from streets surrounding their residences. We also propose to refine estimates of past radio frequency exposure and compare this metric in childhood leukaemia cases and controls.

24.5.2 Pattern recognition

The pattern recognition capability of GIS systems has applications in geology, archaeology, and hydrology which, to my knowledge, have not been used in epidemiology.

Perhaps there is no practical application to epidemiology, but it is worth describing nonetheless.

When could the very shape and patterning of disease cases in space (as opposed to the mere existance of a cluster) suggest a cause? It would have to be a situation where the geographic location of a putative cause was not known, but the very existence of a pattern could suggest one. For example, the stringing out of cases of TB along a connected sequence of streets might point us to a heretofore unsuspected mail man. A linear sinuous distribution of hepatitis A cases might point to a hitherto unsuspected swale which was conducting contaminated sewage to the surface in a series of back yards where children played. The shape of case versus control distribution in space suggests a hypothesis capable of testing, or could be used as a test of, a general hypothesis (e.g. some delivery man is spreading TB or an unsuspected swale is spreading sewage).

The above description relates to spatial pattern recognition. We could look for temporal fluctuations in reported respiratory disease to make inferences about incubation periods and hence the identify of the offending virus. We have been using spectral pattern recognition for identifying crops in use, which is turn may be correlated with the use of certain pesticides.

24.6 CONCLUSION

There are a number of interesting uses of GIS and other geographical analyses in epidemiology. The use of GIS to determine adjacency or estimate exposure simply replaces field activities with computer map activities. Once that activity is completed, the actual analytical procedures are the conventional tabular or logistic regression procedures detecting the presence or absence of association between disease and exposure. The heuristic, clustering and pattern recognition techniques are inherently spatial. The first set of activities has many uses, the last two are, in my opinion, tools whose practical application is still to be proven. The use of ecological studies and the routine use of clustering techniques in cancer and chronic disease epidemiology should be carried out with great caution, if at all.

25

Estimating the Presence and the Degree of Heterogeneity of Disease Rates

M. Martuzzi

International Agency for Research on Cancer

M. Hills

London School of Hygiene (retired)

25.1 INTRODUCTION

Given a study area subdivided into districts, one of the questions often addressed in geographical studies in epidemiology is: Do rates of disease differ from one district to another? Or, in other words, is there *heterogeneity*? Tables or maps of the rates by district and the comparison of rates with a reference rate can be useful, but it is difficult to assess the overall departure from a null hypothesis of constant risk (*homogeneity*) in this way.

This is particularly true when studies are conducted at the small-area scale, because random variability can then be large, and estimated rates often reach extreme values that might appear significant when viewed in isolation.

Tests to assess the evidence against the null hypothesis of constant rates across the study area are available, but they mostly provide only significance levels, which can be difficult to interpret. For example, a small p-value might result from large heterogeneity between few districts (or from a small number of cases), or from a small degree of heterogeneity between many districts (or from many cases). Similarly, when a heterogeneity test yields a non-significant result, this is sometimes erroneously interpreted as supporting homogeneity, and taken to suggest that no further investigation is needed, even though such a result only expresses lack of evidence against homogeneity. Thus, it is desirable to use methods for estimating not only the presence but also the *degree* of heterogeneity.

Disease Mapping and Risk Assessment for Public Health. Edited by A.B. Lawson *et al.*
© 1999 John Wiley & Sons Ltd.

In this chapter two such methods are described. First, the case of rates of rare events such as disease incidence or mortality, following Poisson distributions, is discussed. Secondly, an equivalent method for the prevalence of non-rare conditions, where the binomial model is appropriate, is presented. The methods apply to any dataset consisting of mutually exclusive groups, such as geographical studies, where regions are subdivided in N subareas.

25.2 THE POISSON CASE

To illustrate the method for estimating the degree of heterogeneity in the occurrence of rare events, we analyse perinatal mortality in the (former) North West Thames Health Region (NWTHR), England, in the period 1986–1990. The study area includes part of Greater London, some other urban areas north-west of London, and some rural areas. The resident population at the 1981 census was around 3 million. All births and perinatal outcomes are routinely recorded, and since the post-code of residence is available, it was possible to link the events to the 515 electoral wards of the region, and obtain the number of perinatal deaths (stillbirths plus deaths occurring during the first week of life) and live and stillbirths (the denominators) per ward observed during the study period. For the analyses, the expected number of perinatal deaths in each ward was calculated using the overall sex-specific rates of the region.

There were a total of 2051 observed perinatal deaths (equal to the total number of expected cases because of the way expecteds were calculated), and the number of cases by ward ranged between 0 and 21.

We make the usual Poisson assumption that O_i, the observed number of cases in area i, is Poisson-distributed with mean $\theta_i E_i$, where E_i is the expected number of cases and θ_i is the rate ratio between the rate in the ith ward and the reference rate, adjusted for sex. The Standardised Mortality Ratio (SMR) is O_i/E_i and under the Poisson assumption this is the maximum likelihood estimate (MLE) of θ_i. In our case, however, the SMR is a very unstable estimate of the rate ratio. Its range is from 0 to 3.7, but such a spread overestimates the true heterogeneity of risk because it includes large random fluctuations. In addition, a table showing SMRs and significance tests (or even confidence intervals) might be misleading. For example, one of the wards has 17 observed cases and 6.8 expected, yielding a highly significant SMR of 2.5 ($p < 0.001$), but this could be due to the fact that the most extreme value was selected a posteriori from the entire dataset.

Following a technique used in small-area disease mapping (Clayton and Kaldor, 1987), we assume that the θ_i are drawn from a gamma distribution, with unknown mean μ and variance τ^2 (Martuzzi and Hills, 1995). Gamma distributions are unimodal and skewed to the right and as the mean and variance vary take a large variety of regular shapes. Such a family of distributions seems therefore appropriate to describe the distribution of true area-specific rate ratios. In particular, the variance parameter τ^2, is a convenient measure of heterogeneity. The null hypothesis of no heterogeneity becomes $\tau^2 = 0$, and the larger the value of τ^2 the greater the heterogeneity.

Both μ and τ^2 can be estimated by maximum likelihood. The marginal distribution of counts averaged over the gamma distribution is negative binomial, and the log likelihood function for μ and τ^2 is (Martuzzi and Hills, 1995)

$$\sum_{i=1}^{N}\left(\log\frac{\Gamma(O_i+\mu^2/\tau^2)}{\Gamma(\mu^2/\tau^2)}+\frac{\mu^2}{\tau^2}\log\frac{\mu}{\tau^2}-(O_i+\mu^2/\tau^2)\log(E_i+\mu/\tau^2)\right).$$

To maximise this log likelihood and find the MLE values of μ and τ^2, we can use negative binomial regression or any general-purpose maximisation algorithm. For practical purposes it is convenient to consider the simplified version of the log likelihood function where μ is kept fixed at the value O/E, where $O=\sum O_i$ and $E=\sum E_i$. This log likelihood involves only the τ^2 variance parameter. In our application the constraint $\mu=1$ was made and the log likelihood reached its maximum at $\tau^2=0.034$. Thus, the distribution of the true ward-specific rate ratios is estimated to be a gamma distribution with mean 1 and variance 0.034.

The spread of this distribution can be assessed visually, but it is perhaps preferable to calculate its 5th and 95th percentiles, the limits within which 90% of the true rate ratios lie. These percentiles are 0.72 and 1.32. The null hypothesis $\tau^2=0$ can be tested with a likelihood ratio statistic (Clayton and Hills, 1993). Since for $\tau^2=0$ the marginal distribution of O_i reduces to Poisson, the log likelihood at τ^2 is $O\log\mu-\mu E=-2051$ (this value equals $-E$ when $\mu=1$). The log likelihood at $\tau^2=0.034$ is -2048.03, and referring twice the difference between the two values (5.94) to a table of χ^2 with one degree of freedom gives $p=0.015$.

Simulation methods can also be applied to evaluate the null hypothesis of homogeneity. Since the above analytical method is based on a quadratic approximation to the log likelihood function, simulation methods may be more accurate, especially when we consider few areas, or uneven population distributions, but require large computing power. The method proceeds by repeatedly re-allocating the 2051 cases at random to the 515 wards, each case having a probability of falling in a given ward proportional to the expected cases in that ward. The empirical p-values equal the proportion of times when the simulations exceed the observed value, 0.011 in our case, which is close to the analytical p-value.

Confidence intervals for τ^2 can also be calculated using the log likelihood function (Clayton and Hills, 1993). If a horizontal line is drawn 1.92 units below the maximum, then the two intersection points with the log likelihood curve define the 95% limits. (Note that 1.92 is half of 3.84, the 5th percentile of a χ^2 distribution with one degree of freedom, and must be changed accordingly if different confidence levels are required.) The 95% confidence limits for τ^2 are 0.006 and 0.070, which are in agreement with the significance test.

25.2.1 Profile Log Likelihood

A more precise analysis is based on the two-parameter log likelihood in which the mean μ is not held fixed to a constant value. Since μ is nuisance parameter, the *profile* log likelihood should be used to obtain the confidence interval for τ^2 (Clayton and Hills, 1993). This is defined as

$$l(\hat{\mu}(\tau^2),\tau^2),$$

where $\hat{\mu}(\tau^2)$ is the maximum likelihood estimate of μ for a given value of τ^2. The profile log likelihood normally differs from the log likelihood obtained from the section of the two-parameter surface log likelihood through $\mu = 1$, so that different confidence intervals for τ^2 might be obtained from the profile log likelihood. However, this difference is nearly always small (the two log likelihood functions were almost indistinguishable in the present application), and for practical applications the simplified log likelihood function for τ^2, based on the constraint $\mu = O/E$, can be used.

25.3 BINOMIAL CASE

An analogous methods can be applied to the analysis of prevalence data, when a non-rare disease or condition is under study. In this circumstance the binomial sampling distribution cannot be approximated with a Poisson model, and must be dealt with directly (Martuzzi and Elliott, 1996). The beta-binomial model is analogous to the gamma-Poisson model described above, and can be used along the same lines. As an example, we analyse data collected within the framework of the Small-Area Variations in Air Quality and Health (SAVIAH) study, concerning the prevalence of mild respiratory symptoms among 4395 schoolchildren of the Huddersfield Health Authority, in Northern England. A total of 1306 cases were observed, amounting to an overall prevalence of 29.7%. The presence and degree of heterogeneity is studied using the 64 school catchment areas (SCAs) into which the study area is subdivided. The data are given in the Appendix. The number of prevalent cases per SCA, r_i, follows a binomial distribution with probability π_i, the true prevalence, and n_i, the number of children in the ith SCA. Under this model, the observed prevalence, $p_i = r_i/n_i$, is the MLE of π_i. Its values range between 0% and 63.6%, but again this wide spread reflects large random variability and the instability of the p_i, resulting from the fact that some SCAs have only few children (range of n_i : 1–258).

We assume that the π_i are drawn from a beta distribution, i.e. their probability is proportional to

$$\pi_i^{\alpha-1}(1-\pi_i)^{\beta-1},$$

where $\alpha, \beta > 0$. As in the gamma case, the beta distribution can take a great variety of regular shapes, determined by the two parameters, α and β. The marginal distribution of the r_i obtained by averaging the binomial counts over the beta distributions is of the beta-binomial family, and has parameters α and β which can be estimated maximising the log likelihood function. To this end it is convenient to consider a different parameterisation of the beta distribution, based on the mean $\mu = \alpha/(\alpha+\beta)$ and $\gamma = \alpha + \beta$. The log likelihood function is (Crowder, 1978)

$$l(\mu, \gamma) = \sum_{i=1}^{N} A_i,$$

where

$$A_i = \sum_{k=1, r_i \neq 0}^{r_i} \log(\gamma\mu + r_i - k)$$
$$+ \sum_{k=1, r_i \neq n_i}^{n_i - r_i} \log[\gamma(1-\mu) + n_i - r_i - k]$$
$$+ \sum_{k=1}^{n_i} \log(\gamma + n_i - k).$$

Using the estimates of μ and γ obtained from maximising this likelihood the variance $\tau^2 = \mu(1-\mu)/(\gamma+1)$ can be calculated. As before we begin by setting μ to the fixed value $\sum r_i / \sum n_i$, i.e. the overall prevalence. In the example considered here, the MLE of γ was 95.86, giving $\tau^2 = 0.0021$. To work out percentiles for the prevalences distribution using a computer package, however, it is necessary to use the ordinary parameterisation in $\alpha = \gamma\mu$ and $\beta = \gamma(1-\mu)$, equal to 28.5 and 67.4, respectively. The 5th to 95th percentile range of the prevalence across the study area is from 22.3% to 37.6%.

To test the null hypothesis of constant prevalence across the study area, the log likelihood ratio test can be applied as above. The value of the log likelihood at its maximum is -2667.3, and the value at $\tau^2 = 0, -2674.1$, is given by the binomial log likelihood function

$$\sum_i (r_i \log \mu + (n_i - r_i)\log(1-\mu)),$$

where μ is set to the overall prevalence value. Thus, referring twice the difference of the two log likelihood values (13.6) to a χ^2 table with one degree of freedom, gives a highly significant result ($p < 0.001$), providing strong evidence against homogeneity. Simulation methods can also be used in the binomial case to calculate an empirical p-value based on repeated random sampling.

The 95% confidence interval for the estimated γ, calculated using the likelihood based method described above, was 45.0 to 277.0, or, in terms of τ^2, 0.0045 to 0.00075. In Table 25.1 the resulting 5th and 95th percentile ranges for the prevalence distribution are given. It can be seen that, even at the extremes of the 95% confidence intervals for τ^2, the estimated range of risk, where random variability has been removed, is narrower than the spread of the observed prevalences.

Table 25.1 Percentiles of the beta distribution of the prevalence of respiratory symptoms for different values of γ (or, equivalently, of the variance τ^2). Percentiles of the observed prevalence are shown at the bottom

			Prevalence (%)	
	γ	τ^2	5th percentile	95th percentile
MLE	95.86	0.0021	22.3	37.6
95% lower limit	45.0	0.0045	19.1	41.3
95% upper limit	277.0	0.00075	25.3	34.3
Observed prevalence			16.1	46.9

Although a parameterisation of the log likelihood as a function of τ^2 would allow us to derive the profile log likelihood for τ^2 as was done in the Poisson case, such re-parameterisation is complicated, and little accuracy is lost using the one-parameter log likelihood function above.

25.4 DISCUSSION

The methods described here are a simple way to test for the presence of heterogeneity, and to estimate its extent. Other available methods, such as the chi-squared dispersion test, or the tests proposed by Potthoff and Whittinghill (1966a, b), only provide p-values, which are less informative for descriptive studies. As mentioned in the introduction, large heterogeneity between few areas or based on few cases may produce the same p-value as heterogeneity of small magnitude based on many cases. While the two situations are equivalent in terms of statistical significance, the implications for public health might be different.

When using the gamma or the beta distributions for calculating p-values for the null hypothesis of constant risk, the hypothesis that the variance is equal to zero is tested. Thus, the test is carried out on the boundary of the parameter space and the approximation given by the chi-squared distribution with one degree of freedom might not be entirely correct (Self and Liang, 1987). This potential inaccuracy might be overcome using computer simulations to obtain empirical p-values. However, analytical and empirical p-values are generally close in practice, as is illustrated by the example of perinatal mortality in England. The use of non-parametric distributions is described in Chapter 31 in this volume.

Both in the Poisson and binomial cases the models underlying these methods can be used to obtain empirical Bayes estimates of small-area rate ratios or prevalences. Although such an outcome might be necessary, for example, to prepare geographical maps, often the overall information concerning evidence and degree of heterogeneity is of value. In addition, the set of empirical Bayes estimates is known to be less dispersed than the true rate ratio or prevalence distribution (Louis, 1984), so that 'eyeball' estimates done when looking at a map might underestimate the degree of heterogeneity of the data. In general, results from analyses based on the methods proposed in the chapter, for example a 5th to 95th percentile range, provide useful, synthetic information describing the extent to which risks of disease are likely to differ across a given geographical region. Such information might be of relevance for evaluating the need for public health intervention or for further investigation.

Finally, as a note of caution, we point out that the use of the gamma and beta families as mixing distributions is based on the plausibility of shape and mathematical convenience. There is nothing in the methods that provides any way of criticising the assumption that the true rates follow these distributions.

ACKNOWLEDGEMENTS

Data on perinatal mortality used in this chapter were made available by the North West Thames Health Region Authority, and retrieved from the Small Area Health Statistics

Unit database. SAVIAH (Small-Area Variations in Air quality and Health) was funded by the European Commission. We are grateful to professors Paul Elliott and Dave Briggs for permission to use the data.

APPENDIX

Table A.1 gives the prevalence of mild respiratory symptoms among children living in the Huddersfield Health Authority, Northern England, by school catchment area. Data were collected in the framework of the Small-Area Variations in Air Quality and Health (SAVIAH) study.

Table A.1

Cases (r_i)	Total (n_i)	Cases (r_i)	Total (n_i)	Cases (r_i)	Total (n_i)	Cases (r_i)	Total (n_i)
47	160	24	78	31	90	12	45
70	258	25	77	43	136	66	200
40	175	39	113	26	77	6	18
41	123	14	50	26	86	10	63
33	79	51	111	4	10	24	70
17	62	7	18	30	110	22	68
11	83	26	75	48	198	23	82
6	19	4	12	28	103	7	11
32	95	6	28	18	92	14	58
11	25	19	73	49	104	6	31
33	104	19	58	21	76	8	42
5	29	3	15	32	97	5	10
16	43	18	53	6	17	29	114
19	72	13	44	16	81	4	8
8	36	8	36	5	12	5	16
0	1	5	17	0	1	12	47

26

Ecological Regression with Errors in Covariates: An Application

L. Bernardinelli, C. Pascutto, C. Montomoli and J. Komakec

University of Pavia

W. R. Gilks

MRC Biostatistics Unit

26.1 INTRODUCTION

Ecological analysis studies the geographical variation in disease risk and investigates its association with ecological covariates, i.e. explanatory variables measured at an areal unit level (Walter, 1991a,b; Morgenstein, 1982).

For example, many disease atlases have been produced, mainly cancer atlases (Bernardinelli, *et al.*, 1994) and some of them include information on ecological covariates (Kemp *et al.*, 1985). Further examples include studies relating cardiovascular mortality in different areas to a variety of environmental and socio-economic factors (Cook and Pocock, 1983; Gardner, 1973) and studies relating cancer to dietary intakes (Prentice and Sheppard, 1990).

The simplest approach to ecological analysis uses a multiple regression model for disease risk which only allows for Poisson variation (Clayton and Hills, 1993). More recently it has been observed that the variation not explained by the ecological variables (residual variation) might be substantially in excess of that expected from Poisson sampling theory. Extra-Poisson regression models have been proposed to separate the Poisson sampling variation from the extra-Poisson variability (Pocock *et al.*, 1981; Breslow, 1984).

The current state-of-the-art is to adopt a fully Bayesian approach and Markov chain Monte Carlo (MCMC) methods for model fitting (Bernardinelli and Montomoli, 1992;

Disease Mapping and Risk Assessment for Public Health. Edited by A.B. Lawson *et al.*
© 1999 John Wiley & Sons Ltd.

Besag *et al.*, 1991; Clayton, 1989; Clayton and Bernardinelli, 1992; Bernardinelli *et al.*, 1995a,b). This approach splits the extra-Poisson variation into two components. The first component of variation is simply spatially unstructured extra-Poisson variation, called *heterogeneity*. Modelling the *heterogeneity* variation allows for unmeasured variables that vary between areas in an unstructured way. The second component of variation, called *clustering*, varies smoothly across areas. Modelling the *clustering* variation allows for those unmeasured risk factors that vary smoothly with location. The choice between the *heterogeneity* and the *clustering* model depends upon our prior belief about the size of high/low risk clusters. A cluster size bigger than the area size would lead to a *clustering* model, while a cluster size smaller than the area size would lead to a *heterogeneity* model. Although it is possible to include both terms in the model, this may not be necessary (Bernardinelli *et al.*, 1995a; Clayton and Bernardinelli, 1992). Indeed, for high resolution maps like those in our application the *heterogeneity* component will often be unnecessary, and, if the number of events per area never exceeds one, then is not even identifiable.

By fitting an extra-Poisson Bayesian model, the point estimate of the regression coefficient does not change substantially, but its standard deviation tends to increase with respect to that obtainable via a classical Poisson model (Clayton *et al.*, 1993).

Furthermore, modelling the clustering variation may be thought of as a way of modelling the effect of location. Where the pattern of variation of the covariate has a spatial structure similar to that of disease risk, location acts as a confounder. Modelling the effect of location through the clustering variation causes the estimate of the regression coefficient to be controlled by the effect of location. As a consequence, both the standard error and the point estimate of the regression coefficient change with respect to the classical Poisson model. In particular, the point estimate decreases (Pocock *et al.*, 1981).

The extra-Poisson Bayesian model can be further complicated to allow for measurement error in the ecological covariates. In practice, ecological covariates can rarely be observed directly. Available data may be either imperfect measurements of, or proxies for, the true covariate. Sometimes epidemiological data concerning another disease may be used as a proxy variable. For example, to study the geographical variation of heart disease mortality, an important covariate would be the proportion of smokers living in each area. Such data on smoking would generally not be available, so the prevalence of lung cancer recorded by the cancer registry for each area might be a useful proxy. The simplest approach to this problem would be to estimate the true covariate from the proxy for each area independently, using the proxy estimate in the ecological regression. When the proxy variable is an accurate measure of the true covariate, this approach would be reasonable. However, when the correspondence between the two is not so close, this approach has several disadvantages: not accounting for measurement error causes the point estimate of the regression coefficient to be underestimated and its precision overestimated (Richardson and Gilks, 1993; Bernardinelli *et al.*, 1997). Moreover, when it is reasonable a priori to expect spatial correlation in the true covariate, the Bayesian approach allows us to obtain improved estimates by specifying a spatial smoothing prior on the true covariate (Bernardinelli *et al.*, 1997).

We describe a Bayesian hierarchical–spatial model for ecological regression aimed at investigating the relationship between insulin dependent diabetes mellitus (IDDM) incidence, the proportion of glucose-6-phosphate-dehydrogenase (G6PD) deficient individuals and past prevalence of malaria in Sardinia. Our model is composed of two regression submodels. The first model allows us to estimate the effect of malaria on IDDM risk, while the second model allows us to estimate the effect of malaria on G6PD deficiency. Both models allow also for a common unknown underlying non-malaria factor which we suppose to be responsible for the association between IDDM and G6PD deficiency. Smoothing spatial priors are posited to reduce random geographical variability in the estimates. Measurement error in the ecological covariates is also accounted for.

26.2 BACKGROUND AND DATA

26.2.1 Malaria and IDDM

There is scientific interest in studying the association between IDDM and malaria, since they are both associated with the HLA (Human Leukocyte Antigens) system.

The association between IDDM and the HLA system, known to be involved in controlling immunological responses, has long been established (Todd *et al.*, 1988; Thomson, 1988; Green, 1990). In particular, many studies (Jacob, 1992; Tracey, 1995) have demonstrated an association between IDDM and the HLA loci A, B (class I), DR (class II) (Langholz *et al.*, 1995; Thomas *et al.*, 1995) and the tumor necrosis factor-α gene (TNF-α) localised in the HLA region (class III) (Davies *et al.*, 1994; Tracey *et al.*, 1989). In Sardinia a particular HLA haplotype is associated with IDDM (Cucca *et al.*, 1993).

Malaria is the most important natural selective factor on human populations that has been discovered to date (Jacob, 1992). A West African study showed that an allele of HLA class I and an unusual haplotype of HLA class II are associated with protection from malaria (Hill *et al.*, 1991; Ebert and Lorenzi, 1994). The association between susceptibility to cerebral malaria and the TNF2 allele, a variant located in the promoter region of the TNF-α associated with higher levels of TNF transcription, has also recently been established (McGuire *et al.*, 1994). These elements support the hypothesis that in areas of high endemicity, malaria operates the genetic selection responsible for the influence on the susceptibility to autoimmune diseases (Greenwood, 1968; Wilson and Duff, 1995).

In Sardinia malaria is known to have selected for some serious hereditary diseases such as β-thalassemia, Cooley's disease and favism, the latter caused by glucose-6-phosphate dehydrogenase (G6PD) enzyme deficiency (Bernardinelli *et al.*, 1994). Sardinia is therefore a particularly suitable place for investigating the association between IDDM and malaria.

IDDM incidence in Sardinia is quite atypical of other Mediterranean countries (Muntoni and Songini, 1992). Sardinia has the second highest incidence in Europe (33.2 per 100 000 person years; Songini *et al.*, 1998) after Finland (40 per 100 000; Tuomilehto *et al.*, 1995). A study carried out on the cumulative prevalence of IDDM in 18 year-old military conscripts born in the period 1936–71 showed that the risk for IDDM began increasing with the male birth cohort of 1950 and that the increasing trend is much higher than the one observed in Europe (Songini *et al.*, 1993).

As to malaria, the Sardinian population has a long history of endemicity. Malaria spread gradually all over Sardinia after the Carthaginian conquest, became established after Roman occupation and was a major cause of death in the island until the mid-twentieth century, when it was completely eradicated (Fantini, 1991). Population genetic studies carried out by Piazza *et al.* (1985) suggest that, in the plains of Sardinia where malaria had been endemic, some genetic traits were selected to provide greater resistance to the haemolysing action of *Plasmodium*. In the hilly and mountainous areas, where malaria was almost absent, this adaptation did not occur.

26.2.2 Malaria and G6PD-mutation

Glucose-6-phospate dehydrogenase (G6PD) is the enzyme that catalyses the first reaction of the pentose phosphate pathway, also known as the hexose monophosphate shunt. G6PD deficiency (G6PD-) is responsible for episodes of acute haemolysis provoked by ingestion of oxidising agents such as primaquine, sulphanilamide, chloramphenicol, aspirin, chemical compounds and certain vegetables such as fava beans and can also cause neonatal haemolitic anaemia. G6PD deficiency is therefore a serious hazard to health and can be characterised by high lethality unless promptly and correctly treated. In spite of the natural selection against carriers of G6PD deficiency, this trait is estimated to affect 400 million people worldwide, with particularly high prevalence rates in areas of the world where malaria is or has been endemic (Ganczakowski *et al.*, 1995). Previous studies indicated that carriers of G6PD deficiency and thalassaemic traits have a selective advantage against malaria caused by *Plasmodium falciparum* infections (Siniscalco *et al.*, 1961).

The strongest evidence to support this hypothesis has been the geographical distribution of malaria and the high gene frequencies for these traits. However, since the investigation of a statistically significant correlation between an abnormal haemoglobin trait, enzyme deficiency and malaria show inconsistencies, it is legitimate to question whether the hypothesis that the natural selection by malaria is the sole explanation of these abnormal traits.

The history of external and internal migration in Sardinia during the Carthaginian and Roman epocus points towards the hypothesis that gene flow may have played a role in the distribution of abnormal gene traits (Brown, 1981).

This hypothesis, of course, does not deny the role of malaria in maintaining high gene frequencies in the zones where malaria was widespread. Therefore G6PD deficiency can be considered a genetic adaptation to malaria.

26.2.3 G6PD-mutation and IDDM

A positive association between IDDM and G6PD deficiency has been demonstrated in Iraqi, Indian and Chinese patients raising the interesting question of whether G6PD- is a result or a consequence of diabetes (Saeed *et al.*, 1985). It has been argued that the hexose monophosphate shunt is stimulated by insulin and therefore is depressed in patients affected by IDDM. Although it seems more probable that the enzyme deficiency is a result of IDDM rather than a predisposing cause, the full explanation of the association IDDM—G6PD deficiency is still insufficient (Saha, 1979). A recent study carried out in Sardinia did not demonstrate the above-mentioned association (Meloni *et al.*, 1992).

26.2.4 Graph of conditional independencies

On the basis of the background scientific knowledge summarised above, it seems reasonable to suppose that endemic malaria in Sardinia may have operated a genetic selection on both IDDM and G6PD deficiency. Although the mechanism underlying the relationship between IDDM and G6PD- has not been fully highlighted yet, we think that it is plausible to hypothesise the existence of an association between the two diseases.

Our aim is to study the geographical variation of IDDM in Sardinia, accounting for:

- the spatial correlation in the estimates due to unknown causal factors varying smoothly across areas;
- the effect of malaria selection;
- the hypothesised association with G6PD deficiency.

We also wish to verify the existence of a relationship between malaria selection and G6PD deficiency as reported in the literature.

We can represent our view of the complex relationships between malaria, IDDM and G6PD- using a high-level partially directed conditional independence graph (see Figure 26.1 (a)). In this graph, the parent nodes represent causal factors. Since we are interested in modelling the effect of malaria selection on IDDM and on G6PD deficiency, the node representing malaria is linked to both IDDM and G6PD- with a directed edge. The lack of conditional independence between G6PD- and IDDM given malaria is represented by an undirected edge.

The graph in Figure 26.1 (a) can be transformed into a directed acyclical graph (DAG) by adding a parent node common to IDDM and G6PD- (see Figure 26.1(b)) (Cox and Wermuth, 1996). This node represents a hidden variable, i.e. unknown non-malaria causal factors common to both IDDM and G6PD-. We wish to stress that this new variable is of little epidemiological interest, being only a way to model the relationship between IDDM and G6PD- which has been empirically observed in other studies.

26.3 THE STATISTICAL MODEL

To model the relationships between the quantities represented by the nodes of the DAG in Figure 26.1 (b), we set up an ecological regression model composed of two submodels.

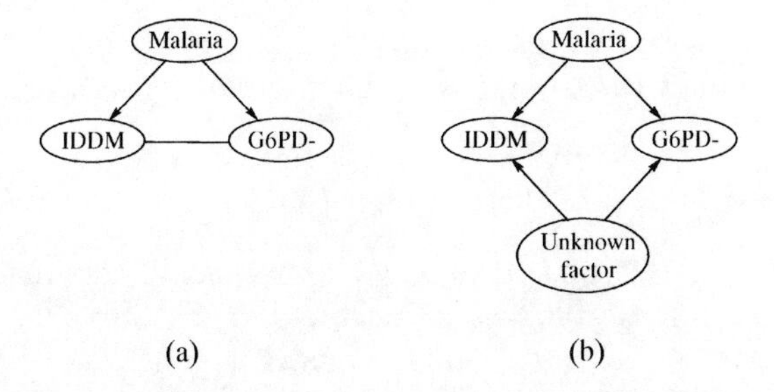

Figure 26.1 High-level view of the relationships between malaria, IDDM and G6PD deficiency. (a) Partially directed conditional independence graph; (b) directed acyclical graph

The dependent variables are IDDM risk in the first submodel and the proportion of G6PD- individuals in the second submodel. In both submodels the ecological covariates are the observed malaria prevalence and a common unobserved non-malaria factor as explained in Section 26.2.4.

Smoothing priors are posited for both the disease risk and the ecological covariates. Measurement error in the covariates is also modelled.

26.3.1 A model for IDDM risk

Suppose we observe d_i events (IDDM cases in our application) in area i as compared with E_i expected from suitable reference rates.

We assume

$$d_i \sim \text{Poisson}(\rho_i E_i), \tag{26.1}$$

where $\{\rho_i\}$ are the area-specific rate ratios (RR) controlled for age (Clayton and Hills, 1993) and $\{E_i\}$ are expected events. If the $\{\rho_i\}$ are not all equal, then the data $\{d_i\}$ will display *extra-Poisson variation.*

To investigate whether extra-Poisson variation is geographically related, we would ideally like to map the true relative risks $\{\rho_i\}$. Since these are unobserved, the most obvious strategy is to estimate ρ_i by the empirical relative risk:

$$\hat{\rho}_i = \frac{d_i}{E_i}, \tag{26.2}$$

which is the maximum likelihood estimate of ρ_i. Mapping the $\{\hat{\rho}_i\}$, however, can be misleading because sampling variability can dominate the map and obscure genuine trends. In particular, areas having exceptionally high or low $\hat{\rho}_i$ will tend to be those with smaller E_i, where sampling variability is most pronounced (var $\hat{\rho}_i = \rho_i/E_i$).

Several strategies for dealing with sampling variability in maps have been proposed. The current state-of-the-art is to adopt a fully Bayesian hierarchical–spatial model (Bernardinelli and Montomoli, 1992; Besag *et al.*, 1991; Clayton, 1989; Clayton and Bernardinelli, 1993). An important feature of this type of model is that the prior distribution for the $\{\rho_i\}$ incorporates spatial correlation, allowing the estimate of ρ_i to formally 'borrow strength' from neighbouring areas. In this way the empirical map is smoothed, and geographical trends and inferences are made more reliable.

Mapping Bayesian estimates of relative risk may reveal geographical trends across the map, or may suggest links with area-specific covariates x_i. To incorporate these covariates into the model, a natural assumption, in conjunction with the Poisson assumption (26.1), would be

$$\log\rho_i = \alpha_i + \beta\mathbf{x}_i, \tag{26.3}$$

where α_i represents the covariate-adjusted area-specific log relative risk, $\beta\mathbf{x}_i = (x_i^1, x_i^2, \ldots, x_i^k)$ the vector of ecological covariates and $\beta = (\beta^1, \beta^2, \ldots, \beta^k)$ the vector of unknown regression coefficients. Such a model is called an *ecological regression model.* Spatial smoothing of the $\{\rho_i\}$ can then be effected via a smoothing prior on the $\{\alpha_i\}$.

26.3.2 A Markov random field prior

In this section we describe a *Markov random field* prior distribution for the $\{\alpha_i\}$ parameters in (26.3). The development follows that in Bernardinelli *et al.* (1995a). This prior will tend to produce similar estimates for α_i and α_j if areas i and j are geographically close.

The Gaussian Markov random field prior we employ assumes, for each area i, that α_i is normally distributed with mean and variance depending on its neighbours. We consider two areas to be 'neighbours' if they share a portion of a boundary.

The conditional prior distribution of α_i given values for $\{\alpha_j, j \neq i\}$ is

$$\alpha_i \sim \mathrm{N}(\mu_{\alpha_i}, \sigma^2_{\alpha_i}), \tag{26.4}$$

where

$$\mu_{\alpha_i} = \frac{\sum_{j \neq i} w_{ij}\alpha_j}{\sum_{j \neq i} w_{ij}}, \tag{26.5}$$

$$\sigma^2_{\alpha_i} = \frac{1}{\gamma_\alpha \sum_{j \neq i} w_{ij}}. \tag{26.6}$$

The *adjacency weights* $\{w_{ij}\}$ are fixed constants. Although other choices are possible, we set $w_{ij} = 0$ unless areas i and j are neighbours, in which case $w_{ij} = 1$.

To ensure that the Gaussian Markov random field model ((26.4),(26.6)) is internally consistent, $\sigma^2_{\alpha_i}$ must depend upon the number of adjacent areas and their adjacency weights. Jointly, the $\{\alpha_i\}$ have an intrinsic multivariate normal prior distribution with *inverse* variance–covariance matrix Λ given by

$$\Lambda_{ij} = \begin{cases} -\gamma_\alpha w_{ij}, & j \neq i, \\ \gamma_\alpha \sum_j w_{ij}, & j = i. \end{cases}$$

The matrix Λ is not of full rank. Thus the prior on the $\{\alpha_i\}$ is improper: adding an arbitrary constant to each α_i will not change the probability (26.4). This need not concern us since the data d_i contain information on the location of the $\{\alpha_i\}$.

The amount of smoothing in the random effects $\{\alpha_i\}$ is controlled by the precision parameter γ_α in (26.6). A small value of γ_α will induce little smoothing, whilst an infinite value would force all the $\{\alpha_i\}$ to be equal. Since we do not wish to impose any fixed amount of smoothing on these parameters, but rather we wish to let the data themselves determine how much smoothing should be induced, we treat γ_α as a model parameter. We chose a $\chi^2_{sf,\nu}$ prior distribution for γ_α, where sf is a scale factor and ν is the degrees-of-freedom parameter.

We then chose a suitable combination of scale factor and degrees of freedom according to our prior belief about the amount of geographical variation across the map.

As discussed by Bernardinelli *et al.* (1995a), given the connection between the precision hyperparameter γ_α and the marginal variability of the area-specific relative risks, it can be shown that the 90% range of variation of the relative risks is approximately $e^{4.7/\gamma_\alpha}$. By expressing the mean and variance of the prior distribution of γ_α in terms of sf and ν (mean$= \nu/sf$; var $= 2\nu/sf^2$), it is possible to choose a combination of sf and ν

leading to a distribution in agreement with the prior belief about the range of the geographical variation.

26.3.3 Measurement error in the prevalence of malaria

When dealing with ecological covariates, available data z_i may be either imperfect measurements of, or proxies for, x_i. The simplest approach to this problem would be to estimate x_i from z_i for each area independently, using this estimate $\hat{x}_i$ in place of x_i in the ecological regression (26.3). When z_i is an accurate measure of x_i, this approach would be reasonable. However, when the correspondence between x_i and z_i is not so close, this approach would have several disadvantages. First, the estimate of the regression coefficient β would probably be underestimated (see, for example, Richardson and Gilks, 1993). Secondly, the precision in parameter estimates or in projections would be overestimated, through failure to take account of uncertainty in the $\{\hat{x}_i\}$. Thirdly, when it is reasonable a priori to expect spatial correlation in the x_i, improved estimates of the $\{x_i\}$ and other unknowns would be obtained through a Bayesian procedure incorporating a spatial smoothing prior on the $\{x_i\}$.

In our application, x_i is related to underlying malaria prevalence, and z_i is the observed number of malaria cases in area *i* at one point in time. We will assume that

$$z_i \sim \text{Binomial}(n_i, \theta_i), \tag{26.7}$$

where n_i is the population size of area *i*, and

$$\log\left(\frac{\theta_i}{1-\theta_i}\right) = x_i. \tag{26.8}$$

Thus the covariate x_i in (26.3) is taken to be the logistic-transformed expectation of z_i.

Since the areal units in our analysis are small, it is reasonable a priori to expect spatial correlation also in the ecological covariates. We used the Markov random field prior described in Section 26.3.2 also for the $\{x_i\}$.

This part of the model accounts for sampling error in the covariate x_i, but there is another source of measurement error we wish to account for in this application.

The hypothesis of interest concerns how genetic adaptation in areas of endemic malaria effects susceptibility to IDDM. Thus the true covariate is the long-term malaria endemicity averaged over many centuries in each commune.

To model the long-term malaria endemicity, we replaced the deterministic relationship in (26.8) by the following stochastic relationship:

$$\log\left(\frac{\theta_i}{1-\theta_i}\right) \sim \text{N}(x_i, \omega). \tag{26.9}$$

We thus introduced an extra layer of uncertainty into the model. This may be interpreted as follows: the true log odds of malaria in commune *i* (i.e. $\log \theta_i/(1-\theta_i)$) represents a single realisation from a latent Normal distribution with mean x_i (i.e. the long-term average endemicity of malaria in commune *i*) and unknown long-term variance, ω. Since the data contain no information by which to estimate ω, we decided to fix its value a priori. We carried out exploratory analyses to select a suitable value for ω as described in Bernardinelli *et al.* (1997) and we chose $\omega = 2.25$.

Having subjectively fixed a value for ω is a potentially controversial aspect of our analysis. However, we examined the sensitivity of the regression coefficient estimate to diferent values of ω in a previous analysis (Bernardinelli *et al.*, 1997). The results did not change qualitatively, but for smaller values of ω the estimates were more similar to those obtained not accounting for long-term measurement error. For larger values of ω the coefficient estimates tended to be larger in absolute value, with wider credible intervals.

26.3.4 A model for the proportion of G6PD- individuals

We assume that the number of G6PD- individuals y_i in area i follows a binomial distribution:

$$y_i \sim \text{Binomial}(m_i, \pi_i), \tag{26.10}$$

where m_i is the number of screened individuals in area i and π_i is the area-specific prevalence of G6PD deficiency.

We used the logistic transformed expectation of y_i,

$$\log\left(\frac{\pi_i}{1-\pi_i}\right) = \xi_i, \tag{26.11}$$

as a second ecological covariate in model (26.3). We specified the same Markov random field prior as described above also on the ξ_i.

26.3.5 Building the full model

Setting up a realistic model to reflect complex aetiological hypotheses and relationships between variables such as those described in Section 26.2 may be quite a difficult task.

To get to the final model, we proceeded in a stepwise way, starting from the fitting of more simple models.

Model 1 (M1)

In a previous analysis fully described in Bernardinelli *et al.*, (1997) we implemented model (26.3) with malaria prevalence as ecological covariate and discussed the measurement error model ((26.7)–(26.9)):

$$\log \rho_i = \alpha_i + \beta_1 x_i. \tag{26.12}$$

Model 2 (M2)

We then investigated the association between IDDM and G6PD- by fitting model (26.3) with the G6PD- proportion as the ecological covariate, as described in

((26.10), (26.11)):

$$\log \rho_i = \alpha_i + \beta_2 \xi_i. \tag{26.13}$$

Model 3 (M3)

As a further step, we fitted model (26.3) with both the ecological covariates defined in Sections 26.3.3 and 26.3.4:

$$\log \rho_i = \alpha_i + \beta_1 x_i + \beta_2 \xi_i. \tag{26.14}$$

We put flat normal prior distributions on all the regression coefficients in models M1, M2 and M3.

Final model

Finally, we set up the full model as follows:

$$\log \rho_i = \alpha_i + \beta x_i + \delta \psi_i, \tag{26.15}$$

$$\log\left(\frac{\pi_i}{1-\pi_i}\right) = \eta x_i + \lambda \psi_i, \tag{26.16}$$

where x_i represents the selective action of endemic malaria and ψ_i represents the underlying unknown causal factor as introduced in Section 26.2.4.

The IDDM and G6PD- models, (26.15) and (26.16), show a symmetry reflecting the structural relationships shown in the DAG in Figure 26.1(b).

We put non-informative normal priors on the regression coefficients β, δ and η and spatial smoothing Markov random field priors on the $\{\alpha_i\}$, $\{x_i\}$ and $\{\psi_i\}$.

We overcame the identification problem concerning δ, λ and the $\{\psi_i\}$ by specifying a proper N(1, 10) prior on λ and a strong prior on the precision parameter of the $\{\psi_i\}$. The prior on the $\{\psi_i\}$ reflects our prior belief on their geographical variation. Our prior assumption was that the highest value of e^{ψ_i} in the map is likely to be around twice the lowest value. Following the method described in Section 26.3.2, we chose $\nu = 20$ and $sf = 2.5$ as *degrees of freedom* and *scale factor* parameters of the prior distribution for the hyperparameter γ_ψ.

In principle, a Bayesian analysis of non-identifiable models is always possible by assigning proper priors for the model unknowns. In practice, a too precise prior may limit Bayesian learning from the data, while a flat or even improper prior may lead to improper posterior or to convergence problems if using Markov chain Monte Carlo methods. In this application, however, we are interested in quantifying the effect of malaria (i.e. the β and η coefficients), for which there are enough data to prevent identifiability problems, rather than the effect of other unknown factors. So the choice of the parameters for the proper prior on λ and on the $\{\psi\}$ is not a major issue.

To give a global view of the final model, the relevant model equations are reported below:

Poisson model for IDDM relative risk:

$$d_i \sim \text{Poisson}(\rho_i E_i).$$

Ecological regression for IDDM relative risk with two ecological covariates:

$$\log \rho_i = \alpha_i + \beta x_i + \delta \psi_i.$$

Binomial model for the proportion of G6PD deficiency:

$$y_i \sim \text{Binomial}(m_i, \pi_i).$$

Logistic ecological regression for G6PD- proportion with the same two ecological covariates:

$$\text{logit}(\pi_i) = \eta x_i + \lambda \psi_i.$$

Model for the first ecological covariate (malaria prevalence):

1. Binomial model for the observed number of malaria cases

$$z_i \sim \text{Binomial}(n_i, \theta_i).$$

2. Second layer of uncertainty

$$\log\left(\frac{\theta_i}{1-\theta_i}\right) \sim \text{N}(x_i, \omega),$$

   where $\omega = 2.25$.

Normal priors for regression coefficients:

$$\beta, \delta, \eta \sim \text{N}(0, 1.0E+5),$$
$$\lambda \sim \text{N}(0, 10).$$

Markov random field priors for the parameters $\{\alpha_i\}$, $\{x_i\}$ and $\{\psi_i\}$:

$$\alpha_i \sim \text{N}(\mu_{\alpha_i}, \sigma^2_{\alpha_i}),$$
$$x_i \sim \text{N}(\mu_{x_i}, \sigma^2_{x_i}),$$
$$\psi_i \sim \text{N}(\mu_{\psi_i}, \sigma^2_{\psi_i}),$$

where

$$\mu_{\alpha_i} = \frac{\sum_{j \neq i} w_{ij}\alpha_j}{\sum_{j \neq i} w_{ij}}, \quad \sigma^2_{\alpha_i} = \frac{1}{\gamma_\alpha \sum_{j \neq i} w_{ij}},$$
$$\mu_{x_i} = \frac{\sum_{j \neq i} w_{ij}x_j}{\sum_{j \neq i} w_{ij}}, \quad \sigma^2_{x_i} = \frac{1}{\gamma_x \sum_{j \neq i} w_{ij}},$$
$$\mu_{\psi_i} = \frac{\sum_{j \neq i} w_{ij}\psi_j}{\sum_{j \neq i} w_{ij}}, \quad \sigma^2_{\psi_i} = \frac{1}{\gamma_\psi \sum_{j \neq i} w_{ij}}.$$

Hyperpriors for the precision parameters:

$$\gamma_\alpha \sim \chi^2_{sf_\alpha, \nu_\alpha},$$
$$\gamma_x \sim \chi^2_{sf_x, \nu_x},$$
$$\gamma_\psi \sim \chi^2_{sf_\psi, \nu_\psi}.$$

26.4 THE DATA

IDDM incidence was calculated from a case registry that has operated in Sardinia since 1989. The incidence data refer to the period 1989–92 and cover the population aged 0–29 years. We let d_i in (26.1) denote the number of IDDM cases in commune $i(i = 1, 2, \ldots, N = 366)$, and E_i denote the expected number of IDDM cases based on Sardinian national rates.

The number of individuals affected by malaria, z_i in (26.7), was recorded for each commune during the period 1938–40 by Fermi (1938, 1940). The population n_i for each commune was taken from the 1936 census.

In the period 1981–82 the Regional Health Service in Sardinia promoted a screening through which about 2000 males affected by G6PD deficiency were identified (15% of participants) (Bernardinelli *et al.*, 1994). We obtained the number of individuals screened for the G6PD mutation in each area (m_i in (26.10)) and the number y_i of G6PD- individuals from this data.

26.5 ESTIMATION

We have specified three arms to the model: a regression submodel for IDDM ((26.1), (26.3)–(26.6)), an analogous regression submodel for G6PD deficiency ((26.10), (26.16)) and a model for long-term malaria endemicity ((26.7)–(26.9)).

The best approach to estimating such a complex model is via Markov chain Monte Carlo methods like Gibbs sampling (Gilks *et al.*, 1996a,b), which allows us to treat the equations representing the three arms of the model as a single large model and hence to estimate all the parameters simultaneously. This can be done using the BUGS software (Spiegelhalter *et al.*, 1995).

26.6 RESULTS

All our models were estimated using BUGS. In each case we ran the Gibbs sampler for 70 000 iterations, discarded the first 60 000 'burn-in' samples and saved one in five sampled values, in order to reduce correlation within each chain. Convergence of the model parameters was checked by looking at the sample traces and using a variety of diagnostics implemented in the CODA (Best *et al.*, 1995) software. Computation took about 13 hours on a Sun Ultra workstation.
Table 26.1 shows the posterior means and 95% Bayesian credible intervals (95% CI) of the regression coefficients in the intermediate models M1, M2 and M3.

The posterior mean of the regression coefficient in model M1 indicates a significant negative association between IDDM and malaria, while the posterior mean of the regression coefficient in model M2 shows no evidence of association between IDDM and G6PD-. The posterior means for the regression coefficients in model M2 did not differ qualitatively from those obtained in the previous two models.

As to the final model ((26.1)–(26.11), (26.15)–(26.16)), the posterior mean estimates of the $\{\rho_i\}$, i.e. the area-specific IDDM relative risks, ranged from 0.512 to 1.76 across

Table 26.1 Posterior mean and 95% credible interval (95% CI) of the regression coefficients in models M1, M2 and M3

	Coefficient	
	β_1	β_2
Model	**Posterior mean (95% CI)**	**Posterior mean (95% CI)**
M1	– 0.060 (– 0.112, – 0.012)	
M2		– 0.410 (– 0.920,0.120)
M3	– 0.058 (– 0.111,0.011)	– 0.086 (– 0.344,0.165)

Table 26.2 Posterior mean, posterior standard deviation (s.d.), numerical standard error of the mean (NSE) and 95% credible interval (95% CI) of the regression coefficients in the final model ((26.1)–(26.16))

Coefficient	Posterior mean	s.d.	NSE	(95% CI)
β	– 0.062	0.028	0.001	(– 0.121, – 0.010)
δ	– 0.006	0.288	0.022	(– 0.591, 0.557)
η	0.044	0.024	0.001	(– 0.002, 0.088)
λ	2.330	0.377	0.028	(1.730, 3.130)

Sardinia, with 5th and 95th percentiles 0.665 and 1.348, respectively. The mean relative risk across areas was 0.995.

The posterior means and standard deviations (s.d.) of the regression cofficients in the final model obtained from the BUGS output are reported in Table 26.2. The numerical standard error of the mean (NSE) were calculated using time series methods to account for correlations within each sample (Best *et al.*, 1995), and the 95% credible intervals (CI) are also reported. The NSEs are a measure of the accuracy of the parameter estimates, which are obtained as means of the sampled values. The NSEs are rather small because we drew a large number of samples.

The regression coefficient β is significantly lower than zero, thus indicating the existence of a negative association between malaria endemicity and IDDM relative risk. The regression coefficient η is not significantly greater than zero, but its credible interval is shifted towards positive values, thus suggesting the existence of a positive association between malaria endemicity and proportion of G6PD-. However, the significantly positive value of the regression coefficient λ indicates that most of the geographical variation in G6PD-proportion is due to non-malaria factors. Finally, the regression coefficient δ indicates that the non-malaria causal factor common to IDDM and G6PD, as defined in Section 26.2.4, has no effect on IDDM.

The Markov random field priors on IDDM risk and on the ecological covariates allowed us to obtain smoothed maps showing a spatially structured variation.

The map in Figure 26.2(a) shows the estimated long-term prevalence of malaria θ_i obtained by fitting the reduced model in 26.3.3; the map in Figure 26.2(b) shows the

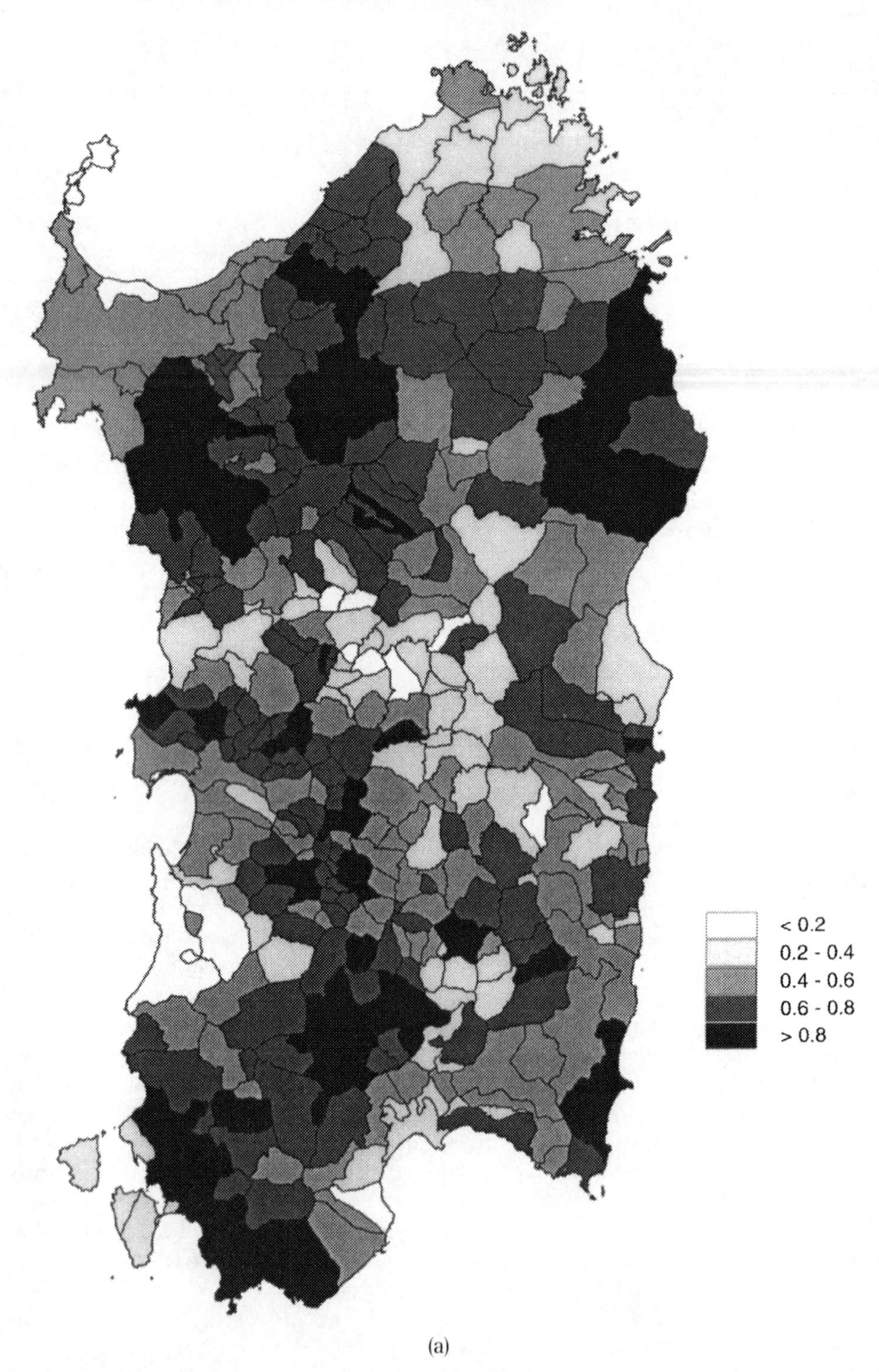

(a)

Figure 26.2 (a) Bayesian estimates of long-term malaria prevalence in Sardinia: proportion of the population affected θ_i

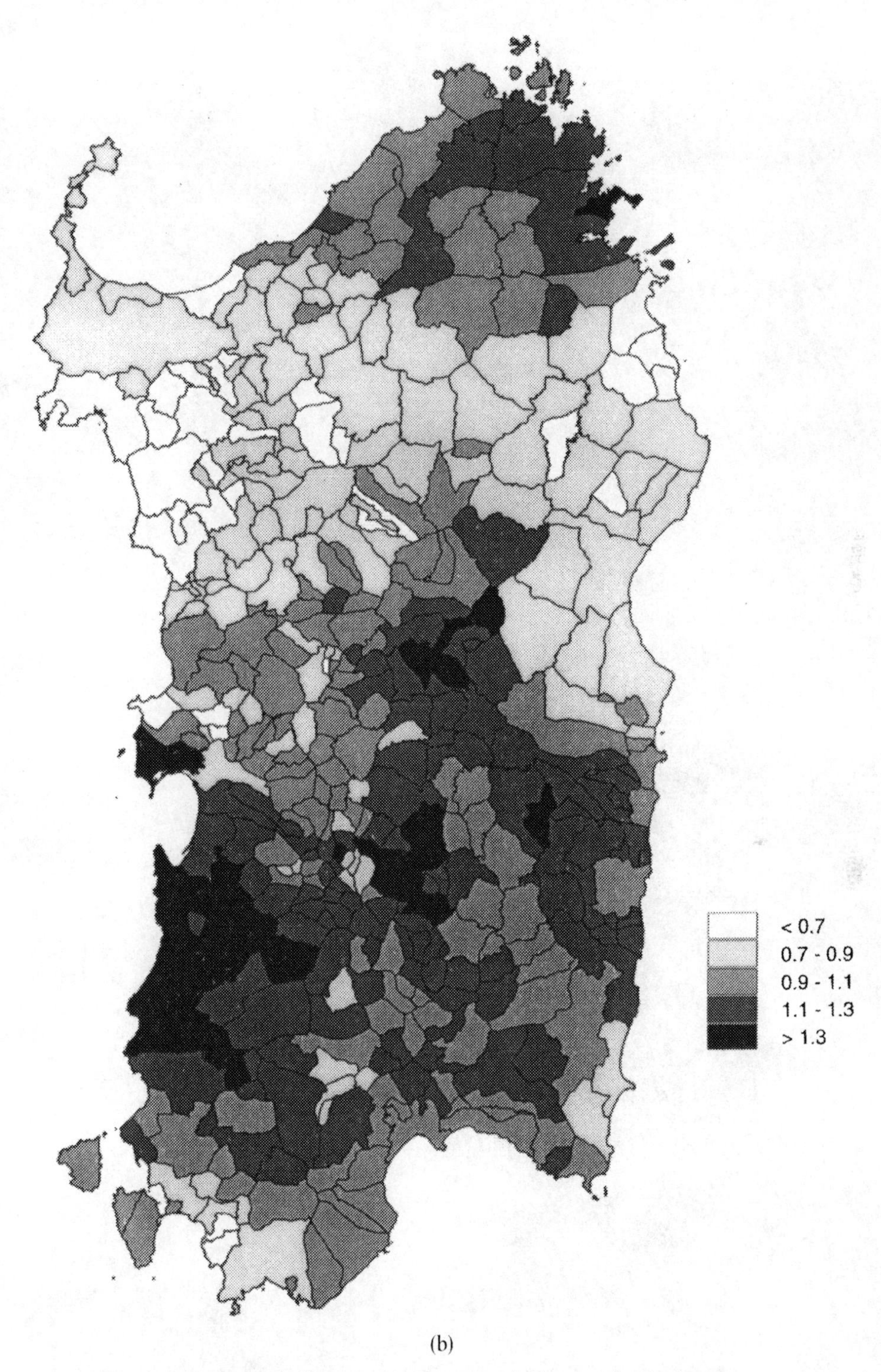

(b)

Figure 26.2 (b) Bayesian estimates of relative risk of IDDM ρ_i

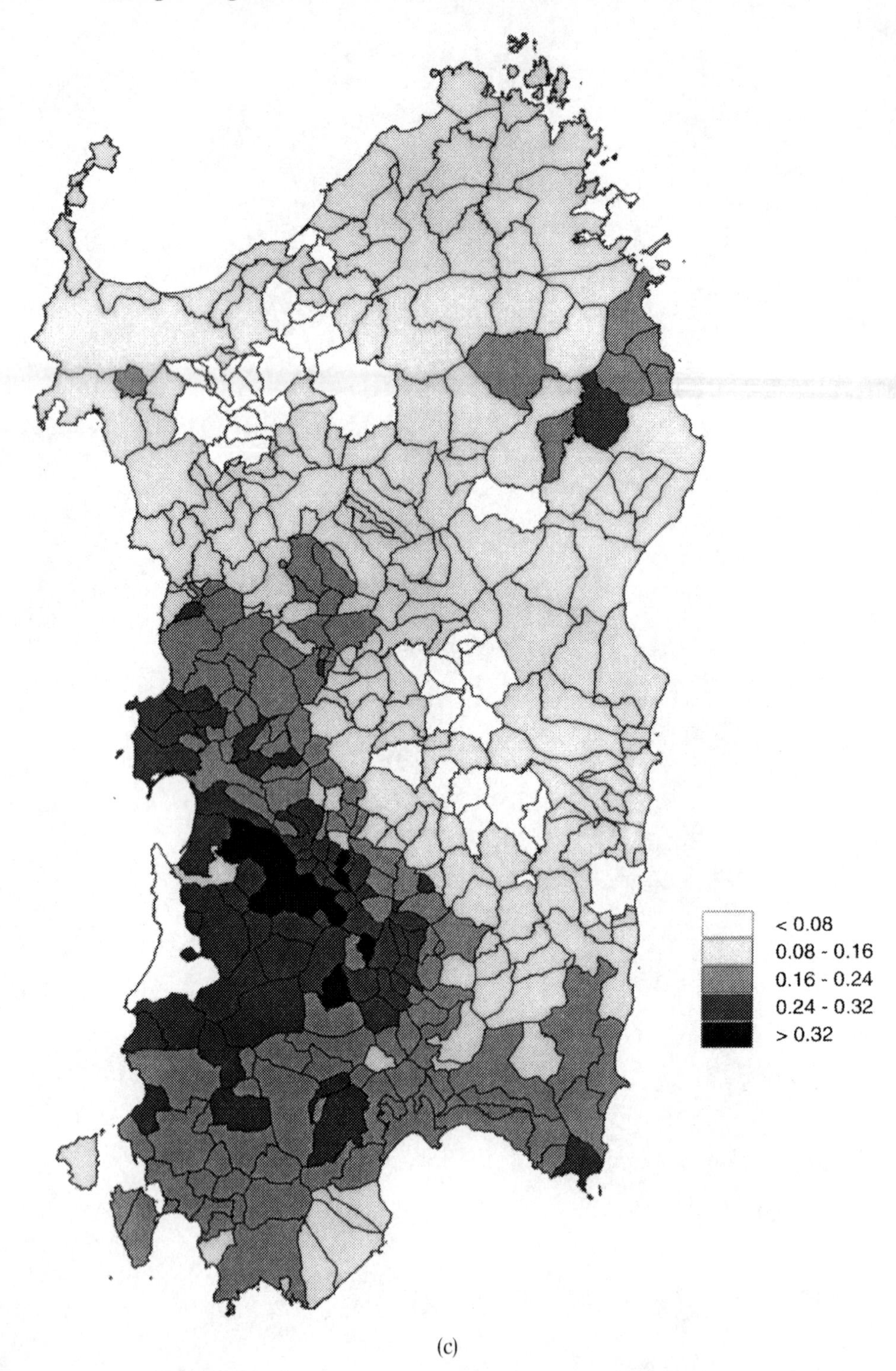

(c)

Figure 26.2 (c) Bayesian estimates of proportion π_i of G6PD-individuals

estimated relative risk of IDDM ρ_i obtained by fitting the reduced model (26.1), (26.4)–(26.6); and the map in Figure 26.2(c) shows the estimated proportion of G6PD- individuals π_i obtained by fitting 26.3.4.

By comparing the maps of IDDM relative risk and of G6PD- proportion with the map of malaria endemicity it is possible to see the direction of the associations of both diseases with malaria that emerges from the estimated regression coefficients. Areas with a low estimated long-term malaria endemicity tend to show higher relative risk for IDDM, while areas with a high estimated long-term malaria endemicity tend to show lower IDDM relative risk.

Considering G6PD deficiency, a positive association with malaria endemicity can be seen in Southern and North-Eastern Sardinia, where low-malaria areas tend to show a lower proportion of G6PD- individuals and vice versa.

Table 26.3 shows the mean over all areas and the corresponding standard deviation of the three components of IDDM log relative risk, i.e. $\log \rho_i$ in model (26.15). βx_i is the component of IDDM log relative risk due to malaria, $\delta\psi_i$ is the component due to a non-malaria factor in common with G6PD deficiency, and α_i represents the area-specific residual component of risk after accounting for the ecological covariates.

The area effects $\{\alpha_i\}$, i.e. the component of IDDM log relative risk not due to malaria or to the non-malaria unknown causal factor in common with G6PD-, show a considerable geographical variation, thus indicating that risk factors other than the ones considered in our model are also responsible for the geographical variation of IDDM.

A perhaps more effective way to show the negative association between IDDM and malaria is to plot $e^{\beta x_i}$, the RR associated with malaria, i.e. the component of the overall RR attributable specifically to the covariate. To do so, we created two groups of areas, according to the estimated endemicity of malaria. We classified the areas belonging to the 25th percentile of estimated endemicity as 'low malaria' and areas belonging to the 75th percentile as 'high malaria'. Figure 26.3(a) shows the plot of $e^{\beta x_i}$ in the two groups of areas.

All low-malaria areas have a relative risk attributable to malaria higher than 1, thus indicating that a low malaria endemicity has the effect of increasing the overall relative risk. On the contrary, all high-malaria areas have a relative risk attributable to malaria lower than 1, thus indicating that a high malaria endemicity has the effect of reducing the overall relative risk.

Similarly, we visualised the association between G6PD deficiency and malaria by plotting the odds of G6PD- associated with malaria ($e^{\psi x_i}$) in the 'high malaria' and 'low malaria' areas (see Figure 26.3(b)).

The positive association between malaria and G6PD deficiency is evident if we observe that all low-malaria areas have an odds of G6PD- attributable to malaria lower

Table 26.3 Mean over all areas and standard deviation (s.d.) of the components of $\log\rho_i$ in model (26.15)

Component of log$_i$	Mean	s.d.
α_i	0.065	0.175
βx_i	– 0.068	0.130
$\delta\psi_i$	– 0.056	0.023
$\log \rho_i$	– 0.059	0.223

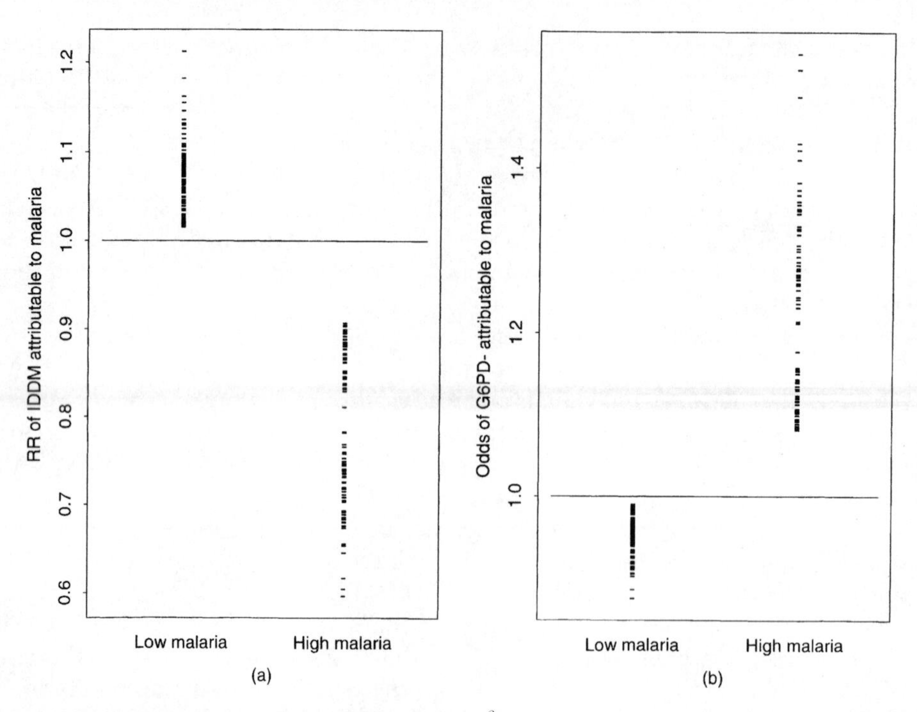

Figure 26.3 (a) Bayesian estimates of the plot of $e^{\beta x_i}$ in (26.15), i.e. the relative risk of IDDM associated with malaria in low- and high-malaria areas. (b) Bayesian estimates of the plot of $e^{\psi x_i}$ in (26.16), i.e. the odds of being G6PD deficient associated with malaria in low- and high-malaria areas

than 1, while all high-malaria areas have an odds of G6PD- attributable to malaria higher than 1.

26.7 DISCUSSION

26.7.1 Substantive conclusions

Sardinians are known to be susceptible to autoimmune diseases. The significant negative association that emerged between long-term malaria endemicity and diabetes relative risk indicates that people living in areas where malaria has been particularly frequent are at less risk of IDDM than those living in areas with a low estimated long-term prevalence of malaria. This is illustrated by the estimated regression coefficient $\beta = -0.062$, which is significantly below zero (Table 26.2).

An alternative way of expressing the negative association between the RR of IDDM and possible genetic selection due to past prevalence of malaria is the relative risk associated with malaria, plotted in Figure 26.3(a), which is consistently lower than 1 in high-malaria areas and vice versa.

A possible interpretation of this finding is that, since malaria has been endemic in the plains of Sardinia for centuries, places with high prevalence of malaria in 1938 are those in which a stronger selection process took place, providing resistance to malaria and also preventing the onset of autoimmune conditions (Jacob, 1992).

The estimated regression coefficient δ, which is almost zero, indicates that there is no evidence of association between IDDM and the component of G6PD- not due to malaria; in other words, IDDM and G6PD deficiency could be considered as conditionally independent given malaria (see Figure 26.1).

The considerable geographical variation of the area effects $\{\alpha_i\}$ indicates that other non-malaria causal factors play a role in determining the geographical variation of IDDM relative risk. This must be taken into account when interpreting the effect of malaria endemicity on IDDM.

Finally, the 95% credible interval of the estimated regression coefficient $\eta = -0.044$ includes zero, but is considerably shifted towards positive values; this indicates that there is a positive though weak association between malaria and G6PD deficiency. This agrees with the results previously reported in the literature and hence supports the hypothesis that the selective action operated by malaria contributed to maintain a higher proportion of G6PD- individuals than expected in normal conditions.

The weakness of the association between G6PD deficiency and malaria, however, might be due to the scarcity of data on the proportion of G6PD-, hence interpretation of this result requires care.

26.7.2 Methodological issues

We have shown that disease maps accounting for covariates measured with error can be constructed using Bayesian hierarchical–spatial models, where spatial smoothing priors are posited for both disease relative risks and underlying covariates. We anticipate that such models will be of particular value when the covariates are themselves incidence or prevalence data for other diseases.

Our choice of prior for malaria prevalence is particularly suitable since it varies between areas in a spatially structured way. Malaria prevalence tends to be higher in low lying and humid regions and lower in the mountains and hills. The spatial prior enables us to obtain a map of the geographical variation of malaria prevalence in which the random variation has been filtered out.

Specifying a spatial smoothing prior for the disease risks (IDDM and G6PD deficiency in our application) represents a way of allowing for unmeasured risk factors (other than malaria) that vary smoothly with location. If the pattern of variation in the covariates is similar to that of disease risk, location may act as a confounder, although it is only a surrogate for other confounding factors. Introducing a spatial prior to model the effect of location thus causes the estimates of the regression coefficients to be controlled for these factors.

By fully acknowledging all potential sources of error and all a priori causal relationships between the variables of interest in our final model, we could investigate different ecological associations simultaneously. We were able to confirm the relationship between malaria prevalence and incidence of IDDM previously found by fitting a less complex model (Bernardinelli *et al.*, 1997). At the same time, our model allowed investigation of the association between malaria endemicity and G6PD deficiency.

Our results show how it is possible to carry out an ecological regression analysis accounting for unprecisely measured covariates and very complex and not fully clarified relationships between the variables of interest. We could achieve this by means of a Bayesian hierarchical–spatial model fitted using Markov chain Monte Carlo.

We wish to emphasise that the results we obtained rely on the assumptions of our model. Indeed, we made the best use of the available data on IDDM malaria and G6PD deficiency to investigate our hypotheses. New data and/or different model assumptions could lead to different results. With models as complex as those considered here it is important to investigate the sensitivity of any conclusions to changes in model specification. In this regard our analysis on IDDM and malaria is continuing.

27

Case Studies in Bayesian Disease Mapping for Health and Health Service Research in Ireland

Alan Kelly

Trinity College Dublin

27.1 INTRODUCTION

Many of the contributions to this volume are essentially methodological in tone; in this chapter the emphasis is firmly on application. Because the author has a direct consultative role with the regional public health bodies in Ireland, such an emphasis is natural. This demands that a prime consideration is to demonstrate the need for, and then ensure the acceptability of, the recent methodological developments for the analysis and presentation of disease and other health outcome measures. In the context of research conducted with a view to public health implementation, it is vital that the relevant professionals be acquainted with the limitations of the heretofore standard approach to disease mapping (using the standardised mortality ratio (SMR)) such as they will have been exposed to through the allied fields of epidemiology and medical geography—see Esteve *et al.*, 1994) and the advantages of adopting recent techniques now being advocated by the statistical community. It must be recalled that these techniques (namely, Bayesian smoothing) are outside the experience of the generality of public health officials, and perceived as highly technical and non-intuitive—features unlikely to encourage easy acceptance.

This chapter illustrates, by way of two case studies (distribution of low birth weight in Dublin and its environs and 'avoidable' deaths from asthma nationally), the advantages of small-area analysis and the consequent requirement for a Bayesian approach to disease mapping. The style of the report is discursive, reflecting the approach adopted by the author in presentations to public health colleagues (Small Area Health Research

Disease Mapping and Risk Assessment for Public Health. Edited by A.B. Lawson *et al.*
© 1999 John Wiley & Sons Ltd.

Unit (SAHRU), 1997a) and therefore avoids a detailed description of the underlying methodology. This has been more than adequately covered by several authors elsewhere. A particularly clear exposition is to be found in Clayton *et al.* (1993), in Cislaghi *et al.* (1995) and again in Mollié (1996) and Chapter 2 in this volume.

27.2 BACKGROUND

The Irish Department of Health's 1995 strategy document—*Shaping a Healthier Future*—in setting out the basis for the proposed developments in health care provision in the near future, repeatedly emphasises the need for relevant and timely information on all aspects of the health care system (Department of Health, 1995). Specific attention is directed to the evaluation of the health care needs of the population, monitoring of the process of care delivery and uptake, and impact in terms of *shifting patterns of mortality and morbidity*. The health strategy document acknowledges the existence of mortality 'black spots' and the need to 'examine variations in the health status of different groups in society' as a basis for the attainment of equity within the system.

It is widely recognised that a potentially fruitful way to represent data on morbidity, mortality, and patterns of health service delivery and uptake is by mapping suitably standardised rates at relevant geographical levels. For example, for state and related administrative purposes, data are routinely generated—or can be identified—variously by small area (District Electoral Division or DED in Ireland), by urban/rural district, by county, or other, larger, geographical unit. Yet, as we illustrate below, real variation in disease and other related measures of health outcome is evident over relatively small geographical distances. This arises because of structural (primarily socio-demographic) differences between areas (with areas comprised of individuals with different risk profiles) leading to *heterogeneity*—and similarities between neighbouring areas sharing similar demographic and environmental profiles leading to area *clustering*. This argues for an analysis directed at a sufficiently disaggregated level to capture adequately substantive differences in underlying risk, whilst of sufficient size to offer realistic opportunities for planners to target intervention.

27.2.1 The traditional approach: consequences for public health

Questions concerning health status, population needs, health care delivery and uptake, are sensitive to both scale and location. Historically, these questions have been addressed at a relatively high level of aggregation; for example, by region or county, with the result that significant variation between smaller areas will have been masked (Kelly and Sinclair, 1997). Yet, while there has been broad recognition of the need for information at a more disaggregated level, two practices have evolved. The first of these, noted already, is to restrict the focus to county-level analysis. This falls within the traditional epidemiological approach and relevant outcomes are typically and adequately expressed in terms of the Standardised Mortality Ratio (SMR)[1] (see Holland, 1991,

[1] If the observed number of events in area i is designated O_i and the expected number (based on regional age and sex specific rates) is E_i, then the SMR is O_i/E_i; this ratio is usually presented as the SMR $\times$ 100.

1993). Of interest will be inter-county comparisons with potential policy and resource implications. The second approach, of recent origin, has been to report SMRs by small area. However, as shown by Clayton and Kaldor (1987) and Clayton *et al.* (1993), reliance on the SMR in the analysis of counts of events in small areas is technically incorrect and maps displaying such SMRs are potentially very misleading. The problem lies with the assumption that the observed number of events in a given small area and in a given time period is Poisson distributed. However, when the true underlying risks are not all equal across areas (an entirely reasonable assumption for most diseases) the observed counts are no longer Poisson distributed—they are said to display *extra-Poisson variation.*[2] In effect, a map of SMRs only partly reflects real differences in risk, because the sampling variability of the risk estimator (which is proportional to population size) may result in the most extreme rates arising in areas having small populations (but often large in physical size and highly visible on the map), producing a disproportionate visual impact based on the *least reliable* data. The problem is exacerbated when considering rare outcomes, such as death from cervical cancer or asthma, that are of considerable public health interest, and the method of presentation is by choropleth map. Mollié (Section 2.2 of Chapter 2 in this volume) discusses and illustrates the problems with mortality data from French 'departments'.

During the last decade, the 'map problem' has been considered afresh from a Bayesian perspective—initially, as the so-called *empirical Bayes* approach discussed in a seminal paper by Clayton and Kaldor (1987) and more recently in terms of a full Bayesian (FB) model offering a more flexible modelling environment (Clayton *et al.*, 1993; Cislaghi *et al.*, 1995; Clayton and Bernardinelli, 1992; Olsen *et al.*, 1996).

27.2.2 The Bayesian approach in brief

In essence, when derived from a reasonable number events (relatively large population size), the Bayesian estimate of the relative risk will be close to the SMR. For less reliable estimates (small population size), the Bayesian estimate is 'shrunk' towards the overall average SMR for the whole area; a plausible a priori solution. However, a better solution involves shrinking the estimate towards the mean of the surrounding areas, because neighbouring areas are likely to share a common aetiological exposure and geographically close areas are, in fact, found to have similar disease rates. 'Borrowing strength' from neighbouring areas in this manner results in smoother and more stable estimates. In practice, 'neighbourhood' is typically defined as areas sharing a common boundary, although alternative definitions are sometimes more appropriate.

Modelling area counts is by means of a random effects Poisson model that decomposes the extra-Poisson variation into two terms: (i) an unstructured component—*relative risk is allowed to vary independently by area*, and (ii) a local spatially structured component—*relative risk is allowed to vary smoothly across neighbouring areas.* Clayton

[2] Assume all areas share the same underlying risk for a specific non-contagious disease, then the observed number of deaths in a given area will have a Poisson distribution. However, in reality, individuals' risks within any small area (e.g. a DED) will vary substantially (individual risks are heterogeneous) with the result that the residual variation will exceed that expected from a Poisson model—this is referred to as *extra-Poisson variation.* Catering for this by allowing relative risks to vary by area is a feature of the Bayesian approach to small-area analysis.

(1994) refers to these components as *heterogeneity* and *clustering*, respectively, and indicates that a measure of the relative contribution of the two types of variability to the total geographical variation may be informative. As in any regression model available area-level covariates may be included in the model. A technical exposition of the Bayesian framework for disease mapping will be found in Mollié (Sections 2.3 to 2.5 of Chapter 2 in this volume).

The estimation of the model parameters is highly computer intensive and requires specialist software employing a Markov chain Monte Carlo algorithm. Suitable software is referenced in Clayton (1994) and Spiegelhalter *et al.* (1997). Determining convergence for such models is important and selected tests have been provided in a library of S-Plus functions by Best *et al.* (1997).

27.3 LOW BIRTH WEIGHT AND AREA DEPRIVATION

27.3.1 Introduction

Low Birth Weight (LBW) is associated with an elevated risk of infant morbidity and neonatal and perinatal mortality. The extent and distribution of LBW is regarded internationally as an important indicator for public health planning (Kelly *et al.*, 1995). It is known that the geographical distribution of LBW is associated with access to, and quality of, local maternity services. It is of interest to determine whether the variation in incidence by small area (i.e. by DED) is related to area deprivation level. A national deprivation index[3] has been developed by SAHRU for public health research and planning in Ireland that will allow any dependency to be investigated. For modelling purposes, each area is assigned a mean-centred deprivation score (range: −2.58 to 5.97), although for mapping and summary purposes the score is converted to an ordinal scale of 1—least deprived to 5—most deprived).

The internally standardised incidence ratios for LBW (SIRs: number of LBWs per DED/number of live births per DED) were computed on the basis of total live births for the 322 DEDs of Dublin County and County Borough (on the east coast of Ireland) for 1988 and 1989.

The total number of LBWs during this two-year period was 1485 from 30 944 live births, giving a rate of 47.9 per 1000 live births. The number of observed LBWs per DED ranged from 0 to 38 with a median of 3.0. Of DEDs, 10% had 1 or 0 cases. The expected LBWs ranged from 0.1 to 36.6 with a median of 3.5. SIRs ranged from 0 to 1000 (median = 96.5) with 18% of DEDs in excess of 150. A map of the raw SIRs is provided (Figure 27.1 (a)). This map shows no clear pattern; neighbouring DEDs apparently differ markedly in levels of relative risk. However, the degree of variation observed is simply not credible. For example, the SIR of 1000 refers to a DED (Lucan North) with one of the smallest populations in the region, having only two births recorded during this period, one of which was an LBW (the expected number was 0.1).

[3] A total of five census-based indicators, widely believed to represent or be a determinant of material disadvantage, were employed in a principal components analysis in the formation of the deprivation index. These are: Unemployment, Low social class, No car, Rented accommodation and Overcrowding. Each of the indicators was first standardised and then subjected to population weighted principal components analysis to determine the appropriate weighting for each variable. Full details can be found in SAHRU (1997b).

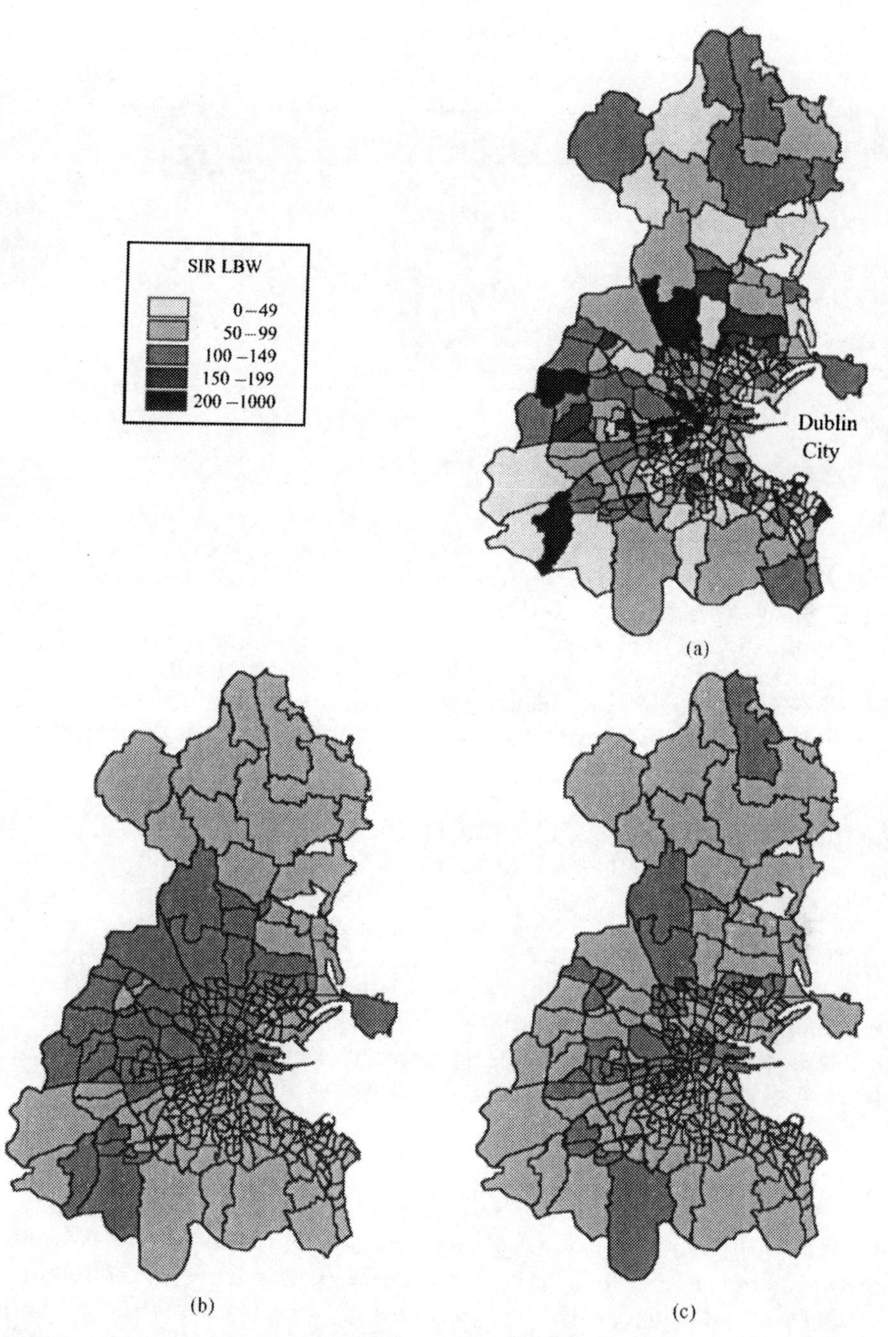

Figure 27.1 Maps of LBW incidence in Dublin County: (a) raw SIR ; (b) Bayesian smoothed SIR by Model 1; and (c) Bayesian smoothed SIR by Model 2

Table 27.1 Summary results of Bayesian models of LBW

	Model 1	Model 2
SIR(min/med/max)	**62/95/146**	**69/92/159**
Heterogeneity component: posterior median (95% CI)	0.001(0.001–0.002)	0.002(0.001–0.012)
Clustering component: posterior median (95% CI)	0.153(0.087–0.291)	0.025(0.004–0.094)
Deprivation coefficient: posterior median (95% CI)	—	0.078(0.050–0.100)

27.3.2 Bayesian models for LBW

Two models have been fitted to the area counts of LBW using Bayesian Ecological Analysis Models (BEAM)—the first (model 1) includes terms for heterogeneity and clustering but without the covariate (the deprivation score), and the second (Model 2) includes the covariate. A summary of both models is presented in Table 27.1 in terms of the estimated minimum SIR, median SIR, and maximum SIR. Also shown are the posterior medians (plus the 95% Bayesian credible interval) for the heterogeneity component (λ^2) and clustering component (κ^2) (in the notation of Mollié in this volume).

Note that for Model 1 the posterior median for the clustering component is two orders of magnitude greater than that for the heterogeneity component, implying strong spatial clustering (as is evident in Figure 27.1 (b)). In Model 2, it is still an order of magnitude larger, even in the presence of the significant covariate, indicating that spatial clustering remains important (see Figure 27.1 (c)). While the heterogeneity component remains small in both models, the decrease in the contribution of the clustering term following the inclusion of the covariate reflects the fact that deprivation shows a strong tendency to spatial clustering (SAHRU, 1997b). As the sign of the covariate's coefficient is positive, relative risk increases with increasing levels of deprivation as anticipated.

27.3.3 Smoothed SIRs for LBW

Figure 27.2 illustrates the extent of smoothing of the raw SIRs on the basis of estimates from Model 1 above (i.e. with no covariate). (NB: the graph is truncated at 400 to facilitate viewing of detail; Bayesian smoothed relative risks are labelled 'FB-SIR'.) The median of the raw SIRs is 96.5, range: 0–1000. The FB-SIRs have a similar median (95), but the range is now 62–146. The latter result is far more credible in public health terms

The scale of shrinkage associated with Model 1 is indirectly a function of population size (or equivalently, the expected counts E_i) and hence uncertainty in the Poisson model. This is made evident in Figure 27.3. Substantial shrinkage (large values of (SIR minus FB-SIR)) occurs for DEDs with small populations (or equivalently, small E_i)

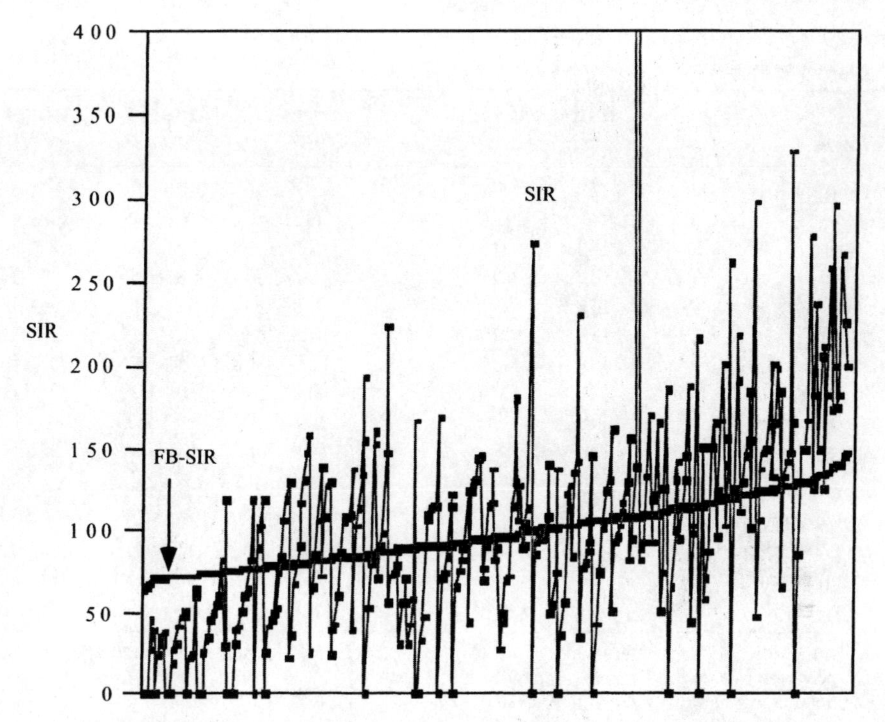

Figure 27.2 Plot of raw SIRs for LBW and Bayesian smoothed SIRs (FB-SIR, according to Model 1) (y-axis truncated at 400; x-axis ordered by increasing FB-SIR)

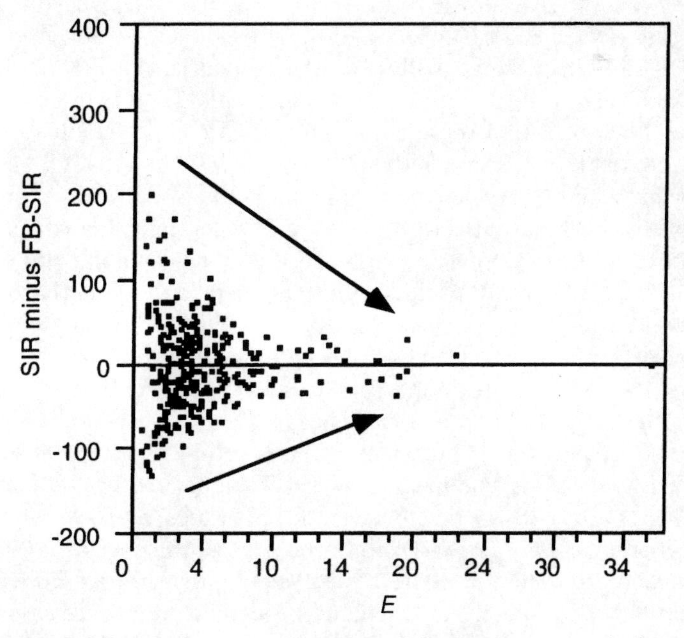

Figure 27.3 Plot of difference between raw SIRs and the Bayesian smoothed SIR (FB-SIR from Model 1) for LBW against the expected counts E (y-axis truncated at 400)

Table 27.2 LBW: raw and Bayesian smoothed SIRs and 95% CIs for selected DEDs ordered according to Bayesian Model 1 [a]

DED name	SIR	Model 1	Model 2	Sig. [d]	SIR 95% CI [b]	Model 1 95% CI [c]	Model 2 95% CI [c]
Arran Quay E	201	149	119	1	87–396	101–214	92–163
Ushers C	227	147	144	0,1,2	117–397	106–205	115–185
Cabra East C	268	143	115	0,1	122–509	106–191	92–147
Arran Quay B	267	142	117		86–624	98–207	92–157
Ushers D	176	142	121		64–383	92–205	94–162
Ballymun B	174	140	149	0,1,2	100–283	101–189	120–187
Kylemore	238	134	131	0,2	123–416	98–182	105–164
Priorswood C	166	129	156	0,2	104–251	91–174	122–196
Swords—Forrest	152	119	99	0	102–219	91–151	80–123
DunLaoghaire—WC	276	100	96	0	110–568	61–156	72–134
Portmarnock North	0	64	75	0	0–60	33–103	50–101
Dundrum-Sandyford	0	67	74	0,1,2	0–66	45–97	56–92
Blackrock-Templehill	0	69	77	2	0–139	41–104	56–97
Balbriggan Urban	48	70	89		13–123	36–115	54–123
Dundrum-Balally	28	71	78	1,2	3–102	46–96	58–96

[a] Model 1: includes terms for heterogeneity and clustering. Model 2: includes terms for heterogeneity, clustering and deprivation (see Table 27.1 and associated text).
[b] 95% Confidence interval based on the Poisson model.
[c] 95% Bayesian credible interval.
[d] Confidence/credible interval does not embrace 100 for model: 0 (SIR); 1 (Model 1); 2(Model 2).

and proportionally less shrinkage for DEDs as the population size increases (i.e. for larger E_i).

Table 27.2 lists selected DEDs (ranked according to estimates—posterior medians—based on Model 1) and the estimated SIRs for various models. The 95% confidence intervals for SIRs and the equivalent empirical Bayesian credible intervals for Models 1 and 2 are given. It will be noted that these latter intervals are considerably narrower than those for SIR and that the interval width for Model 2 is less than that for Model 1. Model 2 confirms the association between area deprivation and LBW incidence and also provides a more precise estimate of the relative risk of LBW for individual areas. Whether or not the respective confidence intervals embrace 100 is indicated in column 5 (Sig.). Of 322 DEDs, 9 had 95% CIs for SIR excluding 100 (Sig. $=0$); only six of these remain 'significantly different from 100' on the basis of either Model 1 (Sig. $=1$) or Model 2 (Sig. $=2$) or both. The majority of DEDs listed in the table with SIRs above 100 are inner-city areas (highly deprived). Those with SIRs below 100 are less deprived suburban areas.

Figure 27.4 gives an overall impression of the contrast between Model 1 and Model 2. This shows a scatterplot of the difference in the point estimates for Model 2 and Model 1 versus the deprivation score. The mean difference is indicated by the horizontal line (this is seen to be about -1.6 units). However, the difference increases with increasing level of deprivation (a spline smoother shows the general trend). The pairwise correlations between point estimates for Model 2 with Model 1, Deprivation score and the raw SIRs are: 0.80, 0.92, 0.34, respectively. Owing to the significant positive association with deprivation, we observe that the point estimates based on Model 2 tend to be lower than Model 1 estimates for lower levels of deprivation and the converse is seen for higher

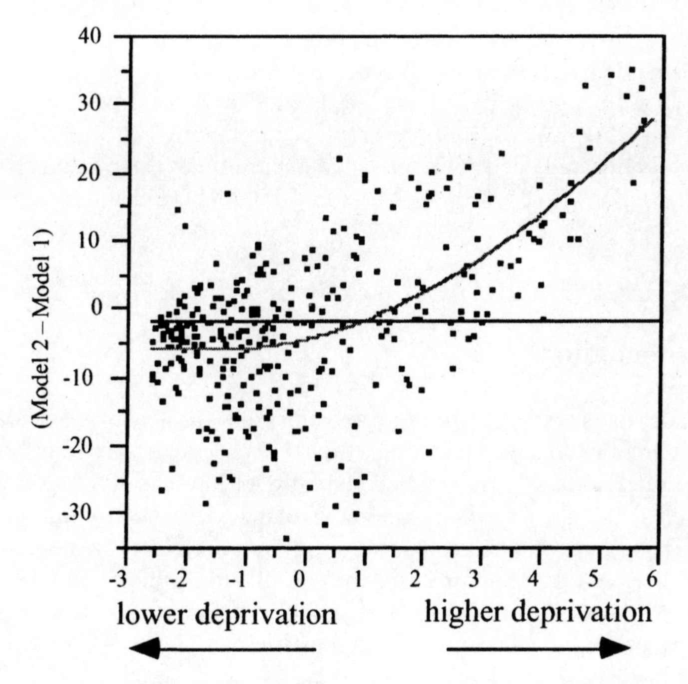

Figure 27.4 Plot of the difference between point estimates of relative risk for Model 2 (with covariate) and Model 1 versus deprivation score. Superimposed are the mean fit (horizontal line) and spline fit

levels of deprivation. The covariate is very influential owing to its tendency to cluster strongly and the association between LBW incidence (noise-free, as in Model 1) and deprivation ($r = 0.6$).

In Summary: LBW incidence is positively associated with deprivation. Higher rates are found in inner-city areas and certain suburban areas immediately to the north and south of the city. This is not an issue of physical access *per se* to maternity services in this instance (three major maternity hospitals are located in the city), but rather it is likely to reflect a lack of 'social access'—in the sense that many of the mothers of LBW infants are late in booking into hospital, they are typically non-attenders at ante-natal clinics, more likely to smoke 'to control weight during pregnancy', and in poorer general health. Areas with relatively high SIRs need to be identified to facilitate targeting of public health measures and improved outreach by hospital maternity services.

27.3.4 Model fitting

Models 1 and 2 were fitted using BEAM and Model 2 was also fitted using BUGS (with negligible differences overall). A 'burn-in' of 5000 iterations was followed by a monitoring run of 10 000, with 1/10 runs saved to file. In view of the large number of parameters to be estimated (all 322 relative risks plus the regression coefficient, plus the random effect variances), a 'burn-in' of 5000 is appropriate. Following the 'burn-in', the number of required iterations was determined by an initial monitoring run of 5000, which was then submitted to the Raftery-Lewis (1992) diagnostic test implemented in CODA. This diagnostic returns the minimum number of runs needed to estimate the model summaries of interest with a given precision. The Heidelberger and Welch (1983) test (also implemented in CODA) was additionally employed following the final run to determine whether convergence had been achieved for each parameter. All parameters passed this test. Typical model run times were 3–4 min on a 250 MHz Pentium II PC.

27.4 AVOIDABLE MORTALITY FOR ASTHMA

27.4.1 Introduction

There is considerable interest within Europe at present in measuring health outcomes that are considered 'avoidable'; for example, mortality for selected conditions in specific age/sex subgroups (Holland, 1991, 1993). Put simply, with proper treatment and access to adequate medical facilities (primary and secondary care), such fatalities should not occur. Where these do occur, health boards and the national Department of Health will need to plan for remedial action. As noted by McColl and Gulliford (1993) managers and health care professionals responsible for distributing resources within the health service need information on health outcomes to ensure that the services provided are effective and located where they are needed.

A series of 15 to 18 indicators has been defined as 'avoidable'—asthma in the 5–44 year age range is one such. Sinclair *et al.* (1995) have shown a general rise in asthma mortality for young adults in Ireland and so it is of interest to investigate the spatial distribution of asthma deaths on a national basis to determine whether there is evidence for urban/rural and regional differences.

Mortality figures for this indicator were compiled between 1986 and 1994 for 65 urban centres (spatially compact) and 23 rural districts (large regions equated with county areas excluding urban centres). Expected numbers of deaths, internally age standardised to 1991 census figures, were computed for each area.

The total number of avoidable asthma deaths during this nine-year period was 163—a very small number overall. The urban/rural breakdown for SMRs is significantly different (Table 27.3). The number of observed deaths per area ranged from 0 to 22. Fully half of the 88 areas had no deaths. The expected number of deaths ranged from 0.06 to 21.6. SMRs ranged from 0 to 1000; the latter refers to Clones urban district with one death observed and 0.1 expected. As noted above with respect to LBW risk, the observed range is not credible.

Table 27.3 Raw SMRs by urban and rural areas

Level	Min	25.0%	Median	75.0%	Max
Rural ($n = 23$)	0	26	77	150	313
Urban ($n = 65$)	0	0	0	103	1000

Wilcoxon Rank Sums test: $\chi^2 = 6.9$, d.f. $= 1$, $p = 0.008$.

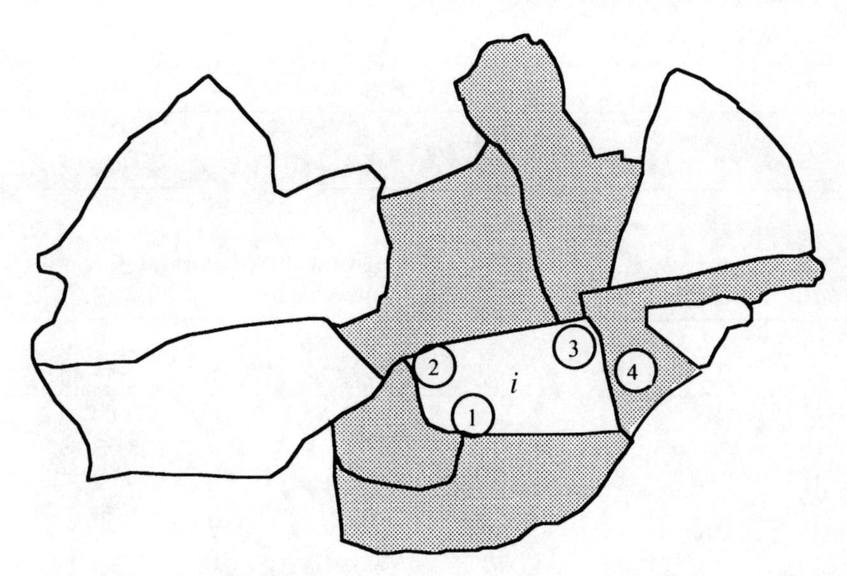

Figure 27.5 Identifying urban and rural neighbours in area i

27.4.2 The 'neighbourhood' problem

'Neighbourhood'—normally defined in terms of polygons with intersecting boundaries—is not feasibly defined in this manner on this occasion. It was found that a mixture of distance and simple contiguity was required. The need for this complication would have been avoided if the focus of analysis had been fixed at county level; however, rates of 'avoidable' mortality were believed a priori to differ between rural and urban areas. All urban centres reside within rural districts. Thus, urban centres within a rural district are considered as neighbours of that district and vice versa. However, not all urban centres within a large district need be neighbours (see Figure 27.5). The district represented by area i in the figure is a neighbour to the shaded districts because it borders on these—this is a standard assumption. There are three urban centres (labelled 1 to 3) in this area. Centres 1 and 2 are neighbours, but neither may be considered a neighbour of Center 3 because the latter exceeds a given threshold distance. Note that Centre 3 would be considered a neighbour of Centre 4. On the basis of considerations of the national grid and demographic factors, only urban centres separated by a maximum distance of 20 km were considered as neighbours. (The robustness of the results to this assumption was tested with the maximum distance set at 15 km and then at 25 km. Differences in SMR point estimates (posterior medians) were negligible.) Due care to ensure symmetry in the neighbourhood relationships is vital before analysis.

27.4.3 Smoothed SMRs

Modelling was undertaken using BUGS (version 0.6). Five thousand iterations were allowed for 'burn-in' before sampling the posterior distribution, followed by a monitoring run of 10 000 iterations. As with the modelling of LBW, the minimum number of

Table 27.4 Summary of Bayesian Smoothed SMRs (FB-SMR) for urban and rural areas

Level	Min	25.0%	Median	75.0%	Max
Rural	64	81	96	111	189
Urban	58	73	94	106	286

Table 27.5 Mortality for asthma (5–44 years) during 1986–1994 for selected urban (U) centers and rural (R) districts. Ranked by Bayesian Model 1[a] SMR value

Name	O	E	SMR	95% CI[b]	Model 1 FB-SMR	95% CI[c]	Sig.[d]
Dundalk (U)	5	1.2	413	132–959	286	120–628	0,1
Louth (R)	6	1.9	313	115–683	189	105–334	0,1
Drogheda (U)	3	1.1	265	53–776	171	75–370	
Kilkenny (U)	2	0.3	556	61–1952	160	51–528	
Laois (R)	8	2.3	335	144–660	137	88–271	0
Thurles (U)	2	0.3	690	76–2430	134	60–361	
Clones (U)	1	0.1	1000	13–5649	133	42–394	
Kerry (R)	1	4.0	25	0–136	69	33–126	
Cobh (U)	0	0.2	0	0–1259	65	24–152	
Cork (R)	5	11.5	43	14–99	64	37–99	0,1
Killarney (U)	0	0.3	0	0–1208	63	15–218	
Macroom (U)	0	0.1	0	0–3510	62	15–203	
Mallow (U)	0	0.2	0	0–1270	61	14–194	
Galway (U)	0	2.5	0	0–145	58	17–143	

[a] Model 1: terms for heterogeneity and clustering.
[b] 95% Confidence interval based on the Poisson model.
[c] 95% Bayesian credible interval.
[d] Confidence/credible interval does not embrace 100 for model: 0 (SIR); 1 (Model 1).

iterations of the Gibbs Sampler required was determined by the Raftery-Lewis diagnostic test and a check on the adequacy of the final model output was determined by means of the Heidelberger and Welch test.

Two Bayesian models were fitted: a simple exchangeable risks model (this is incorporating only an unstructured random effect allowing for extra-Poisson variation and ignoring spatial context) and the random effects Poisson model allowing for both extra-Poisson variation and spatial correlation (see Spiegelhalter *et al.*, 1996, Section 11.2). Examination of the results for the exchangeable risks model suggests that it is inadequate and that spatial effects are clearly evident (not reproduced here). Additionally, the width of the Bayesian credible interval for SMRs was consistently narrower for the full Bayesian model as compared with the exchangeable risks model.

The distribution of the Bayesian smoothed SMRs (FB-SMR) by urban and rural location is summarised in Table 27.4. A Wilcoxon rank sums test indicates that, globally, the difference is no longer statistically significant ($p = 0.47$). Selected results from this model are provided in Table 27.5 and the smoothed SMRs are mapped for urban centres (Figure 27.6(a)) and rural districts (Figure 27.6(b)). Table 27.5 lists only those areas with estimated posterior median SMR above 130 or below 70. As expected, the contrast with the original SMRs is very pronounced at the extremes of the latter with the smoothed values again offering more credible results. The maximum estimated SMR is 286 (Dundalk (U)) followed by Louth (R) and the minimum is 58 (Galway (U)). It is noteworthy that the town of Dundalk is in Country Louth. There is some evidence for an excess of deaths in the urban centres and rural districts in the north-east of the country (labelled

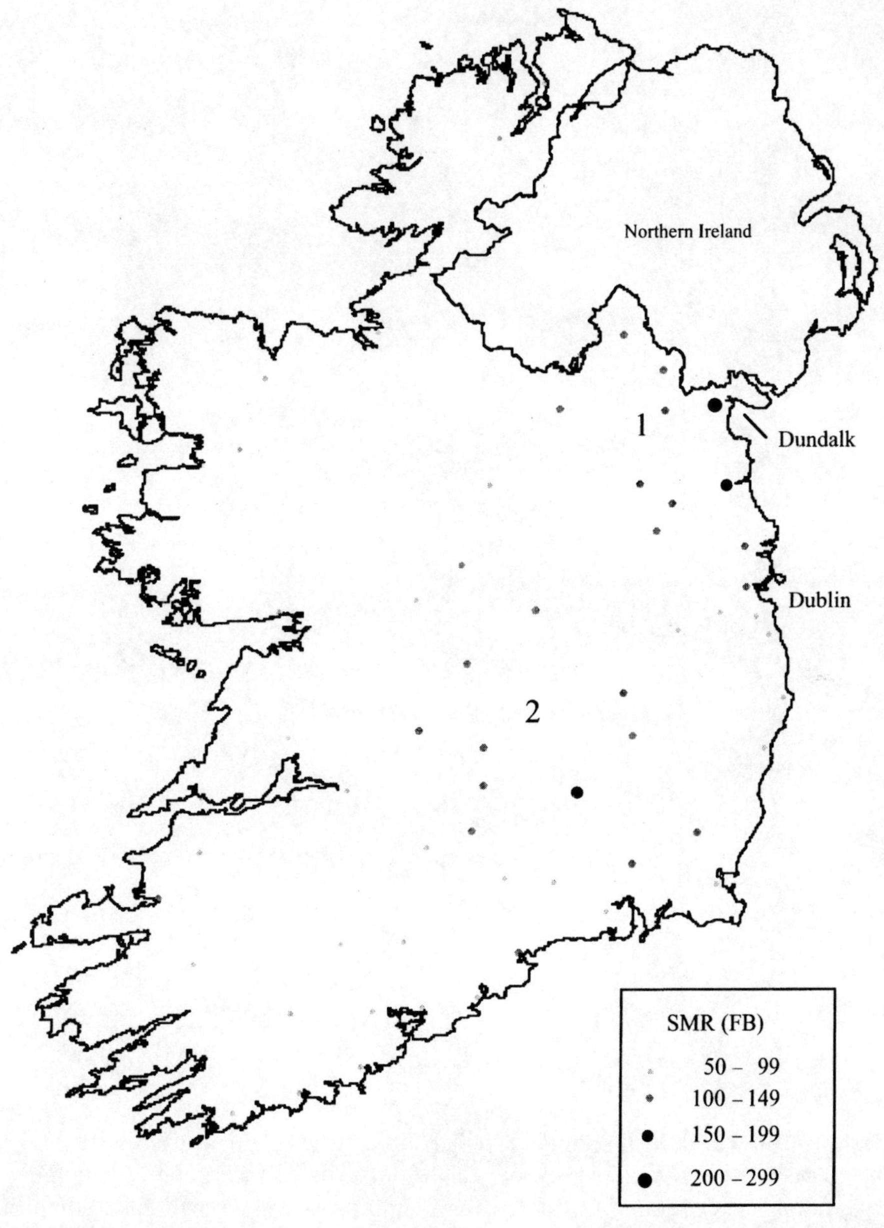

Figure 27.6 (a) Map of Bayesian smoothed SMRs for asthma mortality (5–44 years) (urban centres). (b) Map of Bayesian smoothed SMRs for asthma mortality (5–44 years) (rural districts)

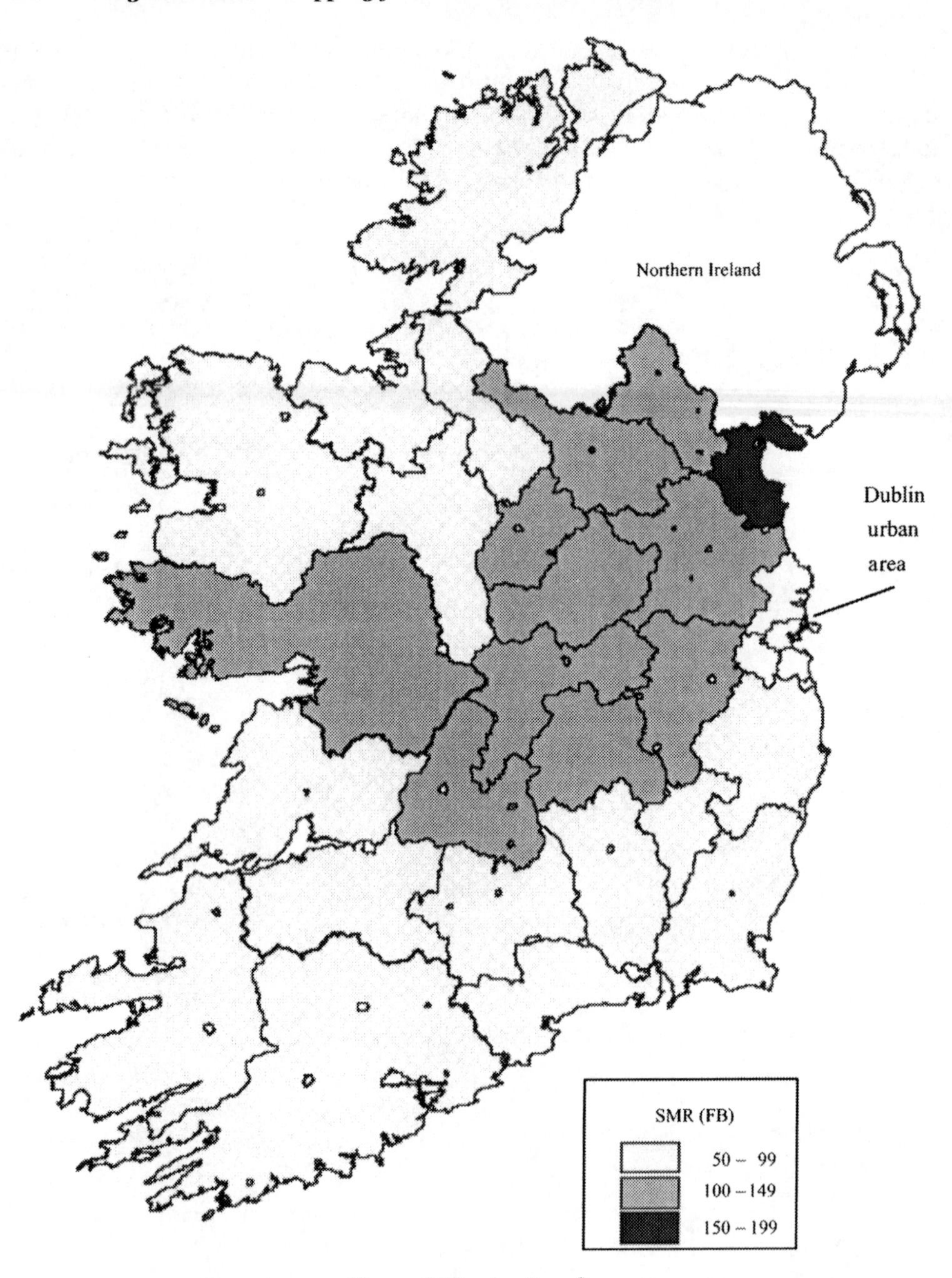

Figure 27.6 *(continued)*

1 in Figure 27.6(a)) and the midlands (labelled 2 in Figure 27.6(a)) and a deficit along the southern, western and northern seaboards, although the 95% CIs generally include 100 with the notable exception of Dundalk (U) and Louth (R) in the north-east. Comparing both maps suggests that urban centres with higher than average SMRs are mostly located within rural districts also with higher than average SMRs.

These results must be related to the quantity and quality of service provision nationally, but firm conclusions from this study must await the results of the analyses of the remaining indicators of 'avoidable' mortality. At that stage, the possible reasons will be considered with the assistance of the departments of public health in each health board.

27.5 CONCLUSION

Analysis and mapping of geographical variation in disease rates is a natural means of enabling spatial patterns and neighbouring clusters or gradients in rates to be discerned readily. Thus, inequalities between districts and regions may be identified and targeted to receive improved health care delivery. Epidemiologists have long found maps useful in generating hypotheses in respect of possible environmental and social or material correlates of disease. Evidence of an unusually high incidence or clustering in space or time of a disease can draw attention to an environmental hazard, or provide an early warning of an outbreak of infectious disease. With the rapid expansion of health service/health outcomes research, decision-makers, health care providers and public health researchers can benefit from ready access to key information suitably analysed by small area and presented in an accessible format, i.e. by map.

The term 'Bayesian Disease Mapping' is generic, and although the majority of published applications to date have been concerned with disease rates, there is no reason that these tools need be confined to analysing disease. Indeed, the potential is considerable for application to other health-related outcomes (e.g. in SAHRU the following areas are being investigated: geographical distribution of attendees at drug clinics, court appearances and sentencing patterns for crime, and health service uptake by households). Recent papers by Spiegelhalter (1998) and Congdon (1998) give additional ideas on other possible applications in the area of health service research.

In conclusion, the demand for information at small-area level for public health research and service planning is increasing rapidly and the advantages of the Bayesian approach will be most evident in such circumstances where data are scarce and raw SMRs (as a measure of relative risk) are liable to be particularly unstable. However, in spite of the obvious (to statisticians!) benefits of modern small-area disease mapping methods, their introduction to, and acceptance by, public health epidemiologists and health planners requires careful preparation and consultation. In SAHRU, we have taken every opportunity to present at special seminars and national conferences and have undertaken to produce a series of reports targeted to public health specialists and the Department of Health. This programme is essential to ensure rapid and broad acceptance of the still novel Bayesian approach.

ACKNOWLEDGEMENTS

SAHRU is part supported by a grant from the Irish Health Research Board. I particularly wish to acknowledge the contribution of my colleague Dr Hamish Sinclair to the work discussed here. I am grateful to the referee for very valuable comments.

28

An Analysis of Determinants of Regional Variation in Cancer Incidence: Ontario, Canada

Stephen D. Walter and S. Martin Taylor

McMaster University

Loraine D. Marrett

Cancer Care Ontario

28.1 INTRODUCTION

This chapter describes an analysis of cancer incidence in Ontario. Our objective was to identify possible determinants of the regional variation in risk that we had observed previously and to identify if any locations or regions had significantly deviant incidence, after taking known risk factors into account. Risk factor information was derived from the Ontario Health Survey, the census, and other sources. Twenty-four types of cancer were investigated. Weighted least squares regressions were fit, with an analysis of the spatial autocorrelation of the residuals. Spatial regressions were also fit, directly incorporating a spatially correlated error structure.

A variety of significant risk factors was identified for the various cancer sites; effects were usually consistent with other results in the aetiological literature. Slightly better fits to the data were usually obtained with the spatial regressions, but there was typically no meaningful change in the risk coefficients. The number of outlier incidence values was close to expectation. They were dispersed geographically, and occurred in a variety of cancer types. The findings provide reassurance that there are no areas with systematically different risk, once known risk factors are accounted for. A few cancers provide limited evidence of residual regional effects that may warrant further investigation.

Disease Mapping and Risk Assessment for Public Health. Edited by A.B. Lawson *et al.*
© 1999 John Wiley & Sons Ltd.

28.2 BACKGROUND AND OBJECTIVES

The purpose of this study was to investigate the determinants of geographical variation in cancer incidence in Ontario. Previous analyses had demonstrated significant spatial patterning of incidence for various cancers in Ontario (Walter *et al.*, 1994). This was a starting point for examining the effects of plausible determinants of geographical variation in incidence, using regression models that incorporated socioeconomic, lifestyle and environmental factors.

Geographical analysis is often used by epidemiologists to maintain surveillance of the regional variation in health, or to identify local anomalies. The many national and regional atlases of mortality and morbidity and associated analyses are good examples of this approach (e.g. Walter and Birnie, 1991; Pukkala, 1989; Pickle *et al.*, 1989, 1996; WHO, 1997; Cislaghi *et al.*, 1990). Some analyses attempting to explain the geographical variation in health have focused on socio-economic indicators and environmental variables such as population density, urbanicity, or latitude (Aase and Bentham, 1996; Charlton, 1996; Kafadar *et al.*, 1996; Sinha and Benedict, 1996; Jones *et al.*, 1992; Nasca *et al.*, 1992; Howe *et al.*, 1993). A smaller number of analyses have examined geographical information on disease-specific risk factors, for example late age at first birth and breast cancer (Sturgeon *et al.*, 1995) and occupational hazards for bladder cancer (Yamaguchi *et al.*, 1991).

Similar techniques have been applied to examine the effect of cancer screening programs (e.g. Lazcano-Ponce *et al.*, 1996) and variation in the utilisation of health care services (Whittle *et al.*, 1991; Roos and Roos, 1982; Paul-Shaheen *et al.*, 1987; Coulter *et al.*, 1988).

Overviews of the use of aggregated health data, using the so-called ecological design, to assess environmental effects are available elsewhere (Walter, 1991a,b; Greenland and Robins, 1994). A related issue is the statistical assessment of geographic patterning in the data, using measures of spatial aggregation or clustering (Walter, 1992a,b, 1994).

Ontario is an especially suitable region for this type of research for several reasons, including:

- the interesting regional patterns in cancer risk previously observed;
- the good quality of the provincial registry data;
- the availability of data on plausible determinants; and
- the relatively large population.

As described later in the chapter in more detail, two regression approaches were adopted: (i) conventional regression models, in which inferences about the possible regional effects are based primarily on a spatial autocorrelation analysis of the residuals; and (ii) spatial regression models that directly incorporate a correlated error structure, based on contiguity criteria.

Our earlier work (Walter *et al.*, 1994) indicated that about one-third of the sex–site combinations examined showed evidence of regional patterns in incidence. We now build on the previous results by identifying plausible risk variables for inclusion in regression models, within the constraints of data availability for Ontario. Our focus was on socio-economic and lifestyle risk factors, but we also included selected occupational variables. The primary purposes of the modelling were to ascertain whether the residuals show spatial patterning, which may indicate the effects of environmental factors warranting further investigation, and to identify specific locations with significantly different risk from the province.

28.3 METHODS

The data ascertained for this project included: age and sex-specific cancer incidence data for various cancer sites, with the Public Health Unit (PHU) as the unit of analysis; and selected primary and derived variables in the Ontario Health Survey (OHS) and the census.

28.3.1 Cancer incidence data

The Ontario Cancer Registry provided the annual numbers of cases of 24 types of cancer by age, sex and PHU, for 1980–91. The time period provides sufficient cases to give relatively stable estimates of the incidence rates in the majority of the PHUs for a wide range of cancer sites. A small number of cases below age 15 or over age 84 were not included, for compatibility with the OHS; cases with unknown residence were also excluded. Population denominators were obtained from the census, with linear interpolation for non-census years. Sex-specific, age-standardised incidence rates (SIRs) were computed, with Ontario as the standard population in the direct method.

28.3.2 Ontario Health Survey (OHS) data

The 1990 OHS is a population-based survey that provides data on social, economic, physical, behavioural, nutritional and other health indicators. Its main instrument was completed by 46 228 individuals aged 15–84, stratified into 44 PHUs, with two small PHUs being combined. It used a complex cluster sample design, incorporating study weights that reflect the probability of selection and non-response at the household and individual levels. The individual study weights were rescaled to analytical weights, which were then used in the aggregation of risk variables within PHUs.

We aggregated the data for six PHUs in Toronto, because of a known lack of specificity in reporting cancer incidence in this area. This resulted in 37 spatial areas for analysis. The selected OHS variables were also standardised (as rates or mean values) to the Ontario age distribution.

28.3.3 Census data

Variables on education, occupation, income, and population density were drawn from the census and used as indicators of socio-economic status and the residential environment. The 1986 census, being approximately mid-way through the cancer incidence period, may represent 'typical' exposure values for the entire period. However, some variables were not available by sex or age, and so could not be sex-specific or age-standardised. Pilot analyses suggested that the regression results might be substantially biased by using unstandardised variables. Accordingly, the data on education and occupation were taken instead from the 1991 census, where standardisation was possible. Data on immigration were also taken from the 1991 census and used for selected cancer sites.

28.3.4 Statistical analysis

The dependent variable in the regressions was the log (base 10) of the directly standardised rate. For each cancer site, a set of candidate independent variables was included in a conventional weighted least squares regression, specified as

$$\boldsymbol{Y} = \boldsymbol{X}\beta + \boldsymbol{U},$$

where $\boldsymbol{Y}$ is an $(n \times 1)$ vector of the values log (SIR), $\boldsymbol{X}$ is an $(n \times p)$ matrix of independent variables, with β being the corresponding $(n \times 1)$ vector of coefficients, and $\boldsymbol{U}$ is the $(n \times 1)$ vector of errors such that $E(\boldsymbol{U}) = 0$ and $E(\boldsymbol{U}\boldsymbol{U}^{\mathrm{T}}) = \sum$. Each PHU was weighted according to its average population for the time period, but the errors were considered to be independent between PHUs, so off-diagonal elements of the variance–covariance matrix $\sum$ were taken as zero. A stepwise backwards elimination algorithm was used, and a significance level of 0.10 in the partial test was required for a given term to be retained in the final model. Candidate variables were based on a review of the aetiological literature. The general socio-demographic census variables were considered as candidate variables for every site. For cancers where both male and female data exist, a further regression, which included the significant variables from both of the sex-specific models, was calculated. This allowed comparisons of effects between sexes.

The regressions were summarised through the regression parameter for each significant term (showing its effect in terms of the original units of measurement), its standard error and partial p-value and the model R^2. The magnitude of each effect was also expressed through the associated relative risk (RR), computed as the expected (multiplicative) change in disease incidence associated with a change of two standard deviations in the independent variable.

Maps of the cancer rates for each site were produced, using unclassed choropleth shading. To examine the fit of the model, the absolute and studentised residuals were also mapped, and their distributions were checked for normality using a Kolmogorov–Smirnov test. To evaluate the spatial patterns, we first calculated the first-order spatial autocorrelation (SAC) in the residuals over all pairs of geographically contiguous PHUs. Specifically, the SAC was estimated through Moran's I-statistic, defined as

$$I = \frac{n \sum_{i=1}^{n} \sum_{j=1}^{n} w_{ij}(y_i - \bar{y})(y_j - \bar{y})}{\sum_{j=1}^{n} w_{ij}(y_i - \bar{y})^2 \left(\sum_{i \neq j} \sum w_{ij}\right)}.$$

The coefficients w_{ij} represent the continguity or closeness of areas i and j. We used a set of binary coefficients such that $w_{ij} = 1$ if areas i and j share a common boundary, and $w_{ij} = 0$ otherwise. The value of the SAC should suggest the magnitude of the regional effects not allowed for in the regression model. Secondly, we carried out a further set of regressions in which the spatial structure of the residuals was modelled directly. The model was defined by

$$\boldsymbol{Y} = \boldsymbol{X}\beta + \boldsymbol{U}, \quad \text{with } \boldsymbol{U} = \rho \boldsymbol{W}\boldsymbol{U} + \varepsilon.$$

Here ε is a vector of independent errors, ρ is the value of the SAC, and $\boldsymbol{W}$ is the matrix of contiguity coefficients w_{ij}. To do this we began with the final model identified in the first conventional regression: we then fit a spatial regression with a fixed value of ρ. This is most conveniently achieved by using spatial differencing and fitting the model $(\boldsymbol{I} - \rho \boldsymbol{W})\boldsymbol{Y} = (\boldsymbol{I} - \rho \boldsymbol{W})\boldsymbol{X}\beta$, as suggested by Bailey and Gatrell (1995). This was repeated with various values of ρ until the mean squared residual error was minimised (to a tolerance of 0.05 in the value of ρ). The INFO-MAP software was used for this purpose (Bailey and Gatrell, 1995). The optimal value of ρ was then taken as a measure of the residual spatial structure in the data, after accounting for the effects of the included independent variables; thus, again the value of ρ should indicate the strength of the regional effects not directly associated with the independent variables in the model. We compared the coefficients and their standard errors for the independent variables in the conventional and spatial models, and also mapped the residuals.

28.4 RESULTS

Table 28.1 summarises the findings for lung cancer. The regression coefficients are directly comparable between sexes, while the associated RR permits comparisons of effects between risk factors. The associated RR involves the distribution of exposure across PHUs, and may be relatively large if exposure rates vary substantially.

Cigarette smoking (pack-years) has a significant effect in both sexes, with an associated RR of approximately 1.2 in males and females. In males, there are additional effects of occupational exposure to fibres, coal tar and metals, and a small association with population density. In females, the occupational effects were not significant, except for a borderline negative effect of occasional exposure to coal tar. There was a negative association with the percentage employed in industry. The R^2 values were 0.84 in males and 0.69 in females.

Table 28.1 also shows the combined models for each sex when the significant terms from both sexes were included. There was little change in the magnitude or significance of the effects. Because more terms were included, the significance of each factor was somewhat reduced, but the qualitative conclusions did not alter substantially.

The SAC for the log incidence rates was quite high, namely 0.516 for males and 0.663 for females, indicating relatively strong regional effects. After computing the regression model, the residual SAC for males was reduced to the -0.059, a value close to zero and suggesting that the regional effects in incidence have been taken account of by the regression variables. Accordingly, there is no advantage in fitting further spatial regressions for males.

In females the residual SAC was 0.291, suggesting that some regional effect still remains. In further spatial regressions, the value of ρ needed to minimise the residual SAC was 0.4. Table 28.2 compares the associated RRs in the standard model and the

Table 28.1 Results of regression analysis: lung cancer

					Combined model			
	Males		**Females**		**Males**		**Females**	
Independent Variable	*b*	Associated RR	*b*	Associated RR	*b*	Associated RR	*b*	Associated RR
Smoking pack-years	0.0197	1.22	0.0213	1.15	0.0193	1.21	0.0221	1.15
Fibres								
(occasional +)	0.0061	1.09	—	—	0.0074	1.12	– 0.0072	(0.98)
(often +)	– 0.0257	0.88	—	—	– 0.0259	0.88	0.0057	(1.01)
Metals (occasional +)	0.0060	1.07	—	—	0.0065	1.08	– 0.0096	(0.97)
Coal tar								
(occasional +)	– 0.0073	(0.93)	– 0.356	(0.95)	– 0.0063	(0.94)	– 0.0387	(0.95)
(often +)	0.0238	1.09			0.0180	1.07	– 0.0129	(0.99)
Population density	0.1742	1.05	—	—	0.1108	1.03	– 0.0132	(1.00)
Percent low income	—	—	0.0026	(1.03)	– 0.0012	(0.95)	0.0027	(1.04)
Percent employed in industry	—	—	– 0.0066	0.85	0.0024	(1.03)	– 0.0059	0.86
R^2 for model		0.84		0.69		0.86		0.71

Note: All terms shown are significant with $p < 0.05$, except terms in parentheses which are significant at $p < 0.10$.

Table 28.2 Comparison of standard and spatial models (lung cancer, females)

Independent variable	Associated RR Standard model	 Spatial model
Smoking, pack-years	1.15	1.10
Coal-tar (occasional)	0.95	0.94
Percent low income	1.03	1.04
Percent employed in industry	0.85	0.89
R^2 for model	0.69	0.76

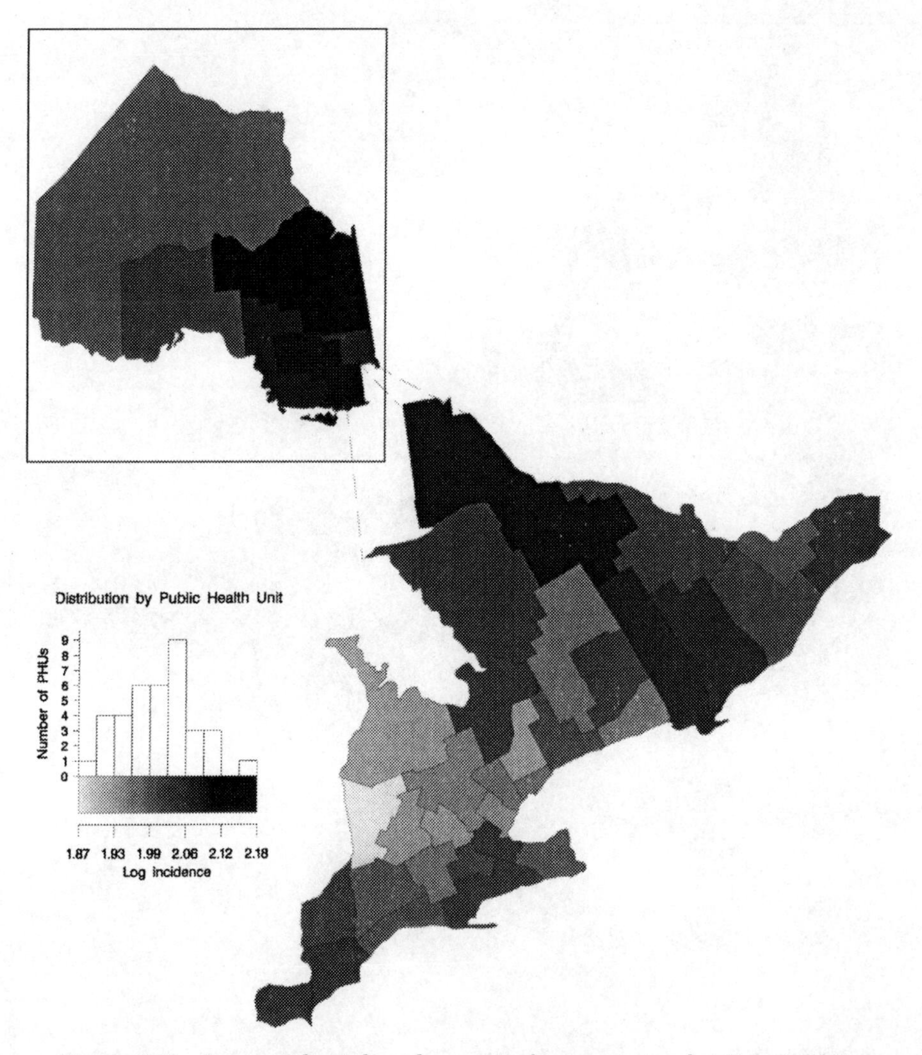

Figure 28.1 Log age-adjusted incidence rates, lung cancer, males, Ontario, 1980

spatial model with $\rho = 0.4$. There is a slight reduction in the smoking effect, but otherwise only small changes are seen. There is a small increase in R^2.

Figures 28.1 and 28.2 map the log incidence rates for males and females. Both show relatively strong regional effects, with elevated risks in the central and northern areas and lower rates in the south-west. In males, the residual pattern was somewhat random, while in females there was a modest suggestion of a remaining regional effect, with higher risk in the east and central parts of the province. Both distributions were approximately normal. For males there were no extreme outliers: all points except one had studentised residuals between $+2$ and -2. The pattern of residuals in females was even less remarkable: the only outlier had a studentised residual of -2.2, suggesting a modest deficit of cases compared with expectation.

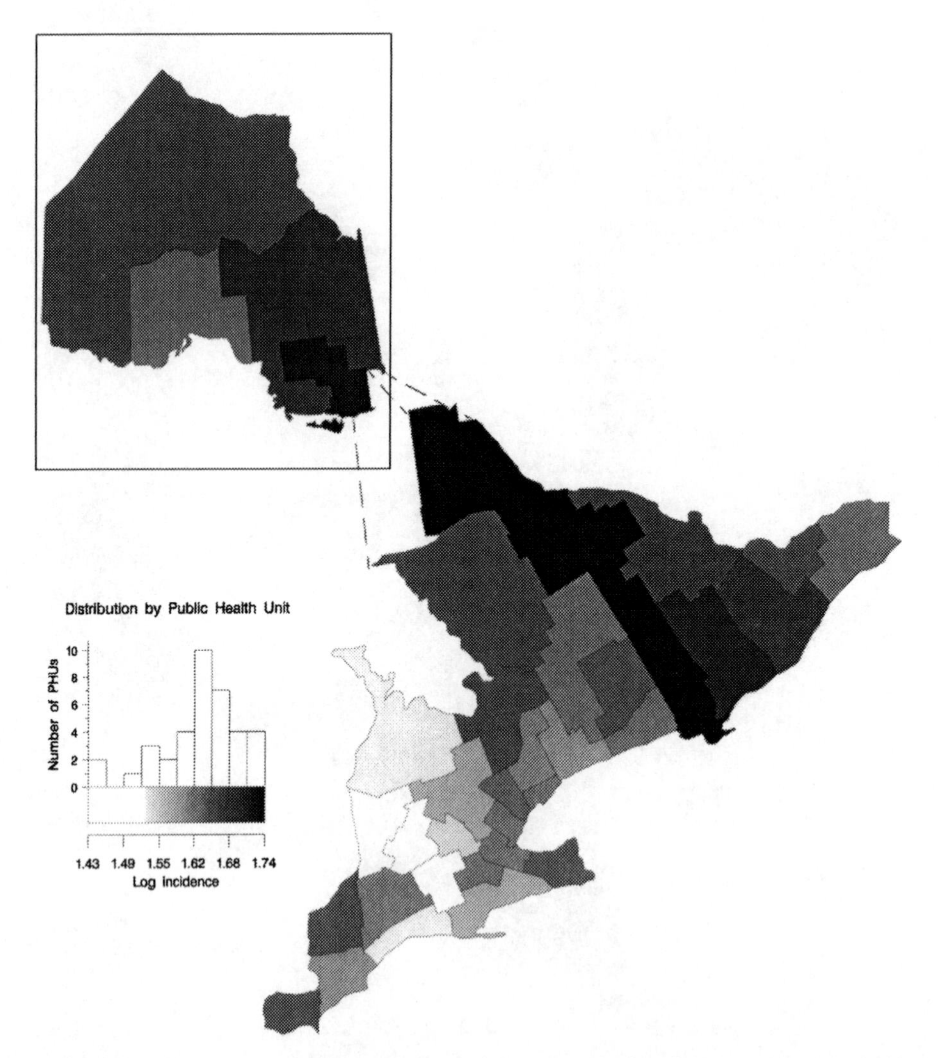

Figure 28.2 Log age-adjusted incidence rates, lung cancer, females, Ontario, 1980–1991

Table 28.3 Summary results for selected cancers: estimated relative risks[a] for significant variables, by sex
(a) Aero-digestive track

Variable		Site											
		Oral cavity		Oesophagus		Stomach		Pancreas		Larynx		Lung	
Class	**Specific**[b]	M	F	M	F	M	F	M	F	M	F	M	F
Socio-demographic	Population density[c]				1.15							1.05	
	Percent households with low income[c]				0.87[d]			0.93		1.17	1.21		1.03[d]
	Percent with i grade 9 education					1.20	1.35	1.29					
	Percent employed in industry	1.40					0.75	0.89[d]					0.85
	Percent semi-skilled/unskilled labourers								1.13				
	Percent immigrants				0.55[d]	1.09	2.78						
	Percent immigrants, ex from US/W. Europe				1.75		0.48						
Smoking	Percent ever-smokers						1.24	1.13		1.11			
	Mean pack-years			1.14					1.16		1.28	1.22	1.15
Diet	Mean fruits/vegetables (servings per day)	1.15							1.12		1.21[d]		
	Mean coffee (cups per day)							0.94[d]					
Alcohol	Percent having <12 drinks per week	1.11[d]	0.88	1.12[d]									
Occupation	Percent glass fibre/asbestos ≥occasionally			1.42		1.17						1.09	
	⩾ often			0.85	0.77	0.90[d]				0.88	0.79	0.88	
	Percent metal fumes ≥occasionally		0.83			1.14[d]						1.07	
	⩾ often					0.85				1.12			
	percent solvents, etc ≥occasionally	0.70		0.75		0.66			0.97				
	⩾ often				1.23								
	Percent coal tar/pitch ≥occasionally					1.20						0.93[d]	0.95[d]
	often											1.09	
Model R^2		0.42	0.45	0.51	0.48	0.83	0.73	0.49	0.50	0.70	0.49	0.84	0.69

(b)

Variable		Site											
Class	**Specific**[b]	**Colorectum** M	F	**Gall-bladder** M	F	**Breast** F	**Cervix uteri** F	**Corpus uteri** F	**Ovary** F	**Prostate** M	**Testis** M	**Thyroid** M	F
Socio-demograhic	Population density[c]						1.06				1.14	1.13	
	Percent households with low income[c]				1.13			1.07				0.81	
	Percent with i grade 9 education					0.95	1.12			0.85	0.84		
	Percent employed in industry							1.10		1.17		0.77	
	Percent semi-skilled/unskilled labourers											1.31	
Smoking	Percent ever-smokers									0.91			0.69
	Mean pack-years						1.16	0.92					1.25[d]
Diet	Mean fat (kg)								0.84				
	Mean fat/total energy (%)	1.09	1.08					1.09	1.14				
Body mass/ physical activity	Percent with BMI 27			0.84				0.88					
	Mean energy expenditure (kcal/kg/day)	1.05											
	Percent inactive	1.06											
Reproductive	Percent with no births					1.03							
	Percent taking female hormones (other than oral contraceptives)							1.12					
Model R^2		0.41	0.44	0.30	0.14	0.40	0.39	0.67	0.23	0.30	0.19	0.51	0.67

(c)

Variable		**Site**													
		Bladder		**Kidney**		**CNS**		**Hodgkin's disease**		**Non-Hodgkin's lymphoma**		**Multiple myeloma**		**Leukaemia**	
Class	**Specific**[b]	**M**	**F**	**M**	**F**	**M**	**F**	**M**	**F**	**M**	**F**	**M**	**F**	**M**	**F**
Socio-demographic	Population density[c]													0.93	0.94
	Percent households with low income[c]		1.12	1.04[d]										1.09	1.12
	Percent with < grade 9 education	0.90	0.90												
	Percent employed in industry				0.88	1.19		0.92[d]		0.86	0.93			1.18[d]	
	Percent semi-skilled/unskilled labourers				1.17					1.08					
Smoking	Mean pack-years	1.10[d]													
Diet	Mean coffee (cups per day)	0.93													
Occupation	Percent solvents, etc: ≥occasionally		0.92			0.81						1.12	1.16	0.85[d]	
	Percent pesticides: ≥occasionally									1.22					
	Percent longest job in farming									0.85					
Model R^2		0.19	0.33	0.10	0.41	0.32	0.00	0.10	0.00	0.40	0.11	0.19	0.17	0.51	0.29

(d)

Variable		**Site**			
		Lip		**Melanoma of skin**	
Class	**Specific**[b]	**M**	**F**	**M**	**F**
socio-demographic	Population density			0.96	
	Percent households with low income	0.81	1.37		0.88
	Percent with < grade 9 education		0.63		
	Percent employed in industry		3.31		
	Percent semi-skilled/ unskilled labourers			0.90	
Smoking	Percent ever-smokers	1.54			
Occupation	longest job in farming	1.78	0.58 [d]		1.11 [d]
Sunlight	Latitude [c]		1.92	0.78	0.80
	Percent involved outdoor activities		0.69		
Model R^2		0.73	0.41	0.59	0.49

[a] Relative risks associated with a 2 s.d. change in the value of each variable.
[b] Sex-specific and age-adjusted unless otherwise specified.
[c] Not sex-specific or age-adjusted.
[d] $0.05 < p < 0.10$; for all other variables, $p < 0.05$.
CNS = Central Nervous System.

Table 28.3 summarises the significant factors in the standard regressions for all cancer sites. The associated RRs are shown; these are comparable between factors for the same analysis in the sense that each effect is shown in terms of its standard deviation for exposure between PHUs. The associated RR for a given factor is directly comparable between cancer sites for the same sex (when included in the model), but comparisons should not be made between sexes for sex-specific variables because of possible differences in their exposure distributions. Cancers with similar sets of candidate variables are grouped together; for many sites, this corresponds roughly to organ systems.

Part (a) of Table 28.3 covers the aero-digestive tract, exclusive of colorectum and gall bladder. Smoking, alcohol, consumption of coffee and of fruit and vegetables, being born outside Canada, and selected occupation exposures are the candidate risk factors. Stomach and lung cancer have the highest SAC and R^2 values in both sexes. Stomach cancer risk varies greatly around the world, with Ontario ranking among the lowest. In our analysis, risk is associated with a high percentage of immigrants, lower mean educational attainment and low socio-economic status in both sexes, and with smoking and less industrial employment in females. Some occupational exposures are associated in males, although not consistently or monotonically.

The cancers in part (b) of Table 28.3 involve a constellation of dietary, body mass and reproductive risk factors. Corpus uteri cancer has a high I-statistic (0.42) and a substantial R^2 value. Consistent with the literature, it is inversely associated with fat intake and positively with the use of female hormones, but the association with body mass is the opposite of expected. Little is known about the causes of prostate cancer, but it is generally more common in those of higher socio-economic status. In our analysis industrial employment is positively associated, and lower educational attainment is inversely related.

The literature suggests a somewhat different set of factors for cervical cancer, including smoking, low socio-economic status and sexual promiscuity. In our data there was evidence of spatial aggregation ($I = 0.38$) and smoking, low educational attainment and high population density were all positively associated with incidence.

Although the risk factors for bladder cancer have been fairly well studied, the variables available in our study did not give high R^2 values (0.19 in males, 0.33 in females), and there was no evidence of spatial aggregation (Table 28.3, part (c)). Smoking, the best-documented factor, was associated with risk, but only in males. Inverse associations were seen with low educational attainment, coffee in males, and occupational exposure to solvents in females, and there was a positive association of low income in females.

For leukaemia (Table 28.3, part (c)), there was some evidence of spatial aggregation ($I = 0.30$ in males, 0.23 in females), with moderate R^2 values (0.51 and 0.29, respectively). Population density was inversely related to risk and percent low income positively related; relative risks for these variables were similar between sexes. In males, exposure to solvents and industrial occupation also predicted risk, the former negatively and the latter positively.

Finally, part (d) of Table 28.3 considers lip cancer and melanoma, both of which are related to sunlight exposure. Although lip cancer shows no evidence of spatial clustering, its model R^2 values were 0.73 for males and 0.41 for females. The female model is based on only 372 cases across the province, so many PHU-level rates will be unstable, and the regression model may not be reliable. In fact, while the male results are consistent with the literature (e.g. increased risks with smoking and long-time farming occupations), the female results are not, with several variables entering the model with coefficients often not in the expected direction. Melanoma, on the other hand, shows strong regional aggregation ($I = 0.39$ in males, 0.41 in females) and high model R^2 (0.59 in males, 0.49 in females). Incidence decreases with increasing latitude in both sexes and increasing percent of unskilled or semi-skilled labourers. it also increases with increasing percent of long-term employment in farming and decreases with percent low income in females. All of these results are consistent with the literature on sun exposure and socio-economic status, where people at higher levels are generally at higher risk.

Table 28.4 summarises the findings from the standard regressions, and shows the case sample sizes on which the regressions are based. The next column indicates the model R^2 in the standard regression. The SAC in the log incidence rate was highest for lung cancer in both males and females, indicating strong spatial aggregation of risk. Moderate regional effects (SAC > 0.2) were found for cervical cancer, male and female leukaemia, male and female melanoma, prostate cancer, male and female stomach cancer, female thyroid cancer, and uterine cancer. The last column shows the SAC for the residuals. As expected, these are typically smaller than the SAC for incidence. Only three residual SACs were greater than 0.2, namely male leukaemia (0.24), female lung cancer (0.29), and female thyroid cancer (0.25); elsewhere they were small or even negative, suggesting that the regional effects had been eliminated through the regressions.

Table 28.4 Summary of standard and spatial regressions (Ontario cancer incidence, 1980–1991)

				Spatial autocorrelation	
Site (ICD 9 Code)	**Sex**	**Number of cases**	**Model R^2**	**Log (incidence)**	**Regression residuals**
Lip	M	1854	0.73	0.14	0.14
(ICD9 140)	F	372	0.41	– 0.12	– 0.22
Oral cancer	M	3781	0.42	– 0.16	– 0.10
(ICD9 141–149)	F	2008	0.45	– 0.09	– 0.21
Oesophagus	M	2808	0.51	0.12	0.03
(ICD9 150)	F	1163	0.48	– 0.02	– 0.22
Stomach	M	6844	0.83	0.38	– 0.16
(ICD9 151)	F	3791	0.73	0.39	– 0.01
Colorectal	M	28052	0.41	0.18	0.04
(ICD9 153–154)	F	25382	0.44	0.24	0.18
Gall-bladder	M	1169	0.30	0.03	0.08
(ICD9 156)	F	1780	0.14	0.10	0.10
Pancreas	M	5141	0.49	0.19	– 0.16
(ICD9 157)	F	4419	0.50	0.11	– 0.07
Larynx	M	3852	0.70	0.19	0.16
(ICD9 161)	F	744	0.49	0.10	– 0.05
Lung	M	33086	0.84	0.52	– 0.06
(ICD9 162)	F	15176	0.69	0.66	0.29
Melanoma	M	4863	0.59	0.39	– 0.04
(ICD9 172)	F	4791	0.49	0.41	0.09
Breast	F	50254	0.40	– 0.06	0.15
(ICD9 174)					
Cervix	F	5826	0.39	0.38	0.13
(ICD9 180)					
Uterus	F	11524	0.67	0.42	– 0.05
(ICD9 182)					
Ovary	F	8342	0.23	– 0.12	0.09
(ICD9 183)					
Prostate	M	20171	0.30	0.34	0.09
(ICD9 185)					
Testis	M	2556	0.19	0.17	0.08
(ICD9 186)					
Bladder	M	12808	0.19	0.11	0.05
(ICD9 188)	F	4170	0.33	– 0.04	– 0.16
Kidney	M	6296	0.10	– 0.15	– 0.19
(ICD9 189)	F	3830	0.41	0.16	0.01

Table 28.4 (*continued*)

Site (ICD 9 Code)	Sex	Number of cases	Model R^2	Spatial autocorrelation: Log (incidence)	Spatial autocorrelation: Regression residuals
Central nervous system (ICD9 191–192)	M	4566	0.32	– 0.14	– 0.16
	F	3786	0	– 0.12	—
Thyroid (ICD9 193)	M	1088	0.51	0.04	– 0.09
	F	3229	0.67	0.35	0.25
Non-Hodgkin's lymphoma (ICD9 200,202)	M	7708	0.40	– 0.08	– 0.12
	F	6385	0.11	0.15	0.07
Hodgkin's disease (ICD9 201)	M	1950	0.10	– 0.12	– 0.10
	F	1500	0	– 0.16	—
Multiple myeloma (ICD9 203)	M	2642	0.19	0.01	– 0.07
	F	2305	0.17	– 0.20	– 0.23
Leukaemia (ICD9 204–8)	M	6913	0.51	0.30	0.24
	F	5085	0.29	0.23	0.13

Further work showed that there were typically only very small changes in the regression coefficients and associated RRs when the spatial model was adopted. The effects were slightly less significant, presumably because of the more complex error structure incorporated in the spatial model. The values of ρ in the spatial regressions (when fitted) were typically similar to those of the SAC for the residuals in the corresponding standard regression. Most of the spatial regressions had higher R^2 values, but the differences from the standard regression were small.

In examining the detailed patterns of residuals, the number of combinations of PHUs with cancer sites is large, so we adopted a stringent definition of an outlier; specifically, we defined a statistical outlier to have a studentised residual greater in absolute value than its 99th percentile value (2.57). Theoretically, if all the relevant determinants had been identified in the regression, so that additional variation was entirely due to chance, then we would expect 1% of the fitted data points to be defined as outliers.

Table 28.5 lists 15 possible outliers. The total number of fitted points is 37 (PHUs) $\times$ 42 (cancer sites) $=$ 1554; therefore, the number of outliers expected by chance alone is approximately 15.5, in close agreement with the observed number. Five points correspond to risk elevations and 10 to deficits; eight are from male data and seven from females. All the site-specific distributions of studentised residuals were approximately normal.

Bearing in mind that the possible outlier points are the most extreme from a set of 1554 points, the five elevated outliers have relatively modest SIR values (1.27, 1.18, 1.27, 1.23 and 1.31). They have no geographic pattern; only one PHU has more one outlier; and

Table 28.5 Possible outliers in regression analysis

Cancer	Sex	PHU	Studentized residual	Number of cases	SIR	SIR (adj)
Lip	F	Niagara	− 3.00	5	0.29	0.47
Oral cancer	M	Timiskaming	− 3.00	6	0.33	0.38
Oesophagus	F	Peterborough	− 3.19	6	0.35	0.52
Colorectal	F	Niagara	− 2.83	1078	0.90	0.88
Gall-bladder	F	York	− 2.70	28	0.59	0.70
Larynx	M	North Western	− 2.79	22	0.62	0.65
Melanoma	M	Hastings	2.59	97	1.27	1.37
		Middlesex	− 2.85	145	0.83	0.81
Prostate	M	Durham	2.77	766	1.31	1.26
Testis	M	Algoma	− 3.10	18	0.50	0.54
		Haliburton	− 2.57	24	0.61	0.61
Kidney	F	Huron	− 3.43	14	0.51	0.52
CNS	M	Sudbury	2.62	124	1.27	1.22
Multiple myeloma	M	Waterloo	2.84	122	1.23	1.43
Leukaemia	F	Halton	3.00	164	1.18	1.36

many types of cancer are involved. The adjusted SIRs represent the cancer incidence for the PHU relative to the province, taking the regression variables into account. These are generally quite similar to the original SIRs. We conclude that if there are indeed true risk differences in these locations, then they are unassociated with the risk factors identified in the regressions.

In summary, given the close agreement of the observed number of outliers to its expectation, their modest SIRs, their dispersed pattern of occurrence, and their lack of grouping by cancer type, the most reasonable interpretation is that most if not all the extreme values are due to chance. In aggregate, the results fail to suggest any particular PHUs with exceptional risk relative to the entire province. There may remain some scope for regionally based studies, given the findings concerning the residual SACs. Particular candidates where both the original incidence rates and the residuals showed stronger SACs (>0.2) are male leukaemia, female lung cancer, and female thyroid cancer.

28.5 CONCLUSIONS

The results presented provide reassurance to public health administrators, in that there do not appear to be any areas in Ontario where the incidence of cancer is exceptionally elevated, once known aetiological factors are taken into account. There are a few cancers with modest regional correlation in the residuals that may be worthy of special study.

Empirically, the data seemed to satisfy the assumption of normality in the residuals. This was probably because of the reasonably large sample sizes and long data collection period. We also found that the results concerning risk factors and their effect sizes were similar using either the standard regression assuming independent errors, or spatial regressions incorporating a spatially correlated error structure. The main differences were slightly larger standard errors for the risk coefficients and less statistical significance in the spatial models.

Additional work is needed to integrate the findings from this analysis with other aetiological literature; this is a complex task, because of the number and diversity of cancer types. Our results are usually consistent but sometimes inconsistent with other literature, so careful interpretation is required. The limited strength of evidence from this study must also be recognised; only relatively weak inferences are possible from ecologically aggregated data to statements about risk for individuals. On the other hand, this study had strength in its wide geographical scope, and in the extent and quality of its data.

ACKNOWLEDGEMENTS

The authors acknowledge the assistance of Derek King, McMaster University, and Vivek Goel, Institute for Clinical Evaluative Sciences, Toronto. The project was partly supported by grant 6606-5034-58 from Health Canada. Dr Walter holds a National Health Scientist award from Health Canada.

29

Congenital Anomalies Near Hazardous Waste Landfill Sites in Europe

H. Dolk, M. Vrijheid, B. Armstrong and the EUROHAZCON Collaborative Group

London School of Hygiene and Tropical Medicine, London

29.1 INTRODUCTION

The EUROHAZCON study is the first European epidemiological study to assess whether the risk of congenital malformation is higher for residents closer to hazardous waste landfill sites than for those farther away.

Waste disposal, whether by landfill or incineration, is one of the foremost environmental concerns today. Knowledge about the potential impact on health is important in deciding on regulation of sites, their siting and remediation. Yet there is little epidemiological evidence on which to base risk assessments. Most studies of pregnancy outcomes among residents near landfill sites have been conducted in North America, from the well-known contamination incident at Love Canal (Vianna and Polan, 1984; Goldman *et al.*, 1985) to more recent assessments around multiple sites (Croen *et al.*, 1997; Geschwind *et al.*, 1992; Marshall *et al.*, 1997; Shaw *et al.*, 1992; Sosniak *et al.*, 1994). Some individual studies have shown raised risks of congenital malformations, but no clear pattern of risk can yet be said to have emerged. There is an extensive literature supporting the potential teratogenicity of many of the chemical classes found in landfill sites (such as heavy metals, pesticides and solvents), but the question is whether nearby residents would be exposed to sufficient doses for there to be any risk, particularly as an individual dose may need to build up to a threshold level for there to be any significant biological effect at all.

Communities close to waste disposal sites are often concerned about the potential health impact, and may link local 'clusters' of adverse health outcomes to exposure to chemicals from nearby sites. Since, even with a random pattern of disease, localised patches of high disease density are bound to occur, it is usually difficult to distinguish

Disease Mapping and Risk Assessment for Public Health. Edited by A.B. Lawson *et al.*
© 1999 John Wiley & Sons Ltd.

clusters derived from the random disease pattern from those where there is a common underlying local cause. Scientifically, and in order to respond to public concern, it is desirable to move beyond *post hoc* cluster investigations, to investigations around waste disposal sites specified a priori.

Residents may be exposed to chemicals from landfill sites through the air or water (Upton, 1989). The air route includes off-site migration of gases, as well as dust and chemicals adhered to dust, especially during periods of active operation of the site. The water route includes contamination of groundwater and surface water, which may contaminate drinking water if local sources are used, or contaminate water used for recreation or household uses. Contamination of air, water or soil may affect locally grown food produce.

Congenital malformations can be divided into those for which there is a pre conceptional mutagenic basis, whether chromosomal or at the level of a single gene, and those that arise from disturbances of in utero development, usually during the organogenetic period in early pregnancy. In this chapter, we consider non-chromosomal malformations.

29.2 METHODS

This report concerns data from seven centres in five European countries (Belgium, Denmark, France, Italy, UK) (Table 29.1), all of which are high-quality, regional, population-based congenital malformation registers. Five of these centres are part of the EUROCAT network of regional registers for the surveillance of congenital anomalies in Europe (EUROCAT, 1991). Three further centres are participating in the study, but two of these register Down's Syndrome only (in Slovenia and the United Kingdom), and one had too little population within the study area around the landfill site to make data analysis meaningful (North-East Italy).

We identified waste landfill sites, located in regions covered by the participating registers, which contained 'hazardous' waste of non-domestic origin, as defined in the EC Directive on Hazardous Waste (ECC, 1991). The EC list includes chemicals such as heavy metals, solvents, pesticides, dioxins. There were twenty one such hazardous waste landfill sites in all participating regions, of which nine closed before the start of the study period and 10 sites were operational for more than 20 years before the end of the study period.

A 7 km zone around each study site was defined as the study area. Where the study areas of two or more study sites overlapped and the sites were within 7 km of each other, the two (or more) study areas were considered as one large study area. Where the sites with overlapping areas were between 7 km and 14 km from each other, the area of overlap was split in such a way that cases and controls were allocated to the nearest site and each study area was considered separately.

The study period began at the start of the malformation register, or, if later, after five years of operation of the nearest landfill site (to allow time for off-site contamination to occur). It ended 31 December 1993 (31 December 1994 for Lyon).

The cases are registered malformed live births, stillbirths and abortions induced following prenatal diagnosis, born within the study period and to a mother resident in a study area, and having one of the malformations on the EUROHAZCON list. This list includes all major malformations, but excludes familial syndromes, neoplasms, metabolic

Table 29.1 Total numbers of cases and controls in EUROHAZCON study areas

Centre	Study area	Number of sites	Study period	Cases	Controls
Funen County (Denmark)	1	1	1987–93	19	44
	2	1	1986–93	28	68
North Thames (West) (UK)	3	1	1990–93	50	124
	4	1	1990–93	10	30
Lyon (France)	5	1	1990–94	35	78
Antwerp (Belgium)	6	1	1990–93	73	160
	7	3	1990–93	35	82
	8	1	1992–93	6	16
Tuscany (Italy)	9	1	1982–93	60	67
	10	1	1982–93	121	138
	11	1	1987–93	45	53
Northern Region (UK)	12	1	1989–93	20	300
	13	4	1986–93	296	740
	14	1	1990–93	23	58
Glasgow (UK)	15	2	1990–91	168	408
Total				1089	2366

diseases and minor malformations. Chromosomal anomalies are excluded from the current analysis.

Controls, two per cases, were randomly selected from all non-malformed live and stillbirths born on the nearest day after the case in the same study area. For convenience, two centres chose to select their controls by taking a random sample from all live births in the same year of birth as the case (Glasgow and Northern Region). In one centre, Tuscany, only one control per case was selected.

Cases and controls were located geographically using addresses or postcodes at birth, with an accuracy of 100 m or less. The distance of residence at birth from the nearest waste site was then used as the surrogate exposure measurement.

29.2.1 Statistical analysis

The association between the proximity to hazardous waste landfill sites and the risk of congenital malformations was investigated using logistic and related binomial regression models (Breslow and Day, 1980). Since individual matching by date of birth was for administrative convenience rather than to control confounding, we carried out an unmatched (unconditional logistic regression) analysis, but included terms for study area and year of birth in all models. The distance from the waste site was first dichotomised into a 0–3 km 'proximate' zone, and a 3–7 km 'distant' zone. These zones were defined a priori on the advice of landfill experts. Information routinely available on

Table 29.2 Odds ratios for maternal age and socio-economic status

	Cases	Controls	OR	95% CI	Trend test *p*-value
Maternal age: all centres					
< 20 years	73	175	0.91	0.68–1.23	
20–24 years	270	615	0.95	0.79–1.15	
25–29 years	391	851	1.00		
30–34 years	232	492	1.02	0.84–1.24	
> 35 years	85	158	1.16	0.87–1.56	0.17
Unknown	38	75			
Socio-economic status					
UK centres: quintiles of small-area deprivation scores					
Affluent: 1	53	167	0.91	0.62–1.34	
2	67	171	1.08	0.75–1.55	
3	100	275	1.00		
4	155	388	1.12	0.84–1.51	
Deprived: 5	290	656	1.25	0.96–1.64	0.04
Unknown	2	3			
Funen County: social class from parental occupation					
High: 1	2	5	1.01	0.17–5.89	
2	2	4	1.22	0.20–7.57	
3	13	33	1.00		
4	18	48	0.95	0.41–2.19	
Low:5	11	20	1.39	0.52–3.69	0.70
Unknown	1	2			
Tuscany: maternal education					
Graduate	8	15	0.58	0.23–1.45	
High School	77	67	1.20	0.76–1.90	
Medium	77	86	1.00		
Elementary	29	56	0.60	0.35–1.04	
None	1	1	1.23	0.08–20.02	0.17
Unknown	34	33			
Lyon: occupational groups					
Professional	1	8	0.20	0.02–1.78	
Intermediate	11	22	0.82	0.31–2.16	
Farmers, craftsmen	4	7	0.98	0.24–3.97	
Workmen	15	26	1.00		
Unemployed	0	8			0.95
Unknown	4	7			
Antwerp: quintiles of average area income					
High income: 1	25	45	1.75	0.81–3.79	
2	23	50	1.73	0.81–3.69	
3	15	58	1.00		
4	21	50	1.59	0.74–3.43	
Low income: 5	28	53	1.84	0.88–3.85	0.92
Unknown	2	2			

socio-economic status (SES) varied greatly between countries participating in the study (Table 29.2). When adjusting for socio-economic status in analyses in which study areas were pooled, SES was therefore separately modelled in each country.

In analyses pooling information over study areas, we analysed the association of risk with distance from a waste site in more detail by grouping more finely and by using distance as a continuous measure in explicit models. As well as standard logistic models in distance and its reciprocal, we fit a model in which excess risk (strictly odds ratio) declines exponentially with distance from the site:

$$\pi/(1-\pi) = \exp(\boldsymbol{\beta}^{\mathrm{T}}\boldsymbol{x})\{1+\alpha\exp(-\gamma d)\},$$

where π is the probability of being a case, d is the distance from the waste site, and $\boldsymbol{x}$ is a vector of possibly confounding covariates. The parameter γ defines the rate of decline in risk with distance, and α defines the maximum risk (right next to the site), relative to being distant from it ($d \to \infty$).

This model is one of a family of 'excess relative risk' models that may be fit using the EPICURE computer package (Preston *et al.*, 1993). These take the form (slightly simplified):

$$R(\boldsymbol{z}_0, \boldsymbol{z}_1, \ldots, \boldsymbol{z}_J) = T_0(\boldsymbol{\beta}_0, \boldsymbol{z}_0)\left[1 + \sum T_j(\boldsymbol{\beta}_j, \boldsymbol{z}_j)\right],$$

where $\boldsymbol{z}_j$ and $\boldsymbol{\beta}_j$ represent vectors of covariates and parameters, respectively, and T_j represents a 'term' comprising in general the product of linear and loglinear 'subterms' $(\boldsymbol{\beta}_{j(1)}^{\mathrm{T}}\boldsymbol{z}_{j(1)}\exp(\boldsymbol{\beta}_{j(2)}^{\mathrm{T}}\boldsymbol{z}_{j(2)}))$. $R(\boldsymbol{z}_0, \boldsymbol{z}_1, \ldots, \boldsymbol{z}_J)$ may represent disease odds, odds ratio, hazard, or hazard ratio at given covariate values. Thus, for this application we have an entirely loglinear term $T_0(\exp(\boldsymbol{\beta}^{\mathrm{T}}\boldsymbol{x}))$ and a single other term T_1 with a linear subterm with a constant only (α), and a loglinear subterm in distance ($\exp(-\gamma d)$). Since this is a case–control study analysed as unmatched, $R(d, \boldsymbol{x})$ represents disease odds. EPICURE implements the maximum likelihood estimation and inference for this model for unmatched or matched (conditional likelihood) case–control data (as well as cohort and case–cohort data).

The development of the EPICURE family of models was motivated by the need to analyse studies of the effects of A-bomb survivors, in order to model the effects of radiation dose with respect to cancer, together with confounders and modifiers (Peirce and Preston, 1985; Pierce *et al.*, 1996). The model we have used also belongs to a family proposed independently specifically for use in case–control studies in the spatial context by Diggle and Rowlingson (1994).

$$\pi/(1-\pi) = \rho\exp(\boldsymbol{\beta}^{\mathrm{T}}\boldsymbol{x})\{1+g(d,\theta)\}$$

(slightly simplifying and changing notation to emphasise the similarities with our formulation). In our formulation, the parameter vector θ has two components α and γ, with $g(d,\theta) = \alpha\exp(-\gamma d)$. Diggle and Rowlingson's term ρ is subsumed in our formulation above as the constant term in $(\boldsymbol{\beta}^{\mathrm{T}}\boldsymbol{x})$, and their nearest specifically illustrated model uses d^2 where we have emphasised d, although we also fit a model using d^2.

Models allowing for effects that varied randomly between study areas (Smith *et al.*, 1995) were explored using the STATA, EGRET, and BUGS packages.

29.3 RESULTS

Fifteen study areas were defined around the 21 landfill sites. Table 29.1 shows the participating centres, study sites, study areas, study periods and numbers of cases and controls on which the current analyses are based. The total number of non-chromosomal cases and controls is 1089 and 2366, respectively. In Table 29.2 the relationship between two potential confounders, maternal age and socio-economic status, and the risk of congenital malformations is shown. Maternal age shows a slight gradient in risk with a higher odds ratio for older, compared with younger, mothers, but this trend is not statistically significant. There was no clear trend in the risk of congenital malformation in relation to socio economic status in any of the centres except in the United Kingdom, where the trend of increasing risk with increasing deprivation was statistically significant ($p = 0.04$). There appears not to be a consistent pattern of more deprived populations living closer to the waste sites (Figure 29.1).

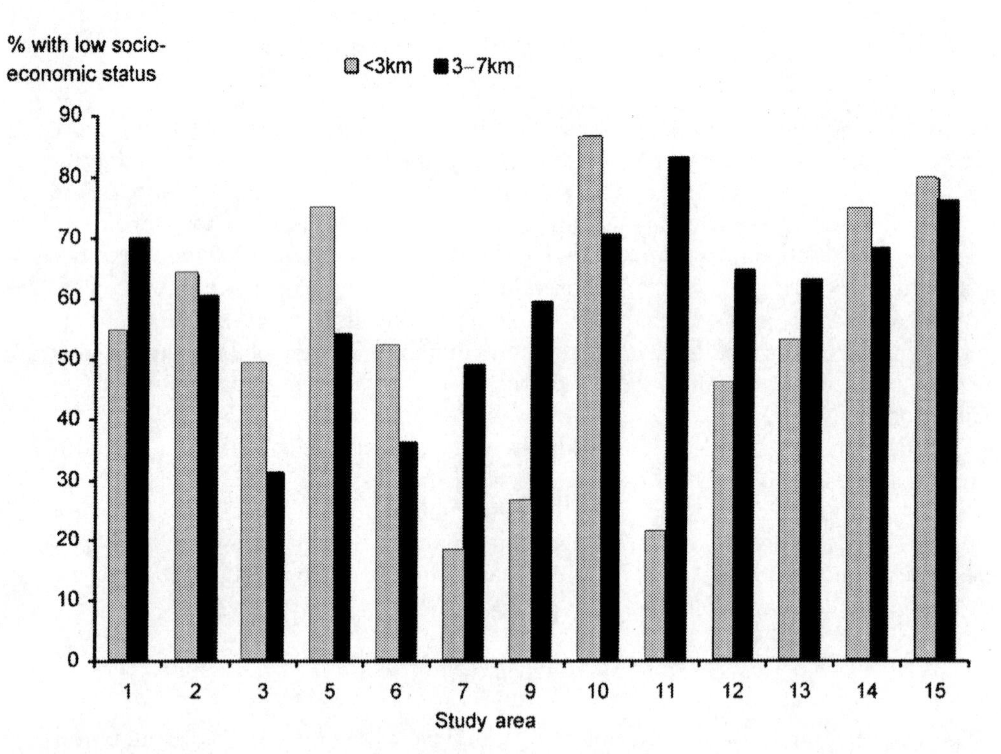

Figure 29.1 Percentage of controls with low socio-economic status close by and farther away from waste sites.

Notes: Areas 1, 2, and 5: % with social class 4 or 5 (from parental occupation);
Areas 6, 7: % in average area income quintiles 4 or 5 (lowest income areas);
Areas 9, 10, 11: % with less then high school eduction (maternal education);
Areas 3, 12–15: % in UK small-area deprivation quintiles 4 or 5 (most deprived areas);
Study area 4 and 8 have not been included in the graph because of small numbers

Table 29.3 Odds ratios for living within 3 km of a hazardous waste landfill site—non-chromosomal anomalies

Distance	Cases	Controls	OR	95% CI	Adj. OR[a]	95% CI
All study areas pooled						
0–3 km	295	511	1.37	1.14–1.63	1.33	1.11–1.59
3–7 km	794	1855				
Study area						
1 0–3 km	7	23	0.49	0.15–1.63	0.43	0.11–1.65
3–7 km	12	21				
2 0–3 km	11	25	1.26	0.47–3.40	1.23	0.41–3.67
3–7 km	17	43				
3 0–3 km	25	59	1.16	0.60–2.26	0.76	0.34–1.69
3–7 km	25	65				
4 0–3 km	6	18	1.12	0.19–6.42	0.83	0.11–6.07
3–7 km	4	12				
5 0–3 km	4	14	0.58	0.17–1.91	0.45	0.13–1.60
3–7 km	31	64				
6 0–3 km	18	21	2.19	1.08–4.45	2.08	0.98–4.41
3–7 km	55	139				
7 0–3 km	11	11	2.92	1.11–7.70	3.93	1.20–12.80
3–7 km	24	71				
8 0–3 km	0	1	0.00		–	
3–7 km	6	15				
9 0–3 km	21	15	2.09	0.92–4.75	1.29	0.48–3.49
3–7 km	39	52				
10 0–3 km	17	15	1.38	0.65–2.94	1.40	0.62–3.15
3–7 km	104	123				
11 0–3 km	28	38	0.65	0.28–1.52	0.72	0.17–2.97
3–7 km	17	15				
12 0–3 km	23	50	1.16	0.67–2.02	1.26	0.71–2.22
3–7 km	97	250				
13 0–3 km	64	113	1.52	1.08–2.15	1.50	1.05–2.13
3–7 km	232	627				
14 0–3 km	1	4	0.63	0.07–6.16	0.94	0.09–9.74
3–7 km	22	54				
15 0–3 km	59	104	1.58	1.07–2.33	1.63	1.09–2.44
3–7 km	109	304				

[a]Adjusted for socio-economic status and maternal age.
Note: The unadjusted odds ratios are not the cross-product ratios from the two-by-two tables because of the stratification by matching variables.

Table 29.3 presents the odds ratios for living within 3 km of a hazardous waste landfill site for each of the 15 study areas and for all study areas pooled, unadjusted and adjusted for maternal age and socio-economic status. The overall adjusted odds ratio was 1.33 (95% CI 1.11–1.59). Adjustment for confounders did not, either for the pooled

Table 29.4 Risk with distance from waste site—modelling of pooled data

Model					**Deviance**	**d.f.**	***p* (model)**
Distance (D) categorised in 1 km bands							
D(km)	Cases	Controls	OR[a]	95% CI			
$<=1$	41	62	1.60	1.03–2.48			
1–2	84	167	1.25	0.92–1.70			
2–3	170	282	1.46	1.15–1.85			
3–4	236	478	1.17	0.95–1.44			
4–5	206	469	1.06	0.86–1.32			
5–7	352	908	1.00				
					4199.8	5	0.025
Logistic regression model[a]							
$\pi/(1-\pi) = \exp(\beta \cdot distance)$			$\beta = -0.08$		4202.2	1	0.001
$\pi/(1-\pi) = \exp(\beta \cdot 1/distance)$			$\beta = 0.32$		4206.5	1	0.012
Exponential excess risk model[a]				95% Cl			
1. $\{1 + \alpha \cdot \exp(-\gamma \cdot \text{distance})\}$			$\alpha = 1.18$ $\gamma = -0.28$	0.38–2.51[b]	4201.7	2	0.004
2. $\{1 + \alpha \cdot \exp(-\gamma \cdot \text{distance}^2)\}$			$\alpha = 0.55$ $\gamma = -0.03$	0.21–1.79[b]	4202.9	2	0.007
Null model[a]					4212.7	0	

[a] Adjusted for maternal age and socio-economic status.
[b] 95% CI estimated keeping γ fixed at its maximum likelihood value, and searching for values of α giving a deviance 3.84 greater than its value at the maximum likelihood estimate.

or individual study areas, substantially change the odds ratio estimates. Adjusted odds ratios for three study areas (7, 13 and 15) showed a statistically significant ($p < 0.05$) increase. The odds ratio for study area 6 borders significance. There was little evidence for heterogeneity in the odds ratios between sites ($p = 0.31$). Of several random effects approaches tried, only Bayes models giving high prior plausibility to large underlying variation suggested substantially different interpretations. A Bayes model with a normal distribution of underlying log odds ratios, and 'non-informative' gamma (0.001, 0.001) prior for the inverse variance of this normal distribution showed (crude) odds ratios distributed about a median of 1.35, with a 95% credible interval (1.07, 1.68).

Dividing subjects into six bands of distance (Table 29.4, Figure 29.2) showed a fairly consistent decrease in risk with distance. Several models using distance as a continuous variable fitted equally well, with the exponential excess model (shown in Figure 29.2) somewhat better than others (Table 29.4). All models showed a statistically significant decreasing risk with distance from the site ($p < 0.05$).

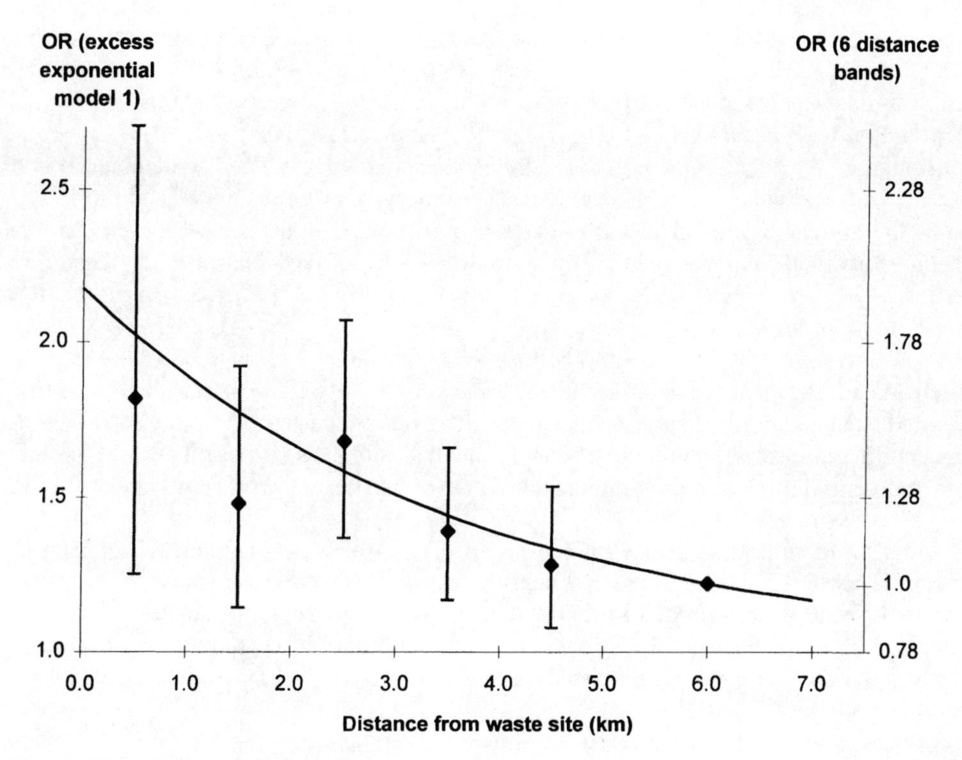

Figure 29.2 Risk with distance from waste site.
Notes: Line shows ORs fitted by exponential excess risk model; diamonds and error bars show ORs and 95% CI for 6 distance bands with 5–7 km band as baseline

29.4 DISCUSSION

Our study has shown a small but statistically significant excess risk of non-chromosomal congenital malformations among residents near (within 3 km of) landfill sites. This excess does not appear to be limited to one or a few sites, and indeed we have no evidence that the risk differs between sites, although our study has limited statistical power to address this issue. The fundamental question is, of course, whether this association is causal, but this cannot be resolved within this single study. Three questions are nevertheless relevant to the interpretation of this excess:

29.4.1 What do we know about potential confounders and sources of bias?

Socio-economic status is the most obvious potential confounder in any spatial analysis of health outcomes. More deprived communities may be both at greater risk of the adverse health outcome, and live closer to industrial sites. In the case of congenital malformations, there is surprisingly little literature to indicate the strength of the relationship between socio-economic status and congenital malformation risk (Hemminki *et al.*, 1980; Knox and Lancashire, 1991; Olsen and Frische, 1993; Olshan *et al.*, 1991). Our own internal analysis has supported a positive association for non-chromosomal malformations with deprivation within the United Kingdom, but little indication of a relationship elsewhere. Although we found differences in the socio-economic profile between residents near and farther from individual sites, no overall pattern emerged for more deprived communities to be living near (within 3 km of) sites. Moreover, adjusting for socio-economic status in our statistical analyses resulted in very little shift in the odds ratios. We therefore conclude that socio-economic status is unlikely to explain the excess in congenital malformation risk found near sites.

A second source of confounding is the possible presence of other industrial sites or environmental exposures near landfill sites. We have not yet exhaustively examined this possibility, but it should be noted that to date there has been very little study of the risk of congenital malformation near any type of industrial site, and our results would have as much potential interest if they implicated other industrial sites as if they implicated the landfill sites under study.

Ascertainment bias, whereby higher case ascertainment occurred close to sites, is a theoretic possibility, but the participating registers had high case ascertainment through the use of multiple sources of information and active case finding, the data were routinely collected blind to the study hypothesis, and an examination of the data by hospital of birth shows that at least hospital-based ascertainment differences are not an explanation for the excess found near sites.

The migration of women between exposure and pregnancy outcome is a further potential source of bias, which would tend to lead to underestimation of any true raised relative risk. Among the chronic effects of exposure, congenital malformations and other adverse pregnancy outcomes are potentially some of the quickest to manifest in terms of the time that elapses between exposure and the detection of the adverse outcome (although for chemicals that bioaccumulate, the length of residence of the mother near the site may be important). Few estimates are available of the proportion of mothers who migrate during pregnancy, but recent figures from England suggest that about one quarter of women change address during pregnancy, of whom half move less than 1 km (Dolk, 1997). We estimate that this would lead to an approximately 10% underestimation of any true excess risk (Armstrong *et al.*, 1996).

29.4.2 To what extent can we distinguish differences in risk according to subgroups of malformations or landfill sites?

Congenital malformations are a very heterogenous set of conditions in terms of pathogenesis and aetiology, and it is thus of obvious interest to establish whether any particular malformations are preferentially linked to either landfill sites in general or to particular chemicals dumped in them. However, we are unable to derive from the literature any very strong a priori hypotheses about which anomalies should show a greater risk in general or in relation to specific chemicals. Furthermore, the landfill sites themselves cannot be classified into clearly differentiated groups according to the likely chemical exposures, both because each site tends to hold a range of chemicals, and because information on the chemicals dumped is incomplete, particularly going back in time when extensive record keeping was not a legal requirement. We established a number of non-mutually exclusive congenital anomaly subgroups (i.e. one child could have more than one anomaly) according to what is known of the epidemiology of these conditions and current practice in surveillance, in order to 'explore' the data, rather than test any hypotheses. Inevitably, these subgroups were a compromise between lumping together heterogeneous conditions, and splitting into multiple subgroups with very few cases in each. Most subgroups exhibited raised odds ratios, with neural tube defects and malformations of cardiac septa and great arteries and veins having odds ratios of nominal statistical significance, and gastroschisis, hypospadias and tracheo-oesophageal fistulas of borderline significance. These results should be regarded as hypotheses to inform further study, but no great weight can be put on any interpretation of the differences in risk between congenital anomalies at this stage.

An analogous problem is distinguishing whether the overall excess risk within 3 km of landfill sites is a general attribute of all sites, or linked to particular sites. Formal testing of heterogeneity in odds ratios did not reveal any evidence of difference between sites, although the statistical power of such an analysis is low. Again, we believe that nothing can essentially be said about differences between individual sites. However, we are in the process of ranking sites according to their general 'hazard potential', using characteristics of their geology, engineering or management that would affect the likelihood of surrounding contamination. We believe that the demonstration of a 'dose-response' effect would strengthen the case for a causal association between the risk of congenital anomaly and residence near sites.

29.4.3 How would interpretation differ if we knew more about the background spatial distribution of the disease, and under what circumstances is more refined spatial modelling of use?

We have used spatial coordinates only to define the distance of cases and controls from the nearest waste site. Having done this, the statistical methods we have used have been standard epidemiological ones, rather than any specifically developed as 'spatial' (with the partial exception of the exponentially declining excess risk model). We believe that these methods have been largely adequate for this study. More explicitly spatial methods would allow one important refinement — allowing for a generalised spatial clustering of abnormalities. If such clustering exists, the finding of an excess near landfill sites is not as unusual as the nominal p-value would suggest. We could apply tests for such clustering and, by characterising it, perhaps in a spatial auto-correlation model, we could make a more appropriate inference on the importance of proximity to a site (Clayton and Bernardinelli, 1992). It may also be that spatial statistical methods would have a part to play in developing more refined indices of exposure.

The problems in interpretation here are not principally statistical, but related to the lack of evidence on exposure near the sites, and on plausible aetiological pathways.

29.4.4 European environmental surveillance

Finally, we would like to consider briefly the implications of this study for the environmental surveillance of congenital malformations at a European level. Environmental problems are now not confined to any one country, and a coordinated policy response is necessary. If science is partly to underpin the policy process, this also needs to be coordinated at a European level. We have shown that it is possible to perform a multicentric study of congenital malformations in relation to a specific environmental point source in Europe. Although this sort of spatially oriented study is only one of many types of research angles needed, it responds to frequently expressed public concerns about spatial clusters, and is therefore valuable from a public health as well as a scientific point of view. Continuation and enlargement of this sort of enterprise requires: (a) more lines of communication being set up between environment departments and analysts of health data, such as congenital malformation registers; (b) a systematic system of control selection and geographical referencing being implemented (or routine post-coding of all births as in the United Kingdom); (c) more attention being given to the establishment of common or comparable European measures of socio-economic status so that socio-economic confounding can be properly included in studies; and (d) a source of funding that recognises the need for environmental surveillance to become part of the general surveillance process which is at present oriented much more towards the traditional concern of detection of clusters in time in relation to the introduction of new drugs.

Membership of the EUROHAZCON Collaborative Group

L. Abramsky	North Thames (West) Congenital Malformation Register, UK
F. Bianchi	Tuscany EUROCAT Register, Italy
E. Garne	Funen County EUROCAT Register, Denmark
V. Nelen	Antwerp EUROCAT Register, Belgium
E. Robert	France Central East Register of Congenital Malformations, France
J.E.S. Scott	Northern Congenital Abnormality Survey, UK
D. Stone	Glasgow EUROCAT Register, UK
R. Tenconi	North-East Italy Register of Congenital Malformations, Italy

ACKNOWLEDGEMENTS

Study co-ordination was Funded by the European Commission DGXII BIOMED programme Concerted Action Contract BMH 1-94-1099.

30

An Analysis of the Geographical Distribution of Leukaemia Incidence in the Vicinity of a Suspected Point Source: A Case Study

Wolfgang Hoffmann

Bremen Institute for Prevention Research and Social Medicine (BIPS)

Peter Schlattmann

Free University Berlin

30.1 INTRODUCTION

Between February 1990 and the end of 1995, six cases of childhood leukaemia were diagnosed among residents of the small rural community of Elbmarsch in Northern Germany. Five of these cases were diagnosed in only 16 months between February 1990 and May 1991 (Dieckmann, 1992). All patients lived in close proximity (500–4500 m) to Germany's largest capacity nuclear boiling water reactor, the 1300 MW (electric) 'Kernkraftwerk Krümmel' (KKK). This plant was commissioned in 1984 and is situated on the northern bank of the Elbe, a major river which separates the Federal States of Schleswig-Holstein and Lower Saxony.

Standardised incidence ratios (SIRs) for childhood leukaemia in a circular area with a radius of 5 km around the plant were 4.60 (95% confidence interval 2.10–10.30) for the time period 1990–1995 and 11.80 (95% CI 4.90–28.30) if the analysis is restricted to the

Disease Mapping and Risk Assessment for Public Health. Edited by A.B. Lawson *et al.*
© 1999 John Wiley & Sons Ltd.

years 1990 and 1991, respectively (Hoffmann *et al.*, 1997a). Since 1 January 1996 until the time of this writing (1998) three additional incident childhood leukaemia cases have been confirmed in the 5-km area around the plant, rendering the magnitude of the childhood leukaemia cluster in the Elbmarsch unprecedented worldwide with respect to its number of cases together with its narrow spatial and brief temporal dimension.

Soon after the cluster had been identified, the governments of Lower Saxony and Schleswig-Holstein established boards of experts to advise on useful investigations and appropriate methods to identify possible causes for the cluster. The board's members' scientific backgrounds included haematology, paediatrics, toxicology, radiobiology, medical physics, geology, virology, statistics, public health, and epidemiology. An extended array of established or suspected risk factors has since been investigated. However, measurements of outdoor and indoor air, soil, drinking water, private wells, milk, vegetables, other garden products and mushrooms for heavy metals, organochlorine compounds, benzene, toluene, and aromatic amines, respectively, did not reveal any clue indicative of unusual contamination (Niedersächsisches Sozialministerium, 1992). An indoor air radon concentration of 610 Bq/m^3 was measured in the home of one case, but not in the homes of the other cases. Moreover, this activity was only slightly above the current recommendation for existing houses in Germany of 500 Bq/m^3(ICRP, 1984; Bundesminister für Umwelt, 1992). A thorough review of medical and hospital records and extensive semi-structured personal interviews with the afflicted families failed to reveal any unusual dose of medical or occupational radiation or exposure to cytostatic or other leukaemogenic drugs. None of the children had a pre-existing medical condition known to be associated with a higher risk of leukaemia. The children had all been born in the local area and most of the parents had lived there for many years prior to the children's births. Despite this residential stability none of the patients' families was found to be related to any of the others and the afflicted children had not had direct contact with each other prior to their diagnoses. Biological samples (breast milk, urine, blood) taken from members of the afflicted families and other inhabitants of Elbmarsch yielded low background values of 2,3,7,8-TCDD, various organochlorines, lead, and cadmium. The prevalence of antibodies against viruses that are discussed as potentially leukaemogenic was below the German average, if any (Niedersächsisches Sozialministerium, 1992).

At this point both boards of experts came to the conclusion that further investigations on the basis of only the diseased children and their families would not shed much more light on the aetiology of the cluster. Instead, a comprehensive retrospective incidence study was suggested which should cover all ages, a large enough study area and a sufficient time period to generate a study base suited for an analytical epidemiological investigation.

30.2 MATERIALS AND METHODS

30.2.1 The retrospective incidence study Elbmarsch

A retrospective incidence study ('Retrospective Incidence Study Elbmarsch'(RIS-E)) was conducted between November 1992 and August 1994. The study region included three counties adjacent to the location of the nuclear power plant, i.e. the counties of

Figure 30.1 Geographical location of the 'Kernkraftwerk Krümmel' in Northern Germany (diameter of the circle is approximately 30 km)

Lüneburg and Harburg in Lower Saxony and the county of Herzogtum Lauenburg in Schleswig-Holstein. The total study population was about 470000.

Ascertainment of cases

Case ascertainment included all leukaemias, malignant lymphomas, and multiple myelomas as well as the myelodysplastic and myeloproliferative syndromes (ICD-9 200–208, 238.7), covering the 10-year period 1984–93. Inclusion criteria were (i) first diagnosis of a target disease in the study period; (ii) place of residence within the study area at the time of first diagnosis; and (iii) German citizenship. Since no epidemiological cancer registry exists in the Federal States of Lower Saxony and Schleswig-Holstein, cases had to be ascertained exclusively from primary data sources.

Data sources included all hospitals, county Departments of Health, and practising physicians in the study area who specialised in family medicine, internal medicine, oncology, or paediatrics. In all hospitals, the Departments of Haematology/Oncology, Internal Medicine, Paediatrics, Radiation Therapy, and Pathology were routinely included in the search. Other departments were included only if patients with target diseases had been treated there. In addition, complete case ascertainment was performed in relevant departments of hospitals and treatment centres (including university hospitals) in adjacent counties outside the study area. Full searches of all single records were performed manually by trained study staff. A standardised set of data was extracted for each respective case (gender, date of birth, date of first diagnosis, full text diagnosis in medical terminology, date of death if deceased, and data source). Extraction was based exclusively on original documents (Hoffmann and Greiser, 1994, 1996).

Original data sources in 56 departments of 13 regional hospitals, four major leukaemia treatment centres, four university hospitals, 130 private practices, and three county health departments were included in the incidence study. Cases were ascertained on average in 2.07 (range 1–7) locations and 2.89 (range 1–13) data sources (Hoffmann *et al.*, 1997b). To warrant completeness of ascertainment, quality control and data validation procedures were routinely implemented over the entire study period. Extended capture–recapture analyses within and between primary data sources did not reveal underascertainment. Particularly, there was no indication of selective ascertainment nor any varying degree of completeness with respect to distance from the plant (Hoffmann and Greiser, 1994). Comparing our data with incidence data from the Saarland Cancer Registry, we found similar numbers of cases for age groups below 65 years for all leukaemia subtypes. Above that, case numbers in our study were generally higher than in the Saarland Cancer Registry. Differences were most pronounced for chronic leukaemias (up to 100% for age groups 75 + for CLL; Stegmaier, 1993; Hoffmann and Greiser, 1994). The complete study area was therefore used as a reference in all geographical analysis.

Analyses presented in this chapter are based on all cases classified as leukaemia. Leukaemia cases were categorised as 'acute' and 'chronic' leukaemia, respectively, according to the Ninth International Classification of Diseases (German Version).

Denominator data/Population figures

To calculate the incidence rates, population figures for all the 217 rural communities of the study area were provided by the State Statistical Offices of Lower Saxony and of Schleswig-Holstein. Data were stratified by five-year age groups, gender, and citizenship for all single years 1984–92. One of the rural communities was uninhabited over the entire study period. Hence all analyses presented in this chapter are based on 216 geographical units representing rural communities which contributed to the denominator. Population data for 1992 were substituted for the year 1993.

Determination of the places of residence of cases

German data confidentiality legislation does not allow the use of full addresses for research purposes. To ensure confidentiality in this study, geographical coding of the residences of all cases was done in a separate procedure by means of collaboration between the owners of the primary data and the land registry offices of the respective counties. After completion of the coding only Gauss–Krueger coordinates with a spatial resolution of approximately 100×100 m were given to the researchers (Grossmann, 1976). After linkage to the abstracted data sets by a unique identification number, these coordinates could subsequently be used in geographical analyses.

For the analyses presented in this chapter, rural communities were used as the unit of observation. Hence all incident cases were assigned to their community of residence at the time of their first diagnosis. Expected cases were then calculated using the total number of incident cases in the study area and each community's respective fraction of age-specific population years at risk.

30.2.2 Statistical methodology

In analysing putative clusters of disease two main strategies may be distinguished; Besag and Newell (1991) describe tests of clustering either as general or focused. Investigators who seek to address 'general clustering' determine whether or not cases are clustered anywhere in the study area, while tests addressing 'focused' clustering assess whether or not cases are clustered around a prespecifed source of hazard, which are frequently called foci. In our case study we simply deal with the point source of the Krümmel nuclear power plant and, hence, are investigating predominantly a focused clustering problem. However, we feel that an analysis of general clustering is still necessary since the use of disease maps based on (empirical) Bayes models allows us to separate signal from noise in the geographical distribution of disease within small areas. This may be highlighted by the fact that due to small expected counts, the variability of the crude SIRs in this study ranges from 0 to 46! Thus the use of variance minimised estimates allows the assessment of extra-Poisson variation which can be seen as indicator of heterogeneity of disease risk.

Our analysis proceeds in two steps. First, we address the presence of general clustering or heterogeneity of disease risk by means of exploratory disease mapping. We then use a focused test in order to investigate the hypothesis of a relationship with the Krümmel nuclear power plant.

30.2.3 Disease Mapping

Here we use mixture models to investigate the hypothesis of heterogeneity of disease risk within the study area. The mixture model approach fits into the framework of empirical Bayes theory (Clayton and Kaldor, 1987). This approach is based on *random effects models*, i.e. models where the distribution of relative risks θ_i between areas is assumed to have a probability density function $g(\theta)$. The O_i are assumed to be Poisson distributed conditional on θ_i with expectation $\theta_i E_i$. The distribution $g(\theta)$ may be either parametric or non-parametric; the parameters of either distribution are estimated from the data in a first step. In a second step, application of Bayes theorem using the prior distribution together with the data allows estimation of the posterior expectation of the relative risk for each individual area.

In the mixture model setting we apply a non-parametric distribution for $g(\theta)$, where we assume that the population under scrutiny consists of subpopulations with different levels of disease risk. Statistically we face the problem of identifying the level of risk for each subpopulation and its corresponding proportion of the overall population. This leads to a random effects model where we assume a *discrete* parameter distribution P for $g(\theta)$ with $P = [\theta_1 \ldots \theta_k; p_1 \ldots p_k]$. P is the discrete probability distribution which gives mass p_j to the parameter θ_j. This model therefore assumes that O_i comes from a non-parametric mixture density of the form:

$$f(o_i, P, E_i) = \sum_{j=1}^{k} p_j f(o_i, \theta_j, E_i), \quad \text{with} \sum_{j=1}^{k} p_j \text{ and}$$

$$p_j \geq 0, i = 1, \ldots, n(\text{number of areas}),$$

where $f(\cdot)$ denotes the Poisson-density with $f(o_i, \theta, E_i) = e^{-\theta E_i}(\theta E_i)^{o_i}/o_i!$ Note that the model consists of the following parameters: the unknown number of components k, the k unknown (relative) risks $\theta_1, \ldots, \theta_k$ and $k-1$ unknown mixing weights $p_1, \ldots, p_k$. The term E_i denotes the population at risk or the expected cases in case SIR's are used. Estimation uses maximum likelihood approaches. Suitable algorithms have been proposed by Böhning *et al.* (1992) and Böhning (1995).

On the basis of this model the posterior probability of belonging to a certain subpopulation can be computed. Application of a maximum rule provides a straightforward method of map construction. The a posteriori expectation of the relative risk for each area may be computed in a similar fashion. For details see Schlattmann and Böhning (1993) and Chapter 31 in this volume.

The null hypothesis of constant disease risk $H_0: \lambda_1 = \lambda_2, \ldots, = \lambda_n$ for all areas within the study area is the special case if $k = 1$. The number of components k may be estimated using the likelihood ratio statistic (LRS) by comparing models with $H_0: k_0 = k$ versus the alternative $H_a: k_a = k + 1$. Note that the regularity conditions for the LRS do not hold and that critical values must therefore be obtained by simulation techniques (Böhning *et al.*, 1994). The calculations in this chapter were performed using Dismap Win (Schlattmann, 1996).

30.2.4 Focused analysis

We start our analysis with the classical approach dividing the study area into circles with different radii, merely for descriptive purposes. Here we chose the following distances in relation to the Krümmel nuclear power plant: 0 to < 5 km, 5 to < 10 km, 10 to < 15 km, 15 to < 20 km, and more than 20 km. For each of these concentric regions we present the SIR, together with a 95% confidence interval based on Byar's approximation (Breslow and Day, 1987) for the Poisson distribution.

However, any such 'concentric region' approach tacitly operationalises 'exposure' through broad categories of distance which are necessarily arbitrary. Consequently, any such analysis does not take advantage of a great deal of distance information that would actually be available in the data. Another disadvantage is the considerable impact that even slight changes in the position of the borders between the concentric regions can have on the categorisation of geographic units, which span over two or more circles (see below).

A more appropriate surrogate would be to investigate the effect of distance to the power plant independent of any predefined categories. The highest possible geographical resolution is achieved in a test statistic that uses the distance d_i of each area i to the power plant. As a result we avoid the bias of choosing arbitrary segments. The distance of the 216 areas to the power plant ranges from 1.3 as the shortest distance and 55.6 km for the most remote community.

Here we apply the score test proposed by Lawson (1993b), Waller *et al.* (1995), and Waller and Lawson (1995) (see also Chapter 19 in this volume). We test the null hypothesis $H_0: \lambda_1 = \lambda_2, \ldots, = \lambda_n$ of constant disease risk within the study region versus the alternative hypothesis $H_1: \mathrm{E}(O_i): E_i\lambda(1 + g_i\varepsilon),\ i = 1, n, \varepsilon > 0$. This hypothesis implies that the number of cases increases in proportion to the exposure of individuals in area i relative to the focus, where $i = 1, \ldots, n$. The test statistic is

$$U = \sum_{i=1}^{n} g_i(O_i - \mathrm{E}(O_i)),$$

where $E(O_i)$ is given by the expected cases E_i. Under the null hypothesis the expectation is $E(U) = 0$ and the variance is given by

$$\mathrm{var}(U) = \sum_{i=1}^{n} g_i^2 E_i,$$

where $U^* = U/\sqrt{var}(U)$ has an asymptotic standard normal distribution.

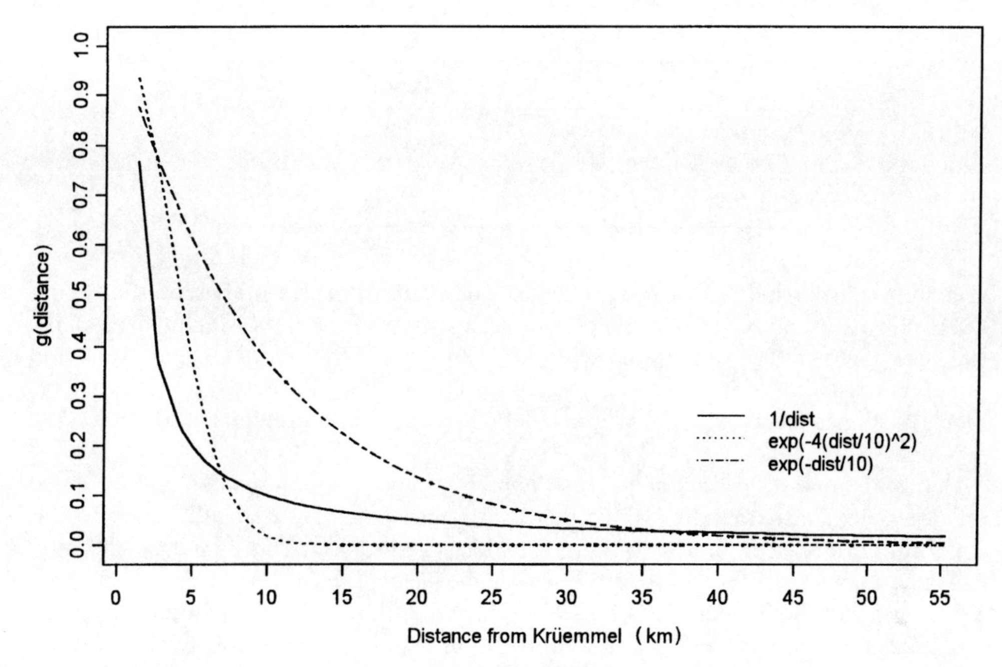

Figure 30.2 Comparison of three models of distance-based surrogates

To parameterise the distance exposure relationship g_i, we use the simple inverse distance

$$g_i = 1/d_i, \quad i = 1, \ldots, n, \tag{30.1}$$

and two exponential functions of distance d_i (Tango, 1995):

$$g_i = \exp(-d_i/\tau), \quad i = 1, \ldots, n, \tau > 0, \tag{30.1}$$

$$g_i = \exp(-4(d_i/L)^2), \quad i = 1, \ldots, n, L > 0. \tag{30.3}$$

We have chosen $\tau = 10$ to allow for large clusters; here most of the weight is put on a radius of 20 km. We also allow for a small cluster asuming a threshold of 10 km, thus we define $L = 10$. The resulting distance-based surrogates for exposure are shown in Figure 30.2. Clearly model (30.3) gives most weight to the vicinity of the focus, followed by model (30.1) and (30.2).

All statistical tests and the calculations required to obtain confidence intervals for the SIRs were programmed in Fortran 77 using the GNU-Fortran (1997) Compiler.

30.3 RESULTS

30.3.1 Incidence data base

Results of the analysis of the original dataset are presented elsewhere (Hoffmann and Greiser, 1994, 1996). Here we have updated the dataset with information from an

Table 30.1 Case population

	Males	Females	Total
Acute leukaemias	133	100	233
Chronic leukaemias	197	186	383
Total	330	286	616

ongoing extension of the previous incidence study. This extended study is presently conducted by the Bremen Institute of Prevention Research and Social Medicine to establish the case population for a major case–control study on risk factors for leukaemia and malignant lymphoma. The extended study area completely covers the area of the previous incidence study, and all primary data sources of the previous study are being revisited.

Hence all analyses presented in this Chapter are based on an updated and validated incidence data base including all incident leukaemia cases between 1984 and 1993 in three counties adjacent to the Krümmel nuclear power plant. This data base presently contains complete information on 616 leukaemia cases (Table 30.1).

The fact that the incidence database is updated regularly in the ongoing study, so far has caused few changes compared with the figures published previously. Changes are predominantly due to the reclassification of cases with leukaemia following myelodysplastic stages as 'myelodysplasia' rather than 'leukaemia', in accordance with the definition of the present study. In the previous study, these cases were coded according to what their physicians had labelled the 'main clinical diagnosis' of a patient.

Moreover, as expected, we occasionally obtain additional information on some of the cases, i.e. those who were already ascertained in the previous study. In a few instances additional evidence has now proved that, for example, a diagnosis that was still only 'rule out' in one source in the previous study (an exclusion criterion), was confirmed in another source, which, however, had hitherto been unknown to us. In theory, two case records which did not match in the record linkage of the previous study, nevertheless could later turn out to belong to the same patient. However, since this constellation requires a mistake in the date of birth, the place of residence or even the gender of a patient in the primary clinical documentation, it is encountered extremely rarely.

Altogether, as a result of these changes, 19 leukaemia cases (3.1%) have changed their respective categories compared with the results presented earlier.

A second modification refers to the geographical regions. To provide the most precise distance data for the various geographical methods presented herein, the distance to the nuclear power plant of every rural community has been recalculated, this time without rounding. On the basis of this recalculation 14 rural communities, all of which were very close to one of the borders between the concentric regions, were reassigned to adjacent circles.

30.3.2 Disease mapping

We start with the mixture model analysis of the regional distribution of acute leukaemias for men and women. Using the package Dismap Win (Schlattmann, 1996) we

Table 30.2 Mixture model analysis (acute leukaemia in men)

Components $\hat{k}$	Weight $\hat{p}_j$	Relative risk $\hat{\theta}_j$	λ_k
$k=2$	0.96	0.96	– 266.39
Men	0.44	2.15	
$k=1$	1	0.99	– 266.5

Table 30.3 Mixture model analysis (chronic leukaemia)

	Men		Women	
Weight $\hat{p}_j$	0.60	0.40	0.75	0.25
Relative risk $\hat{\theta}_j$	0.72	1.36	0.75	1.71

obtain a homogeneous solution with $\hat{k} = 1$ and relative risk estimate $\hat{\theta} = 1$ for women (not shown in Table 30.2). For men we obtain an initial two-component solution (Table 30.2). However, the improvement in the log likelihood λ due to inclusion of the second component is only marginal. Hence we conclude for acute leukaemia that there is no deviation from a constant risk within the study region in separate analyses for men and women.

For chronic leukaemias we obtain a two-component solution for both men and women. For men, we obtain a two-component solution with a log likelihood of – 165.9 compared with a log likelihood of – 168.18 for the homogeneous solution. The value of the likelihood ratio statistic (LRS) is 4.56, with a 95% simulated critical value of 4.01 (Schlattmann, 1993). Hence we accept the two-component model. For women, we obtain a two component solution with a log likelihood of – 188.58 compared with a log likelihood of – 191.5 for the homogeneous solution. Thus we obtain a value for the LRS of 5.84 and accept the heterogeneous two-component solution (Table 30.3).

The maps for chronic leukaemia in men and women are in shown in Figures 30.3 and 30.4. Clearly, for chronic leukaemias there is heterogeneity of disease risk present. For chronic leukaemia in men, 60% of the regions have a risk of 0.72 compared with the whole area and 40% have a higher relative risk of 1.36. In women, we find 75% of the areas with a relative risk of 0.77 and 25% with a risk of 1.71 (Table 30.3). Some of the areas with higher relative risk are located close to the power plant, but areas with higher risk are not constrained to the focus.

30.3.3 Focused analysis

For the focused analysis we start with the traditional descriptive approach of calculating the SIR for each concentric region together with a 95% confidence interval (Tables 30.4 and 30.5). Hence, in the analysis based on concentric regions we find an excess risk for chronic leukaemia for men within the first circle and an excess risk for women within the third circle.

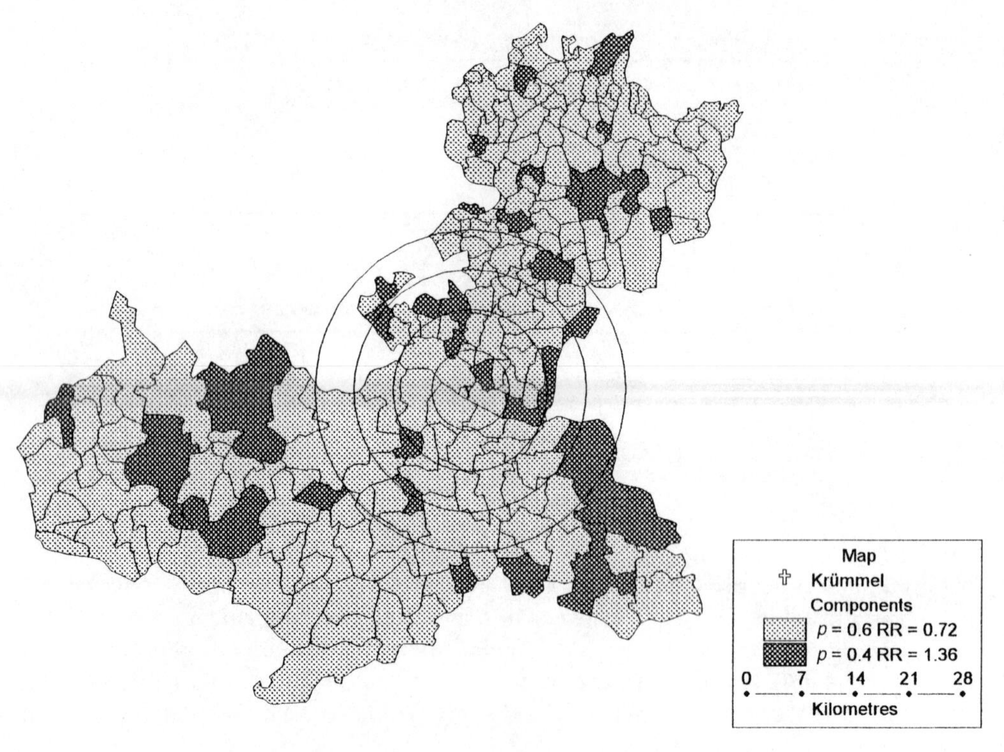

Figure 30.3 Chronic leukaemia in men

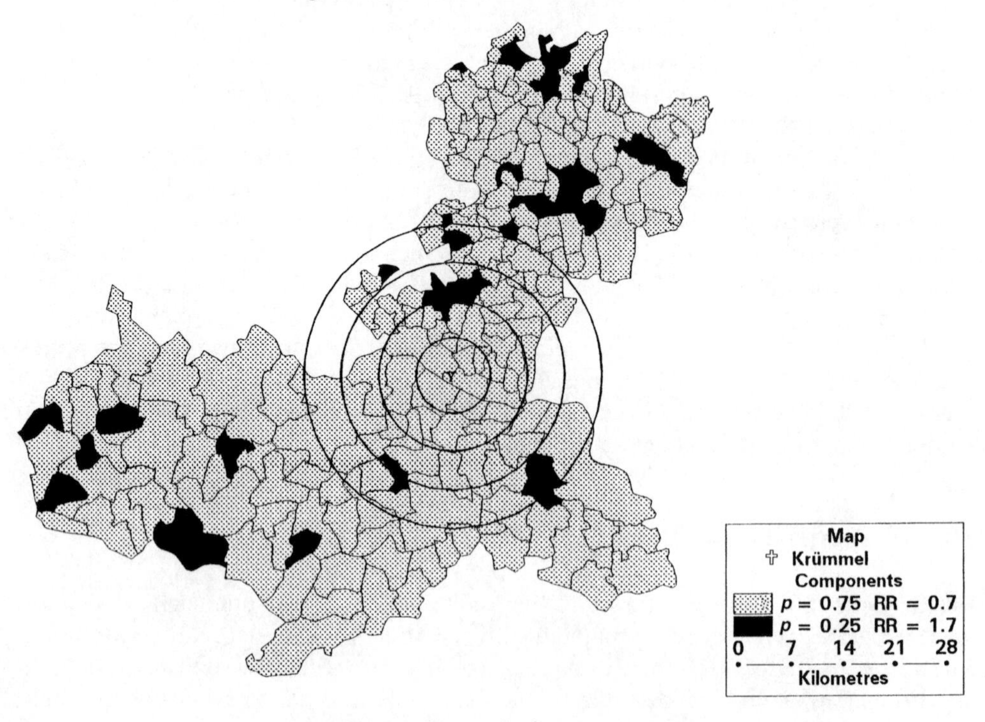

Figure 30.4 Chronic leukaemia in women

Table 30.4 Acute leukaemia for men and women

	Men				Women			
Distance	O_i	E_i	**SIR**	**95%CI**	O_i	E_i	**SIR**	**95%CI**
< 5 km	11	8.16	1.35	0.67, 2.41	5	6.31	0.79	0.26, 1.85
5–< 10 km	11	6.91	1.59	0.79, 2.85	5	5.10	0.98	0.32, 2.29
10–< 15 km	20	22.52	0.763	0.54, 1.37	16	16.94	0.94	0.54, 1.53
15–< 20 km	20	22.19	0.89	0.55, 1.39	18	17.86	1.01	0.60, 1.53
≥20 km	71	73.2	0.901	0.75, 1.22	56	53.79	1.04	0.78, 1.35

O_i: observed cases; E_i: expected cases; SIR: standardised incidence ratio; 95%CI: 95% confidence interval; km: kilometres

Table 30.5 Chronic leukaemia for men and women

	Men				Women			
Distance	O_i	E_i	**SIR**	**95%CI**	O_i	E_i	**SIR**	**95%CI**
< 5 km	20	11.74	1.70	1.03, 2.63	5	6.31	1.11	0.59, 1.90
5–< 10 km	9	10.23	0.88	0.40, 1.67	5	5.10	1.06	0.51, 1.96
10–< 15 km	26	32.35	0.80	0.52, 1.18	16	16.94	1.39	1.01, 1.88
15–< 20 km	24	33.25	0.72	0.46, 1.07	18	17.86	0.83	0.57, 1.19
≥20 km	118	109.41	1.07	0.89, 1.29	56	53.79	0.92	0.74, 1.12

See table 30.4 for abbreviations.

Table 30.6 Results of U^* for the different types of leukaemia in men and women

Diagnostic category	$U^*: g_i = 1/d_i$	$U^*: g_i = exp(-d_i/10)$	$U^*: g_i = exp(-4(d_i/10)^2)$
Acute leukaemia (men)	0.83, $p = 0.21$	0.88, $p = 0.19$	1.21, $p = 0.14$
Acute leukaemia (women)	0.71, $p = 0.24$	– 0.18, $p = 0.57$	0.61, $p = 0.27$
Chronic leukaemia (men)	0.13, $p = 0.45$	0.41, $p = 0.33$	1.35, $p = 0.09$
Chronic leukaemia (women)	0.24, $p = 0.40$	0.85, $p = 0.20$	0.14, $p = 0.44$

In Table 30.6 we present the results of the Waller/Lawson test, based on the distances of each individual geographical unit to the suspected point source using different types of the distance–exposure relationship.

According to Table 30.6 there is no clear-cut relationship between the distance to the nuclear power plant and the risk of leukaemia. Regardless of the assumed structure of the exposure, neither a simple linear decline with distance nor an exponential relationship shows a significant effect. A non-significant trend for chronic leukaemia in males is conceivable only with the third model, which gives the highest weight to the direct vicinity. This corresponds to the result in the innermost concentric region.

30.4 DISCUSSION

Appropriate operationalisation of 'point-source hypotheses' and quantification of the respective environmental 'exposure' have been a challenge for statisticians and epidemiologists for many years. Suspected or real childhood leukaemia clusters in the vicinity of nuclear installations were among the most typical applications.

As of this writing (1998) the question of whether or not emissions of nuclear power plants could induce leukaemias and other types of cancer in the population living in the immediate vicinity of these plants is a matter of intense debate among epidemiologists and in the general public.

30.4.1 Traditional methodological approaches

The majority of examples published so far have used simple geographical approaches of dividing the neighbourhood of the respective point source into concentric regions, which mostly, but not always, had equidistant radii (e.g. $0-<5$ km, $5-<10$ km, $10-<15$ km, or the respective number of miles, etc.). Subsequently, the number of cases and the population at risk were ascertained separately for each concentric region. Either incidence density (ID) rates, standardised incidence density (SID) rates or relative measures such as the standardised incidence rate ratio (SIR) were calculated to compare the circular regions and to conduct statistical testing.

Most of the applications of this simple approach have not supported the presence of any measurable risk.

However, these negative studies should not be interpreted as proof of the absence of risk (for a critical discussion of this misconception, see Chapter 22 in this volume). In some instances negative findings could well have been a consequence of the inherent conceptional and methodological limitations of this approach (Gardner, 1989; Wakeford and Binks, 1989; Shleien *et al.*, 1991; MacMahon, 1992). The results of simple geographical incidence studies nevertheless have often proven helpful for local health professionals in the context of risk communication to the public.

Despite this crudeness, there is a number of examples of positive findings as well. Like most of the negative studies, these investigations have predominantly focused on childhood malignancies. In particular, increased risks for childhood leukaemia in the vicinity of nuclear power plants were observed in England and Wales (Baron, 1984; Black, 1984; Gardner and Winter, 1984b; Urquhart *et al.*, 1984, 1986; Barton *et al.*, 1985; Forman *et al.*, 1987; Roman *et al.*, 1987; Ewings *et al.*, 1989), Scotland (Heasman *et al.*, 1984, 1986; Hole and Gillis, 1986), the United States (Johnson, 1981; Clapp *et al.*, 1987; Crump *et al.*, 1987; Goldsmith, 1989a,b; Hatch *et al*, 1991), Canada (McLaughlin *et al.*, 1993), Germany (Grosche *et al.*, 1987; Dieckmann, 1992; Michaelis *et al.*, 1992; Hoffmann *et al.*, 1993; Prindull *et al.*, 1993), France (Viel *et al.*, 1993, 1995), and recently Japan (Iwasaki *et al.*, 1995; Hoffmann *et al.*, 1996). Some of these examples have stimulated interesting hypotheses and further research, including refined exposure assessment (Wing *et al.*, 1997a,b; Schmitz-Feuerhake *et al.*, 1997) and analytical epidemiological studies (Gardner *et al.*, 1990a, Urqhart *et al.*, 1991; Morris and Knorr, 1996; Pobel and Viel, 1997).

30.4.2 Distance-based approaches

One of the most important methodological problems of the traditional approach is bias due to the loss of geographical information. This bias results from the use of broad categories, arbitrary choice of the radii, the problem of multiple testing for numerous circles, and the problem of extra-Poisson variation, which cannot be addressed using traditional methods.

Some of these problems can be avoided using the whole dataset. Some authors have tried to model the effect of distance in loglinear models (see, for example Cook-Mozzafari *et al.*, 1989).

In addition to this approach an increasing number of authors try to address the effect of distance to the point source as a surrogate exposure measure by applying either the score test used here or Stone's Poisson maximum distance test. The score test has been applied in the context of leukaemia and waste sites (Waller and Turnbull, 1993) and in the context of leukaemia and nuclear power plants (Waller *et al.*, 1995). The latter analysis also tried to address 'general' and 'focused' clustering as in the present analysis.

Stone's test has been used even more frequently in order to evaluate a distance–risk of disease relationship with regard to nuclear power plants and leukaemia. Bithell and Stone (1988) did not find an association between leukaemia incidence in the Sizewell area, whereas Viel *et al.* (1995) report a distance-based gradient for the La Hague plant.

In this chapter we have discussed the use of disease mapping methods and focused tests for risk assessment. Clearly, our database is limited to three counties (Landkreise), and results cannot readily be generalised to other locations. From our point of view a map of the whole country with the same spatial resolution would be desirable to reduce the problem of selection bias and *post hoc* analysis, given that an (empirical) Bayes approach is used, since in this case inference for an individual area is performed in the light of the whole dataset.

On the other hand, the incidence dataset, which we used for the first time in Germany, has provided high resolution geographical data for adult leukaemias and has been most intensely validated for completeness of case ascertainment and data accuracy.

The use of map-based techniques allows us to investigate the presence of extra-Poisson variation, which would indicate heterogeneity of disease risk. For acute leukaemias in men and women, we found no indication of extra-Poisson variation or heterogeneity of disease risk within the study area. Clearly this is in accordance with a negative finding for the effect of distance, which otherwise would have produced extra-Poisson variation.

Using the traditional approach we found a significant excess of leukaemia risk for men for chronic leukaemias in the 5 km radius and for women in the 15–20 km radius. Likewise, for this diagnostic category we found heterogeneity of disease risk using disease mapping methods. No trend was observed over concentric regions for neither diagnostic category and sex, respectively. Application of the Poisson maximum test by Stone confirms the absence of any significant trend over the concentric regions (results not shown). Since there is extra-Poisson variation present for the chronic leukaemia data, we also computed covariate adjusted mixture models (Schlattmann *et al.*, 1996). Here we included the above-defined distance functions, which act as an exposure surrogate, as a covariate into the model. We also followed a suggestion by Lawson (1993b) to allow for directional effects. Neither a homogeneous log linear model nor a covariate adjusted mixture model revealed a significant effect of distance to the power plant (result not shown) or significant directional effects.

Thus, on the level of individual geographical regions there is only a slight indication of a trend in the SIR with respect to distance from the nuclear power plant, if any, which is restricted to males and well below statistical significance.

From a methodological point of view we should mention that there is certainly a need to address edge effects, as discussed in Chapter 6 in this volume, since our study area is close to the city of Hamburg and countries of the former GDR. For both of these adjacent areas there are no data available.

30.5 CONCLUSION

At the present stage, our results do not indicate any increased risk for the diagnostic category of acute leukaemias for either males or females of all ages in the vicinity of the nuclear power plant and within the whole study region. Likewise, there is no increased risk for chronic lymphatic leukaemia in females. For males, however, an increased SIR was observed in the 5 km region around the plant. Despite considerable modifications in the coding of diagnoses and the assignment of rural communities to the concentric regions, this result is in accordance with a result published earlier (Hoffman and Greiser, 1996). No statistically significant trend with distance, however, was found based on individual geographical units.

Looking at the maps of the study region we find heterogeneity of disease risk for chronic leukaemia in men and women. Some of the areas close to the power plant fall into the higher risk category with an excess risk of about 32%, but the high risk areas are different for women and men and generally are not constrained to the vicinity of the power plant.

No conclusive explanation of these findings is possible without additional data. Clearly this heterogeneity of disease risk is due to an *unobserved* heterogeneous distribution of risk factors. There might be a multitude of different exposure patterns present within the study area. One could think of a heterogeneous distribution of pre-existing conditions, which increase leukaemia risk, such as therapeutic radiation or antineoplastic chemotherapy. Our findings could also be related to different life-style patterns such as smoking or to environmental factors such as the use of herbicides or other pesticides on farms and greenhouses or possibly the indoor use of insecticides (Greiser *et al.*, 1995; Hostrup *et al.*, 1997). Along the same line of reasoning, the increased risk in the vicinity of the power plant could be generated by any of the factors mentioned before or by either direct radiation or by radioactive nuclides which are emitted close to the ground, e.g. vented directly from the reactor building. Another possibility would be ground contamination through locally confined irrigation with contaminated water or specific patterns of food production. Currently we do not have sufficient information on the spatial distribution of any of these parameters.

In principle, inclusion of the predominant wind direction in the model would be an option. However, there is a serious drawback to this. According to available environmental surveillance data, emissions from the plant are highly variable over time. Hence the direction of the distribution of any released radionuclides is dependent on the wind direction around the specific time of the release which might well be different from the main or average direction.

The pattern of areas with increased incidence of chronic leukaemias is different in males and females. This difference could well be attributable to chance. However, alternative explanations include gender-specific differences in sensitivity for the induction of chronic leukaemias, possibly due to a different pattern of histological subtypes within this broad and heterogeneous category. In a more general sense this could explain the observed heterogeneity of chronic leukaemia risk for both genders in terms of regional differences in the proportion of more 'susceptible' histological subtypes. However, case numbers are insufficient to allow for a simultaneous analysis of gender and histological subgroup of leukaemia.

Moreover, the estimation of distance to any point source uses people's residences rather than their actual geographical position at any given time. Hence even while distance and the direction of someone's home at some specific time of radionuclide emission might indicate exposure, this might or might not be true for the person living there. The person could well have been at work or even on vacation many miles away from his/her home when the radioactive plumes passed by. Exposure can also be underestimated in this approach. Workers in the nuclear industry are likely to live in close proximity to the plant where they are employed. Gardner *et al.* have found that fathers of children with leukaemia residing in the vicinity of BNFL Sellafield were not only more likely to work on the site but also on average had accumulated higher radiation doses compared with other fathers (Gardner *et al.*, 1990a,b).

The problem is even worse considering the fact that residence is usually operationalised as 'residence at the time of the first diagnosis of a target disease' in retrospective incidence studies. Since initiation of any cancer occurs many years before clinical diagnosis, a person's most recent residence's distance to a point source may or may bear any significance with respect to exposure at the most relevant time period.

In conclusion, at the present stage all these scenarios are purely speculative. Given the ecological nature of the data, any inference as to causality is of course impossible. In the next step we will analyse our data using methods that refine the spatial resolution and take full advantage of the available geographical resolution of the cases' residences. Thus, we will apply methods such as extraction mapping (Lawson and Williams, 1994; Viel *et al*, 1995) which are based on kernel density estimates that allow the use of individual case locations. An alternative method would be the use of case–control methods (see, for example Chapter 20 in this volume). Clearly, owing to the need to fulfil the above-mentioned data protection requirements, there is the problem of tentatively imprecise locations, as discussed by Jacquez (1994b). Thus, we will also follow the suggestions given in this chapter to quantify the impact of imprecise locations.

Currently a major analytical epidemiological case–control study is in progress in an extended study area in which a multitude of potential risk factors for leukaemias and lymphomas will be evaluated. This study will collect information on risk factors as well as on confounders and, hence, provide a unique opportunity for a geographical case–control analysis.

31

Lung Cancer Mortality in Women in Germany 1995: A Case Study in Disease Mapping

Peter Schlattmann and Dankmar Böhning

Free University Berlin

Allan Clark and Andrew B. Lawson

University of Aberdeen

31.1 INTRODUCTION

From a public health point of view the investigation of the regional distribution of lung cancer mortality in women may be fruitful for a number of reasons. First, in contrast to men the incidence and mortality of lung cancer for women in Germany is on the rise (Schön *et al.*, 1995). The major risk factor for lung cancer is tobacco smoking (Tomatis *et al.*, 1990); thus the mortality of lung cancer is closely related to the increasing prevalence of smoking in women. Estimates of the attributable risk of lung cancer due to smoking range from 57% to 83% in women. Thus lung cancer is frequently addressed as a disease that belongs to the category of avoidable death, i.e. deaths which may be avoided by medical and/or preventive intervention (Holland, 1993). Since the prognosis of patients suffering from lung cancer is poor and until today the effectiveness of screening measures is controversial (Chamberlain, 1996), the major preventive action available are smoking cessation programmes as a means of primary prevention. Thus identification of high risk areas in disease maps could provide targets for preventive action such as smoking cessation programmes.

Besides smoking there are several other risk factors for lung cancer established or under discussion which are worth mentioning from a public health perspective. The

Disease Mapping and Risk Assessment for Public Health. Edited by A.B. Lawson *et al.*
© 1999 John Wiley & Sons Ltd.

well-known urban/rural difference in lung cancer incidence and the detection of known carcinogens in the atmosphere gave rise to the hypothesis that long-term, exposure to air pollution may have an effect on lung cancer risk. A recent review by Katsouyanni and Pershagen (1997) summarised the evidence that ambient air pollution may have an effect on lung cancer risk. The estimated relative risk of ambient air pollution was up to 1.5. However, owing to the difficulties in exposure assessment the effect of air pollution on lung cancer is till controversial. Thus identification of high risk areas in disease maps may be a starting point for further analytical studies.

Another risk factor under discussion is indoor radon. Studies of underground miners exposed to radioactive radon and its decay products have found that exposure increases the risk of lung cancer. Consequently, when radon was found to accumulate in houses, there was concern about the public health impact from exposure to a known carcinogen. Because of differences between working in underground mines and living in houses, estimates are subject to major uncertainties. Numerous case–control studies were launched to assess directly the lung cancer risk from indoor radon. Some studies report positive or weakly positive findings, while others report no increased risk (Lubin and Boice, 1997). Thus disease maps combined with geological information could provide starting points for analytical studies investigating the risk of radon.

Finally, there are several protective agents such as selenium or antioxidants like vitamins E or A under discussion (Blot, 1997). Thus identification of high risk areas could be a starting point for intervention trials, introducing chemoprevention using minerals and/or vitamins. And, of course, the identification of low risk areas could provide hypotheses for unknown protective factors.

As a result, the implications from disease maps of lung cancer mortality in women are manifold. The common denominator of the ideas above is to display the heterogeneity of disease risk in maps. Our case study will present and compare the results of several methods for disease mapping using mortality data from Germany in 1995.

31.2 THE DATA

The establishment of population-based cancer registries is still under development in Germany. Only mortality data are routinely available. Thus we use for our analysis mortality data for lung cancer from women in the year 1995. The data were coded according to the 9th revision of the International Classification of Diseases ICD-9 as ICD 162.

When constructing disease maps one of the first steps is the choice of the spatial resolution. Frequently the spatial resolution is limited by the availability of the data. In Germany, unfortunately there is no central database accessible as a source for small-area health data. Data even on a spatial resolution of 'Landkreise' are not routinely available (Landkreise refers to small areas such as counties). Such data can only be obtained by directly addressing the census bureaus (Statistische Landesämter) of the 16 states of Germany. As a result, the collection of health data on a small-area level such as the Landkreise is quite tedious and expensive. Only because of financial support by the European Union, were we able to collect data on the spatial resolution of the 439 Landkreise and cities of Germany for the year 1995.

Once the spatial resolution has been defined an appropriate epidemiological measure has to be chosen. Define the observed count in the ith small area as O_i. The corresponding expected cases in that region are denoted E_i. One frequently used measure of relative risk is the Standardised Mortality Ratio, $\text{SMR}_i = O_i/E_i$, where the expected cases E_i are calculated based on a reference population. For our data we used the age-specific lung cancer mortality rates for women in Germany for the year 1995 (Statistisches Bundesamt, 1997) as the reference population. The necessary population data of the individual area were taken from the database 'Statistik regional' (Statistische Ämter, 1997) as well as the boundary file of Germany (DFLR, 96).

Here we can use the SMR_i of the individual region as an estimate for the relative risk of that area compared with the whole country.

31.3 THE METHODS

31.3.1 Traditional methods

Once the spatial resolution and the epidemiological measure are defined, a suitable mapping method has to be chosen. A common approach for the construction of thematic maps in epidemiology is the choropleth method (Howe, 1990). This method implies categorising each area and then shading or colouring the individual regions accordingly.

One of the traditional approaches to categorisation is based on the percentiles of the SMR distribution. Most cancer atlases use this approach, usually based on quartiles, quintiles or sextiles (Walter and Birnie, 1991). Figure 31.1 shows the map of lung cancer mortality based on the quartiles of the SMR distribution.

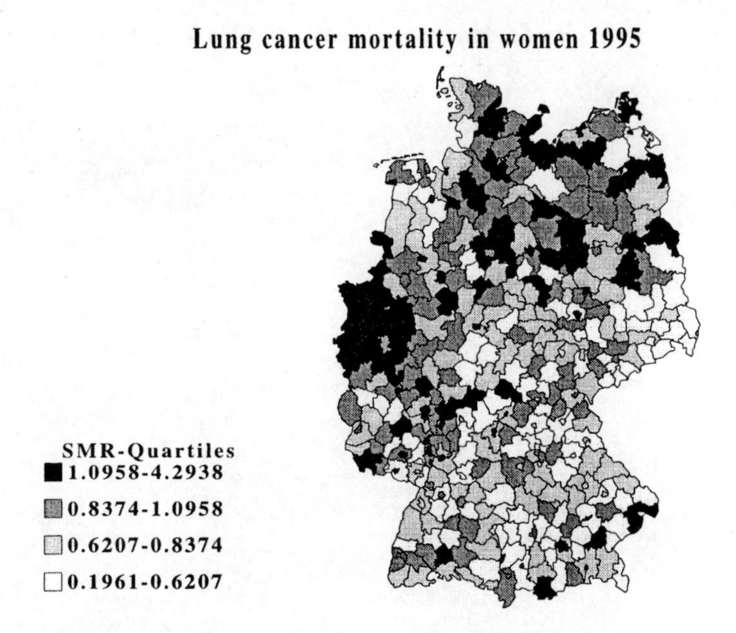

Figure 31.1 Map based on the quartiles of the SMR distribution

Note the high variability of the data, with relative risk estimates ranging from 0.196 to 4.294. This would indicate a relative risk that is up to four times higher in high risk areas or six times lower in low risk areas. But in the worst case this variability reflects only random fluctuations due to different population size and corresponding small counts.

Thus another frequently used approach is based on the assumption that the observed cases O_i of the individual region follow a Poisson distribution with

$$O_i \sim \text{Po}(\theta E_i), \text{ with density } f(O_i, \theta, E_i) = \frac{e^{-(\theta E_i)}(\theta E_i)^{O_i}}{O_i!},$$

where again E_i denotes the expected cases in the ith region. Computation of the p-value is done under the null hypothesis $\theta = 1$ or based on the maximum-likelihood estimator

$$\hat{\theta} = \frac{\sum_{i=1}^{n} O_i}{\sum_{i=1}^{n} E_i},$$

where the latter is called the adjusted null hypothesis (n is the number of areas). This is a minimal model for the relative risk, namely that a constant risk is assumed and a significant result would indicate departure from constant risk. Figure 31.2 shows the probability map using the adjusted null hypothesis. In terms of interpretation both maps reflect the above-mentioned urban/rural difference in mortality. Common to both maps

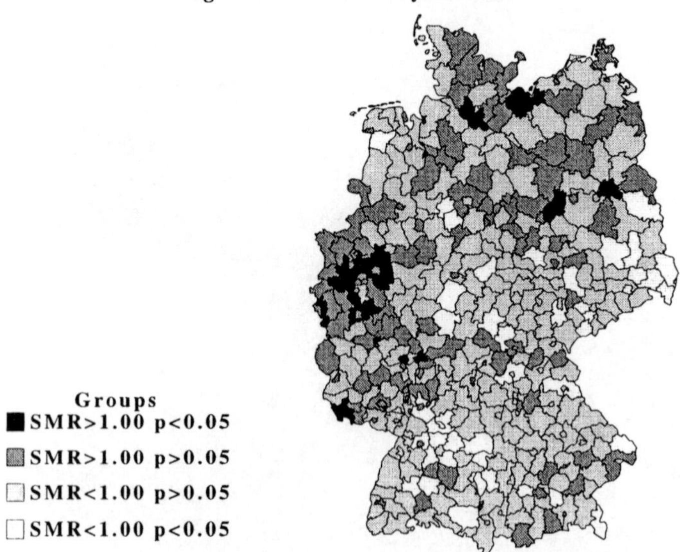

Figure 31.2 Map based on Poisson probabilities

is an excess risk in large urban areas such as Berlin, Hamburg and the Ruhrgebiet, just to mention a few. There seems also to be an east–west and a north–south gradient.

However, probability maps can suffer from the presence of artefacts that are unobserved in the data. Again, population size is a possible confounder; large areas tend to have significant results. Also, we have the problem of multiple testing, even adjusting for the number of comparisons does not lead to a consistent estimate of heterogeneity of the data (Schlattmann and Böhning, 1993).

31.3.2 The empirical Bayes approach

The parametric empirical Bayes approach

To circumvent the problems associated with percentile and probability maps, frequently random effect models are used. These are models where the distribution of relative risks θ_i between areas is assumed to have a probability density function $g(\theta)$. The O_i are assumed to be Poisson distributed conditional on θ_i with expectation $\theta_i E_i$.

Several parametric distributions like the gamma distribution or the log normal distribution have been suggested for $g(\theta)$; for details see Clayton and Kaldor (1987) or Mollié and Richardson (1991) and Chapter 2 in this volume. Among these parametric distributions describing the heterogeneity involved in the population values θ_i the Gamma distribution has been used several times for epidemiological purposes (see Chapter 25 in this volume). Note that in a Bayesian sense we can also think of $g(\theta)$ as a parametric prior distribution on θ_i. The very nature of empirical Bayesian procedures is that the parameters of the prior distribution are estimated from the data; in the case when the

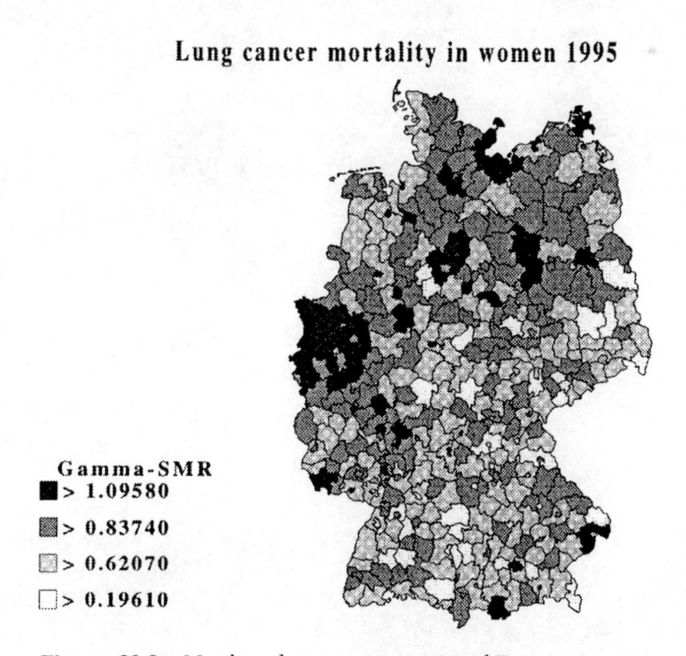

Figure 31.3 Map based on gamma empirical Bayes estimates

θ_i are assumed to be gamma distributed, with $\theta_i \sim \Gamma(\alpha, \nu)$, the parameters α and ν have to be estimated from our data. This estimation has to be done iteratively here we have applied the algorithm proposed by Clayton and Kaldor (1987), which is also implemented in DismapWin (Schlattmann, 1996). Applying Bayes' Theorem we obtain the posterior expectation of the relative risk of the individual area as

$$\widehat{\mathrm{SMR}}_i = \mathrm{E}(\hat{\theta}_i|\hat{\alpha}, \hat{\nu}, E_i, O_i) = \frac{O_i + \hat{\nu}}{E_i + \hat{\alpha}}.$$

Figure 31.3 shows the map based on this empirical Bayes approach. Note that the range of the posterior SMRs is now reduced. The lowest estimated risk is now 0.4 and the highest estimated risk is 3.12. Thus we removed from the data random variability due to the small counts. To ensure comparability of our maps we used the same grey scale as in our first percentile map. Thus we see immediately that we are now dealing with a smoothed map with less extremes in the relative risk estimates. Again we observe excess risk in metropolitan areas.

The non-parametric empirical Bayes approach

Now let us assume that our population under scrutiny consists of subpopulations with different levels of disease risk θ_j. Each of these subpopulations with disease risk θ_j represents a certain proportion p_j of all regional units. Statistically, this means that the mixing distribution reduces to a finite mass point distribution. Here we face the problem of identifying the level of risk for each subpopulation and the corresponding proportion of the overall population. One can think of this situation as a *hidden* (*or latent*) structure, since it remains unobserved to which subpopulation each area belongs. These subpopulations may have different interpretation. For example, they could indicate that an important covariate has not been taken into account. Consequently, it is straightforward to introduce an unobserved or latent random vector Z of length k consisting of only zeros besides one 1 at some position (say the Jth) which then indicates that the area belongs to the Jth subpopulation. Taking the marginal density over the unobserved random variable Z we are led to a discrete semiparametric mixture model. If we assume a non-parametric parameter distribution for

$$P = \begin{bmatrix} \theta_1 & \dots & \theta_k \\ p_1 & \dots & p_k \end{bmatrix}$$

for the mixing density $g(\theta)$ (whose MLE can be shown to be always discrete), we obtain the mixture density as the weighted sum of Poisson densities for each area i:

$$f(O_i, P, E_i) = \sum_{j=1}^{k} p_j f(O_i, \theta_j, E_i), \quad \text{with} \sum_{j=1}^{k} p_j = 1 \text{ and } p_j \geq 0, j = 1, \dots, k.$$

The model consists of the following parameters: the number of components k, the k unknown relative risks $\theta_1, \dots, \theta_k$ and $k-1$ unknown mixing weights $p_1, \dots, p_{k-1}$. There are no closed-form solutions available; to find the maximum likelihood estimates suitable algorithms are given by Böhning *et al* (1992). These algorithms are implemented in the package DismapWin (Schlattmann, 1996). Estimating the unknown parameters

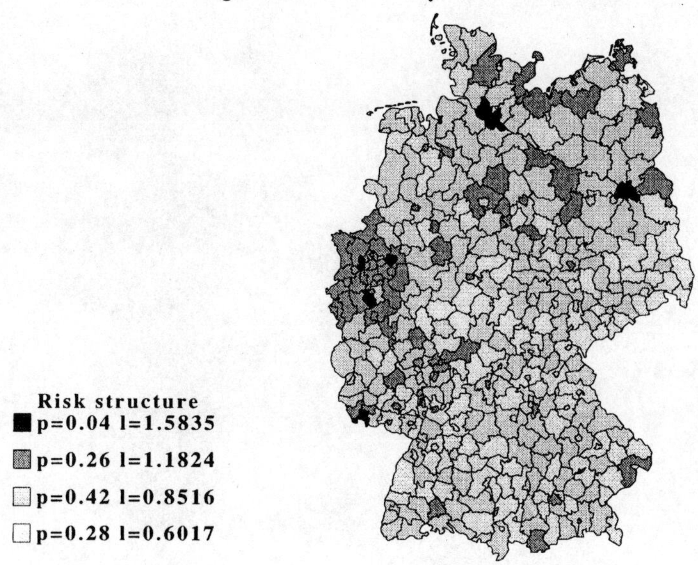

Figure 31.4 Map based on the mixture model

and applying Bayes theorem enables us to construct a map based on the posterior density. Details on this approach can be found in Chapter 4 in this volume.

The map based on this approach is shown in Figure 31.4

We obtain a four-component mixture model with DismapWin as an appropriate solution for our data. The category with the highest risk has a weight of 0.04% with a relative risk of 1.5; the next category has a weight of 0.26% and a slightly elevated risk of 1.13. The next category has a weight of 0.42% and a relative risk of 0.85 and the category with the lowest relative risk of 0.6 has a weight of 28% of all regions. It should be pointed out that one of the advantages of the non-parametric mixture approach is that—as a by-product—a number of colours or greyshading patterns to be used in the map are provided.

We can compute the posterior expectation for this model as well:

$$\widehat{\mathrm{SMR}}_i = \mathrm{E}(\hat{\theta}_i | O_i, \hat{P}, E_i) = \frac{\sum_{j=1}^{k} \hat{\theta}_j \hat{p}_j f(O_i, \hat{\theta}_j, E_i)}{\sum_{l=1}^{k} \hat{p}_l f(O_i, \hat{\theta}_l, E_i)}.$$

Using the same categorisation as in the SMR percentile map we obtain the map based on the posterior expectation of the SMR_i as shown in Figure 31.5.

The appearance of the map looks very similar to that in Figure 31.3, the map based on the parametric empirical Bayes approach. Comparing Figures 31.3 and 31.5 with Figure 31.4, i.e. the map constructed according to the mixture posterior classification, we find

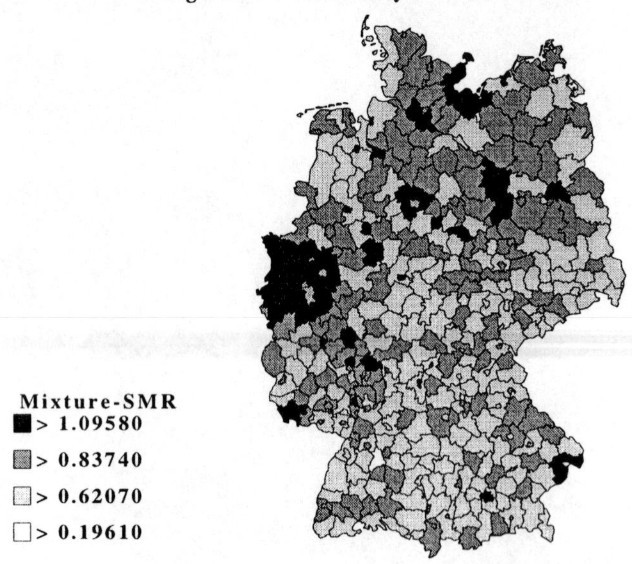

Figure 31.5 Map based on the mixture model relative risk estimates

different patterns. The similarity of Figure 31.3 and 31.5 is mainly based on the fact that they use the same method of classification using the greyscale of the percentile map of the crude SMR. The major gain of these two maps is that they remove random variability from the map. The map based on the mixture classification, however, now introduces a different pattern, since it not only provides shrinkage of the estimators, but also provides an estimate of the underlying risk structure.

Also, the range of the estimated relative risks based on the non-parametric empirical Bayes approach is even lower than in the parametric empirical Bayes approach. Thus the mixture model approach provides a higher degree of shrinkage than the approach based on the gamma distribution. The lowest posterior relative risk is 0.51, and the highest posterior relative risk is 1.58. Again we find an excess risk in metropolitan areas. We also observe a lower risk in the east and the south of the country.

31.3.3 Full Bayesian analysis

In this section we demonstrate the use of a fully Bayesian modelling approach to the analysis of the German cancer data. A form of this approach was first proposed by Besag *et al.* (1991). Using the notation of the previous sections, we define the Poisson likelihood for a realisation $\{O_i\}, i = 1, \ldots, n$, of counts in n small areas as

$$L = \prod_{i=1}^{n} \frac{e^{-\theta_i E_i} (\theta_i E_i)^{O_i}}{O_i!}.$$

Lung cancer mortality in women 1995

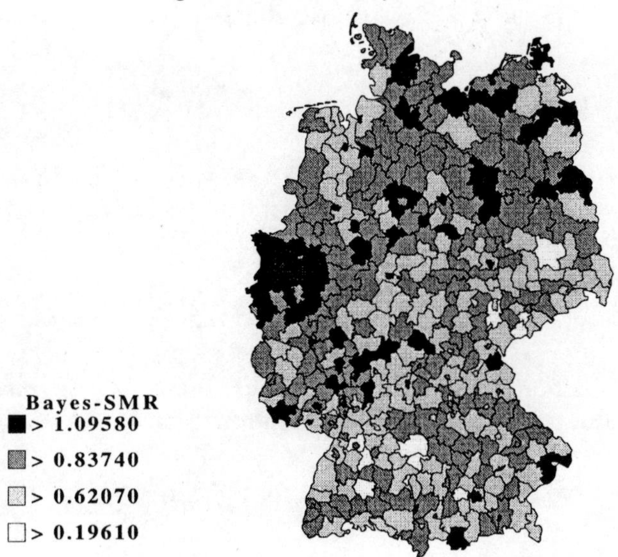

Figure 31.6 Map based on the full Bayes model

Here

$$\theta_i = \exp\{t_i + u_i + v_i\},$$

where E_i are again the expected cases for the ith small area, and we have a log linear link between the Poisson expectation and the terms t_i, u_i, and v_i. These model terms represent different types of variation that could be considered in the model. The first term represents trend in the rates across the study region and can be thought of as *long-range* variation. In our example, we do not include trend variation, although it is straightforward to do so in any particular application. The second and third terms (u_i, v_i) represent types of random effect or heterogeneity, which can be included if there is thought to be any extra random structure in the counts which may remain unexplained by the other model components. This extra structure could be due to inherent extra variation not captured by the Poisson likelihood model. Uncorrelated extra variation is sometimes called uncorrelated heterogeneity (see Chapter 1 in this volume). In addition, there could also exist autocorrelated variation, which is often termed 'correlated heterogeneity'. In our model we represent correlated heterogeneity by u_i and uncorrelated heterogeneity by v_i. Because we wish to apply a fully Bayesian analysis to the dataset, we assume that all parameters in our model have prior distributions. In fact, the heterogeneity terms are random effects, and since we have no other external support for their estimation (than the dataset) we need to make distributional assumptions to allow us to properly distinguish their form.

The prior distributions employed here are those specified by Besag and coworkers. The correlated random effect has an intrinsic singular Gaussian prior distribution:

$$p_i(u_i|\ldots) \propto \frac{1}{\sqrt{\beta}}\left\{-\sum_{j\in\partial_i} w_{ij}(u_i - u_j)^2\right\},$$

where $w_{ij} = 1/2\beta, \forall ij$. The neighbourhood ∂_i is assumed to be the areas with a common boundary with ith area. The uncorrelated heterogeneity (v_i) is defined to have a Gaussian prior distribution:

$$p(\mathbf{v}) \propto \sigma^{-n/2}\exp\left\{-\frac{1}{2\sigma}\sum_{i=1}^{n} v_i^2\right\}.$$

These prior distributions have parameters which must also be considered to have (hyper)prior distributions. Both β and σ are assumed to have improper inverse exponential hyperpriors:

$$\text{prior}\,(\beta, \sigma) \propto \mathrm{e}^{-\varepsilon/2\beta}\mathrm{e}^{-\varepsilon/2\sigma}, \quad \sigma, r \succ 0,$$

where ε is taken as 0.001. These prior distributions penalise the absorbing state at zero, but provide considerable indifference over a large range. The trend parameters have been assumed to have uniform prior distributions on suitable ranges, since we have little prior preference for their values.

Once the prior distributions are specified, we must consider the evaluation of the full posterior distribution (P_0), which combines the Poisson likelihood and all the prior distributions. To sample parameter values from P_0 we employ a Markov chain Monte Carlo (MCMC) method. We have employed a Metropolis–Hastings algorithm to sample all parameters. This algorithm allows for the iterative evaluation of the proposed new parameters via the use of posterior ratios. Convergence of the algorithm was assessed by the Geweke criteria, based on the log posterior surface, and chains with separate start values were examined. Cowles and Carlin (1996) and Brooks (1998) discuss the variety of methods available for convergence checking of this algorithm. Convergence was achieved in this example by 35 000 iterations.

Our analysis of the lung cancer mortality data for women has led to the production of a posterior expected relative risk map (Figure 31.6). This map represents a summary of the final converged relative risks for the dataset from the sampling algorithm.

The resulting map shows some marked features. First, the inclusion of a correlation term has smoothed the map and produced many patches of similar risk level. This is commonly found when autocorrelation is included in such an analysis. The main features displayed on the map relating to difference in lung cancer incidence are: (i) a large concentration of elevated risk in the western area of Germany and (ii) a noticeable north–south gradient, with elevated risks in the northern region. Further analysis of the mapped data could include the explicit modelling of a trend surface which could capture the north–south gradient, if such an estimated surface were required.

However, comparing these maps with those based on the empirical Bayes methods, especially the map based on the posterior expectation of the relative risks based on the mixture model, we find a similar pattern. But again the mixture model approach induces the highest degree of shrinking.

31.3.4 Discussion and conclusion

Our case study shows some marked features. All Bayesian methods (empirical and full) applied produce smoothed maps, indicating that random variability has been extracted from the data. It is now more commonly accepted that percentile maps should be accompanied by smoothed maps. From our point of view a Bayesian methodology should be applied in order to obtain smoothed maps. The major issue here is to decide between empirical and full Bayesian analysis. For a detailed discussion of this topic, see Chapter 1 in this volume. In addition, although not pursued here, it is also important in applications to assess goodness-of-fit via the use of global measures such as AIC and BIC, or point-wise via residual analysis. This can lead to the choice of models which may not be favoured by prior considerations. In addition, the assessment of point-wise variances or standard errors for relative risks on maps should be employed when the assessment of individual regions is a focus of attention. Currently, the authors are involved in a simulation study which investigates and compares empirical and full Bayesian methods for disease mapping.

From a practical point of view, especially from the viewpoint of the public health practioner, the empirical Bayes approaches described here have several advantages. First, they provide relatively easy computation and implementation. Secondly, with packages such as DismapWin, there is free software available, which directly produces maps based on these methods. Thirdly, empirical Bayes methods do not require a difficult convergence diagnostic such as the full Bayesian approach. This relative simplicity is mainly due to the fact that the empirical Bayes methods described here model unstructured heterogeneity of disease risk and ignore structured heterogeneity. Certainly the full Bayes approach offers the most flexible approach to the data, since any aspect of structured and unstructured heterogeneity and trend can easily be modelled. Also, complete inference for any part of the model may be obtained from the posterior distribution. Inference for the mixture model approach, for example, has to rely on resampling methods, which introduces some complexity, but again may be done with DismapWin.

All Bayesian methods determine the presence of heterogeneity of disease risk in one way or the other. From an epidemiological point of view this is an important feature, since only in the case of heterogeneity of disease risk is further action required. The non-parametric mixture model approach offers the additional feature that it may be used to classify the individual region; this of particular importance when disease maps are used in order to investigate the potential hazardous effects of point sources.

From a public health point of view our case study provides some interesting results. As might have been expected, we find an excess risk in urban areas of the western part of the country. Without additional data no further causal assumptions can be made; this excess could be either due to a different prevalence of smoking in the western part or might be related to ambient air pollution. Further ecological studies would be an interesting task. Currently neither data on smoking habits nor on environmental pollution are available to investigate this question further.

References

Aase, A. and Bentham, G. (1996) Gender, geography and socio-economic status in the diffusion of malignant melanoma risk. *Social Science and Medicine*, **42:** 1621–1637. (366)

Abel, U. and Becker, N. (1987) Geographical clusters and common patterns in cancer mortality of the Federal Republic of Germany. *Archives of Environmental Health*, **42:** 51–57. (171)

Ahlbom, A., Axelson, O., Stöttrup-Hansen, E., Hogstedt, C., Jensen, U. and Olsen, J. (1990) Interpretation of 'negative' studies in occupational epidemiology. *Scandinavian Journal of Work, Environment and Health*, **16:** 153–157. (305)

Aitken, M. (1996) A general maximum likelihood analysis of overdispersion in generalised linear models. *Statistics and Computing*, **6:** 251–262. (10)

Albers, J. M., Kreis, I. A., Liem, A. K. and van Zoonen, P. (1996) Factors that influence the level of contamination of human milk with poly-chlorinated organic compounds. *Archives of Environmental Contamination and Toxicology*, **30:** 285–291. (304)

Alexander, F. E. and Boyle, P. editors (1996) *Methods for Investigating Localized Clustering of Disease*. IARC Scientific Publication 135. International Agency for Research on Cancer, Lyon. France. (100, 176)

Alexander, F. E. Cartwright, R. A. and McKinney, P. M. (1988) A comparison of recent statistical techniques of testing for spatial clustering: preliminary results. In P. Elliot, editor, *Methodology of Enquiries into Disease Clustering*. School of Hygiene and Tropical Medicine, London, pp. 23–33. (171)

Alexander, F. E., Wray, N., Boyle, P., Coebergh, J. W., Draper, G., Bring, J., Levi, F., Kinney, P. A. M., Michaelis, J., Peris-Bonet, R., Petridou, E., Pukkala, E., Storm, H., Terracini, B. and Vatten, L. (1996) Clustering of childhood leukaemia: a European study in progress. *Journal of Epidemiology and Biostatistics*, **1:** 13–24. (109)

Alt, K. W. and Vach, W. (1991) The reconstruction of genetic kinship in prehistoric burial complexes —problems and statistics. In H. H. Bock and P. Ihm, editors, *Classification, Data Analysis and Knowledge Organization*. Springer-Verlag, Berlin. (115)

Altman, D. and Bland, J. (1995) Absence of evidence is not evidence of absence. *British Medical Journal*, **311:** 485. (298)

American Journal of Epidemiology, **132:** S1–S202 (1990). (111)

Andersen, P. K., Borgan, Ø., Gill, R. D. and Keiding, N. (1992) *Statistical Models Based on Counting Processes*. Springer-Verlag, New York. (273)

Anderson, N. H. and Titterington, D. M. (1997) Some methods for investigating spatial clustering, with epidemiological applications. *Journal of the Royal Statistical Society, Series A*, **160:** 87–105. (105, 108, 115–16, 253)

Anonymous (1997) Adult blood lead epidemiology and surveillance—United States, first quarter 1997, and annual 1996. *Morbidity and Mortality Weekly Report*, **46:** 643–647. (304)

Anselin, L. (1995) Local indicators of spatial association—LISA. *Geographical Analysis*, **27:** 93–115. (154, 178)

Anto, J., Sunyer, J. and Rodriguez-Roisin, R. (1989) Community outbreaks of asthma associated with inhalation of soybean dust. *New England Journal of Medicine*, **320:** 1097–1102. (100)

Armstrong, B. G., Gleave, S., Wilkinson, P. (1996) The impact of migration on disease rates in areas with previous environmental exposures. *Epidemiology*, **7:** S88. (392)

Aronowitz, S. (1987) *Science as Power: Discourse and Ideology in Modern Society*. University of Minnesota Press, Minneapolis. (295)

Athas, W. F. and Key, C. R. (1993) *Los Alamos Cancer Rate Study: Phase I, Final Report*. New Mexico Department of Health. (143)

Axelson, O. (1994) Dynamics of management and labor in dealing with occupational risks. In M. A. Mehlman and A. Upton, editors, *The Identification and Control of Environmental and Occupational Diseases*, Vol. XXIII. Princeton Scientific Publishing, Princeton, pp. 587–603. (305)

Axelson, O. (1995) Cancer risks from exposure to radon in homes. *Environmental Health Perspectives*, **103**, Supplement 2: 37–43. (306)

Baddeley, A. and Lieshout, M. (1993) Stochastic geometry models in high-level vision. In K. Mardia, editor, *Statistics and Images*. Carfax, Abingdon, pp. 233–258. (134, 236)

Baddeley, A. and Møller, J. (1989) Nearest-neighbour Markov point processes and random sets. *International Statistical Review*, **57:** 89–121. (236)

Bailey, T. C. and Gatrell A. C. (1995) *Interactive Spatial Data Analysis*. Longman, Harlow. (220, 221, 223, 369)

Baker, R. D. (1996) Testing for space–time clusters of unknown size. *Journal of Applied Statistics*, **23:** 543–554. (149)

Barbone, F., Bovenzi, M., Cavallieri, F. and Stanta, G. (1995) Air pollution and lung cancer in Trieste, Italy. *American Journal of Epidemiology*, **141:** 1161–1169. (289, 290)

Baron, J. A. (1984) Cancer mortality in small areas around nuclear facilities in England and Wales. *British Journal of Cancer*, **50:** 815–824. (406)

Barris, Y. I., Sahin, A. and Ozesmin, M. (1978) An outbreak of pleural mesotheliomas and chronic fibrosing pleurisy in the village of Karain. *Thorax*, **33:** 181–192. (100)

Barton, C. J., Roman, E., Ryder, H. M. and Watson, A. (1985) Childhood leukaemia in West Berkshire (letter). *The Lancet*, **2:** 1248–1249. (406)

Baser, M. E. (1992) The development of registries for surveillance of adult lead exposure, 1981 to 1992. *American Journal of Public Health*, **82:** 1113–1118. (304)

Becker, N., Frentzel-Beyme, R. and Wagner, G. (1984) *Krebsatlas der Bundesrepublik Deutschland*. Springer-Verlag Berlin/Heidelberg/New York/Tokyo. (170)

BEIR V, Committee on the Biological Effects of Ionizing Radiations (1990) *Health Effects of Exposure to Low Levels of Ionizing Radiation*. National Academy Press, Washington. (304)

Berman, M. and Turner, T. R. (1992) Approximating point process likelihoods with GLIM. *Applied Statistics*, **41:** 31–38. (240)

Bernardinelli, L. and Montomoli, M. (1992). Empirical Bayes versus fully Bayesian analysis of geographical variation in disease risk. *Statistics in Medicine*, **11:** 983–1007. (21, 26, 33, 218, 329, 334)

Bernardinelli, L. Clayton, D. and Montomoli, C. (1995a) Bayesian estimates of disease maps: how important are priors? *Statistics in Medicine*, **14:** 2411–2431. (11, 26, 29, 61, 204, 218, 330, 335)

Bernardinelli, L., Clayton, D., Pascutto, C., Montmoli, C. and Ghislandi, M. (1995b) Bayesian analysis of space–time variation in disease risk. *Statistics in Medicine*, **14:** 2433–2443. (11, 26, 29, 61, 204, 218, 330, 335)

Bernardinelli, L., Maida, A., Marinoni, A., Clayton, D., Romano, G., Montomoli, C., Fadda, D., Solinas, G., Castiglia, P., Cocco, P. L., Ghislandi, M., Berzuini, C., Pascutto, C., Nerini, M., Styles, B., Capocaccia, R., Lispi, L. and Mallardo, E. (1994) *Atlante della mortalità per tumore in Sardegna* (*Atlas of Cancer Mortality in Sardinia*, 1983–1987). FATMA-CNR. (329, 331, 340)

Bernardinelli, L., Pascutto, C., Best, N. G. and Gilks, W. R. (1997) Disease mapping with errors in covariates. *Statistics in Medicine*, **16:** 741–752. (185, 196, 198, 204, 330, 337, 347)

Bertazzi, P. A., Pesatori, A. C., Consonni, D., Tironi, A., Landi, M. T. and Zocchetti, C. (1993) Cancer incidence in a population accidentally exposed to 2,3,7,8-tetrachlorodibenzo-para-dioxin. *Epidemiology*, 4: 398–406. (304)

Besag, J. (1974) Spatial interaction and the statistical analysis of lattice systems (with discussion). *Journal of the Royal Statistical Society, Series B*, **36:** 192–236. (23, 59, 63, 188, 206)

Besag, J. (1975) Statistical analysis of non-lattice data. *The Statistician*, **24:** 179–195. (206, 210)

Besag, J. (1986) On the statistical analysis of dirty pictures. *Journal of the Royal Statistical Society, Series B*, **48:** 259–302. (20)

Besag, J. (1989) Towards Bayesian image analysis.*Journal of Applied Statistics*, **16:** 395–407. (20, 24, 196)

Besag, J. and Clifford, P. (1989) Generalized Monte Carlo significance tests. *Biometrika*, **76:** 633–642. (110, 240)

Besag, J. and Green, P. J. (1993) Spatial statistics and Bayesian computation. *Journal of the Royal Statistical Society, Series B*, **55:** 25–37. (31, 89, 240)

Besag, J. and Knorr-Held, L. (1998) Modelling risk from a disease in time and space. *Statistics in Medicine*, **17:** 2045–2060. (11)

Besag, J. and Kooperburg, C. (1995) On conditional and intrinsic autoregressions. *Biometrika*, **82:** 733–746. (195, 207)

Besag, J. and Mollié, A. (1989) Bayesian mapping of mortality rates. *Bulletin of the International Statistical Institute*, **53:** 127–128. (24, 196)

Besag, J. and Newell, J. (1991) The detection of clusters in rare diseases. *Journal of the Royal Statistical Society, Series A*, **154:** 143–155. (105, 109, 111–14, 144, 149, 171, 232, 244, 257, 398)

Besag, J., Green, P., Higdon, D. and Mengersen, K. (1995) Bayesian computation and stochastic systems (with discussion). *Statistical Science*, **10:** 3–66. (64)

Besag, J., York, J. and Mollié, A. (1991) Bayesian image restoration with two applications in spatial statistics. *Annals of the Institute of Statistical Mathematics*, **43:** 1–59. (9, 23, 31, 33, 59, 63–65, 87, 90, 105–107, 124, 128, 134, 135, 186, 195, 204, 207, 220, 221, 269, 330, 334, 418)

Best, N. G., Cowles, M. K. and Vines, S. K. (1995) *CODA: Convergence Diagnosis and Output Analysis for Gibbs Sampling Output*, Version 0.30. Medical Research Council Biostatistics Unit, Cambridge. (340, 341)

Best, N. G., Ickstadt, K. and Wolpert, R. L. (1998a) *Spatial Poisson Regression for Health and Exposure Data Measured at Disparate Spatial Scales*. Technical report, Imperial College, Department of Epidemiology and Public Health, Imperial College School of Medicine at St Mary's, London. (196, 200)

Best, N. G., Spiegelhalter, D. J., Thomas, A. and Brayne, C. E. G. (1996) Bayesian analysis of realistically complex models. *Journal of the Royal Statistical Society, Series A*, **159:** 323–342. (197)

Best, N. G., Waller, L. A., Thomas, A., Conlon, E. M. and Arnold, R. A. (1998b) Bayesian models for spatially correlated disease and exposure data. In J. Bernardo, J. Berger, A. Dawid, and A. Smith, editors, *Bayesian Statistics 6*. Oxford University Press, Oxford. (to appear). (196, 200)

Best, N., Cowles, K. C. and Vines, K. (1997) *CODA., Version 0.40*. MRC Biostatistics Unit, Cambridge. (352)

Biggeri, A., Barbone, F., Lagazio, C., Bovenzi, M. and Stanta, G. (1996) Air pollution and lung cancer in Trieste: spatial analysis of risk as a function of distance from sources. *Environmental Health Perspectives*, **104:** 750–754. (272, 275, 285)

Biggeri, A., Braga, M. and Marchi, M. (1993) Empirical Bayes *Interval* estimates: an application to geographical epidemiology. *Journal of the Italian Statistical Society*, **3:** 251–267. (25)

Bithell, J. (1990) An application of density estimation to geographical epidemiology. *Statistics in Medicine*, **9:** 691–701. (6, 234, 242, 243, 247, 249, 255, 275)

Bithell, J. (1995) The choice of test for detecting raised disease risk near a point source. *Statistics in Medicine*, **14:** 2309–2322. (144, 244, 254, 259, 260)

Bithell, J. and Stone, R. (1989) On statistical methods for analysing the geographical distribution of cancer cases near nuclear installations. *Journal of Epidemiology and Community Health*, **43:** 77–83. (242, 243, 407)

Bithell, J. F. (1998) Geographical analysis. In P. Armitage and T. Colton, editors, *Encyclopedia of Biostatistics*, Vol. 2. Wiley, Chichester, pp. 1701–1716. (247)

Bithell, J. F. and Draper, G. J. (1996) Uranium-235 and childhood leukaemia around Greenham Common airfield. Paper read to the Committee on the Medical Aspects of Radiation in the Environment. (250)

Bithell, J. F., Dutton, S. J., Draper, G. J. and Neary, N. M. (1994) The distribution of childhood leukaemias and non-Hodgkin's lymphomas near nuclear installations in England and Wales. *British Medical Journal*, **309:** 501–505. (249, 250)

Black, D. (1984) *Investigation of the Possible Increased Incidence of Cancer in West Cumbria: Report of the Independent Advisory Group*. Her Majesty's Stationery Office, London. (406)

Blot, W. J. (1997) Vitamin and mineral supplementation and cancer risk: international chemoprevention trials. *Proceedings of the Society of Experimental Biology and Medicine*, **216:** 291–296. (412)

Blot, W. J., Harrington, J. M., Toledo, A., Hoover, R., Heath, C. W. and Fraumeni, J. F. (1978) Lung cancer after employment in shipyards during World War II. *New England Journal of Medicine*, **304:** 620–624. (100)

Bond, V. P., Wielopolski, L. and Shani, G. (1996) Current misinterpretations of the linear no-threshold hypothesis. *Health Physics*, **70:** 877–882. (297)

Brenner, H., Savitz, A. D., Jöckel, K. H. and Greenland, S. (1992) Effects of nondifferential exposure misclassification in ecologic studies. *American Journal of Epidemiology*, **135:** 83–93.

Breslow, N. and Clayton, D. (1993) Approximate inference in generalised linear mixed models. *Journal of the American Statistical Association*, **88:** 9–25. (9, 16, 61, 64, 65, 87, 186, 187, 204, 219, 221, 241, 269)

Breslow, N. and Day, N. (1987) *Statistical Methods in Cancer Research, Volume 2: The Design and Analysis of Cohort Studies*. International Agency for Research on Cancer, Lyon. (7, 235, 243, 260)

Breslow, N. E. (1984) Extra-Poisson variation in log-linear models. *Applied Statistics*, **33:** 38–44. (329)

Breslow, N. E. and Day, N. E. (1975) Indirect standardization and the multiplicative model for rates with reference to the age adjustment of cancer incidence and relative frequency data. *Journal of Chronic Diseases*, **28:** 289–303.

Breslow, N. E. and Day, N. E. (1980) *Statistical Methods for Cancer Research: Volume 1*. International Agency for Research in Cancer, Lyon. (385)

Brooks, S. (1998). Markov chain Monte Carlo and its application. *The Statistician*, **47:** 69–100. (420)

Brown, L. M., Pottern, L. M. and Blot, W. J. (1984) Lung cancer in relation to environmental pollutants emitted from industrial sources. *Environmental Research*, **34:** 250–261. (288)

Brown, P. (1992) Popular epidemiology and toxic waste contamination: lay and professional ways of knowing. *Journal of Health and Social Behaviour*, **33:** 267–281. (295)

Brown, P. J. (1981) New considerations on the distribution of malaria, thalassemia and glucose-6-phosphate dehydrogenase deficiency in Sardinia. *Human Biology*, **53:** 367–382. (332)

Bundesforschungsanstalt für Landeskunde und Raumordnung (BfLR) (1996). Digital boundary file of Germany.

Bundesminister für Umwelt, Naturschutz und Reaktorsicherheit (Federal Ministry for the Environment, Nature Protection and Nuclear Safety), editor (1992) *Die Exposition durch Radon und seine Zerfallsprodukte in Wohnungen in der Bundesrepublik Deutschland und deren Bewertung*. Veröffentlichungen der Strahlenschutzkommission, Bd. 19. Gustav Fischer Verlag, Stuttgart, Jena, New York (in German). (396)

Böhning, D. (1995) A review of reliable maximum likelihood algorithms for semiparametric mixture models. *Journal of Statistical Planning and Inference*, **47:** 5–28. (53, 399)

Böhning, D., Dietz, E. and Schlattmann, P. (1998) Recent developments in C.A.MAN. *Biometrics*, **52**(2), 525–536. (11)

Böhning, D., Dietz, E., Schaub, R., Schlattmann, P. and Lindsay, B. G. (1994) The distribution of the likelihood-ratio for mixtures of densities from the one parameter exponential family. *Annals of the Institute of Mathematical Statistics*, **46:** 373–388. (59)

Böhning, D., Schlattmann, P. and Lindsay, B. G. (1992) C. A. MAN—computer assisted analysis of mixtures: statistical algorithms. *Biometrics*, **48:** 283–303. (11, 53, 399, 416)

Carlin, B. P. and Louis, T. A. (1996) *Bayes and Empirical Bayes Methods for Data Analysis.* Chapman & Hall, London. (31, 32)

Carrat, F. and Valleron, A. J. (1992) Epidemiological mapping using the 'kriging' method: application to an influenza-like illness epidemic in France. *American Journal of Epidemiology*, **135:** 1293–1300. (7, 235)

Carstairs, V. (1981) Small area analysis and health service research. *Community Medicine*, **3:** 131–139. (6, 140)

Carstairs, V. and Morris, R. (1991) *Deprivation and Health in Scotland.* Aberdeen University Press, Aberdeen. (199)

Centre for Disease Control (1981) Pneumocystis pneumonia—Los Angeles. *Mortality and Morbidity Weekly Report*, **30:** 250–252. (100)

Centres for Disease Control (1990) Guidelines for investigating clusters of health events. *Morbidity and Mortality Weekly Report*, **39:** 1–23. (151)

Chamberlain, J. (1996) Screening for cancers of other sites: lung, stomach, oral and neuroblastoma. In J. Chamberlain, and S. Moss, editors, *Evaluation of Cancer Screening.* Springer-Verlag, Berlin, pp. 136–156. (411)

Chapman, S. (1994) What is public health advocacy? In S. Chapman and D. Lupton, editors, *The Fight for Public Health: Principles and Practice of Media Advocacy.* BMJ Publishing Group, London, pp. 3–22. (299)

Charlton, J. (1996) Which areas are healthiest? *Population Trends*, **83:** 17–24. (366)

Chen, R., Connelly, R. R. and Mantel, N. (1993) Analyzing post alarm data in a monitoring system, in order to accept or reject the alarm. *Statistics in Medicine*, **12:** 1807–1812. (145)

Chenguri, L. (1987) *The Population Atlas of China.* Oxford University Press, Hong Kong. (162)

Chia, S. E., Chan, O. Y., Sam, C. T. and Heng, B. H. (1994) Blood cadmium levels in non-occupationally exposed adult subjects in Singapore. *Science of the Total Environment*, **145:** 119–123. (304)

Chikuni, O., Nhachi, C. F., Nyazema, N. Z., Polder, A., Nafstad, I. and Skaare, J. U. (1997) Assessment of environmental pollution of PCBs, DDT and its metabolites using human milk of mothers in Zimbabwe. *Science of the Total Environment*, **199:** 183–190. (304)

Christiansen, C. L. and Morris, C. N. (1997) Hierarchical Poisson Regression Modeling. *Journal of the American Statistical Association*, **92:** 618–632. (31)

Cislaghi, C., Biggeri, A., Braga, M., Lagazio, C. and Marchi, M. (1995) Exploratory tools for disease mapping in geographic epidemiology. *Statistics in Medicine*, **14:** 2363–2381. (9, 218, 350, 351)

Cislaghi, C., Decarli, A., La Vecchia, C., Mezzanotte, G. and Vigotti, M. A. (1990) Trends surface models applied to the analysis of geographical variations in cancer mortality. *Revue Epidemiologie et de Santé Publique*, **38:** 57–69. (366)

Clapp, R., Cobb, S., Chan, C. and Walker, B. (1987) Leukaemia near Massachusetts nuclear power plant (letter). *The Lancet*, **2:** 1324–1325. (406)

Clayton, D. (1994) *BEAM: Bayesian Ecological Analysis Models*. MRC Biostatistics Unit, Cambridge. (92, 351–52)

Clayton, D. (1996) Generalized linear mixed models. In W. Gilks, S. Richardson and D., Spiegelhalter, editors, *Markov Chain Monte Carlo in Practice*. Chapman & Hall, London, pp. 275–301. (163)

Clayton, D. and Bernardinelli, L. (1992) Bayesian methods for mapping disease risk. In P. Elliot, J. Cuzick, D. English and R. Stern, editors, *Geographical and Environmental Epidemiology: Methods for Small-Area Studies*. Oxford University Press, London, pp. 205–220. (21, 24, 31, 33, 64, 105, 107, 194, 196, 204, 330, 351, 393)

Clayton, D. and Hills, M. (1993). *Statistical Models in Epidemiology*. Oxford University Press, Oxford. (220, 323, 329, 334)

Clayton, D. G. (1989) Hierarchical model in descriptive epidemiology. *Proceedings of the XIVth International Biometrics Conference*, pp. 201–213. (330, 334)

Clayton, D. G. and Kaldor, J. (1987) Empirical Bayes estimates of age-standardised relative risks for use in disease mapping. *Biometrics*, **43:** 671–681. (10, 22, 23, 25, 31, 33, 51, 61, 106, 186, 194, 204, 218, 220, 242, 247, 322, 351, 399, 415, 416)

Clayton, D., Bernardinelli, L. and Montomoli, C. (1993) Spatial correlation and ecological analysis. *International Journal of Epidemiology*, **22:** 1193–1201. (24, 183, 185, 186, 350, 351)

Cliff, A. D. and Ord, J. K. (1981) *Spatial Processes. Models and Applications*. Pion, London. (109, 115, 149, 153, 154, 171)

Cochran, W. G. (1954) Some methods of strengthening the common chi square tests. *Biometrics*, **10:** 417–451. (282)

Cohen, B. (1990) Ecological versus case–control studies for testing a linear no-threshold dose–response relationship. *International Journal of Epidemiology*, **19:** 680–684. (192, 199)

Colhoun, H., Ben-Shlomo, Y., Dong, W., Bost, L. and Marmot, M. (1977) Ecological analysis of collectivity of alcohol consumption in England: importance of average drinker. *British Medical Journal*, **314:** 1164–1168. (301)

Collings, B. J. and Margolin, B. H. (1985) Testing goodness of fit for the Poisson assumption when observations are not identically distributed. *Journal of the American Statistical Association*, **80:** 411–418. (109)

Congdon, P. (1998) A multilevel model for infant health outcomes: maternal risk factors and geographic variation. *Journal of the Royal Statistical Society, Series D*, **47:** 159–182. (363)

Cook, D. G. and Pocock, S. J. (1983) Multiple regression in geographical mortality studies, with allowance for spatially correlated errors. *Biometrics*, **39:** 361–371. (189, 329)

Cook-Mozaffari, P. J., Darby, S. C., Doll, R., Forman, D., Hermon, C., Pike, M. and Vincent, T. (1989) Geographical variation in mortality from leukaemia and other cancers in England and Wales in relation to proximity to nuclear installations, 1969–78. *British Journal of Cancer*, **59:** 476–485. (407)

Coulter, A., McPherson, K. and Vessey, M. (1988) Do birth women undergo too many or too few hysterectomies? *Social Science and Medicine* **27:** 987–994. (366)

Cowles, M. K. and Carlin, B. (1996) Markov chain Monte Carlo convergence diagnostics: a comparative review. *Journal of the American Statistical Association*, **91:** 883–904. (27, 28, 240, 420)

Cox, D. R. and Isham, V. (1980) *Point Processes*. Chapman & Hall, London. (272)

Cox, D. R. and Lewis, P. A. W. (1976) *Statistical Analysis of Series of Events*. Chapman Hall, London. (240)

Cox, D. R. and Wermuth, N. (1996) *Multivariate Dependencies—Models, Analysis and Interpretation*. Chapman & Hall, London. (333)

Cressie, N. (1992) Smoothing regional maps using empirical Bayes predictors. *Geographical Analysis*, **24:** 75–95. (31, 61)

Cressie, N. (1993) *Statistics for Spatial Data*, rev. edn. Wiley, New York. (31, 65–67, 87, 154, 209, 223, 273)

Cressie, N. A. C. (1996) Bayesian and constrained inference for extremes in epidemiology. *Epidemiology Proceedings of the American Statistical Association, Joint Meetings 1995*. ASA, Alexandria, pp. 11–17. (247)

Cressie, N. and Chan, N. H. (1989) Spatial modelling of regional variables. *Journal of the American Statistical Association*, **84:** 393–401. (62, 64, 242)

Croen, L. A., Shaw, G. M., Sabonmatsu, L., Selvin, S. and Buffler, P. A. (1997) Maternal residential proximity to hazardous waste sites and risk for selected congenital malformations. *Epidemiology*, **8:** 347–354. (318, 383)

Croner, C., J. Sperling, *et al.* (1996) Geographic Information Systems (GIS): new perspectives in understanding human health and environmental relationships. *Statistics in Medicine*, **15:** 1961–1977. (165)

Cronkite, E. P. and Musolino, S. V. (1996) The linear no-threshold model: is it still valid for the prediction of dose-effects and risks from low-level radiation exposure (guest editorial). *Health Physics*, **70:** 775–776. (297)

Crowder, M. J. (1978) Beta-binomial anova for proportions. *Applied Statistics*, **27:** 34–37. (324)

Crump, K. S., Ng, T.-H, and Cuddihy, R. G. (1987) Cancer incidence patterns in the Denver metropolitan area in relation to the Rocky Flats Plant. *American Journal of Epidemiology*, **126:** 127–135. (406)

Cucca, F., Muntoni, F., Lampis, R., Frau, F., Argiolas, L., Silvetti, M., Angius, E., Cao, A., DeVirgiliis, S. and Congia, M. (1993) Combinations of specific DRB1, DQA1, DQB1 haplotypes are associated with insulin-dependent diabetes mellitus in Sardinia. *Human Immunology*, **37:** 859–894. (331)

Cuzick, J. and Edwards, R. (1990) Spatial clustering for inhomogeneous populations, *Journal of the Royal Statistical Society, Series B*, **52:** 73–104. (105, 108, 115, 132, 134, 149, 151, 153, 154, 177, 255, 278, 279)

Czeizel, A. E., Elek, C. and Gundy, S. (1993) Environmental trichlorfon and cluster of congenital abnormalities. *The Lancet*, **341:** 539–542. (100)

Damber, L. (1986) Lung cancer in males: an epidemiological study in northern Sweden with special regard to smoking and occupation. *Thesis*. University of Umeå, Umeå. (291)

Davies, J. L., Kawaguchi, Y., Bennet, S. T.,Copeman, J. B., Cordell, H. J., Pritchard, L. E., Reed, P. W., Gough, S. C. L., Jenkins, S. C., Palmer, S. M., Balfour, K. M., Rowe, B. R., Farral, M., Barnett, A. H., Bain, S. C. and Todd, J. A. (1994) A genome-wide search for human type 1 diabetes susceptibility genes. *Nature*, **371:** 130–136. (331)

Deely, J. J. and Lindley, D. V. (1981) Bayes empirical Bayes. *Journal of the American Statistical Association*, **76:** 833–841. (25)

Dellaportas, P. and Stephens, D. A. (1995) Bayesian analysis of errors-in-variables regression models. *Biometrics*, **51:** 1085–1095. (197)

Dempster, A. P., Laird, N. M. and Rubin, D. B. (1977) Maximum likelihood from incomplete data via the EM algorithm. *Journal of the Royal Statistical Society, Series B*, **39:** 1–38. (25, 57)

Department of Health (1995) *Shaping a Healthier Future: A Strategy for Effective Health Care in the 1990s*. Government Publications Office, Dublin. (350)

Devine, O. and Louis,T. (1994) A constrained empirical Bayes estimator for incidence rates in areas with small populations. *Statistics in Medicine*, **13:** 1119–1133. (10, 25, 242)

Devine, O. J., Louis,T. A. and Halloran, M. E. (1994a) Empirical Bayes estimators for spatially correlated incidence rates. *Environmetrics*, **5:** 381–398. (25, 31, 33)

Devine, O. J., Louis, T. A. and Halloran, M. E. (1994b) Empirical Bayes methods for stabilizing incidence rates before mapping. *Epidemiology*, **5:** 622–630. (25, 31, 33)

Dickson, D. (1984) *The New Politics of Science*. Pantheon Books, New York. (295)

Dieckmann, H. (1992). Häufung von Leukämieerkrankungen in der Elbmarsch. *Gesundheitswesen* **10:** 592–596 (in German). (395, 406)

Dietz, E. (1992) Estimation of heterogeneity — A GLM-Approach. In L. Fahrmeir, F. Francis, R. Gilchrist and G. Tutz, editors, *Advances in GLIM and Statistical Modelling*, Lecture Notes in Statistics Springer, Berlin, pp. 66–72. (57)

Diggle, P. (1983) *Statistical Analysis of Spatial Point Patterns*. Academic Press, London. (223, 234, 272, 273)

Diggle, P. (1985) A kernel method for smoothing point process data. *Applied Statistics*, **34:** 138–147. (234)

Diggle, P. (1989) Contribution of the 'cancer near nuclear installations' meeting. *Journal of the Royal Statistical Society, Series, A*, **152:** 367–369. (126, 235, 237)

Diggle, P. (1990) A point process modelling approach to raised incidence of a rare phenomenon in the vicinity of a prespecified point. *Journal of the Royal Statistical Society, Series A*, **153:** 349–362. (104, 125, 126, 236, 241, 278, 283)

Diggle, P. (1993) Point process modelling in environmental epidemiology. In V. Barnett and K. Turkman, editors, *Statistics for the Environment 1*, Wiley, New York. (120, 121, 151, 242)

Diggle, P. and Chetwynd, A. (1991) Second-order analysis of spatial clustering for inhomogeneous populations. *Biometrics*, **47:** 1155–1163. (105, 108, 112, 132)

Diggle, P. and Elliott, P. (1995) Disease risk near point sources: statistical issues in the analysis of disease risk near point sources using individual or spatially aggregated data. *Journal of Epidemiology and Community Health*, **49:** S20–S27. (241, 271, 286)

Diggle, P. and Milne, R. K. (1983) Negative binomial quadrat counts and point processes. *Scandinavian Journal of Statistics*, **10:** 257–267. (242)

Diggle, P. and Rowlingson, B. (1994) A conditional approach to point process modelling of elevated risk. *Journal of the Royal Statistical Society, Series A*, **157:** 433–440. (125, 126, 236, 274, 275, 283, 387)

Diggle, P., Morris, S., Elliott, P. and Shaddick, G. (1997) Regression modelling of disease risk in relation to point sources. *Journal of the Royal Statistical Society, Series A*, **160:** 491–505. (237, 238, 241, 283)

Diggle, P., Moyeed, R. and Tawn, J. (1998) Model-based Geostatistics. *Journal of the Royal Statistical Society, Series C*, **47**, 299–350. (235)

Divino, F., Frigessi, A. and Green, P. J. (1998) *Penalized Pseudolikelihood Inference in Spatial Interaction Models with Covariates*. Technical Report of the Norwegian Computing Centre. (59, 189, 190)

Dockery, D. W. and Pope, C. A. III. (1994) Acute respiratory effects of particulate air pollution. *Annual Reviews of Public Health*, **15:** 107–132. (305)

Dockery, D. W., Pope, C. A., Xu, X. *et al.* (1993) Mortality risks of air pollution: a prospective cohort study. *New England Journal of Medicine* **329:** 1753–1759. (289)

Dolk, H. (1997) The influence of migration in small area studies of environment and health—migration during pregnancy. *The ONS Longitudinal Study Update*, No. 27; June, pp. 6–8. (392)

Doll, R. (1989) The epidemiology of childhood leukemia. *Journal of the Royal Statistical Society, Series A*, **152:** 341–351. (261)

Douglas, J. B. (1979) *Analysis with Standard Contagious Distributions.* International Cooperative Publishing House, Fairland. (106)

Duncan, O. D., Cuzzort, R. P. and Duncan, B. (1961) *Statistical Geography.* The Free Press, New York. (182)

Ebert, D. and Lorenzi, R. (1994) Evolutionary Biology, Parasites and Polymorphisms. *Nature*, **369:** 705–706. (331)

Efron, B. and Morris, C. N. (1973) Stein's estimation rule and its competitors—an empirical Bayes approach. *Journal of the American Statistical Association*, **68:** 117–130. (63)

Elliot, P., Hill, M., Beresford, J. *et al.* (1992a) Incidence of cancers of the larynx and lung near incinerators of waste solvents and oils in Great Britain. *Lancet*, **339:** 854–858. (218, 237, 238, 242, 243, 244, 291)

Elliott, P. (1995) Investigation of disease risks in small areas. *Occupational and Environmental Medicine*, **52:** 785–789. (237, 238, 242)

Elliott, P., Cuzick, J., English, D. and Stern, R., editors (1992b) *Geographical and Environmental Epidemiology: Methods for Small-Area Studies.* Oxford University Press, London. (218, 237, 238, 242, 243, 244)

Elliott, P., Martuzzi, M. and Shaddick, G. (1995) Spatial statistical methods in environmental epidemiology: a critique. *Statistical Methods in Medical Research*, **4:** 137–159. (242, 259)

Elliott, P., Shaddick, G., Kleinschmidt, I., Jolley, D., Walls, P., Beresford, J. and Grundy, C. (1996) Cancer incidence near municipal solid waste incinerators in great britain. *British Journal of Cancer*, **73:** 702–710. (238, 244)

English, D. (1992) Geographical epidemiology and ecological studies. In *Geographical and Environmental Epidemiology: Methods for Small-Area Studies*, Oxford University Press, Oxford, pp. 3–13. (193)

Escobar, M. D. (1994) Estimating normal means with a Dirichlet process prior. *Journal of the American Statistical Association*, **89:** 268–277. (46)

Esman, N. A. and Marsh, G. M. (1996) Applications and limitations of air dispersion modelling in environmental epidemiology. *Journal of Exposure Analysis and Environmental Epidemiology*, **6:** 339–353. (237, 240)

Esteve, J., Benhamou, E. and Raymound, L. (1994) *Statistical Methods in Cancer Research: Vol. IV, Descriptive Epidemiology*. IARC Scientific Publications, Lyon. (349)

EUROCAT (1991) *Eurocat Registry Descriptions 1979–1990*. Office for official Publications of the European Communities, Luxembourg. (384)

European Communities Council (1991) Council directive of 12 December 1991 on hazardous waste (91/689/EC). *Official Journal of the European Communities L 377/20, 31.12.91*. (384)

Ewings, P. D., Bowies, C., Phillips, M. J. and Johnson, S. A. N. (1989) Incidence of leukaemia in young people in the vicinity of Hinkley Point nuclear power station, 1959–1986. *British Medical Journal*, **299:** 289–293. (406)

Fantini, B. (1991) La lotta antimalarica in Italia fra controllo ed eradicazione: l' esperimento Sardegna. *Parassitologia*, **33:** 11–23. (331)

Faust, K. and Romney, A. K. (1985) The effect of skewed distributions on matrix permutation tests. *British Journal of Mathematical and Statistical Psychology*, **38:** 152–160. (154)

Fermi, C. (1938) Provincia di Nuoro. Malaria, danni economici. Risanamento e proposte per il suo risorgimento. *Gallizzi*, **2:** 1–311. (340)

Fermi, C. (1940) Provincia di Cagliari. Malaria, danni economici. Risanamento e proposte per il suo risorgimento. *Gallizzi*, **3:** 1–610. (340)

Ferrándiz, J. and López, A. (1996) *On Maximum Likelihood Estimation and Model Choice in Spatial Regression Models*. Technical Report 4-96, Dep. d'Estadística i I.O., Universitat de València. (209)

Ferrándiz, J., López, A., Llopis, A. , Morales, M. and Tejerizo, M. L. (1995) Spatial interaction between neighbouring counties: cancer mortality data in Valencia (Spain). *Biometrics*, **51:** 665–678. (59, 187, 188, 204, 205, 211)

Fischhoff, B., Bostrom, A. and Quadrel, M. J. (1993) Risk perception and communication. *Annual Reviews of Public Health*, **14:** 183–203. (307, 308)

Fisher, R. A. (1949) *The Design of Experiments*. Hafner, New York. (166)

Forman, D., Cook-Mozaffari, P., Darby, S., Davey, G., Stratton, I., Doll, R. and Pike, M. (1987) Cancer near nuclear installations. *Nature*, **329:** 499–505. (406)

Freeman, M. F. and Tukey, J. W. (1950) Transformations related to the angular and square root. *Annals of Mathematical Statistics*, **21:** 607–611. (62)

Ganczakowski, M., Town, M., Bowden, D. K., Vulliamy, T. J., Kaneko, A., Clegg, J. B., Weatherall, D. J. and Luttazzo, M. (1995) Multiple glucose 6-phosphate dehydrogenase-deficient variants correlate with malaria endemicity in the Vanuatu Archipelago (Southwestern Pacific). *American Journal of Human Genetics*, **56:** 294–301. (332)

Gardner, M. J. (1973) Using environment to explain and predict mortality. *Journal of the Royal Statistical Society*, **136:** 421–440. (329)

Gardner, M. J. (1989) Review of reported increases of childhood cancer rates in the vicinity of nuclear installations in the UK. *Journal of the Royal Statistical Society Series, A*, **152:** 307–325. (233, 244, 406)

Gardner, M. J. and Winter, P. D. (1984a) Mapping small area cancer mortality: a residential coding story. *Journal of Epidemiology and Community Health*, **38:** 81–84. (286)

Gardner, M. J. and Winter, P. D. (1984b) Mortality in Cumberland during 1959–1978 with reference to cancer in young people around Windscale (letter). *The Lancet*, 217–218. (406)

Gardner, M. J., Hall, A. J., Snee, M. P., Downes, S., Powell, C. A. and Terrell, J. D. (1990b) Methods and basic data of case–control study of leukaemia and lymphoma among young people near Sellafied nuclear plant in West Cumbria. *British Medical Journal*, **300:** 429–434. (409)

Gardner, M. J., Snee, M. P., Hall, A. J., Powell, C. A., Downes, S. and Terrell, J. D. (1990a) Results of a case–control study of leukaemia and lymphoma among young people near Sellafied nuclear plant in West Cumbria. *British Medical Journal*, **300:** 423–429. (406, 409)

Gbary, A. R., Philippe, D., Ducis, S. and Beland, F. (1995) Distribution spatiale des sieges anatomiques choisis de cancer au Quebec. *Social Science and Medicine*, **41:** 863–872. (177)

Geary, R. C. (1954) The contiguity ratio and statistical mapping. *The Incorporated Statistician*, **5:** 115–145. (169)

Geiger, H. J., Rush, D. and Michaels, D. (1992) *Dead Reckoning: A Critical Review of the Department of Energy's Epidemiologic Research.* Physicians for Social Responsibility, Washington. (296)

Gelfand, A. E. (1996). Model determination using smapling-based methods. In W. R. Gilks, S. Richardson and D. J. Spiegelhalter, editors, *Markov Chain Monte Carlo in Practice.* Chapman & Hall, London, pp. 145–161. (213)

Gelman, A (1996) Inference and monitoring convergence. In W. Gilks, S. Richardson and D. Spiegelhalter, editors, *Markov Chain Monte Carlo in Practice.* Chapman & Hall, London, pp. 131–143. (28)

Gelman, A. and Price, P. N. (1999) All maps of parameter estimates are misleading. *Statistics in Medicine*, to appear. (75)

Gelman, A. and Rubin, D. B. (1992a) A single sequence from the Gibb sampler gives a false sense of security. In J. M. Bernardo, J. O. Berger, A. P. Dawid and A. F. M. Smith, editors, *Bayesian Statistics*, vol 4. Oxford University Press, Oxford, pp. 625–631. (28)

Gelman, A. and Rubin, D. B. (1992b) Inference from iterative simulation using multiple sequences. (with discussion). *Statistical Science*, **7:** 457–511. (209)

Gelman, A., Carlin, J., Stern, H. and Rubin, D. (1995) *Bayesian Data Analysis.* Chapman & Hall, London. (31, 119, 127, 131–2, 240)

Geman, S. and Geman, D. (1984) Stochastic relaxation, Gibbs distributions and the Bayesian restoration of images. *IEEE Transactions on Pattern Analysis and Machine Intelligence*, **6:** 721–741. (27)

George, P. (1991) *Automatic Mesh Generation: Applications to Finite Element Methods.* Wiley, New York. (139)

Geschwind, S., Stolwijk, J., Bracken, M., Fitzgerald, E., Stark, A., Olsen, C. and Melius, J. (1992) Risk of congenital malformations associated with proximity to hazardous waste sites. *American Journal of Epidemiology*, **135:** 1197–1207. (383)

Getis, A. (1992) Spatial interaction and spatial autocorrelation: a cross-product approach. *Environment and Planning A*, **23:** 1269–1277. (153)

Getis, A. and Ord, J. K. (1996) Local spatial autocorrelation statistics: an overview. In P. Lougley and M. Batty, editors, *Spatial Analysis: Modelling in a GIS Environment.* Geoinformation International, Cambridge. (154, 178)

Geweke, J. (1992) Evaluating the accuracy of sampling-based approaches to the calculation of posterior moments. In J. M. Bernardo, J. O. Berger, A. P. Dawid and A. F. M. Smith, editors, *Bayesian Statistics 4.* Oxford University Press, Oxford.

Geyer, C. (1996) Estimation and optimization of functions. In W. R. Gilks, S. Richardson and D. J. Spiegelhalter, editors, *Markov Chain Monte Carlo in Practice.* Chapman & Hall, London, pp. 241–258. (204)

Geyer, C. J. and Thompson, E. A. (1992) Constrained Monte Carlo maximum likelihood calculations (with discussion). *Journal of the Royal Statistical Society, Series B*, **54:** 657–699. (204)

Ghosh, M. (1992) Constrained Bayes estimates with applications. *Journal of the American Statistical Association*, **87:** 533–540. (36, 74, 75)

Ghosh, M., Natarajan, K., Stroud, T. and Carlin, B. (1998) Generalized linear models for small-area estimation. *Journal of the American Statistical Association*, **93:** 273–282. (9)

Gilks, W. R. (1996) Full conditional distributions. In W. Gilks, S. Richardson and D. Spiegelhalter, editors, *Markov Chain Monte Carlo in Practice*. Chapman & Hall, London, pp. 75–88. (26)

Gilks, W. R. and Wild, P. (1992) Adaptive rejection sampling. *Applied Statistics*, **41:** 337–348. (27)

Gilks, W. R., Clayton, D. G., Spiegelhalter, D. J., Best, N. g., McNeil, A., Sharples, L. D. and Kirby, A. J. (1993). Modelling complexity: applications of Gibbs sampling in medicine (with discussion). *Journal of the Royal Statistical Society, Series B*, **55:** 39–102. (219)

Gilks, W. R., Richardson, S. and Spiegelhalter, D. J. (1996a) *Markov Chain Monte Carlo in Practice*. Chapman & Hall, London. (11, 26, 31, 32, 90, 124, 127, 240, 340)

Gilks, W. R., Richardson, S. and Spiegelhalter, D. J. (1996b) Introducing Markov Chain Monte Carlo. In W. Gilks, S. Richardson and D., Spiegelhalter, editors, Chapman & Hall, London, pp. 1–19. (11, 26, 31, 32, 90, 124, 127, 240, 340)

Goldman, L., Paigen, B., Magnant, M. M. and Highland, J. H. (1985) Low birth weight, prematurity and birth defects in children living near the hazardous waste site, Love Canal. *Hazardous Waste and Hazardous Material*, **2:** 209–223. (383)

Goldsmith, J. R. (1989a) Childhood leukaemia mortality before 1970 among populations near two US nuclear installations (letter). *The Lancet*, **1:** 793. (406)

Goldsmith, J. R. (1989b) Childhood leukaemia mortality before 1970 among populations near two US nuclear installations. *The Lancet*, **2:** 1443–1444. (406)

Goldstein, H. (1995) *Multilevel Statistical Models*. *Edward Arnold*, London. (186, 187, 217, 219, 221, 227)

Goldstein, H. (1996a) *Likelihood Computations for Discrete Response Multilevel Models*. Technical Report, Multilevel Models Project, Institute of Education, London. (221)

Goldstein, H. (1996b) Consistent estimator for multilevel generalized linear models using an iterated bootstrap. *Multilevel Modelling Newsletter*, **8:** 3–6. (187)

Goldstein, H. and Rasbash, J. (1992) Efficient computational procedures for the estimation of parameters in multilevel models based on iterative generalised generalised least squares. *Computational Statistics and Data Analysis*, **13:** 63–71. (221)

Goldstein, H. and Spiegelhalter, D. J. (1996) League tables and their limitations: statistical issues in comparisons of institutional performance (with discussion). *Journal of the Royal Statistical Society, Series A*, **159:** 385–443. (36, 37)

Goldstein, H., Healy, M. and Rasbash, J. (1994). Multilevel time series models with applications to repeated measures data. *Statistics in Medicine*, **13:** 1643–1655. (223)

Gori, G. B. (1995) Re: Epidemiology faces its limits. *Science*, **269:** 1327–1328. (298)

Green, A. (1990) The role of genetic factors in the development of insulin-dependent diabetes mellitus. *Current Topics in Microbiology and Immunology*, **164:** 3–16. (331)

Green, P. J. (1995) Reversible jump MCMC computation and Bayesian model determination. *Biometrika*, **82:** 711–732. (107, 123–25)

Green, P. J. and Silverman, B. W. (1994) *Non Parametric Regression and Generalized Linear Models*. Chapman Hall, London. (189)

Greenland, S. (1992). Divergent biases in ecologic and individual-level studies. *Statistics in Medicine*, **11:** 1209–1223. (182, 185, 186)

Greenland, S. and Brenner, H. (1993) Correcting for non-differential misclassification in ecologic analyses. *Applied Statistics*, **42:** 117–126. (182)

Greenland, S. and Morgenstern, H. (1989) Ecological bias, confounding and effects modification. *International Journal of Epidemiology*, **18:** 269–274. (182, 199)

Greenland, S. and Robins, J. (1994) Ecologic studies—biases, misconceptions and counterexamples. *American Journal of Epidemiology*, **139:** 747–759. (182, 366)

Greenwood, B. M. (1968) Autoimmune disease and parasitic infections in Nigerians. *The Lancet*, **380–382**, August 17. (331)

Greiser, E., Schäfer, I., Hoffmann, W., Schill, W., Hilbig, K. and Weber, H. (1995) Risk factors of leukaemias and malignant lymphomas in three counties around a toxic waste dump. In R. Frentzel-Beyme, U. Ackermann-Liebrich, P. A. Bertazzi, E. Greiser, W. Hoffmann and J. Olsen, editors, *Environmental Epidemiology in Europe 1995. Proceedings of an International Symposium* European Commission, Directorate General V, Bremen, Germany, pp. 153–170. (408)

Griffith, D. A. (1983) The boundary value problem in spatial statistical analysis. *Journal of Regional Science*, **23:** 377–387. (85)

Grimson, R. C. (1991) A versatile test for clustering and a proximity analysis of neurons. *Methods of Information in Medicine*, **30:** 299–303. (149)

Grimson, R. C., Wand, K. C. and Johnson, P. W. C. (1981) Searching for hierarchical clusters of disease: spatial patterns of sudden infant death syndrome. *Social Science and Medicine*, **15D:** 287–293. (172)

Grosche, B., Hinz, G., Tsavachidis, C. and Kaul, A. (1987) *Analyse der Leukämiemorbidität in Bayern in den Jahren 1976–1981.* Tiel II. *Risikofaktoren, regionale Verteilung und epidemiologische Aspekte.* Bayerisches Staatsministerium für Landesentwicklung und Umweltfragen (Hrsg.), Institut für Strahlenhygiene des Bundesgesundheitsamtes, Neuherberg (in German). (406)

Grossmann, W. (1976) *Geodätische Rechnungen and Abbildungen in der Landesvermessung. 3. Auflage.* Vermessungswesen bei Konrad Wittwer, Stuttgart, pp. 182–190, 254 (in German). (398)

Gurevicius, R. (1987) Methodological aspects and basic descriptive data regarding the epidemiology of lung cancer in Lithuania. *Thesis.* Institute of Oncology, Vilnius. (288)

Haining, R. (1990) *Spatial Data Analysis in the Social and Environmental Sciences.* Cambridge University Press, Cambridge. (152, 153, 154)

Hardings, S. (1991) *Whose Science? Whose Knowledge?* Cornell University Press, Ithaca. (295)

Hastie, T. J. and Tibshirani, R. J. (1990) *Generalized Additive Models.* Chapman & Hall, New York. (285)

Hatch, M. C., Wallenstein, S., Beyea, J., Nieves, J. W. and Susser, M. (1991) Cancer rates after the Three Mile Island nuclear accident and proximity of residence to the plant. *American Journal of Public Health*, **81:** 719–724. (406)

Haybrittle, J., Yuen, P. and Machin, D. (1995) Multiple comparisons in disease mapping: Letter to the editor. *Statistics in Medicine*, **14:** 2503–2505. (234)

Heasman, M. A., Kemp, I. W., MacLaren, A., Trotter, P., Gillis, C. R. and Hole, A. J. (1984) Incidence of leukemia in young persons in the West of Scotland (letter). *The Lancet*, **1:** 1188–1189. (406)

Heasman, M. A., Kemp, I. W., Urquhart, J. D. and Black, R. (1986) Childhood leukemia in northern Scotland (letter). *The Lancet*, **1:** 266. (406)

Heidelberger, P. and Welch, P. (1983) Simulation run length control in the presence of an initial transient. *Operations Research*, **31:** 1109–1144. (358)

Heikkinen, J. and Penttinen, A. (1995) *Bayesian Smoothing in Estimation of the Pair Potential Function of Gibbs Point Processes.* Technical Report 17, Department of Statistics, University of Jyvaskyla. (205)

Hemminki, K., Mutanen, P., Luoma, K. and Saloniemi, I. (1980) Congenital malformations by the parental occupation in Finland. *International Archives of Occupational and Environmental Health*, **46:** 93–98. (392)

Hertz-Picciotto, I. (1996) Comment: toward a coordinated system for the surveillance of environmentel health hazards. *American Journal of Public Health*, **86:** 638–641. (151)

Hill, A. B. (1965) The environment and disease: association or causation? *Proceedings of the Royal Society of Medicine* **58:** 295–300. (297)

Hill, A. V. S., Allsopp, C. E. M., Kwiatkowski, D., Anstey, N. M., Twumasi, P., Rowe, P. A., Bennett, S., Brewster, D., McMichael, A. J. and Greenwood, B. M. (1991) Common West African HLA antigens are associated with protection from severe malaria. *Nature*, **352:** 595–600. (331)

Hills, M. and Alexander, F. (1989) Statistical methods used in assessing the risk of disease near a source of possible environmental pollution: a review. *Journal of the Royal Statistical Society, Series A*, **152:** 353–363. (231, 233, 242)

Hirotsu, C. (1983) Defining the pattern of association in two-way contingency tables. *Biometrika*, **70:** 579–589. (278, 279, 281)

Hirotsu, C. (1993) Beyond analysis of variance techniques: some applications in clinical trials. *International Statistical Review*, **61:** 183–201. (280)

Hoffmann, W. and Greiser, E. (1994) *Retrospektive Inzidenzstudie Elbmarsch. Inzidenz von Leukämien, malignen Lymphomen, multiplen Myelomen und von verwandten Erkrankungen in den Landkreisen Herzogtum Lauenburg, Harburg und Lüneburg, 1984–1993*, Technical Report. Bremer Institut für Präventionsforschung und Sozialmedizin (BIPS) (in German). (397, 397–8, 398, 401)

Hoffmann, W. and Greiser, E. (1996) Increased incidence of leukemias in the vicinity of the Krümmel nuclear power plant in Northern Germany. In R. Frentzel-Beyme, U. Ackermann-Liebrich, P. A. Bertazzi, E. Greiser, W. Hoffmann and J. Olsen, editors, *Environmental Epidemiology in Europe 1995. Proceedings of an International Symposium*, European Commission, Directorate General V, Bremen, Germany, pp. 185–206. (397, 401, 408)

Hoffmann, W., Dieckmann, H., Dieckmann, H. and Schmitz-Feuerhake, I. (1997a) A cluster of childhood leukemia near a nuclear reactor in Northern Germany. *Archives of Environmental Health*, **52:** 275–280. (396)

Hoffmann, W., Kranefeld, A. and Schmitz-Feuerhake, I. (1993) Radium-226-contaminated drinking water: hypothesis on an exposure pathway in a population with elevated childhood leukemia. *Environmental Health Perspectives*, **101:** Supplement 3: 113–115. (406)

Hoffmann, W., Kuni, H. and Ziggel, H. (1996). Leukemia and lymphoma mortality in the vicinity of nuclear power stations in Japan 1973–1987 (letter). *Journal of Radiological Protection*, **16:** 213–215. (406)

Hoffmann, W., Schäfer, I. and Greiser, E. (1997b) Retrospektive Fallerhebung in ökologischen und analytischen epidemiologischen Studien: Quantitative Analyse der Vollständigkeit und Validität primärer und sekundärer Datenquellen. In M. P. Baur, R. Fimmers and M. Blettner, editors, *Medizinische Informatik, Biometrie und Epidemiologie GMDS '96*, MMV Medizin Verlag, München, pp. 479–485 (in German). (397)

Hogan, M. D., Chi, P. Y., Hoel, D. G. and Mitchell, T. J. (1979) Association between chloroform levels in finished drinking water supplies and various site-specific cancer mortality rates. *Journal of Environmental Pathology and Toxicology*, **2:** 873–887. (304)

Hole, D. J. and Gillis, C. R. (1986) Childhood leukemia in the west of Scotland (letter). *The Lancet*, **2:** 524–525. (406)

Holland, W. W. (1991) *European Community Atlas of 'Avoidable Death', 2nd edn. Volume 1, Oxford University Press*, Oxford. (350, 358)

Holland, W. W., editor (1993) *European Community Atlas of 'Avoidable Death'*, 2nd edn. Oxford University Press, Oxford. (351, 358, 411)

Hostrup, O., Witte, I., Hoffmann, W., Greiser, E., Butte, W. and Walker, G. (1997) *Biozidanwendungen im Haushalt als möglicher Risikofaktor für die Gesundeit der Raumnutzer*, Technical Report. Biozidberatungsstelle der AG Biochemie/Toxikologie der Carl-von-Ossietzky Universität Oldenburg und Bremer Institut für Präventionsforschung und Sozialmedizin (BIPS), Oldenburg (in German). (408)

Howe, G. M. (1990) Historical evolution of disease mapping in general and specifically of cancer mapping. In P. Boyle, C. S. Muir and E. Grundmann, editors, *Cancer Mapping*. Springer-Verlag, Berlin, pp. 1–21. (49, 413)

Howe, H. L., Keller, J. E. and Lehnherr, M. (1993) Relation between population density and cancer incidence, Illinois, 1986–1990. *American Journal of Epidemiology*, **138:** 29–36. (366)

Hubbard, R. (1990) *The Politics of Women's Biology*. Rutgers University Press, New Brunswick. (295)

Hutchinson, S. J. (1995) Using density estimation to construct relative risk functions from areal data. *M.Sc. Dissertation*. University of Oxford. (254)

Härdle, W. (1991) *Smoothing Techniques: With Implementation in S. Springer-Verlag*, New York. (5, 127)

Ickstadt, K. and Wolpert, R. L. (1996) Spatial correlation or spatial variation? A comparison of Gamma/Poisson hierarchical models. Working paper 96–01, Institute of Statistics and Decision Sciences, Duke University. (204)

ICRP (International Commission On Radiological Protection) (1984) *Principles for Limiting Exposure of the Public to Natural Sources of Radiation*. ICRP Publication No. 39. Pergamon Press, Oxford. (396)

Inskip, H., Beral, V., Fraser, P. and Haskey, P. (1983) Methods for age-adjustment of rates. *Statistics in Medicine*, **2:** 483–493. (6, 126)

International Agency for Research on Cancer (IARC) (1985). *Scottish Cancer Atlas*. IARC Scientific Publication, **72**. IARC, Lyon. (38)

Ishimaru, T., Hishino, T., Ichimaru, M., Okada, H., Tomiyyasu, T., Tsuchimoto, T. and Yamamoto, T. (1971) Leukemia in atomic bomb survivors, Hiroshima and Nagasaki, 1 October 1950–30 September 1966. *Radiation Research*, **45:** 216–233. (100)

Iwasaki, T., Nishizawa, K. and Murata, M. (1995) Leukemia and lymphoma mortality in the vicinity of nuclear power stations in Japan, 1973–1987. *Journal of Radiological Protection*, **25:** 271–288. (406)

Jacob, C. O. (1992) Tumor necrosis factor α in autoimmunity: pretty girl or old witch? *Immunology Today*, **13:** 122–125. (331, 346)

Jacquez, G. M. (1994a) Stat! Statistical software for the clustering of health events. BioMedware, Ann Arbor. (117)

Jacquez, G. M. (1994b) Cuzick and Edwards' Test when exact locations are unknown. *American Journal of Epidemiology*, **140:** 58–65. (409)

Jacquez, G. M. (1996a) Disease cluster statistics for imprecise space–time locations. *Statistics in Medicine*, **15:** 873–885. (152, 153)

Jacquez, G. M. (1996b) A *k*-nearest neighbor test for space-time interaction. *Statistics in Medicine*, **15:** 1935–1949. (154, 164)

Jacquez, G. M. (1997) Method for measuring a degree of association for dimensionally referenced data. In *United States Patent Application*. BioMedware, Inc., USA, pp. 20. (149, 157)

Jacquez, G. M. (1999) *Spatial Randomisation Methods*. BioMedware Press, Ann Arbor, in press.

Jacquez, G. M. and Waller, L. A. (1998) The effect of uncertain locations on disease cluster statistics. In H. T. Mowerer and R. G. Congalton, editors, *Quantifying Spatial Uncertainty in Natural Resources: Theory and Applications for GIS and Remote Sensing*. Ann Arbor Press, Chelsea. (152, 155, 164)

Jacquez, G. M., Grimson, R., Waller, L and Wartenberg, D. (1996a). The analysis of disease clusters. Part II: Introduction to techniques. *Infection Control and Hospital Epidemiology*, **17:** 385–397. (151)

Jacquez, G. M., Waller, L. A., Grimson, R. and Wartenberg, D. (1996b). The analysis of disease clusters. Part I: State of the art. *Infection Control and Hospital Epidemiology*, **17:** 319–327. (151)

Janerich, D. (1991) Can stress cause cancer? *American Journal of Public Health*, **81:** 687–688. (298)

Jansson, B., Tholander, M. and Axelson, O. (1989) Exposure to radon in Swedish dwellings—attitudes and elimination. *Environment International*, **15:** 293–297. (308)

Jenks, S. (1994) Researchers to comb Long Island for potential cancer factors. *Journal of the National Cancer Institute*, **86:** 88–89. (143)

Johnson, C. J. (1981). Cancer incidence in an area contaminated with radionuclides near a nuclear installation. *Colorado Medicine*, **78:** 385–392. (406)

Johnson, N. L. and Kotz, S. (1969) *Distribution in Statistics: Discrete Distributions*. Houghton Mifflin, Boston. (280)

Joiner, M. C. (1994) Evidence for induced radioresistance from survival and other end points: an introduction. *Radiation Research*, **138:** S5–S8. (297)

Jolley, D. J., Jarman, B. and Elliott, P. (1992) Socio-economic confounding. In *Geographical and Environmental Epidemiology: Methods for Small-Area Studies*, Oxford University Press, Oxford, pp. 115–124. (199)

Jones, M. E., Shugg, D., Dwyer, T., Young, B. and Bonett, A. (1992) Interstate differences in incidence and mortality from melanoma. A re-examination of the latitudinal gradient. *Medical Journal of Australia*, **157:** 373–378. (366)

Journal of *the Royal Statistical Society, Series A*, **152:** 305–384 (1989). (111)

Järup, L., Carlsson, M. D., Elinder, C. G., Hellström, L., Persson, B. and Schütz, A. (1995) Enzymuria in a population living near a cadmium battery plant. *Occupational and Environmental Medicine*, **52:** 770–772. (304)

Kafadar, K., Freedman, L. S., Goodall, C. R. and Turkey, J. W. (1996) Urbanicity-related trends in lung cancer mortality in US counties: white females and white males, 1970–1987. *International Journal of Epidemiology*, **25:** 918–931. (366)

Kass, R. E. and Steffey, D. (1989) Approximate Bayesian inference in conditionally independent hierarchical models (parametric empirical Bayes models). *Journal of the American Statistical Association*, **84:** 717–726. (74)

Katsouyanni, K. and Pershagen, G. (1997) Ambient air pollution exposure and cancer. *Cancer Causes and Control*, **8:** 284–291. (287, 289, 291, 412)

Katsouyanni, K., Trichopoulos, D., Kalandidi, A., Tomos, P. and Riboli, E. (1991) A case–control study of air pollution and tobacco smoking in lung cancer among women in Athens. *Preventive Medicine*, **20:** 271–278. (290)

Kelly, A. and Sinclair, H. (1997) Deprivation and health: identifying the black spots. *Journal of Health Gain*, **1** (2): 13–14. (350)

Kelly, A., Kevany, J., de Onis, M. and Shah, S. M. (1995) Maternal anthprometry and pregnancy outcomes: a WHO collaborative study. *Bulletin of the World Health Organization*, Special Supplement, **73**. (352)

Kelsall, J. and Diggle, P. (1995) Non-parametric estimation of spatial variation in relative risk. *Statistics in Medicine*, **14:** 2335–2342. (6–8, 110, 234, 254, 255)

Kemp, I., Boyle, P., Smans, M. and Muir, C. (1985) *Atlas of Cancer in Scotland, 1975–1980, Incidence and Epidemiologic Perspective*. IARC Scientific Publication No. 72. International Agency for Research on Cancer, Lyon. (223, 329)

Key, T. (1995). Risk factors for prostate cancer. *Cancer Surveys*, **23:** 63–76. (223)

Kharrazi, M. (1998) *Pregnancy Outcomes Around the B. K. K. Landfill, West Covina, California: An Analysis by Address*. California Department of Health Services, Berkeley, California. (316)

Kinlen, L. J. (1995) Epidemiological evidence for an infective basis in chidhood leukaemia. *British Journal of Cancer*, **71:** 1–5. (99, 122, 144)

Kjuus, H., Langård, S. and Skaerven, R. (1986) A case-referent study of lung cancer, occupational exposures and smoking. Etiologic fraction of occupational exposures. *Scandinavian Journal of Work, Environment and Health*, **12:** 210–215. (303)

Knorr-Held, L. and Besag, J. (1998) Modelling risk from a disease in time and space. *Statistics in Medicine*, **17:** 2045–2060. (196)

Knox, E. G. (1964) The detection of space–time interactions. *Applied Statistics*, **13:** 25–29. (108, 149, 151, 154)

Knox, E. G. (1989) Detection of clusters. In P. Elliott, editor, *Methodology of Enquiries into Disease Clustering*, pp. 17–20. London School of Hygiene and Tropical Medicine, London. (105, 122, 129)

Knox, E. G. and Lancashire, R. J. (1991) Frequencies and social variations. In *Epidemiology of Congenital Malformations*. HMSO, London. (392)

Koopman, J. S. and Longini, I. M. (1994) The ecological effects of individual exposures and non linear disease dynamics in populations. *American Journal of Public Health*, **84:** 836–842. (192)

Kuk, A. Y. C. (1995) Asymptotically unbiased estimation in generalised linear models with random effects. *Journal of the Royal Statistical Society, Series B* **57:** 395–407. (221)

Kulldorff, M. (1997) A spatial scan statistic. *Communications in Statistics—Theory and Methods*, **26:** 1481–1496. (109, 146, 147, 149)

Kulldorff, M. (1998) Statistical methods for spatial epidemiology: tests for randomness. In A. Gatrell and M. Loytonen, editors, *GIS and Health in Europe*. Taylor & Francis, London. (112)

Kulldorff, M. and Nagarwalla, N. (1995) Spatial disease clusters: detection and inference. *Statistics in Medicine*, **14:** 799–810. (109, 113, 116, 146)

Kulldorff, M., Athas, W. F., Feuer, E. J., Miller, B. A. and Key, C. R. (1998a) Evaluating cluster alarms: a space–time scan statistic and brain cancer in Los Alamos. *American Journal of Public Health*, **88:** 1377–1380. (147)

Kulldorff, M., Feuer, E. J., Miller, B. A. and Freedman, L. S. (1997) Breast cancer clusters in Northeast United States: a geographic analysis. *American Journal of Epidemiology*, **146:** 161–170. (104, 143)

Kulldorff, M., Rand, K. and Williams, G. (1996) SaTScan, version 1.0, program for the space and time scan statistic. National Cancer Institute, Bethesda. (117)

Kulldorff, M., Rand, K., Gherman, G., Williams, G. and DeFrancesco D. (1998b) SaTScan version 1.0, Software for the spatial and space–time scan statistic. National Cancer Institute, Bethesda. (147)

Kvåle, G., Bjelke, E. and Heuch, I. (1986) Occupational exposure and lung cancer risk. *International Journal of Cancer*, **37:** 185–193. (303)

Lagazio, C., Marchi, M. and Biggeri, A. (1996) The association between risk of disease and point sources of pollution: a test for case-control data. *Statistica Applicata*, **8:** 343–356. (280, 281)

Lai, M.-S., Hsueh, Y.-M., Chen, C.-J., Shyu, M.-P., Chen, S.-Y. and Kuo, T.-L. (1994) Ingested inorganic arsenic and prevalence of diabetes mellitus. *American Journal of Epidemiology*, **139:** 485–491. (304)

Laird, N. M. and Louis, T. A. (1987) Empirical Bayes confidence intervals based on bootstrap samples. *Journal of the American Statistical Association*, **82:** 739–750. (25)

Laird, N. M. and Louis, T. A. (1989) Empirical Bayes ranking methods. *Journal of Educational Statistics*, **14:** 29–46. (36, 37)

Lancaster P. and Salkauskas, K. (1986) *Curve and Surface Fitting: An Introduction*. Academic Press, London. (7)

Lanes, S. (1985) Causal inference is not a matter of science. *American Journal of Epidemiology*, **122:** 550. (295)

Langford, I. H. (1994) Using empirical Bayes estimates in the geographical analysis of disease risk. *Area*, **26:** 142–149. (219, 220)

Langford, I. H. (1995) A log-linear multi-level model of childhood leukaemia mortality, *Journal of Health and Place*, **1:** 113–120. (219)

Langford, I. H. and Lewis, T. (1998). Outliers in multilevel models. *Journal of the Royal Statistical Society, Series A.*, **101:** 121–160. (227)

Langford, I. H., Bentham, G. and McDonald, A.-L. (1998). Multilevel modelling of geographically aggregated health data: a case study on malignant melanoma mortality and UV exposure in the European Community. *Statistics in Medicine*, **17:** 41–57. (218)

Langholz, B., Tuomilehto-Wolf, E., Thomas, D., Pitkaniemi, J., Tuomilehto, J. and the DiMe Study Group (1995) Variation in HLA-associated risks of childhood insulin-dependent diabetes mellitus in the Finnish population: I. Allele effects at A, B, and DR loci. *Genetic Epidemiology*, **12:** 441–453. (331)

Last, J. (1996) Professional standards of conduct for epidemiologists. In S.S. Coughlin and T. L. Beauchamp, editors, *Ethics and Epidemiology*. Oxford University Press, New York, pp. 53–75. (295)

Last, J. M. (1993) Global change: ozone depletion, greenhouse warming, and public health. *Annual Reviews of Public Health*, **14:** 115–136. (304)

Latour, B. (1987) *Science in Action: How to follow Scientists and Engineers Through Society*. Harvard University Press, Cambridge. (295)

Lawson, A. B. (1989) Contribution to the 'cancer near nuclear installations' meeting. *Journal of the Royal Statistical Society, Series A*, **152:** 374–375. (104, 126, 235)

Lawson, A. B. (1992a) GLIM and normalising constant models in spatial and directional data analysis. *Computational Statistics and Data Analysis*, **13:** 331–348. (239, 240)

Lawson, A. B. (1992b) On composite intensity score tests. *Communications in Statistics—Theory and Methods*, **22:** 3223–3235. (240)

Lawson, A. B. (1993a) A deviance residual for heterogeneous spatial Poisson processes. *Biometrics*, **49:** 889–897. (241)

Lawson, A. B. (1993b) On the analysis of mortality events around a prespecified fixed point. *Journal of the Royal Statistical Society, Series A*, **156:** 363–377. (233–35, 239, 241–44)

Lawson, A. B. (1994) Using spatial Gaussian priors to model heterogeneity in environmental epidemiology. *The Statistician*, **43:** 69–76. (219, 237, 243, 247)

Lawson, A. B. (1995) Markov chain Monte Carlo methods for putative pollution source problems in environmental epidemiology. *Statistics in Medicine*, **14:** 2473–2486. (105, 107, 119, 123, 124, 125, 135)

Lawson, A. B. (1996a) Markov chain Monte Carlo methods for spatial cluster processes. In *Computer Science and Statistics: Proceedings of the Interface*, Volume 27, pp. 314–319. (59, 134, 135, 238, 241)

Lawson, A. B. (1996b) Spatial interaction between neighboring counties (Letter). *Biometrics*, **52:** 370. (59, 134, 135, 238, 241)

Lawson, A. B. (1996c) Use of deprivation indices in small area studies: letter. *Journal of Epidemiology and Community Health*, **50:** 689–690. (59, 134, 135, 238, 241)

Lawson, A. B. (1999) Cluster modelling of disease incidence via MCMC methods. *Journal of Statistical Planning and Inference*, to appear. (119)

Lawson, A. B. and Clark, A. (1999) Markov chain Monte Carlo methods for clustering in case event and count data in spatial epidemiology. In Statistics and Epidemiology: Environment and Health, Editors Halloran, E. and Greenhouse, J. Springer Verlag, New York. (135)

Lawson, A. B. (1997) Some spatial statistical tools for pattern recognition. In A. Stein, F. W. T. P. de Vries, and J. Schut, editors, *Quantitative Approaches in Systems Analysis, Volume 7*. C. T. de Wit Graduate School for Production Ecology, pp. 43–58. (9, 107)

Lawson, A. B. (1998a) Spatial modelling of cluster object and non-specific random effects, with application in spatial epidemiology. In L. Giergl, A. Cliff, A.-J. Valleron, P. Farrington and M. Bull, editors, *GEOMED'97: Proceedings of the International Workshop on Geomedical Systems.*, pp. 141–156. Teubner Verlag. (4, 119, 236)

Lawson, A. B. (1998b) Statistical maps. In P. Armitage and T. Colton, editors, *Encyclopedia of Biostatistics*, volume 6. Wiley, Chichester, pp. 4267–4271. (4, 119, 236)

Lawson, A. B. and Cressie, N. (1999) Spatial statistical methods for environmental epidemiology. In C. R. Rao and P. K. Sen, editors, *Handbook of Statistics: Bio-Environmental and Public Health Statistics*. North-Holland, Amsterdam, to appear. (103, 233)

Lawson, A. B. and Harrington, N. W. (1996) The analysis of putative environmental pollution gradients in spatially correlated epidemiological data. *Journal of Applied Statistics* (special issue on Statistics in the Environment edited by J. Jeffers), **23:** 301–310. (241, 242)

Lawson, A. B. and Viel, J.-F. (1995) Tests for directional space–time interaction in epidemiological data. *Statistics in Medicine*, **14:** 2383–2392. (96)

Lawson, A. B. and Waller, L. (1996) A review of point pattern methods for spatial modelling of events around sources of pollution. *Environmetrics*, **7:** 471–488. (4, 5, 85, 125, 151, 223, 233, 235, 240, 243)

Lawson, A. B. and Williams F. L. R. (1994). Armadale: a case study in environmental epidemiology. *Journal of the Royal Statistical Society, Series A*, **157:** 285–298. (100, 125, 126, 128, 130, 132, 140, 183, 219, 236, 240, 241, 275, 409)

Lawson, A. B. and Williams, F. L. R. (1993) Application of extraction mapping in environmental epidemiology. *Statistics in Medicine*, **12:** 1249–1258. (6, 234)

Lawson, A. B., Biggeri, A. and Lagazio, C. (1996) Modelling heterogeneity in discrete spatial data models via MAP and MCMC methods. In A. Forcina, G. Marchetti, R. Hatzinger and G. Galmacci, editors, *Proceedings of the 11th International Workshop on Statistical Modelling*. Graphos, Citta di Castello, pp. 240–250. (9, 10, 105, 107, 119, 124, 189, 220, 237, 238, 243, 245)

Lazcano-Ponce, E. C., Rascon-Pacheco, R. A., Lozano-Ascencio, R. and Velasco-Mondragon, H. E. (1996) Mortality from cervical carcinoma in Mexico: impact of screening, 1980–1990. *Acta Cytologica*, **40:** 506–512. (366)

Le, N. D., Petkan, A. J. and Rosychuk, R. (1996) Surveillance of clustering near point sources. *Statistics in Medicine*, **15:** 727–740. (109, 241)

Lee, Y. and Nelder, J. A. (1996) Hierarchical generalized linear models (with discussion). *Journal of the Royal Statistical Society, Series B*, **58:** 619–678. (204)

Letourneau, E. G., Krewski, D., Choi, N. W. *et al* (1994) Case–control study of residential radon and lung cancer in Winnipeg, Manitoba Canada. *American Journal of Epidemiology*, **140:** 310–322. (291)

Levins, R. and Lewontin, R. (1985) *The Dialectical Biologist.* Harvard University Press, Cambridge. (295)

Liang, K. Y. and Zeger, S. L. (1986) Longitudinal data analysis using generalized linear models. *Biometrika*, **73:** 13–22. (187)

Llopis, A. (1985) Estudio quimico-sanitario de las aguas de consumo público de la provincia de Valencia. PhD thesis, Universitat de Valéncia. (211)

Lloyd, O. L. (1978) Respiratory cancer clustering associated with localized industrial air pollution. *Lancet*, **1:** 318–320. (288)

Lloyd, O. L. (1982) Mortality in a small industrial town. In A. Gardner, editor, *Current Approaches to Occupational Health—2*, Wright, pp. 283–309. (129)

Lloyd, O. L., Smith, G., Lloyd, M. M., Holland, Y. and Gailey, F. (1985) Raised mortality from lung cancer and high sex ratios of births associated with industrial pollution. *British Journal of Industrial Medicine*, **42:** 475–480. (288)

Louis, T. A. (1984) Estimating a population of parameter values using Bayes and empirical Bayes methods. *Journal of the American Statistical Association*, **79:** 393–398. (36, 75, 326)

Louis, T. A. (1991) Using empirical Bayes methods in biopharmaceutical research. *Statistics in Medicine*, **10:** 811–829. (25)

Lubin, J. H. and Boice, J. D. Jr (1997) Lung cancer risk from residential radon: meta-analysis of eight epidemiologic studies. *Journal of the National Cancer Institute*, **89:** 49–57. (306, 412)

Luckey, T. (1980) *Hormesis with Ionizing radiation.* CRC Press, Boca Raton. (297)

Lumley, T. (1995) Efficient execution of stone's likelihood ratio tests for disease clustering. *Computational Statistics and Data Analysis*, **20:** 499–510. (244)

Macbeth, R. (1965) Malignant disease of the paranasal sinuses. *Journal of Laryngology and Orology*, **79:** 592–612. (100)

MacMahon, B. (1992) Leukemia clusters around nuclear facilities in Britain (comment). *Cancer Causes and Control*, **3:** 283–288. (406)

Magder, L. S. and Zeger, S. (1996) A smooth nonparametric estimate of a mixing distribution using mixtures of Gaussians. *Journal of the American Statistical Association*, **91:** 1141–1151. (46)

Mallet, A. (1986) A maximum likelihood estimation method for random coefficient regression models. *Biometrika*, **73:** 645–656. (57)

Manly, B. F. J. (1991) *Randomization and Monte Carlo Methods in Biology*. Chapman & Hall, London. (153, 154)

Mantel, N. (1967) The detection of disease clustering and a generalised regression approach. *Cancer Research*, **27:** 209–220. (108, 149, 151, 153, 154)

Mantel, N. and Haenszel, W. (1959) Statistical aspects of the analysis of data from retrospective studies of disease. *Journal of the National Cancer Institute*, **22:** 719–748. (282)

Manton, K. G., Stallard, E., Woodbury, M. A., Riggan, W. B., Creason, J. P. and Mason, T. J. (1987) Statistically adjusted estimates of geographic mortality profiles. *Journal of the National Cancer Institute*, **78:** 805–815. (22, 23)

Manton, K. G., Woodbury, M. A. and Stallard, E. (1981) A variance components approach to categorical data models with heterogeneous cell populations: analysis of spatial gradients in lung cancer mortality rates in North Carolina counties. *Biometrics*, **37:** 259–269. (9, 10, 22, 23)

Manton, K. G., Woodbury, M. A., Stallard, E., Riggan, W. B., Creason, J. P. and Pellom, A. C. (1989) Empirical Bayes procedures for stabilizing maps of U.S. cancer mortality rates. *Journal of the American Statistical Association*, **84:** 637–650. (22, 61)

Marshall, E. G., Gensburg, L. J., Deres, D. A., Geary, N. S. and Cayo, M. R. (1997) Maternal residential exposure to hazardous wastes and risk of central nervous system and musculoskeletal birth defects. *Archives of Environmental Health*, **52:** 416–425. (383)

Marshall, R. (1991a) A review of methods for the statistical analysis of spatial patterns of disease. *Journal of the Royal Statistical Society, Series A*, **154:** 421–441. (105, 111, 112, 151, 153)

Marshall, R. (1991b) Mapping disease and mortality rates using empirical Bayes estimators. *Applied Statistics*, **40:** 283–294. (9, 10, 105, 111, 112, 151, 153, 242)

Martuzzi, M. and Elliott, P. (1996) Empirical Bayes estimation of small area prevalence of non-rare conditions. *Statistics in Medicine*, **15:** 1867–73. (324)

Martuzzi, M. and Hills, M. (1995) Estimating the degree of heterogeneity between event rates using likelihood. *American Journal of Epidemiology*, **141:** 369–374. (51, 322, 323)

McColl, A. J. and Gulliford, M. C. (1993) *Population Health Outcome Indicators for the NHS: A Feasibility Study*. Faculty of Public Health Medicine, Royal College of Physicians, London. (358)

McCullagh, P. and Nelder, J. A. (1989) *Generalized Linear Models*, 2nd edn. Chapman Hall, London. (194, 221)

McCulloch, C. E. (1997) Maximum likelihood algorithms for Generalized Linear Mixed Models. *Journal of the American Statistical Association*, **92:** 162–170. (204)

McGuire, W., Hill, A. V. S., Allsopp, C. E. M., Greenwood, B. M. and Kwiatowski, D. (1994) Variation in the TNF—a promoter region associated with susceptibility to cerebral malaria. *Nature*, **371:** 508–511. (331)

McLaughlin, J. R., Clarke, E. A., Nishri, E. D. and Anderson, T. W. (1993) Childhood leukemia in the vicinity of Canadian nuclear facilities. *Cancer Causes and Control*, **4:** 51–58. (406)

McShane, L. M., Albert, P. S. and Palmatier, M. A. (1997) A latent process regression model for spatially correlated count data. *Biometrics*, **53:** 698–706. (204)

Meloni, T., Pacifico, A., Forteleone, G. and Meloni, G. F. (1992) Association of G6PD deficiency and diabetes mellitus in northern Sardinian subjects. *Haematology*, **77:** 94–100. (332)

Michaelis, J., Keller, B., Haaf, G. and Kaatsch, P. (1992) Incidence of childhood malignancies in the vicinity of West German nuclear power plants. *Cancer Causes and Control*, **3:** 255–263. (406)

Mielke, P. W. (1978) Clarification and appropriate inferences for Mantel and Valand's nonparametric multivariate analysis technique. *Biometrics*, **34:** 277–282. (154)

Miettinen, O. S. (1985) *Theoretical Epidemiology, Principles of Occurrence Research in Medicine*. Wiley, New York. (304)

Mills, P. K., Abbey, D., Beeson, W. L. and Petersen, F. (1991) Ambient air pollution and cancer in California Seventh-day Adventists. *Archives of Environmental Health*, **46:** 271–280. (289)

Mollié, A. (1990) Représentation Géographique des Taux de Mortalité: Modélisation Spatiale et Méthodes Bayésiennes. Thèse de Doctorat, Université Paris VI. (24)

Mollié, A. (1996) Bayesian mapping of disease. In W. Gilks, S. Richardson and D. J. Spiegelhalter, editors, *Markov Chain Monte Carlo in Practice*. Chapman & Hall, London, pp. 359–379. (26, 194, 205, 207, 350)

Mollié, A. and Richardson, L. (1991) Empirical Bayes estimates and cancer mortality rates using spatial models. *Statistics in Medicine*, **10:** 95–112. (25, 51, 61, 219, 415)

Moolgavkar, S. H. and Luebeck, E. G. (1996) A critical review of the evidence on particulate air pollution and mortality. *Epidemiology*, **7:** 420–428. (305)

Moolgavkar, S. H. and Venzon, D. J. (1987) General relative risk regression models for epidemiologic studies. *American Journal of Epidemiology*, **126:** 949–961. (283)

Morales, M., Llopis, A., Tejerizo, M. L. and Ferrándiz, J. (1993) Concentration of nitrates in drinking waters and its relation with trends cancer mortality in Valencia Community and Spain. *Journal of Environmental Pathology, Toxicology and Oncology*, **12:** 229–236. (211)

Moran, P. A. P. (1948) The interpretation of statistical maps. *Journal of Royal Statistical Society, Series B*, **10:** 243–251. (112, 169)

Moran, P. A. P. (1950) Notes on continuous stochastic phenomena, *Biometrika*, **37:** 17–23. (154, 169)

Morgenstein, H. (1982) Uses of ecologic analysis in epidemiologic research. *American Journal of Public Health*, **72:** 1336–1344. (329)

Morgenstern, H. (1998) Ecologic studies. In K. Rothman and S. Greenland, editors, *Modern Epidemiology*. Lippincott-Raven, New York. (182)

Morris, C. N. (1983) Parametric empirical Bayes inference: theory and applications (with discussion). *Journal of the American Statistical Association*, **78:** 47–65. (23, 25)

Morris, C. N. and Christiansen, C. L. (1996) Hierarchical models for ranking and for identifying extremes, with applications (with discussion). In J. M. Bernardo, J. O. Berger, A. P. Dawid and A. F. M. Smith, editors, *Bayesian Statistics*, Volume 5. Oxford University Press, London, pp. 277–296. (37)

Morris, M. S. and Knorr, R. S. (1996) Adult leukemia and proximity-based surrogates for exposure to pilgrim Plant's nuclear emissions. *Archives of Environmental Health*, **51:** 266–274. (406)

Motykiewicz, G., Perera, F. P., Santella, R. M., Hemminki, K., Seemayer, N. H. and Chorazy, M. (1996) Assessment of cancer hazard from environmental pollution of Silesia. *Toxicology Letters*, **88:** 169–173. (304)

Mowerer, H. T., Czaplewski, R. L. and Hamre, R. H. (1997) *Spatial Accuracy Assessment in Natural Resources and Environmental Sciences*. Arbor Press, Ann Arbor. (161)

Mumford, J. L., He, X. Z., Chapman, R. S. *et al.* (1987) Lung cancer and indoor air pollution in Xuan Wei, China. *Science*, **235:** 217–220. (314)

Munasinghe, R. L. and Morris, R. D. (1996) Localization of disease clusters using regional measures of spatial autocorrelation. *Statistics in Medicine*, **15:** 893–905. (178)

Muntoni, S. and Songini, M. (1992) High incidence rate of IDDM in Sardinia. *Diabetes Care*, **15:** 1317–1322. (331)

Nair, V. N. (1986) Testing in industrial experiments with ordered categorical data (with discussion). *Technometrics*, **28:** 283–311. (279)

Nair, V. N. (1987) Chi-squared type tests for ordered alternatives in contingency tables. *Journal of the American Statistical Association*, **82:** 283–291. (278, 279)

Nasca, P. C., Mahoney, M. C. and Wolfgang, P. E. (1992) Population density and cancer incidence differentials in New York State, 1978–82. *Cancer Causes and Control*, **3:** 7–15. (366)

Neutra, R. R. (1990) Counterpoint from a cluster buster. *American Journal of Epidemiology*, **132**, Supplement: 1–8. (233, 314, 315)

Niedersächsisches Sozialministerium (Lower Saxony Ministry of Social Affairs), editor (1992) *Kinderleukämie in der Elbmarsch*, Berichtsband. Niedersächsisches Soziallministerium, Hannover (in German). (396)

Nitta, H. and Maeda, K. (1982) Personal exposure monitoring to nitrogen dioxide. *Environment International*, **8:** 243–248. (290)

Norstrom, T. (1989) The use of aggregate data in alcohol epidemiology. *British Journal of Addiction*, **84:** 969–977. (301)

Oden, N. (1995) Adjusting Moran's *I* for population density. *Statistics in Medicine*, **14:** 17–26. (105, 109, 112, 176)

Oden, N., Jacquez, G. and Grimson, R. (1996) Realistic power simulations compare point and are area-based disease cluster tests. *Statistics in Medicine*, **15:** 783–806. (172)

Ogata, Y. (1988) Statistical models for earthquake occurrences and residual analysis for point processes. *Journal of the American Statistical Association*, **83:** 9–27. (241)

Ohno, Y. and Aoki, K. (1979) A test of significance for geographic clusters of disease. *International Journal of Epidemiology*, **8:** 273–281. (171)

Ohno, Y. and Aoki, K. (1981) Cancer deaths by city and county in Japan (1969–71): a test of significance for geographic clusters of disease. *Social Science and Medicine*, **15:** 251–258. (171)

Olsen, J. and Frische, G. (1993) Social differences in reproductive health. *Scandinavian Journal of Social Medicine*, **21:** 90–97. (392)

Olsen, S., Martuzzi, M. and Elliot, P. (1996) Cluster analysis and disease mapping—why, when and how? A step by step guide. *British Medical Journal*, **313:** 863–865. (351)

Olshan, F., Baird, P. A. and Lo, K. H. (1991) Socioeconomic status and the risk of birth defects. *American Journal of Epidemiology*, **134:** 778–779. (392)

Openshaw, S. (1991) A new approach to the detection and validation of cancer clusters: a review of opportunities, progress and problems. In F. Dunstan and J. Pickles, editors, *Statistics in Medicine*. Clarendon Press, Oxford, pp. 49–63. (151)

Openshaw, S., Charlton, M., Wymer, C. and Craft, A. (1987) A mark I grographical analysis machine for the automated analysis of point data sets. *International Journal of Geographical Information Systems*, **1:** 335–358. (109)

Ott, W. R. (1988) Human activity patterns: a review of the literature for estimating time spent indoor, outdoors and in transit. *In Research Planning Conference of Human Activity Patterns*. Las Vegas. (290)

Panopsky, M. A. and Dutton, J. A. (1984) *Atmospheric Turbulence*. Wiley, New York. (237, 240, 241)

Paul-Saheen, P., Clark, J. D. and Williams, D. (1987) Small area analysis: a review and analysis of North American literature. *Journal of Health Politics, Policy and Law*, **12:** 742–809. (366)

Penttinen, A. (1984) *Modelling Interaction in Spatial Point Patterns: Parameter Estimation by the Maximum Likelihood Method*. Volume 7 of *Jyvaskyla Studies in Computer Science, Economics and Statistics*. University of Jyvaskyla, Jyvaskyla. (204)

Pershagen, G. (1985) Lung cancer mortality among men living near an arsenic-emitting smelter. *American Journal of Epidemiology*, **122:** 684–694. (288, 290, 291, 303)

Pershagen, G. (1990) Air pollution and cancer. In H. Vainio, M. Sorsa and A. McMichael, editors, *Complex Mixtures and Cancer Risk*. IARC Scientific Publications No 104. International Agency for Cancer Research, Lyon, pp. 240–251. (287)

Pershagen, G. and Nyberg, F. (1995) Air Pollution and Lung Cancer in an Area Near a Smelter. Report 1/95. Institute of Environmental Medicine, Stockholm (in Swedish). (288)

Pershagen, G. and Simonato, L. (1993) Epidemiological evidence on outdoor air pollution and cancer. In L. Tomatis, editor, *Indoor and Outdoor Air Pollution and Human Cancer*. Springer-Verlag, Berlin, pp. 135–145. (288)

Pershagen, G., Åkerblom, G., Axelson, O., Clavensjö, B., Damber, L., Desai, G., Enflo, A., Lagarde, F., Mellander, H., Svartengren, M. and Swedjemark, G. A. (1994) Residential radon exposure and lung cancer in Sweden. *New England Journal of Medicine*, **330:** 159–164. (306)

Piantadosi, S., Byar, D. P. and Green, S. B. (1988) The ecological fallacy. *American Journal of Epidemiology*, **127:** 893–904. (182, 199)

Piazza, A., Mayr, W. R., Contu, L., Amoroso, A. *et al.* (1985) Genetic and population structure of four Sardinian villages. *Annals of Human Genetics*, **4:** 47–63. (332)

Pickle, L. W., Mason, T. J. and Fraumeni, J. F. Jr (1989) The new United States cancer atlas. *Recent Results in Cancer Researcg*, **114:** 196–207. (366)

Pickle, L. W., Mungiole, M., Jones, G. K. and White, A. A. (1996) *Atlas of United States Mortality*. National Center for Health Statistics, Hyattsville. (31, 366)

Pierce, D. A. and Preston, D. L. (1985) Analysis of cancer mortality in the A-bomb survivor cohort. In *Proceedings of the 45th Session of the International Statistical Institute, pp. 557–570 International Statistical Institute*, Amsterdam. (387)

Pierce, D. A., Shimuzu, Y., Preston, D. L., Vaeth, M. and Mabuchi, K. (1996) Studies of the mortality of atomic bomb survivors. Report 12. Part 1. Cancer 1950–1990. *Radiation Research*, **146:** 1–27. (387)

Plummer, M. and Clayton, D. (1996) Estimation of population exposure in ecological studies. *Journal of the Royal Statistical Society, Series B*, **58:** 113–126. (183, 185, 199)

Pobel, D. and Viel, J-F. (1997) Case–control study of leukemia among young people near LaHague nuclear reprocessing plant: the environmental hypothesis revisited. *British Medical Journal*, **314:** 101–106. (406)

Pocock, S. J., Cock, D. G. and Beresforod, S. A. A. (1981) Regression of area mortality rates on explanatory variables: what weighting is appropriate? *Applied Statistics* **30:** 286–295. (329, 330)

Pocock, S. J., Cook, D. G. and Shaper, A. G. (1982) Analysing geographic variation in cardiovascular mortality: methods and results. *Journal of the Royal Statistical Society, Series A*, **145:** 313–341. (189)

Pocock, S. J., Delves, H. T., Ashby, D., Shaper, A. G. and Clayton, B. E. (1998) Blood cadmium concentrations in the general population of British middle-aged men. *Human Toxicologys*, **7:** 95–103. (304)

Pope, C. A., Thun, M. J., Namboodiri, M. M. *et al.* (1995) Particulate air pollution as predictor of mortality in a prospective study of US adults. *American Journal of Respiratory and Critical Care Medicine*, **151:** 669–674. (289)

Potthoff, R. F. and Whittinghill, M. (1966a) Testing for homogeneity I: the binomial and multinomial distributions. *Biometrika*, **53:** 167–182. (109, 326)

Potthoff, R. F. and Whittinghill, M. (1996b) Testing for homogeneity II: the poisson distribution. *Biometrika*, **53:** 183–90. (109, 326)

Prentice, R. and Sheppard, L. (1995) Aggreagate data studies of disease risk factors. *Biometrika*, **82:** 113–125. (183, 291)

Prentice, R. L. and Sheppard, L. (1990) Dietary fat and cancer: consistency of the epidemiologic data, and disease prevention that may follow from a practical reduction in fat consumption. *Cancer Causes and Control*, **1:** 81–97. (329)

Press, W. H., Flannery, B. P., Teukolsky, S. A. and Vetterling, W. T. (1986) *Numerical Recipes: The Art of Scientific Computing*. Cambridge University Press, Cambridge. (74)

Preston, D. L., Lubin, J. H. and Pierce, D. A. (1993) *EPICURE User's Guide*. Hirosoft International Corp Seattle. (387)

Prindull, G. Demuth, M. and Wehinger, H. (1993) Cancer morbidity rates of children from the vicinity of the nuclear power plant of Wurgassen (FRG). *Acta Haematologica*, **90:** 90–93. (406)

Public Citizen Health Research (PCHR) Group (1997) Who poisoned the children? *Health Letter*, **13:** 3–6. (100, 143)

Pukkala, E. (1989) Cancer maps of Finland: an example of small area-based mapping. *Recent Results in Cancer Research*, **114:** 208–215. (366)

Raftery, A. and Lewis, S. (1992) How many iterations in the Gibbs sampler? In: J. M. Bernardo, J. O. Berger, A. P. Dawid and A. F. M. Smith, editors, *Bayesian Statistics 4*, Oxford University Press, pp. 763–774. (358)

Ranta, J. (1996) Detection of overall space–time clustering in a non-uniform distributed population. *Statistics in Medicine*, **15:** 2561–2572. (112)

Raubertas, R. F. (1988) Spatial and temporal analysis of disease occurrence for detection of clustering. *Biometrics*, **44:** 1121–1129. (105)

Rezvani, A., Mollié, A., Doyon, F. and Sancho-Garnier, H. (1997) *Atlas de la Mortalité par Cancer en France: 1986–1993*. INSERM, Paris. (16)

Rhind, D., Evans, I and Visvalingam, M. (1980) Making a national atlas of population be computer. *The Cartographic Journal*, **17:** 1–11. (162)

Richardson, S. (1992) Statistical methods for geographical correlation studies. In P. Elliott, J. Cuzick, D. English and R. Stern, editors, *Geographical and Environmental Epidemiology: Methods for Small-Area Studies*, Oxford University Press, London, pp. 181–204. (199, 200, 204)

Richardson, S. and Gilks, W. R. (1993a) A Bayesian approach to measurement error problems in epidemiology using conditional independence models. *American Journal of Epidemiology*, **138:** 430–442. (197, 330, 336)

Richardson, S. and Gilks, W. R. (1993b) Conditional independence models for epidemiological studies with covariate measurement error. *Statistics in Medicine*, **12:** 1703–1722. (197, 330, 336)

Richardson, S. and Green, P. (1997) On Bayesian analysis of mixtures with an unknown number of components (with discussion). *Journal of the Royal Statistical Society, Series B*, **59:** 731–792. (59)

Richardson, S. and Hemon, D. (1990) Ecological bias and confounding (letter). *International Journal of Epidemiology*, **19:** 764–776. (199)

Richardson, S., Guihenneuc, C. and Laserre, V. (1992) Spatial linear models with autocorrelated error structure. *The Statistician*, **41:** 539–557. (204)

Richardson, S., Monfort, C., Green, M., Draper, G. and Muirhead, C. (1995) Spatial variation of natural radiation and childhood leukaemia incidence in Great Britain. *Statistics in Medicine*, **14:** 2487–2501. (196)

Richardson, S., Stucker, I. and Hemon, D. (1987) Comparisons of relative risks obtained in ecological and individual studies: some methodological considerations. *International Journal of Epidemiology*, **16:** 111–120. (182, 186)

Ripley, B. (1981) *Spatial Statistics*. Wiley, New York. (7, 84)

Ripley, B. (1988) *Statistical Inference for Spatial Processes*. Cambridge University Press, Cambridge. (84, 86, 225)

Roberts, G. (1996) Markov chain concepts related to sampling algorithms. In W. Gilks, S. Richardson and D. Spiegelhalter, editors, *Markov Chain Monte Carlo in Practice*. Chapman & Hall, London, pp. 45–57. (27)

Roberts, G. and Polson, N. (1994) On the geometric convergence of the Gibbs sampler. *Journal of the Royal Statistical Society B*, **56:** 377–384. (27)

Robinson, W. S. (1950) Ecological correlation and the behavior of individuals. *American Sociological Review*, **15:** 351–357. (182)

Roman, E., Beral, V., Carpenter, L., Watson, A., Barton, C., Ryder, H. and Aston, L. D. (1987) Childhood leukemia in the West Berkshire and Basingstoke and North Hampshire District Health Authorities in relation to nuclear establishments in the vicinity. *British Medical Journal*, **294:** 597–602. (406)

Roos, N. P. and Roos, L. L. (1982) Surgical rate variations: do they reflect the health or socioeconomic characteristics of the population? *Medical Care*, **20:** 945–958. (366)

Rosenbaum, P. R. and Rubin, D. B. (1984) Difficulties with regression of age-adjusted rates. *Biometrics*, **40:** 437–443. (186)

Ross, A. and Davis, S. (1990) Point pattern analysis of the spatial proximity of residences prior to diagnosis of persons with Hodgkin disease. *American Journal of Epidemiology*, **132**, Supplement 1: S53–S62. (286)

Rossiter, J. E. (1991) Epidemiological applications of density estimation. *D.Phil. thesis*. University of Oxford. (254, 255)

Rothman, K. and Greenland, S. (1998) *Modern Epidemiology*, 2nd edn. Lippincott, Boston. (274, 286)

Rothman, K. J. (1990) A sobering start to the cluster busters' conference. *American Journal of Epidemiology*, **132**, Supplement: S6–S13. (233)

Rushton, G. and Lolonis, P. (1996) Exploratory spatial analysis of birth defects in an urban population. *Statistics in Medicine*, **15:** 717–726. (155, 161)

Saeed, Th. Kh., Hamamy, H. A. and Alwan, A. A. S. (1985) Association of glucose 6-phosphate dehydrogenase deficiency with diabetes mellitus. *Diabetic Medicine*, **2:** 110–112. (332)

Saha, N. (1979) Association of glucose 6-phosphate dehydrogenase deficiency with diabetes mellitus in ethnic groups of Singapore. *Journal of Medical Genetics*, **16:** 431–434. (332)

SAHRU (1997a) *Small Area Analysis in Health and Health Service Research: Principles and Application*. Technical Report No. 1, Small Area Health Research Unit, Department of Community Health and General Practice, Trinity college, Dublin. (349–50)

SAHRU (1997b) *A National Deprivation Index for Health and Health Services Research.* Technical Report No. 2, Small Area Health Research Unit, Department of Community Health and General Practice, Trinity College, Dublin. (354)

Sans, S., Elliott, P., Kleinschmidt, I., Shaddick, G., Pattenden, S., Walls, P., Grundy, C. and Dolk, H. (1995) Cancer incidence and mortality near the Baglan Bay petrochemical works, South Wales. *Occupational and Environmental Medicine,* **52:** 217–224. (238, 244)

Sasaki, S. (1993) An ecological study of the relationship between dietary fat intake and breast cancer mortality. *Preventive Medicine,* **22:** 187–202. (301)

Satterthwaite, F. E. (1946) An approximate distribution of estimate of variance components. *Biometrics Bulletin,* **2:** 110–114. (280)

Schlattmann, P. (1993) Statistische Methoden zur Darstellung der räumlichen Verteilung von Krankheiten unter besonderer Berücksichtigung von Mischverteilungen. *Dissertation,* Free University Berlin (in German). (55)

Schlattmann, P. (1996) The computer package DismapWin. *Statistics in Medicine,* **15:** 931. (53, 402, 416)

Schlattmann, P. and Böhning, D. (1993) Mixture models and disease mapping. *Statistics in Medicine,* **12:** 1943–1950. (11, 51, 53, 219, 399, 415)

Schlattmann, P. and Böhning, D. (1997) Contribution to the discussion of the paper by S. Richardson and P. Green: 'On Bayesian Analysis of Mixtures with an Unknown Number of Components'. *Journal of the Royal Statistical Society, Series B,* **59:** 781–782. (59)

Schlattmann, P. and Böhning, D. (1999) Spatially dependent mixture models. *Statistics in Medicine* (in revision). (59)

Schlattmann, P., Böhning, D. and Dietz, E. (1996) Covariate adjusted mixture models with the program DismapWin. *Statistics in Medicine,* **15:** 919–929. (11, 57, 407)

Schlaud, M., Seidler, A., Salje, A., Behrendt, W., Schwartz, F. W., Ende, M., Knoll, A. and Grugel, C. (1995) Organochlorine residues in human breast milk: analysis through a sentinel practice network. *Journal of Epidemiology and Community Health,* **49**, Supplement 1: 17–21. (304)

Schmidtmann, I. and Zoellner, I. (1993) Results of various methodological approaches to cluster detection in the German registry of childhood malignancies. In J. Michaelis *et al.*, editors, *Medizinische Forschung 6, Epidemiology of Childhood Leukaemia.* Gustav Fischer Verlag, Stuttgart, pp. 189–202. (172)

Schmitz-Feuerhake, I., Dannheim, B., Heimers, A., Oberheitmann, B., Schröder and Ziggel, H. (1997) Leukemia in the proximity of a German boiling-water nuclear reactor: evidence of population exposure by chromosome studies and environmental radioactivity. *Environmental Health Perspectives,* **105**, Supplement: I 1499–1504. (406)

Schoenberg, J. B., Klotz, J. B., Wilcox, H. B. *et al* (1990) Case–control study of residential radon and lung cancer among New Jersey women. *Cancer Research,* **50:** 6520–6524. (290)

Schwab, M., Colome, S. D., Spengler, J. D. *et al.* (1990) Activity patterns applied to pollutant exposure assessment: data from a personal monitoring study in Los Angeles. *Toxicology and Industrial Health,* **6:** 517–532. (290)

Schwartz, S. (1994) The fallacy of ecological fallacy: the potential misuse of a concept and the consequences. *American Journal of Public Health,* **84:** 819–823. (192)

Schön, D, Bertz, J. and Hoffmeister, H, editors (1995) *Bevölkerungsbezogene Krebsregister in der Bundesrepublik Deutschland,* Band, 3, RKI-Schriften. MMV-Verlag, München, pp. 233–239. (411)

Scott, D. W. (1992) *Multivariate Density Estimation.* Wiley, London. (249)

Searle, S., Cassella, G. and McCulloch, C. (1992) *Variance Components.* Wiley, New York. (240)

Self, S. G. and Liang, K. Y. (1987) Asymptotic properties of maximum-likelihood estimators and likelihood ratio tests under nonstandard conditions. *Journal of the American Statistical Association,* **82:** 605–610. (326)

Shaw, G., Schulman, J., Frisch, J. D., Cummins, S. K. and Harris, J. A. (1992) Congenital malformations and birthweight in areas with potential environmental contamination. *Archives Environmental Health* **47:** 147–154. (383)

Shen, W. and Louis, T. A. (1997) Empirical Bayes estimation via the smoothing by roughening approach. Abstract, Joint Statistical Meetings, Anaheim, p. 342. (46)

Shen, W., and Louis, T. A. (1998) Triple-goal estimates in two-stage, hierarchical models. *Journal of the Royal Statistical Society, Series B*, **60:** 455–471. (32, 36, 37)

Shleien, B., Ruttenber, A. J. and Sage, M. (1991) Epidemiologic studies of cancer in populations near nuclear facilities. *Health Physics*, **61:** 699–713. (406)

Shy, C. (1997) The failure of academic epidemiology: witness for the prosecution. *American Journal of Epidemiology*, **145:** 470–484. (299)

Sibson, R. (1980) The Dirichlet tesselation as an aid in data analysis. *Scandinavian Journal of Statistics*, **7:** 14–20. (234)

Silverman, B. D. (1986) *Density Estimation for Statistics and Data Analysis*. Chapman & Hall, London. (132, 249, 276)

Sinclair, H., Allwright, S. and Prichard, J. (1995) Secular trends in mortality from asthma in children and young adults: Republic of Ireland, 1970–91. *Irish Journal of Medical Science*. (358)

Sinha, T. and Benedict, R. (1996) Relationship between latitude and melanoma incidence: international evidence. *Cancer Letters*, **99:** 225–231. (366)

Siniscalco, M., Bernini, L., Latte, B. and Motulsky, A. G. (1961) Favism and thalassaemia in Sardinia and their relationship to malaria. *Nature*, **190:** 1179–1180. (332)

Smans, M. (1989) Analysis of spatial aggregation. In P. Boyle, C. S. Muir and E. Grundmann, editors, *Cancer Mapping. Recent Results in Cancer Research*. Springer-Verlag, Heidelberg, pp. 83–86. (171)

Smith, A. F. M. and Roberts, G. O. (1993) Bayesian computation via the Gibbs Sampler and related Markov chain Monte Carlo methods. *Journal of the Royal Statistical Society, Series B*, **55:** 3–23. (200)

Smith, D. and Neutra, R. (1993) Approaches to disease cluster investigations in a state health department. *Statistics in Medicine*, **12:** 1757–1762. (151)

Smith, G. H., Williams, F. L. R. and Lloyd, O. L. (1987) Respiratory cancer and air pollution from iron foundries in a Scottish town: an epidemiological and environmental study. *British Journal of Industrial Medicine*, **44:** 495–802. (288)

Smith, T., Spiegelhalter, D. and Thomas, A. (1995) Bayesian approaches to random-effects meta-analysis: a comparative study. *Statistics of Medicine*, **14:** 2685–2699. (387)

Smouse, P. E., Long, J. C. and Sokal, R. R. (1986) Multiple regression and correlation extensions of the Mantel test for matrix correspondence. *Systematic Zoology*, **35:** 627–632. (154)

Snow, J. (1854) *On the Mode of Communication of Cholera*, 2nd edn. Churchill Livingstone, London. (3, 100)

Songini, M., Bernardinelli, L., Clayton, D., Montomoli, C., Pascutto, C., Ghislandi, M., Fadda, D., Bottazzo, G. F. and the Sardinian IDDM Study Groups. The Sardinian IDDM Study (1998) 1. Epidemiology and geographical distribution of IDDM in Sardinia during 1989 to 1994. *Diabetologia*, **41:** 221–227. (331)

Songini, M., Loche, M. and Muntoni, S. (1993) Increasing prevalence of juvenile onset Type 1 (insulin dependent) diabetes mellitus in Sardinia: the military service approach. *Diabetologia*, **36:** 457–552. (331)

Sosniak, W., Kaye, W. and Gomez, T. M. (1994) Data linkage to explore the risk of low birthweight associated with maternal proximity to hazardous waste sites from the National Priorities List. *Archives of Environmental Health*, **49:** 251–255. (383)

Spiegelhalter, D. (1998) Bayesian graphical modelling: a case-study in monitoring health outcomes. *Journal of the Royal Statistical Society, Series C*, **47:** 115–133. (363)

Spiegelhalter, D. J., Thomas, A. and Best, N. G. (1996) Computation on Bayesian graphical models. In *Bayesian Statistics* 5. Clarendon Press, Oxford. (200, 360)

Spiegelhalter, D. J., Thomas, A., Best, N. G. and Gilks, W. R. (1995) *BUGS: Bayesian Inference Using Gibbs Sampling, Version 0.50.* Medical Research Council Biostatistics Unit, Cambridge. (200, 219, 340)

Spiegelhalter, D., Thomas, A., Best, N. and Gilks, W. (1997) *BUGS: Bayesian Inference Using Gibbs Sampling* (v0.60). MRC Biostatistics Unit, Cambridge, UK (BUGS (v0.50) Manual, Examples Volume 2, p. 40.) (BUGS homepage: `http://www.mrc-bsu.cam.ac.uk/bugs/Welcome.html`). (352)

Staessen, J. and Lauwerys, R. (1993) Health effects of environmental exposure to cadmium in a population study. *Journal of Human Hypertension*, **7:** 195–199. (304)

Statistics in *Medicine*, **12:** 1751–1968 (1993). (111)

Statistics in *Medicine*, **14:** 2289–2501 (1995). (111)

Statistics in *Medicine*, **15:** 681–952 (1996). (111)

Statistische Ämter der Länder und des Bundes (1997) Statistik regional. Wiesbaden. (413)

Statistisches Bundesamt (1997) *Todesursachen.* Fachserie 12. Reihe 4. Wiesbaden. (413)

Steel, D. G. and Holt, D. (1996) Analysing and adjusting aggregation effects: the ecological fallacy revisited. *International Statistical Review*, **64:** 39–60. (182)

Stegmaier, C. (1993) Neuauswertung maligner lymphatischer und hämatologischer Neubildungen 1984–1993. Persönliche Mitteilung (in German). (398)

Stern, H.S., and Cressie, N. (1996) Bayesian and constrained Bayesian inference for extremes in epidemiology. In *1995 Proceedings of the Epidemiology Section, American Statistical Association*, pp. 11–20. (37)

Stone, R. (1988) Investigations of excess environmental risks around putative sources: statistical problems and a proposed test. *Statistics in Medicine* **7:** 649–660. (144, 243, 244, 278)

Sturgeon, S. R., Schairer, C., Gail, M., McAdams, M., Brinton, L. A. and Hoover, R. N. (1995) Geographic variation in mortality from breast cancer among white women in the United States. *Journal of the National Cancer Institute*, **87:** 1846–1853. (366)

Susser, M. (1994a) The logic in ecological: I. The logic of analysis. *American Journal of Public Health*, **84:** 825–829. (192)

Susser, M. (1994b) The logic in ecological: II. The logic of design. *American Journal of Public Health* **84:** 830–835. (192)

Svensson, C., Pershagen, G. and Klominek, J. (1989) Lung cancer in women and type of dwelling in relation to radon exposure. *Cancer Research*, **49:** 1861–1865. (291)

Swan, S. H., Waller, K., Hopkins, B., Windham, G. C., Fenster, L., Schaefer, C. and Neutra, R. (1998) A prospective study of spontaneous abortion. *Epidemiology*, **9:** 126–133. (316)

Taguchi, G. (1974) A new statistical analysis for clinical data, the accumulating analysis, in contrast with the chi-squared test. *The Newest Medicine*, **29:** 806–813. (280)

Tango, T. (1984) The detection of disease clustering in time. *Biometrics*, **40:** 15–26. (112, 242)

Tango, T. (1995) A class of test for detecting 'general' and 'focused' clustering of rare diseases. *Statistics in Medicine*, **14:** 2323–2334. (105, 109, 112–15, 148, 177, 178, 242, 259)

Tango, T. (1997) Letter to the editor on 'Detection of overall space–time clustering in a non-uniformly distributed populations' by Ranta *et al. Statistics in Medicine*, **16:** 2621–2623. (112)

Tango, T. (1998) Letter to the editor on 'Adjusting Moran's I for Population Density' by Oden, N. *Statistics in Medicine*, **17:** 1055–1062. (112)

Tanner, M. A. (1996) *Tools for Statistical Inference*, 3rd edn. Springer-Verlag, New York. (89, 91, 127, 138)

Taubes, G. (1995) Epidemiology faces its limits. *Science*, **269:** 164–169. (297, 298)

Thacker, S. B., Stroup, D. F., Parrish, R. G. and Anderson, H. A. (1996) Surveillance in environmental public health: Issues, systems and sources. *American Journal of Public Health*, **86:** 633–638. (151)

Thomas, D. C. (1985) The problem of multiple inference in identifying point-source environmental hazards. *Environmental Health Perspectives*, **62:** 407–414. (234, 244)

Thomas, D., Pitkaniemi, J., Langholz, B., Tuomilehto-Wolf, E., Tuomilehto, J. and the DiMe Study Group (1995) Variation in HLA-associated risks of childhood insulin-dependent diabetes mellitus in the Finnish population: II. Haplotype effects. *Genetic Epidemiology*, **12:** 441–453. (331)

Thomson, G. (1988) HLA disease associations: models for insulin dependent diabetes mellitus and the study of complex human genetic disorders. *Annual Reviews of Genetics*, **22:** 31–50. (331)

Tiefelsdorf, M. (1998) Some practical applications of Moran's *I*'s exact conditional distribution. *Papers in Regional Science*, **77:** 101–129. (178)

Tierney, L. (1996) Introduction to general state-space Markov chain theory. In W. Gilks, S. Richardson and D. Spiegelhalter, editors, *Markov Chain Monte Carlo in Practice*. Chapman & Hall. London, pp. 59–74. (27)

Tierney, L. and Kadane, J. B. (1986) Accurate approximations for posterior moments and marginal densities. *Journal of the American Statistical Association*, **81:** 82–86. (26)

Tobler, W., Deichmann, U., Gottsegen, J. and Maloy, K. (1995) The global demography project. National Center for Geographic Information and Analysis, Santa Barbara. (155, 164, 165)

Todd, J. A., Bell, J. I. and McDevitt, H. O. (1988) A molecular basis for genetic susceptibility in insulin dependent diabetes mellitus. *Trends in Genetics*, **4:** 129–134. (331)

Tomatis, L., Aitio, A., Heseltine, E., Kaldor, J., Mier, A. B., Parkin, D. M. and Riboli, E. (1990) *Cancer: Causes, Occurrence and Control*. IARC Scientific Publications. Lyon, pp. 169–181. (411)

Tracey, K. J. (1995) TNF and Mae West or: death from too much of a good thing. *The Lancet*, **345:** 75. (331)

Tracey, K. J., Vlassara, H. and Cerami, A. (1989) Cachetin/tumour necrosis factor. *The Lancet*, 1122–1126, May 20. (331)

Traven, N., Talbott, E. and Ishii, E. (1995) Association and causation in environmental epidemiology. In E. Talbott and G. Vraun, editors, *Introduction to Environmental Epidemiology*. CRC Press, Boca Raton, pp. 39–46. (297)

Tsutakawa, R. K. (1985) Estimation of cancer mortality rates: a Bayesian analysis of small frequencies. *Biometrics*, **41:** 69–79. (21, 61)

Tsutakawa, R. K. (1988). Mixed model for analyzing geographic variability in mortality rates. *Journal of the American Statistical Association*, **83:** 37–42. (10, 22)

Tsutakawa, R. K., Shoop, G. L. and Marienfeld, C. J. (1985) Empirical Bayes estimation of cancer mortality rates. *Statistics in Medicine*, **4:** 201–212. (22, 26)

Tuomilehto, J., Virtala, E., Karvonen, M. *et al.* (1995) Increase in incidence in insulin-dependent diabetes mellitus among children in Finland. *International Journal of Epidemiology*, **24:** 984–992. (331)

Turnbull, B., Iwano, E., Burnett, W., Howe, H. and Clark L. (1990) Monitoring for clusters of disease: application to leukaemia incidence in upstate New York. *American Journal of Epidemiology*, **132:** Supplement: 136–143. (108, 113, 243)

Upton, A. C. (1989) Public health aspects of toxic chemical disposal sites. *Annual Review of Public Health*, **10:** 1–25. (384)

Urquhart, J. D., Black, R. J., Muirhead, M. J., Sharp, L., Maxwell, M., Eden, O. B. and Jones, D. A. (1991) Case–control study of leukemia and non-Hodgkin's lymphoma in children in Caithness near the Dounreay nuclear installation. *British Medical Journal*, **302:** 687–692. (406)

Urquhart, J., Cutler, J. and Burke, M. (1986) Leukemia and lymphatic cancer in young people near nuclear installations (letter). *The Lancet*, **1:** 384. (406)

Urquhart, J., Palmer, M. and Cutler, J. (1984) Cancer in Cumbria: The Windscale connection. *The Lancet*, **1:** 217–218. (406)

US Environmental Protection Agency (1996) *Air Quality Criteria for Particulate Matter*, Vol. III. EPA/600/p-95/0016F. (304)

Vedal, S. (1997) Ambient particles and health: lines that divide. *Journal of Air and Waste Management Association*, **47:** 551–581. (305)

Vianna, N. and Polan, A. (1984) Incidence of low birth weight among Love Canal residents. *Science*, **226**: 1217–1219. (383)

Viel, J-F., Pobel, D. and Carré, A. (1995) Incidence of leukemia in young people around the La Hague nuclear waste reprocessing plant: a sensitivity analysis. *Statistics in Medicine*, **14**: 2459–2472. (144, 234, 241, 406)

Viel, J.-F., Richardson, S., Danel, P., Boutard, P., Malet, M., Barrelier, P., Reman, O. and Carré, A. (1993) Childhood leukemia incidence in the vicinity of La Hague nuclear-waste reprocessing facility (France). *Cancer Causes and Control*, **4**: 341–343. (406, 407, 409)

Vine, M. F., Degnan, D. and Hanchette, C. (1997) Geographic information systems: their use in environmental epidemiologic research. *Environmental Health Perspectives*, **105**: 598–605. (311)

Vose, D. (1996) *Quantitative Risk Analysis, A Guide to Monte Carlo Simulation Modelling*. Wiley and Sons, New York. (166)

Wakeford, R. and Binks, K. (1989) Childhood leukemia and nuclear installations. *Journal of the Royal Statistical Society, Series* A, **152**: 61–86. (406)

Waller, L. (1996) Statistical power and design of focused clustering studies. *Statistics in Medicine*, **15**: 765–782. (244)

Waller, L. A. and Jacquez, G. M. (1995) Disease models implicit in statistical tests of disease clustering. *Epidemiology*, **6**: 584–590. (257)

Waller, L. A. and Lawson, A. B. (1995) The power of focused tests to detect disease clustering. *Statistics in Medicine* **14**: 2291–2308. (244, 258, 269, 400)

Waller, L. A., Carlin, B. P. and Xia, H. (1997a) Structuring correlation within hierarchical spatio-temporal models for disease rates. In T. G. Gregoire, D. R. Brillinger, P. J. Diggle, E. Russek-Cohen, W. G. Warren and R. D. Wolfinger, editors, *Modelling Longitudinal and Spatially Correlated Data: Lecture Notes in Statistics 122*, Springer Verlag, New York, pp. 309–319. (31, 33)

Waller, L. A., Carlin, B. P., Xia, H. and Gelfand, A. (1997b) Hierarchical spatio-temporal mapping of disease rates. *Journal of the American Statistical Association*, **92**: 607–617. (31, 33)

Waller, L. A., Turnbull, B., Clark, L. and Nasca, P. (1994) Spatial pattern analyses to detect rare disease clusters. In N. Lange, L. Ryan, L. Billard, D. Brillinger, L. Conquest and J. Greenhouse, editors, *Case Studies in Biometry*. Wiley New York, pp. 3–23. (259, 260, 266)

Waller, L. and Turnbull (1993) The effect of scale on tests for disease clustering. *Statistics in Medicine*, **12**: 1869–1884. (407)

Waller, L., Turnbull, B., Clark, L. and Nasca, P. (1992) Chronic disease surveillance and testing of clustering of disease and exposure: application to leukaemia incidence and TCE-contaminated dumpsites in upstate New York. *Environmetrics*, **3**: 281–300. (144, 242, 259, 260)

Waller, L., Turnbull, W., Hjalmars, U., Gustafsson, G. and Andersson, B. (1995) Detection and assessment of clusters of disease: an application to nuclear power plant facilities and childhood leukaemia in Sweden. *Statistics in Medicine*, **14**: 3–16. (144, 400, 407)

Walter, S. (1994) A simple test for spatial pattern in regional health data. *Statistics in Medicine*, **13**: 1037–1044. (149)

Walter, S. D. (1991a) The ecologic method in the study of environmental health. I. Overview of the method. *Environmental Health Perspective*, **94**: 61–65. (329, 366)

Walter, S. D. (1991b) The ecologic method in the study of environmental health. II. Methodologic issues and feasibility. *Environmental Health Perspective*, **94**: 67–73. (329, 366)

Walter, S. D. (1992a) The analysis of regional patterns in health data. I. Distributional considerations. *American Journal of Epidemiology*, **136**: 730–741. (366)

Walter, S. D. (1992b) The analysis of regional patterns in health data. II. The power to detect environmental effects. *American Journal of Epidemiology*, **136**: 742–759. (366)

Walter, S. D. (1993a) Visual and statistical assessment of spatial clustering in mapped data, *Statistics in Medicine*, **12**: 1275–1291. (178)

Walter, S. D. (1993b) Assessing spatial patterns in disease rates. *Statistics in Medicine*, **12**: 1885–1894. (172)

Walter, S. D. (1994) A simple test for spatial pattern in regioinal health data. *Statistics in Medicine*, **13:** 1037–1044. (366)

Walter, S. D. and Birnie, S. E. (1991) Mapping mortality and morbidity patterns: an international comparison. *International Journal of Epidemiology*, **20:** 678–689. (51, 366, 413)

Walter, S. D., Birnie, S. E., Marrett, L. D.,Taylor, S. M., Reynolds, D., Davies, J., Drake, J. J. and Hayes, M. (1994) Geographic variation of cancer incidence in Ontario. *American Journal of Public Health*, **84:** 367–376. (366)

Wartenberg, D. and Greenberg, M. (1990a) Detecting disease clusters: the importance of statistical power. *American Journal of Epidemiology*, **132:** *Supplement* S156–S166. (258)

Wartenberg, D. and Greenberg, M. (1990b) Space–time models for the detection of clusters of disease. In R.W. Thomas, editor, *Spatial Epidemiology*, volume 21. Pion, London. (153)

Wartenberg, D. and Greenberg, M. (1997) Characterizing cluster studies: a review of the literature. Manuscript. (103)

Webster, R., Oliver, M. A., Muir, K. R. and Mann, J. R. (1994) Kriging the local risk of a rare disease from a register of diagnoses. *Geographical Analysis*, **26:** 168–185. (235)

Whittemore, A., Friend, N., Brown, B. and Holly, E. (1987) A test to detect clusters of disease. *Biometrika*, **74:** 631–635. (105, 108, 112, 242)

Whittle, J., Steinberg, E. P., Anderson, G. F. and Herbert, R. (1991) Accuracy of Medicare claims data for estimation of cancer incidence and resection rates among elderly Americans. *Medical Care*, **29:** 1226–1236. (366)

Williams, F. and Lloyd, O. (1988) The epidemic of respiratory cancer in the town of Armadale: the use of long-term epidemiological surveillance to test a causal hypothesis. *Public Health*, **102:** 531–538. (129)

Wilson, A. G. and Duff, G. W. (1995) Genetic traits in common diseases support the adage that autoimmunity is the price paid for eradicating infectious diseases. *British Medical Journal*, **310:** 1482–1483. (331)

Wing, S. (1998) Whose epidemiology, whose health? *International Journal of Health Services* **28:** 241–252. (295, 299)

Wing, S., Richardson, D. and Armstrong, D. (1997a) Re: Science, public health, and objectivity: research into the accident at Three Mile Island. *Environmental Health Perspectives*, **105:** 567–570. (296)

Wing, S., Richardson, D., Armstrong, D. and Crawford-Brown, D. (1997b) A reevaluation of cancer incidence near the Three Mile Island nuclear plant: the collision of evidence and assumptions. *Environmental Health Perspectives*, **105:** 52–57. (406)

Winn, D. M., Blot, W. J., Shy, C. M., Pickle, L. W., Toledo, A. and Fraumeni, J. F. (1981) Snuff dipping and oral cancer among women in the southern United States. *New England Journal of Medicine*, **304:** 745–749. (100)

World Health Organisation (1997) *Atlas of Mortality in Europe: Subnational Patterns 1980–1981 and 1990–1991*. WHO Regional Publications, European Series, No. 75, Copenhagen. (35, 366)

Wynder, E. (1996) Invited commentary: response to Science article, 'Epidemiology faces its limits'. *American Journal of Epidemiology*, **143:** 747–749. (295, 296)

Wynder, E. (1997) Re: "Invited commentary: response to Science article, 'Epidemiology faces its limits'". *American Journal of Epidemiology*, **145:** 476. (296)

Xia, H., Carlin, B. P., and Waller, L. A. (1997) Hierarchical models for mapping Ohio lung cancer rates. *Environmetrics*, **8:** 107–120. (31)

Xu, Z. Y., Blot, W. J., Xiao, H. P. *et al.* (1989) Smoking, air pollution and the high rates of lung cancer in Shenyang, China. *Journal of the National Cancer Institute, JNCI* **81:** 1800–1806. (290)

Yamaguchi, N., Watanabe, S., Okubo, T. and Takahashi, K. (1991) Work-related bladder cancer risks in male Japanese workers: estimation of attributable fraction and geographical correlation analysis. *Japanese Journal of Cancer Research*, **82:** 624–631. (366)

Yasui, Y. and Lele, S. (1997) A regression method for spatial disease rates: an estimating function approach. *Journal of the American Statistical Association*, **92:** 21–32. (187, 204)

Yu, M. C. (1986) Cantonese salted fish as a cause of naso pharyngeal carcinoma. *Cancer Research*, **6:** 1956–1961. (314)

Zelterman, D. (1987) Goodness-of-fit tests for large sparse multinomial distributions. *Journal of the American Statistical Association*, **82:** 624–629. (242)

Zeman, P. (1997) Objective assessment of risk maps of tick-borne encephalitis and Lyme borrellosis based on spatial patterns of located cases. *International Journal of Epidemiology*, **26:** 1121–1130. (316)

Zoellner, I. (1991) *Statistische Methoden zur Analyse räumlicher Konzentrationen—Anwendung auf die Verteilung von Krebsfällen in Deutschland*. Dissertation (Ph.D. thesis), University of Dortmund. (172)

Appendix

Disease Mapping and Risk Assessment for Public Health Decision Making: Report on a WHO/Biomed2 International Workshop

INTRODUCTION

In recent years many new methods have been developed in the field of disease mapping, and a number of health-oriented institutions from many countries have undertaken the production of atlases of diseases and mortality. Despite advances in methodology and increasing data availability, no systematic evaluation of the available techniques, with regard to their use in public health and decision making, has been done so far. Besides

---

The workshop was held at the WHO European Centre for Environment and Health, Rome, Italy on 2–4 October 1997.

disease mapping and descriptive studies, a growing interest has emerged on the evaluation of putative sources of risk. Initiated by the querelle on the risk of childhood leukaemia around nuclear plants to the recent claims on raised risk of asthma in areas with high traffic pollution, a variety of methods for the analysis of case event clustering and their relation to sources of noxious agents have appeared in both statistical and epidemiological literature. Furthermore, some countries host research centres devoted to the investigation of claims regarding spontaneous clusters of disease. Thus, it was felt that it was urgently needed to clarify the conditions of use and the merit of different techniques to address such questions and to inform public health response in an appropriate way.

OBJECTIVES

The European Initiative in Disease Mapping and Risk Assessment[1] and the WHO European Centre for Environment and Health, Rome Division organised an international workshop with the following aims:

- to review and assess the current development of methods of data analysis to be used in geographical epidemiological studies;
- to provide an evaluation of the application of each of the available approaches for public health use;
- to reach a consensus upon a list of recommendations on the use of the techniques that are most appropriate to orient public health policy decisions.

Specific areas of interest include: disease mapping and its role in health surveillance and public health resource allocation; ecological analyses and their controversial use in aetiologic research; the role of cluster detection in epidemiology; and the analysis of risk around putative sources, with special emphasis on environmental causes of diseases.

37 temporary advisers, 9 observers and 4 WHO officers from Belgium, Canada, France, Germany, Italy, Ireland, Japan, The Netherlands, Norway, Spain, Sweden, United Kingdom and the United States were invited to attend the workshop (see Annex 1 for a complete list of names and addresses).

Professor Benedetto Terracini was elected Chairman of the meeting (unanimously), Professor Noel Cressie Vice-Chairman (with one dissenting vote) and Dr Marco Martuzzi rapporteur (unanimously).

A total of 36 working papers, listed in Annex 2, were presented and discussed in five sessions:

1. Disease Mapping, chaired by Noel Cressie (USA)
2. Clustering, chaired by Julian Besag (USA)
3. Ecological Analyses, chaired by Arnoldo Frigessi (Italy/Norway)

---

[1] In 1996, a European Initiative in Disease Mapping and Risk Assessment (DMRA, project coordinator A.B. Lawson, University of Abertay Dundee) was funded by the European Union under the second Biomed Programme, section on Risk factors of occupational and environmental diseases. Belgium (E. Lesaffre, University of Leuven), France (J.F. Viel, University of Besançon), Germany (D. Böhning, Free University, Berlin), Italy (A. Biggeri, University of Florence) and United Kingdom (A.B. Lawson, University of Abertay Dundee) are the participating countries. The objectives of the initiative are to provide a review of current disease mapping methods in member countries; to review the available methods for assessment of geographical variations in disease; to assess, via Europe-wide applications, the most appropriate spatial methods.

4. Risk assessment around putative sources, chaired by Göran Pershagen (Sweden)
5. Health Surveillance and applications, chaired by Benedetto Terracini (Italy).

Topics of discussion and conclusions of these sessions are summarised below.

Following the five sessions, three parallel working groups, chaired by Annibale Biggeri (Italy), Tony Fletcher (UK) and Jean-François Viel (France), addressed a series of questions regarding geographical analyses and their applications in public health. Each working group drafted a list of recommendations for disease mapping and risk assessment for public health decision making. Finally, in a meeting held in plenary, the conclusions reached separately by the three working groups were re-discussed, a consensus was sought and the workshop's conclusions and recommendations were finalised. These are reported below, in form of answers to the questions discussed by the working groups. A draft of the present report was circulated for comments after the meeting with all participants before publication.

The working papers presented during the workshop have been peer-reviewed both by participants in the workshop and by external consultants and will be collected and published in a book jointly edited by the DMRA initiative and the WHO European Centre for Environment and Health.

SUMMARY OF FIVE PARALLEL SESSIONS (CONTRIBUTED PAPERS)

Session 1: Disease mapping
Chair: Noel Cressie

Eight papers were presented: Lawson et al. (LEA); Cressie and Stern (CS); Besag and Knorr-Held (BK); Böhning and Schlattman (BS); Louis (L); Mollié (MO); Martuzzi and Hills (MH); and Lawson, Dreassi and Biggeri (LDB).

Disease mapping has two common uses: smoothing away noise to draw maps and assessing specific hypotheses concerning incidence. The presentations and discussion were almost exclusively concerned with the merits of various statistical models and analyses for disease incidence rates of small areas. However, one paper (LEA) did discuss the situation where case-event data (i.e., point patterns of case locations) are available.

In discussion, there was considerable enthusiasm for the development of case-event models and associated data analysis. However, it was recognised that in practice exact locations are often ambiguous (e.g., home versus workplace). One might attempt to build a point-level model for this ambiguity. Or, one might aggregate the point-level, case-event data into small-area counts, accepting the possibility of ecological bias caused by the aggregation.

For the most part, the session was devoted to modelling and analysis of small-area count data. It was thought that the standard Poisson assumption on count data is appropriate. While some of the papers mentioned hypothesis tests on the set of area-specific rates or risk, all papers put some form of mixing distribution on them. For example, BS assumed that the risks are i.i.d. (independently and identically distributed) according to a discrete distribution; their interest was in classification of the small areas. Also, MH put a continuous gamma distribution for the risks; their interest was in the posterior distribution of its variance and how different it was from zero (the case of constant risk). In all other papers BK, L, MO and CS used multivariate log-normal distributions. One question of interest (largely unresolved) was whether these models are consistent at different levels of aggregation.

Spatial hierarchical models are very useful for disease mapping. There was not agreement about model choice at the second level of the hierarchy, but there was agreement that such mixing distributions are needed. With regard to statistical analysis there were two dominant approaches presented, both developed in the framework of Bayesian statistics. The empirical Bayes method is where one attempts to estimate parameters of "prior" distributions using observed marginal distributions. The second approach is the fully Bayesian approach, where the prior and "posterior" distributions are obtained via Markov Chain Monte Carlo (MCMC) computations (after assessing appropriate convergence). MO mapped smoothed rates using posterior means; BK have a dynamic temporal component and mapped posterior medians; L estimated the edf and associated ranks using squared error loss; CS estimated extrema with special loss functions and developed Bayesian diagnostics; LDB considered data outside the region of interest as missing and mapped the resulting posterior means.

In conclusion, the Bayesian approach (empirical or full) is appealing because almost any question can be asked and addressed. Incorporating spatial dependence requires an extra one or two parameters and is worthwhile insurance against mis-specification of regression effects. If it is found that the spatial component is not needed, one might do a reanalysis without it, to obtain a more parsimonious model. Finally, we should realise that if we are asked to smooth and map, there is a good chance that our smoothed estimates will be used later (perhaps inappropriately) to assess specific hypotheses. The paper by L makes this problem clear and gives an example of this important issue.

Session 2: Clustering

Chair: Julian Besag

There were six talks in the sessions, given by (in order) Drs Jacquez, Kulldorff, Lawson, Schmidtman, Tango and Zöllner. With the exception of a paper on MCMC methods, based on Bayesian statistics, all the approaches adopted a frequentist standpoint.

The resolution of the data which the authors had in mind ranged from county level, where data are reasonably reliable but correspondingly the analysis is rather coarse, to census enumeration district (ED) and even case level, for both of which there is a greater danger of database problems caused, for example, by incompatibility between numerators (the cases) and denominators (those at risk). Potential incompatibilities include the fact that cases occur over a substantial period of time but that the at risk population may be measured on a single census day and hence be susceptible to substantial migration effects (e.g. new housing projects); and that location may be ambiguous (e.g., it can defined as place of residence, birth, work or school).

In the talks, the purpose of a geographical analysis was to provide a p-value for evidence of overall clustering in the data, based on one or more test statistics assessed with respect to a particular reference distribution; or to identify apparent clusters presumably often with the intention of follow-up analysis by case-control or other methods if an identifiable putative cause was subsequently suggested; or both of these.

Various different approaches to such investigations were suggested and some authors included limited comparisons but there was no consensus on what might be a "best" method. This is not unreasonable: a method of analysis should be appropriate to the eventual purpose, to knowledge about the disease aetiology, to the form of the available data, and so on. Calculation of p-values depended on analytical approximations or, more commonly, Monte Carlo simulation. There was some support for more extensive power studies against meaningful alternatives.

As regards p-values, several speakers and discussants noted that for some tests, both ancient and modern, the calculations recommended in the literature are generally invalid; this is a quite common fault for tests of clustering in cancer atlases. For example, in Moran's test and similar ones, the traditional Gaussian approximation or simple randomisation analysis, applied to incidence rates or ranks is typically incorrect: even if risk is constant, there should be a geographical pattern of incidence rates if high (e.g. urban) and low (e.g. rural) populations show a geographical pattern, as they usually do. The reason is that high and low rates predominate in the low population zones and this induces the pattern. Speakers noted that the incorrect methodology can be replaced by Monte Carlo versions based e.g. on multivariate hypergeometric distributions, though these become highly computationally intensive when large populations are involved. Sometimes the incorrect results are adequate in practice but this needs careful monitoring. The existence of the problem requires more publicity but was not the focus of the session.

Practical examples were shown by most speakers and also by the chairman.

SESSION 3: ECOLOGICAL ANALYSIS

Chair: Arnoldo Frigessi

Papers were presented by Drs Bernardinelli (Italy), Best (UK), Divino (Italy), Braga (Italy), Ferrandiz (Spain), Frigessi (Italy/Norway), Kelly (Ireland) and Langford (UK) and were mainly concerned with hierarchical Bayesian modelling and modelling spatial interaction in the presence of informative covariates measured with or without error.

Several case studies have been described in order to introduce and illustrate old and new methods, including, among others, avoidable mortality for asthma; the association between malaria and diabetes; mapping of prostate and lung cancer mortality rates; and ultraviolet light as a possible cause of skin cancer.

The talks and the discussion focussed on the following issues:

1. Bayesian complex models allow a realistic and structured description of nature, in particular through incorporation of cause/effects features and measurement errors in covariates. However, statistical tools for model validation and criticism do not yet seem fully adequate, and sensitivity studies are important. Co-operation between statisticians and other fields of expertise is needed for building reliable models.
2. Models may seem sometimes too large and overparametrised; covariates may act as confounders; and there is a risk of overadjustment to watch out for.
3. The introduction of spatial smoothing and interaction among estimated spatial parameters are now standard in the presence of covariates. There seem to be three main rationales behind the introduction of terms describing spatial features into models:

   a) When we believe that measured covariates and the selected model are appropriate, the spatial part is introduced in order to check a posteriori that it is not significant.

If this is the case, the selected model explains the phenomena under study well enough.

b) The spatial part is significant and reflects what is not explained by the covariates included in the regression model. It may capture spatial features related to covariates that cannot be measured (e.g., diet) or even defined. The presence of such spatial residual leaves something unexplained from the epidemiological viewpoint. However, keeping a spatial part in the model seems generally cautious.
c) Presence of strong spatially structured effects. The spatial part is a fundamental feature of the data under study. For example, in the case of infectious diseases, vicinity is a truly explanatory factor. Such informative spatial models can also be introduced in order to investigate certain hypotheses, for example the presence of an infective agent in the aetiology of childhood leukaemia.

4. Two new modelling inferential ideas were introduced:

   a) a new way to handle data collected on different scales/grids, avoiding the aggregation of information to the least detailed scale; and
   b) a new way of doing non-parametric inference in spatial auto models with covariates, when covariates modulate spatial interaction. Non parametric methods are computationally feasible and results can be usefully compared to those obtained from parametric models.

5. There should be concern over the possible misinterpretation of disease maps. Maps without clear information on the underlying assumptions, models and approaches should not be delivered to public health authorities without careful consideration.
6. Disease maps are often interpreted "macroscopically", i.e., looking for overall spatial trends or large areas at increased risk, with no interest in the resolution at which detailed information is conveyed, such as the exact boundaries among areas with different level of risk. When this is likely to be the case, models should try to incorporate higher level features like templates, in the spirit of Bayesian image analysis.

Session 4. Risk Assessment Around Putative Sources
Chair: Göran Pershagen

Papers were presented by Drs Biggeri (Italy), Bithell (UK), Armstrong (*in lieu* of Dr Dolk, UK), Maul (France), Pershagen (Sweden), and Waller (USA).

One paper presented several methods for case-control analysis of risk around putative sources. An example was provided in which the area under study was divided into circular annuli with the postulated sources of pollution in the centre. This suggested that lung cancer risk was related to distance of residence to city centre, as well as to an incinerator in a multivariate analysis including individual information on smoking, occupation and air particulate levels.

Another paper described some exploratory methods for testing the uniformity of a risk surface over a specified region using a relative risk function. This implies that the relative risk at any given point represents the risk relative to a population-weighted average for the region as a whole. Using as an example childhood cancer data from England, it was discussed how far to go in analysis and presentation of results from such exploratory activities in situations where there is no significant heterogeneity in the data.

One presentation dealt with the power of focused score tests under mis-specified cluster models. Focused tests assess whether cases are clustered around some pre-specified potential sources of hazard in the study area. Mis-specified score tests imply that tests defined for one type of cluster model are applied to clustering generated through a different model. One problem, which was discussed, is that proximity is often not a good proxy for true exposure and the importance of accurate environmental exposure measurements was stressed.

Sequential monitoring of low event rates was discussed in one paper. The intention is to consider time intervals in the appearance of cases, making it possible to perform a stepwise assessment of the putative source while monitoring the data in a prospective design. One problem is to adjust the critical threshold so as to maintain type I error rate under a pre-specified level of significance during the entire study. Another concern arises when the phenomenon under study is rare since this may lead to a less reliable model fitting and invalid asymptotic distribution of standard test statistics. The methodology was applied to leukaemia incidence near a French nuclear reprocessing plant and did not indicate spatial-temporal clustering within the space-time window under investigation.

A multinational European study on congenital anomalies near hazardous waste landfill sites was also described. This involved 21 sites in 15 areas and a total of more than 1000 cases and 2000 controls. Overall there appeared to be a 30% increase in risk associated with residence within 3 km of the sites following adjustment for socio-economic status, maternal age and year of birth. Various types of bias were discussed such as confounding by socio-economic status, other industrial sites, differences in hospital ascertainment and migration but it was considered unlikely that this explained the results. The problems in interpretation were regarded as not principally statistical but related to the lack of evidence on exposure and plausible aetiologic pathways.

Finally, one paper focussed on the methodology for assessing lung cancer risks near point emission sources. The occurrence of lung cancer shows substantial geographic variation and ecological analyses have provided useful hints for explaining these findings, but are generally not sufficient for assessing causal relationships. For example, detailed evaluation of the role of ambient air pollution requires data on important risk factors for lung cancer, particularly regarding smoking, to adequately assess confounding and interactions. Poor characterisation of exposure is a major problem. Ideally measurements of air pollutants should cover time periods relevant for disease aetiology and the role of individual differences in activity patterns and mobility for exposure should be assessed. International collaboration is desirable in studies of lung cancer near point emission sources, in view of the often small populations with excessive exposure near each site and the possibility of combining data from several sites.

SESSION 5: HEALTH SURVEILLANCE AND APPLICATIONS
Chair: Benedetto Terracini

Papers were presented by Drs Axelson (SWE), Becker (GER), Fletcher (UK), Neutra (USA), Terracini (ITA), Viel (FRA), and Walter (CAN)

Three presentations described geographical analyses of the occurrence of a number of cancer types in Germany and in Ontario and of pleural cancer in Italy. In Ontario, the availability of databases on the distribution of social, economic, behavioural, nutritional and other health indicators allowed to consider known risk factors in the analyses. This greatly reduced the regional correlation of the residuals, thus increasing the specificity of the identification of areas requiring further special investigation.

The other four presentations related to the interaction of statisticians/epidemiologists investigating geographical patterns of disease and the rest of the society. Members of the group felt that one should be very cautious about using risk estimates obtained from aggregated data in order to estimate risk at the individual level. Possible exceptions are unusual circumstances of highly specific associations with high attributable risks (e.g. asbestos and mesothelioma).

Emphasis was given to the following points:

1. Maps can be understood by lay people provided they include any caveat or other consideration for their interpretation (including absolute numbers). Although scientific uncertainties can be shared with the lay people, scientists should not miss any opportunity to remind that lack of evidence for an effect does not correspond to evidence of lack of effect.
2. Studies aiming at assessing the effectiveness of different forms of communication of findings to the lay people are needed.
3. The pros and cons of several categories of geographical analyses deserve attention:

    a) Mapping of crude rates, standardised incidence ratios (SIR) or even case numbers;
    b) Mapping of smoothed and model-adjusted rates and SIRs;
    c) Procedures for detecting or locating clusters of disease, either at the individual level or at the level of counties or greater;
    d) Ecological studies, i.e. studies of rates in a series of populations as a function of the prevalence of risk factors in those populations, either without accounting for spatial autocorrelation or accounting for it by several methods;
    e) Case-control or cohort studies using procedures accounting for the effects of any spatial autocorrelation;
    f) Procedures using computer mapping to calculate:

        - the location and extent of exposure to pollution (e.g. numbers of schools near agricultural fields);
        - the number of persons at risk of some exposure;
        - exposures to agents like traffic fumes or electro-magnetic fields;

    g) Procedures using computer mapping to recognise a pattern of disease location suggesting a particular mode of disease transmission (e.g. linearity of disease excess suggests a polluted water grid).

It was noted that the workshop provided detailed discussion of the technical issues and applications of the first five categories above (points a through e). However it is noted that in practical applications the choice of methods for addressing points f) and g) is also important.

4. Geographical analyses of pre-clinical endpoints - when feasible - have a potential for informing public health action.
5. Whenever geographical data on more than one risk factor are available, the combined effects should be investigated in terms of effect modification as well as confounding.

6. When prioritising interventions, mere numerical comparisons between risks are incorrect, in that they ignore other values, such as the difference between imposed vs. self-inflicted risks.
7. Several episodes world-wide indicate that the media may encounter difficulties in communicating to the public, in a balanced way, epidemiological findings and their interpretation in terms of reliability and risk estimates. The collection of case histories in this respect may help systematise and perhaps prevent misunderstandings between scientists and the lay press.
8. There is a need for a better knowledge of the societal context (including politicians, activists, unions, etc.) in which geographical studies are carried out and for analysing and publishing episodes in which the epidemiologists' work has been impaired.

CONCLUSIONS AND RECOMMENDATIONS

It was agreed that "geographical analyses" refer to studies designed to exploit information on the spatial location in the data.

1. *When is it appropriate to use geographical analyses in public health decision making?*

Geographical analyses are appropriate when outcomes or exposures or a combination of both have a spatial structure. Studies of this nature can assist in public health decision making.

In particular, geographical analyses of the distribution of risk factors can be useful in prioritising preventive measures. Disease mapping is useful for health service provision and targeting interventions if avoidable risk factors are known. Geographical studies of disease and environmental exposures may in some cases be sufficient by themselves to justify action, for example if the exposure-disease association is specific, the latency is short and the exposure is spatially defined.

2. *If appropriate, what is the methodology of choice?*

No methodology of choice can be recommended in general. Analytical methods should be selected on the basis of the structure of the data to be analysed and of the hypotheses to be investigated (e.g., individual or aggregated data, presence of putative foci of risk). In most circumstances, however, it might be helpful to envisage a first level of descriptive analysis, to be followed by more specific, and problem-dependent, analyses involving parameter estimation and hypothesis testing. These will often be based on multivariate techniques and statistical modelling.

3. *Is a disease cluster, with no prior hypothesis, a sufficient cause for public health action?*

The reporting of any disease cluster, even in the absence of a hypothesis defined *a priori*, should never be ignored but critically evaluated. The process of decision making following cluster detection should be informed by considerations concerning the plausibility of any post-hoc hypothesis, its relevance in public health terms, the feasibility of possible preventive measures, and the resources needed for further investigation. While further evaluation is needed of the conditions under which public health action following cluster detection should be taken, it is noted that the simple statistical evidence of a localised excess is not sufficient to warrant intervention. Governments and other agencies should proceed with caution when communicating the occurrence of possible disease clusters.

4. *Is it possible to depict a screening device for geographical clusters, in the context of public health surveillance?*

Public health surveillance, for example from population registries, may be valuable in responding to cluster reports. It is technically possible to devise tools that systematically screen for geographical clusters, allowing for the problem of multiple testing. However, for aetiologic research, screening programmes are likely to be fruitful only in special circumstances, where highly specific exposure-disease associations and large attributable risks are involved. If surveillance is undertaken, it should be based on a clear protocol on the action to be taken if notable clusters are detected.

5. *Is it possible to use geographical methods to monitor public health interventions?*

In some circumstances geographical analyses have been used for monitoring the impact and the effectiveness of public health intervention. This might be more informative when dealing with public health measures with immediate or short-term effects.

6. *Are the new methods proposed of added value in generating hypotheses?*

Traditional methods for investigating spatial patterns of disease/exposure can be valuable in several circumstances to inform public health action, provided the data are valid, accurate and complete. However, recently developed methods of analysis, designed to deal with the spatial component of the data, have the potential to provide results that improve or correct those obtained using conventional methods, especially with small area or individual case location data. The presence of underlying geographical patterns can be more easily detected and their properties better described. In addition, such new methods have a greater potential in generating new hypotheses from geographically referenced health data.

7. *Do geographical analyses contribute to the strengthening of the evidence of the causal nature of an association?*

Geographic analyses with no information at the individual level are vulnerable to bias. However, while individually based epidemiological studies are in general needed to demonstrate the causal nature of an exposure-disease association, geographical analyses can help strengthen the available evidence. For some spatially distributed exposures such as environmental pollution, geographical studies are appropriate designs and can provide useful evidence in assessing causality, especially when the appropriate time scale is accounted for.

8. *How to communicate to decision makers the results of geographical analyses?*

Providing public health decision makers with results of geographical analyses, without addressing the underlying assumptions and discussing the implications, should be avoided. Communication of results should be based first on simple tabulations of the data and complemented by analyses addressing clearly specified hypotheses.

If maps are to be used, they should be accompanied by appropriate indices of uncertainty and variability along with some word of caution on interpretation. It is important that overall evaluations of the findings and conclusions be given.

9. *Is it possible to sketch a cost-utility evaluation of geographical methodologies in public health?*

Some technologies, such as geographic information systems (GIS), used for geographical analyses require data availability and can involve substantial financial investments. It is therefore appropriate to identify the direct benefits associated with this investment. These benefits include the capability of: undertaking rapid screening of apparent disease clusters to address the need for a ad-hoc study; bringing together the available information on geographically-based factors and/or confounders potentially of relevance; targeting or delivering public health intervention or environmental health controls more efficiently; focussing attention or investigations on the areas where intervention might be most beneficial; contributing to a surveillance scheme when needed.

10. *To what extent does data quality affect the different geographical methods?*

As in all epidemiological studies, high data quality is crucial in geographical analyses. At the small area level, however, even relatively minor inconsistencies might have a large impact on the findings, especially with routinely collected data (e.g. population estimates derived from census data).

Annex 1
Participants

TEMPORARY ADVISERS

Dr Ben Armstrong

Environmental Epidemiology Unit, London School of Hygiene and Tropical Medicine, Keppel Street, London WC1E 7HT, United Kingdom

Dr Olav Axelson

Department of Occupational Medicine, University Hospital, S-58185 Linkoping, Sweden

Dr Nikolaus Becker

Deutsches Krebsforschungszentrum Abteilung Epidemiologie, Im Neuenheimer Feld 280, D-69120 Heidelberg, Germany

Dr Luisa Bernardinelli

Dipartimento di Scienze Sanitarie Applicate e Psicocomportamentali, Università di Pavia, Via Bassi 21, 27100 Pavia, Italy

Dr Julian Besag

Department of Statistics, University of Washington, Box 354322, Seattle, WA 98195 USA

Dr Nicky Best

Department of Epidemiology & Public Health, Imperial College, School of Medicine at St Marys, Norfolk Place, London W2 1PG, United Kingdom

Dr Annibale Biggeri

Dipartimento di Statistica "G. Parenti", Università di Firenze, Viale Morgagni, 5950134 Firenze, Italy

Dr John Francis Bithell

Lecturer in Statistics, Department of Statistics, 1 South Parks Road, Oxford OX1 3TG, United Kingdom

Dr Dankmar Böhning

Department of Epidemiology, Free University Berlin, Fabeck Str. 60-62, Haus 562, 14195 Berlin, Germany

Dr Cesare Cislaghi

Istituto di Biometria, Via Venezian 1, 20133 Milano, Italy

Dr Allan Clark

School of Informatics, University of Bell Street, Abertay, Dundee DD1 1HG, United Kingdom

Dr Noel A. C. Cressie

Department of Statistics, Iowa State University, Ames, Iowa IA 50011-1210, USA

Dr Fabio Divino

Dipartimento Statistico Universitario, Università di Firenze, Viale Morgagni, 59, 50134 Firenze, Italy

Dr Emanuela Dreassi

Dipartimento Statistico Universitario, Università di Firenze, Viale Morgagni, 5, 50134 Firenze, Italy

Dr Juan R. Ferrandiz

Universidad de Valencia, Departamento de Estadistica e I.O., Facultad de Matematicas, Dr. Moliner 50, E-46100 Burjassot, Spain

Dr Tony Fletcher

Head of Unit, Environmental Epidemiology Unit, Department of Public Health & Policy, London School of Hygiene & Tropical Medicine, Keppel Street, London WC1E 7HT, United Kingdom

Dr Arnoldo Frigessi

Norwegian Computing Center, Gaudstadalléen, 23, P.O. Box 114, Blindern, N-0314 Oslo, Norway

Mr Geoffrey M. Jacquez

BioMedware, 516 North State Street, Ann Arbor, Michigan MI 48104-1236, USA

Dr Alan Kelly

Lecturer in Biostatistics, Department of Community Health and General Practice, Trinity College, Dublin 2, Ireland

Dr Martin Kulldorff

Biometry Branch, DCPC National Cancer Institute, EPN 344, 6130 Executive Blvd, Bethesda, Maryland 20892-7354, USA

Dr Corrado Lagazio

Dipartimento di Statistica "G. Parenti", Università di Firenze, Viale Morgagni, 59, 50134 Firenze, Italy

Dr Ian H. Langford

School of Environmental Sciences, University of East Anglia, Norwich NR4 7TJ, United Kingdom

Dr Andrew B. Lawson

School of Informatics, University of Bell Street, Abertay, Dundee DD1 1HG, United Kingdom

Dr Emmanuel Lesaffre

Biostatistical Centre for Clinical Trials, UZ St. Rafael, Kapucijnenvoer, 353000 Leuven, Belgium

Dr Thomas A. Louis

Professor and Head of Biostatistics, University of Minnesota, School of Public Health, 420 Delaware St. SE, Box 303, Minneapolis, MN 55455, USA

Dr Marco Marchi

Dipartimento Statistico Universitario, Università di Firenze, Viale Morgagni, 59, 50134 Firenze, Italy

Dr Annie Mollié

INSERM U351, Institut Gustave Roussy, 39 rue Camille Desmoulins, 94805 Villejuif Cedex, France

Dr Raymond Neutra

California Department of Health Services, Division of Environmental and Occupational Disease Control, 5801 Christie Avenue, Suite 600, Emeryville CA 94608, USA

Dr Goran Pershagen

Insitute of Environmental Medicine, Karolinska Institute, PO Box 210, S-17177 Stockholm, Sweden

Dr Peter Schlattman

Department of Epidemiology, Free University Berlin, Fabeckstr. 60-62, 14195 Berlin, Germany

Dr Irene Schmidtmann

Institut für Medizinische Statistik und Dokumentation, Klinikum der Johannes Gutenberg-Universität, 55101 Mainz, Germany

Dr Toshiro Tango

Director, Division of Theoretical Epidemiology, The Institute of Public Health, 4-6-1 Shirokanedai, Minato-ku, Tokyo 108, Japan

Dr Benedetto Terracini

Dipartimento di Scienze Biomediche e Oncologia Umana, Università di Torino, Via Santena, 710126 Torino, Italy

Dr Jean-François Viel

Department of Public Health, Faculty of Medicine, 2 Place Saint Jacques, 25030 Besançon, France

Dr Lance A Waller

Division of Biostatistics, University of Minnesota, Box 303 Mayo Building, 420 Delaware St. SE, Minneapolis, MN 55455-0392, USA

Dr Stephen Walter

McMaster University, Dept of Clinical Epidemiology and Biostatistics, HSC-2C16,1200 Main St W, Hamilton, Ontario L8N 3Z5, Canada

Dr Iris K. Zöllner

Landesgesundheitsamt Baden Württemberg Referat Epidemiologie, Hoppenlaustrasse 7, 70174 Stuttgart, Germany

OBSERVERS

Dr Mario Braga

Agenzia per i Servizi Sanitari Regionali, Palazzo Italia, Piano VIII, Piazza G. Marconi, 25, 00144 Roma, Italy

Dr Francesca Gallo

Direzione Centrale delle Statistiche su Popolazione e Territorio, Istituto Nazionale di Statistica (ISTAT), Via Adolfo Ravà, 150, 00144 Roma, Italy

Dr Francesco Forastiere

Dirigente, Unitá Operative Epidemiologiche, Osservatorio Epidemiologlco della Regione Lazio, Via di S. Costanza, 53, 00198 Roma, Italy

Dr Wolfgang Hoffmann

Universität Bremen, Postfach 330 440, D28334 Bremen, Germany

Dr Corrado Magnani

Servizio Epidemiologia Tumori, Università di Torino, Via Santena, 7, 10126 Torino, Italy

Dr Paola Michelozzi

Osservatorio Epidemiologico della Regione Lazio, Via di Santa Costanza, 53, 00198 Roma, Italy

Dr Bart Ostro

Air Pollution Epidemiology Unit, California Environmental Protection Agency, 2151 Berkley Way, Annex 11, Berkley, California 94704 USA

Dr Adele Seniori Costantini

Occupational Epidemiology Branch, Centro per lo Studio per la Prevenzione Oncologia (CSPO), Via di San Salvi, 12, 50134 Firenze, Italy

WORLD HEALTH ORGANIZATION

International Agency for Research on Cancer

Dr Marco Martuzzi

Unit of Radiation and Cancer

European Centre for Environment and Health

Dr Roberto Bertollini

Director, ECEH Rome Division

Dr Alexander Kuchuk

Manager, EHIS, ECEH Bilthoven Division

Ms Candida Sansone

Programme Assistant, ECEH Rome Division

Index

References to figures are shown in *italics*, to maps are <u>underlined</u>, to tables are <u>*underlined italics*</u>. A page number followed by an italic *n* indicates that the reference is in a footnote.